Urban Metabolism

Yan Zhang

Urban Metabolism

Theory, Methods and Applications

科学出版社
Science Press

Springer

Yan Zhang
School of Environment
Beijing Normal University
Beijing, China

ISBN 978-981-19-9122-6 ISBN 978-981-19-9123-3 (eBook)
https://doi.org/10.1007/978-981-19-9123-3

Jointly published with Science Press
The print edition is not for sale in China mainland. Customers from China mainland please order the print book from: Science Press.

Translation from the Chinese Simplified language edition: "Chengshidaixie: lilun, fangfa he yingyong" by Yan Zhang, © Science Press 2020. Published by Science Press. All Rights Reserved.

This Springer imprint is published by the registered company Springer Nature Singapore Pte Ltd.
The registered company address is: 152 Beach Road, #21-01/04 Gateway East, Singapore 189721, Singapore

Preface

With the acceleration of urbanization and the improvement of people's living standards, resource consumption, and pollution emissions are increasing. However, due to gaps in our understanding of social and economic development, it has not been possible to establish a complete analytical framework that provides guidance on how to reduce resource consumption and increase waste recycling. This has led to serious urban ecological and environmental problems. Finding solutions to these problems has attracted considerable attention from academic researchers, government departments, and city managers. I wrote this book on urban metabolism with the goal of exploring innovations in this field of research to reveal the progress that has been made both in China and around the world.

Urbanization has become an irresistible wave that has swept the world. According to the United Nations' *2018 Revision of World Urbanization Prospects*, the global urbanization rate in 1950 was only 30%, but by 2014 had increased to 1.8 times that level, reaching 54%. By 2050, two out of every three people on Earth are expected to be urbanites. This rapid urbanization has led to increasingly prominent ecological and environmental problems. The United Nations Environment Program's *Cities and Climate Change* (https://www.unep.org/explore-topics/resource-efficiency/what-we-do/cities/cities-and-climate-change) notes that cities account for 67% of the world's energy consumption, 70% of the greenhouse gas emissions, and three-fourths of consumption of natural resources. The *Global Metro Monitor 2018* reports that China's urbanization process is unprecedented, and continues to increase faster than in any other country. The impacts of this process are enormous, and understanding them has become an important field in urban research. The continuous evolution of China's urban scale and expansion will have a profound impact on the global society, economy, ecology, environment, and many related factors. To reduce or mitigate the adverse effects of rapid urban development and define more eco-friendly future directions, urban metabolism researchers are providing both new research methods and new perspectives, making this field a hotspot in urban and ecological research both in China and around the world.

Urban metabolism research is based on an analogy in which cities are compared with giant organisms, so that their metabolic processes can be analyzed using the tools developed by metabolic researchers. Just as the human body takes in nutrients, transforms them into a range of forms, and excretes waste, urban systems also require input materials and energy, and after consuming these resources, distribute new products and discharge waste. Human metabolism supports growth (e.g., bones, muscles), and cities also grow (e.g., infrastructure, buildings). A human body that is too thin or too fat can experience metabolic disorders, leading to a loss of vitality. A too-thin body is equivalent to a city that lacks a sufficient supply of materials and energy of the right forms (malnutrition), or that has high consumption but insufficient transformation of the consumed materials into energy or infrastructure (i.e., low efficiency), and may show incomplete or inadequate infrastructure. In contrast, excessive body weight is equivalent to the accumulation of large amounts of urban material and to high energy consumption to produce the in-use stock of materials, and this can increase material and energy consumption, resulting in serious ecological and environmental problems. For example, excess nutrition can lead to urban bloating (e.g., resource depletion, serious pollution, traffic congestion).

These examples show how urban metabolism can provide intuitive insights into urban problems. To quantify these problems, researchers have developed important measurement indexes to support urban planning and design with the goal of achieving sustainable urban development. There is broad academic consensus about the importance of the tools provided by this field of research. Since Abel Wolman's pioneering research in 1965, considerable progress has been made in the technical methods, models, practical application, and other aspects of urban metabolism. This progress is also providing important theoretical support for efforts to achieve "zero-waste cities" in China. In 2018, the State Council of the People's Republic of China issued a work plan for a pilot project that would begin in 2019 to establish a zero-waste city (http://www.chinadaily.com.cn/a/201901/24/WS5c48f80aa3106c65c34e62bc.html).

In this book, I will comprehensively discuss the concepts, technical and theoretical frameworks, research subjects, research methods, and practical applications of urban metabolism research. It results from my 15 years of urban metabolism research. I wrote this book with assistance from Li Shengsheng, Xia Linlin, Zheng Hongmei, Li Juan, Li Yaoguang, Fu Chenling, Zhang Xiaolin, Zhang Jinyun, Wu Qiong, Li Yanxian, Wang Xinjing, Xu Dongxiao, and Liu Ningyin. I am very grateful for their assistance.

I am also very grateful to the National Key Research and Development Program Project "Development of Ecological Security Protection Techniques for the Urban Agglomeration Area in Beijing-Tianjin-Hebei" (2016YFC0503005) for funding this book and the research it required, and additional support from the National Natural Science Foundation of China Innovative Research Group Project "Watershed Water Environment, Water Ecology and Integrated Management" (51721093).

As urban metabolism is a relatively new subject, its theoretical framework is not yet mature, either in China or elsewhere. In addition, the complexity of urban ecological problems, the diversity of urban types and characteristics, and inconsistencies in how a healthy urban metabolism is defined will increase the difficulty of this study. Thus, this book should be considered to be a preliminary exploration of urban metabolism concepts, theory, frameworks, research methods, and their practical application. There are still many problems to be solved, including the challenge of obtaining sufficient high-resolution data to fully understand a city's metabolism. The book nonetheless represents an important first step for promoting this research, and provides a reference for future urban metabolism research, urban planning, and urban management both in China and around the world.

As is the case in describing any immature field of research, this book will inevitably include mistakes and omissions. I encourage readers to criticize and correct the book's contents, thereby improving the usefulness of future versions. I hope that this book can promote the in-depth study of urban metabolism, and focus the attention of all sectors of Chinese and global society on urban ecological problems and the urban research methods that have been developed to solve them.

Beijing, China
June 2021

Yan Zhang

Contents

Part I
Theoretical Framework

This section of the book focuses on the connotations of urban metabolism research and its associated areas of research, the basic theory, and the resulting theoretical framework for this research. Using the CiteSpace (http://cluster.cis.drexel.edu/~cchen/citespace/) literature analysis tool, I mined the literature to define the knowledge structure, context, and evolution of urban metabolism research. By so doing, I was able to comprehensively summarize the subject's background, key insights from related fields, and the foundation for urban metabolism research. On this basis, I describe the importance of the study of urban metabolism and establish a technical framework and research paradigm for the research based on the theories of urban complex ecosystems, thermodynamics, and systems ecology. In subsequent chapters, I will demonstrate how this framework supports real-world studies of urban ecosystems and management of these ecosystems to improve their sustainability.

Chapter 1
Connotations of Urban Metabolism

1.1 The Concept of an Urban Organism and Ecosystem

To benefit from the analogies with ecosystems and organisms, which have been rigorously studied by researchers, it is first necessary to understand the relationship between the two. On the one hand, a city resembles an ecosystem because it is composed of multiple components (e.g., sectors such as the manufacturing industry and agriculture) that resemble living organisms because they exchange materials, energy, and information both with each other and with their environment. Ecosystems don't really have a metabolism, but do have flows of materials and energy that resemble the metabolic flows in an organism. On the other hand, cities are not living bodies with a beating heart, bones, muscles, and nerves, and they don't have a simple metabolism that provides materials and energy to different parts of the body. The purpose of comparing cities with ecosystems and organisms is to use familiar terminology that makes it easier to understand the complex flows within an urban ecosystem or organism (i.e., its metabolism). In this chapter, I will review how the organism metaphor has been applied to the formation and development of cities, and will explore the similarities between cities and living organisms in terms of their structure and functions.

Chengshi is the Chinese word for a city. It is composed of two parts: *cheng* refers to a defensive area surrounded by walls and a moat, and *shi* refers to a center for the distribution of materials, energy, information, capital, and people, as well as a place where trade occurs (i.e., a market). Cities were originally centers of agriculture. Subsequently, three new social divisions evolved (animal husbandry, manufacturing of specialized products, and commerce) and cities became increasingly complex. As human society evolved and became increasingly complex, the ability of humans to exploit nature increased, providing an incentive to concentrate on activities such as animal husbandry and agriculture in a defined area to improve labor productivity, leading to the development of early small settlements that became the embryo for a market. At the same time, the exchange of goods between groups of people (e.g., professions such as herding and crop cultivation) created the material conditions for

Y. Zhang, *Urban Metabolism*,
https://doi.org/10.1007/978-981-19-9123-3_1

private ownership. As society continued to evolve, the use and improvement of metal tools led to the evolution of craftsmanship and a form of manufacturing industry that was distinct from agriculture. The efficiency permitted by this separation led to the rapid development of commodity production, an increasingly obvious gap between the rich and the poor, and the emergence of settlements dominated by the crafts and by a manufacturing industry. This, in turn led to the continuous growth and development of markets, which function as areas in which exchanges of materials, energy, and information occur.

These markets were supported by the emergence of merchants who specialized in trading of various commodities, accompanied by the separation of mental and physical labor. As some merchants grew rich, the continuous accumulation of commercial capital led to continuing maturation of the market. However, the emergence of private ownership led to conflict between early rulers or governments and private owners over which group should control resources. To increase their wealth, some settlements or merchants provoked conflict with others, leading to a need to create a city with walls and a moat to protect its market (i.e., *chengshi*). This represents an early example of the concept of an organism (the city) that is clearly separated from its external environment.

Cities usually developed in areas with suitable natural conditions, such as alluvial plains near bodies of water. By taking advantage of steadily evolving social structures and technologies, cities allowed humans to exploit natural resources more efficiently, thereby allowing the cities to expand and become more complex. Because of the advantages offered by cities, they have become the main home of the world's population. The rapid development of cities has led to an increasingly high density of material flow, energy flow, population flow, and information flow, accompanied by rapid turnover (i.e., transformation) of resources. Because cities function as open systems, they depend on their external environment for resources they cannot obtain internally; as a result, large external inputs are required, and many byproducts of their consumption (including waste) are exported from the city. Unlike living organisms, the materials consumed by cities tend to flow along linear paths rather than circular flows that would lead to a closed system, which is more common in nature. The massive inputs of materials and energy and the resulting discharge of waste have increased the pressure on the city's internal and external environments, resulting in problems such as excessive consumption of resources and energy, pollution, and deterioration of environmental quality. Acuto et al. (2018) showed that cities occupy only 2%–3% of the world's land area, but create more than 75% of the world's GDP, consume about 75% of the world's natural resources and nearly 67% of the world's fossil fuels, and contribute more than 70% of the world's greenhouse gas emissions. The land, fresh water, food, energy, and infrastructure within cities can no longer support such large artificial structures. Problems for the ecological environment caused by the global urbanization process become concentrated in these small areas and have a profound impact on the health of the urban ecosystems and on sustainable development of their regions and the world.

In this context, humans must answer two crucial questions: as hybrid artificial and natural systems, can cities better mimic organisms and natural ecosystems by using resources more efficiently to mitigate environmental problems? If this is possible, how can it be achieved?

To answer these questions, researchers have focused on the city itself both as the source of the problem and the location where the problem must be solved. To accomplish the goal of making cities more environmentally sustainable, it's first necessary to understand the laws that govern their internal processes by taking advantage of the insights provided by ecological science. To do so, researchers consider cities as if they are giant living organisms ("superorganisms") with a complex circulatory system or as an ecosystem, with complex interactions among its components. Based on this metaphor, it becomes possible to analyze the city's metabolic mechanisms and detect imbalances that cause ecological and environmental problems, which are analogous to metabolic disorders (Zhang 2019; Kennedy et al. 2007; Newman 1999).

The first step in analyzing an urban metabolism is to understand the many similarities that cities have with biological organisms and ecosystems. Cities cannot be simply and rigidly separated into walls and moats (that is, there are no clear barriers between the city and its external environment) and markets (flows of materials, energy, and information); instead, the internal and external environments are defined arbitrarily, and differ from the discrete barriers found in a living organism. Under this framework, urban roads, buildings, factories and other artificial structures are analogous to the flesh, bones, organs, and blood of a living organism; that is, they accumulate and transfer materials, energy, and information both within the city and between the city and its external environment. As in an ecosystem or living organism, urban material and energy flows are no longer linear, but rather travel along networks of internal connections that support various metabolic functions. Different cities are analogous to different ecosystems; for example, some are predominantly agricultural and others are predominantly industrial. Material circulation, energy conversion, and waste excretion are carried out by means that resemble those of autotrophic and heterotrophic organisms. Cities also follow natural evolutionary processes such as growth, shrinkage, and self-renewal, and conduct self-regulation and self-reproduction during this evolution (Huang et al. 2006).

1.2 Multi-level Similarity of Urban Systems to Organisms

Like ecosystems and organisms, cities have a distinctive structure. Even though most of a city is composed of non-biological elements, it also has natural components (e.g., green spaces, bodies of water), so its structure has many similarities with a living organism. If we consider the human body as a reference, cities and bodies are similar in terms of their structural hierarchy and functional mechanisms.

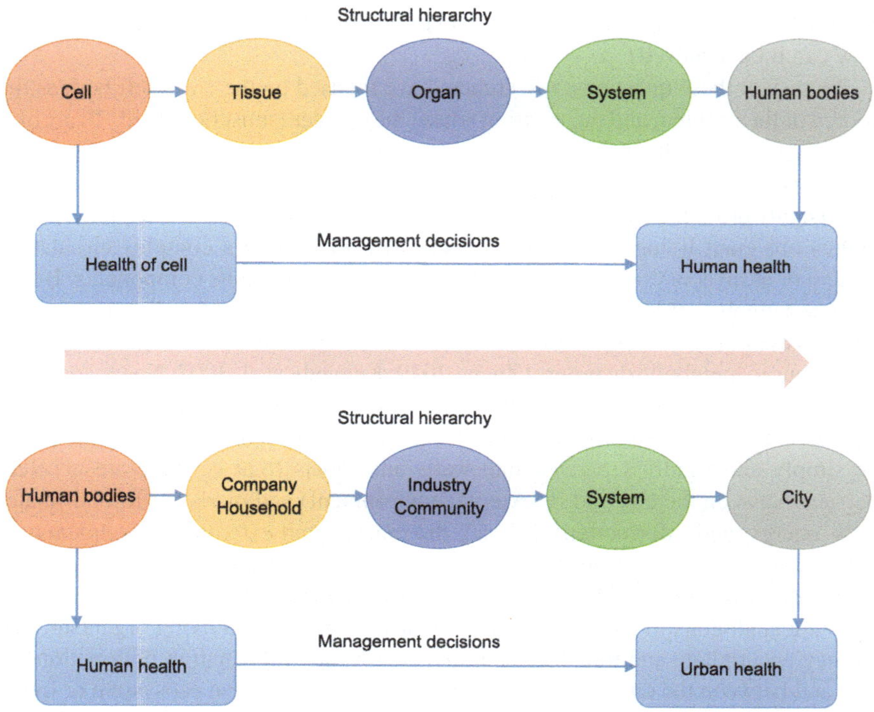

Fig. 1.1 Comparison of the structural hierarchy of human bodies (top) and cities (bottom)

1.2.1 Similarity of the Structural Hierarchy

Figure 1.1 summarizes the similarity of the structural hierarchy between cities and human bodies. For example, the overall urban system represents a body that is composed of organs (e.g., energy creation structures such as power plants or material transformation structures such as factories), and each organ is composed of cells (e.g., the workers in an industrial sector). This hierarchical structure defines the flows within and between key components of the urban metabolism. Metabolic processes at different structural levels must be organized to guarantee stable and orderly operation of the overall living organism, and this is also true of a city. Where the stability or order is disrupted, the urban organism functions less efficiently or even malfunctions, representing an "urban metabolic disorder".

1.2.2 Similarity of the Functional Mechanisms

Based on the structural hierarchy described in the previous section, we can also define specific functional mechanisms such as digestion (systems that obtain raw materials and transform them into energy and nutrients), excretion systems that remove waste, and a circulatory system that distributes resources and waste throughout the system. Just as the lungs take in oxygen and the heart distributes it throughout the body, the city's circulatory system (e.g., roads, railways, and pipelines) is defined by the city's ability to transport materials, energy, and information, so understanding this system is essential to understanding the flows within a city. Railways and roads are analogous to a body's major veins and arteries, whereas the finely subdivided streets at a local level are equivalent to capillaries. The city's government functions as the system's brain, since it coordinates flows of information and defines policies that will shape the flows of materials and energy through other components of the system. Although the brain has not yet been included in studies of urban metabolism, it obviously exerts an important control over the other flows. Interestingly, Chinese traditional medicine focuses on the flows of energy (*qi*) within the human body, and therefore provides a better metaphor for understanding metabolic disorders than Western medicine, which tends to examine parts of the overall system in isolation.

The human digestive system breaks down food into easily absorbed nutrients that can be used to construct, enlarge, or renew the body's components or to provide energy for those components. The equivalent functional structures in a city convert resources such as metallic and non-metallic minerals and fossil fuels (i.e., raw materials) into processed materials that can be used by the city's component systems. Cities also perform functions similar to those of the liver, kidneys, and intestine by detoxifying waste and recovering useful components, thereby emulating the circular (recycling) flows in a living organism. Figure 1.2 illustrates these and other components of the urban superorganism.

Urban forest, grassland, bodies of water, wetlands, and other natural components represent areas that both provide materials and energy for the city and absorb some of its waste. In this sense, they also play the purification roles of urban organs that resemble the liver, kidneys, and lungs. In this role, they are assisted by sectors that recycle or treat and release urban waste. Thus, both natural and socioeconomic components of urban systems perform indispensable functions treating the waste produced by the urban organism, and participate actively in urban renewal and growth processes.

Urban growth and other forms of metabolism cannot occur without large flows of materials and energy, and the urban transportation infrastructure provides the flows required by these functions. The railways and roads represent the infrastructure equivalent of veins and arteries, and transportation vehicles are analogous to muscles. Buildings represent one example of the parts of the city's "body" that are created by these flows.

Just as in a human body, urban growth results from immobilization of the transported materials in the form of buildings (analogous to bones) and transportation infrastructure such as roads (analogous to veins and arteries). Where these structures are permanent or semi-permanent, they become part of the city's "stock" of materials. Subsequently, these structures evolve and form new connections with other structures in a way analogous to cell division (i.e., population growth) and the formation of new structures (e.g., the emergence of new businesses). Through these processes,

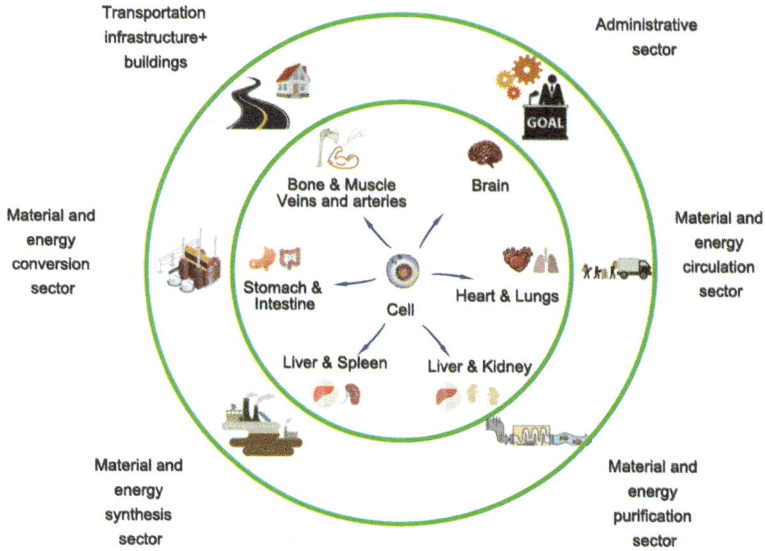

Fig. 1.2 Comparison of the functional mechanisms of an urban organism with the corresponding structures in a human body

urban organisms become increasingly complex in their behaviors and activities, and a correspondingly complex theoretical framework must be developed to describe and explain this complexity. Such a framework provides insights into problems with the urban organism. For example, if the transportation infrastructure grows too slowly, rapidly growing parts of the city are starved of materials and energy, much the same way that a blocked or narrowed artery can prevent the flow of blood or oxygen to parts of the human body. Such problems highlight the challenge of how to turn an inflexible gray area of reinforced concrete into a flexible "living" city capable of continuing growth (Jacobs 2011).

1.3 Evolution of the Concept of an Urban Metabolism

The urban metabolism concept originated from the analogy between cities and natural ecosystems. In 1965, American water treatment expert Abel Wolman proposed the concept of a metabolism for humanity's social and economic systems in an attempt to understand the impact of urban development on the environment and the limitations of the environment's ability to supply resources and minimize environmental pollution during urban development (Wolman 1965). His goal was to seek effective ways to quantify urban health and develop standards that would optimize a city's development. Wolman regarded the city as equivalent to a living organism or ecosystem with metabolic processes, and to accomplish this, he described urban metabolism as the processes responsible for flows of materials, energy, manufactured products, and waste. Taking a hypothetical city in the United States with a population of 1 million as an example, he found that larger cities experience three serious urban metabolic problems: water shortages, inadequate sewage treatment, and air pollution. He noted that by measuring the substances (such as food, clothing, fuel, electricity, and building materials) needed by urban residents for their life, work, and entertainment, it became possible to examine the relationships between material flows and urban ecological problems. Based on this analysis, it became possible to propose engineering and other tools to assist public decision-making.

Wolman's research was a breakthrough that opened up new areas of research. Many scholars extended his work and expanded the connotations of urban metabolism in terms of the processes, mechanisms, influences, and measurements of these factors. These researchers mainly focused on the data collection process (Kennedy et al. 2007; Decker et al. 2000), interactions among the components of the system (Zucaro et al. 2014), evaluation of the load imposed on the environment (Warren-Rhodes and Koenig 2001), and opportunities for improvement (Newman 1999).

Urban metabolism research focused on the processes involved in transforming raw materials, fuel, and water into components of the urban environment such as buildings, human biomass, and waste (Decker et al. 2000). It therefore collects data on all technologies and socioeconomic processes involved in urban development, energy production, and waste disposal (Kennedy et al. 2007). This approach provides a framework for describing the interactions between humans and nature and analyzing the relationships between a city's inputs and outputs, represented by flows from and to the surrounding environment (Barles 2007; Kennedy et al. 2007; McDonald and Patterson 2007; Odum 1996).

This research represented a fundamental change in how cities are seen. Cities are no longer planned and managed as if they were a purely architectural space, but are instead seen as living bodies with metabolic processes that can be controlled by their managers. All metabolic actors interact through dynamic transformations of flows, and particularly recycling (circular) flows, jointly nourishing the urban organism's living body. These concepts strongly emphasize the social and economic activities involved in the metabolic processes; however, they do not restrict analysis of the human processes to considerations of technology, but instead regard nature as integral to human life rather than as something separate. Urban metabolism should therefore consider all metabolic processes, including both those related to human technologies and those related to natural processes. Failure to integrate human social and economic activities with natural processes results in a misleading description of the urban organism. That approach does not account for how human activities distort or change natural metabolic processes; despite the large and dominant impacts of technological metabolic processes, natural processes remain an essential part of a city's metabolism.

Researchers have disagreed over the definition and meaning of the term urban metabolism (Pincetl et al. 2014; Bettencourt 2013). Bohle (1994), one of the early researchers, noted that the metaphor was necessary to support careful research on the social structure and natural and technological processes to quantify how they function. Fischer-Kowalski (1998) suggested that the concept of metabolism was an appropriate metaphor, since it emphasized the flows of materials and energy through the urban socioeconomic system. However, Warren-Rhodes and Koenig (2001) and Golubiewski (2012) noted that the metaphor is imperfect because there are also significant differences despite the many similarities between living organisms and cities. For example, cities are not a single organism, unlike humans, animals, and plants. Instead, they function more like an ecosystem (Pincetl et al. 2014; Golubiewski 2012). Nonetheless, so long as we remember that the analogy is not perfect, and use it for facilitating understanding, important insights can be achieved by comparing urban metabolism with a living organism's input of materials (food), transformation of those materials (digestion), and output of waste (excretion) (Kennedy et al. 2012). By simulating how the metaphor of urban metabolism applies to the flows into, within, and out of an ecosystem, it becomes easier to understand the complexity of these flows (Zhang et al. 2006). The organism and ecosystem metaphors both provide insights that can help managers identify metabolic disorders that result from pressure on the resource supply and pollution emission from each component of the system and pressure on the links among the components, and help urban planners and designers try to create a more sustainable city by relieving these pressures and healing the associated disorders (Kennedy et al. 2012).

Although a city is dominated by its abiotic parts, urban metabolism research recognizes that these parts change dynamically, and thus have characteristics that resemble a biological system. The resulting insights can reveal ways to reduce pressure on the city's ecological environment. Despite being a metaphor, urban metabolism is an important concept because of how it can guide us to develop new and more powerful ways to understand the origins and destinations of materials and energy. It also lets us develop metabolic indices such as throughput and metabolic efficiency that can reveal the resource utilization problems responsible for pollution of the environment, which, in turn, lets us take effective measures to solve these problems and guides us towards sustainable urban development.

Although many scholars have expanded and deepened the concepts of urban metabolism from different perspectives since Wolman's breakthrough, they continued to focus on overall resource inputs and waste outputs. However, this description treated cities as "black boxes", whose internal processes were invisible. As a result, researchers failed to account for conversion and transformation of materials within a city, and as a result, could not explain the reasons for dynamic changes in inputs and outputs over time. To explain those changes, it was necessary to "open the black box" to reveal what was happening inside. Writing in the *Encyclopedia of Ecology*, I noted (Zhang 2019) that it's not possible to study urban metabolism without being able to observe the metabolic processes that result from interactions between the socioeconomic and natural components of the system.

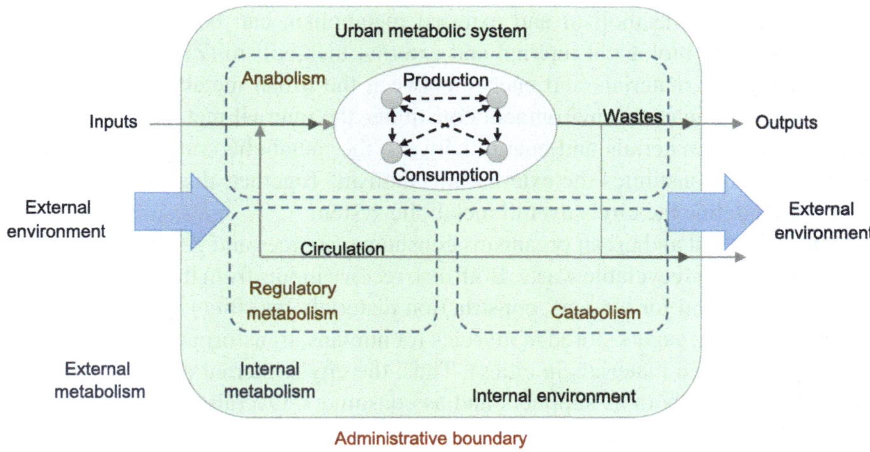

Fig. 1.3 Illustration of the phases of urban metabolic processes

1.4 Urban Metabolic Processes

1.4.1 Metabolic Phases

The material and energy exchanges between an organism and its environment, as well as the transformations of materials and energy within an organism, can be called a metabolism. This metabolic category includes anabolism (constructive growth through assimilation), catabolism (breaking materials down into their components), and regulatory metabolism (controlling the rate of anabolic and catabolic processes) (Costa 2008). From the perspective of an organism's metabolism, urban metabolism can be described using key words such as external metabolism (focused on interactions with the city's external environment) and internal metabolism (focused on interactions within the city), synthetic metabolism (creating new structures), decomposition (removal of old or failed structures), and regulation (balancing the rates of these processes). Based on this overview, we can divide the metabolism of urban organisms into different metabolic phases (Fig. 1.3).

1.4.2 External and Internal Flows

The internal environment of the urban metabolic system and the external environment can be compared to the internal and external environments of the human body. Because cities don't necessarily have a distinct transition between their internal and external environment, I have defined a city's internal environment for the purposes of this book as the area within its administrative boundary. Therefore, the processes of urban internal metabolism and external metabolism can be defined by referring to human or biological internal and external metabolism (Zhang et al. 2006). The exchange of materials and energy between the urban metabolic components within the city's internal environment constitutes the internal metabolism, whereas the exchange of materials and energy between the metabolic components and the external region constitutes the external metabolism. Together, these two metabolic components define the city's overall metabolic system.

Both biological and urban organisms consume resources and generate recyclable byproducts and unrecyclable waste. Both also receive inputs from their external environment (e.g., food for humans, construction materials for cities) and their internal environment (e.g., sugars stored in muscles for humans, transformation of raw materials into processed materials in cities). Thus, the city's internal and external environments function both as suppliers and as consumers. Overall, the categories and forms of internal and external metabolism in cities are very similar to those in living organisms.

Urban nutrients include resources such as energy, land, water, minerals, and biological resources. They are the "food" on which the urban organism depends for its survival and health (i.e., for its stability and sustainability). The vitality of the urban organism depends on the types of "food" that it consumes, and how, as well as on the degrees of absorption and digestion (i.e., the uptake and transformation efficiencies) and on the smoothness of excretion. The amount, throughput, rate, and properties of the city's internal and external nutrients are important indicators of its health.

1.4.3 Anabolism, Catabolism, and Regulatory Metabolism

Living organisms obtain nutrients from their environment through anabolism (assimilation of nutrients into new or existing structures) and catabolism (dissimilation of existing structures to release nutrients). When the organism's internal and external environments are constantly changing, these changes affect the organism's metabolism, and the organism uses regulatory mechanisms to mitigate the effects of these changes. The process of adjusting the metabolic intensity to adapt to changes in its environment, which comprises changes in the direction and rate of the adjustments, is called regulatory metabolism.

The nutrients obtained by urban organisms from the organism's internal and external environments are transformed into intermediate products and products the city needs to support its metabolic activity. At the same time, urban organisms decompose the byproducts, including waste, and discharge the resulting substances into the external environment if they cannot be recycled internally. Anabolism and catabolism are therefore opposing metabolic phases, balanced by the organism's regulatory metabolism. Reuse and recycling of materials are supported by the organism's circulatory system (which moves materials from where they are surplus to where they are needed) and by its regulatory system (which adjusts the rates of these processes).

1.4.4 Metabolic Linkages

Analysis of urban metabolic processes has evolved from a primarily linear perspective (with separate inputs and outputs) (Wolman 1965) to quasi-cyclic (with some outputs becoming inputs) (Girardet 1996) and finally to networks (with transformations and cycling) (Zhang et al. 2009) (Fig. 1.4). The urban metabolic processes can be summarized as inputs, transformations, outputs, and cycling, and this framework can be used to trace the flows of materials and energy through an urban metabolism.

Nutrients (metabolites) flow through three key components of the urban system: arterial components that carry nutrients or energy to components of the system, consumer components that consume the nutrients or energy and generate waste, and venous components that carry away waste. The input and output flows result from interactions between the arterial components and the city's internal and external

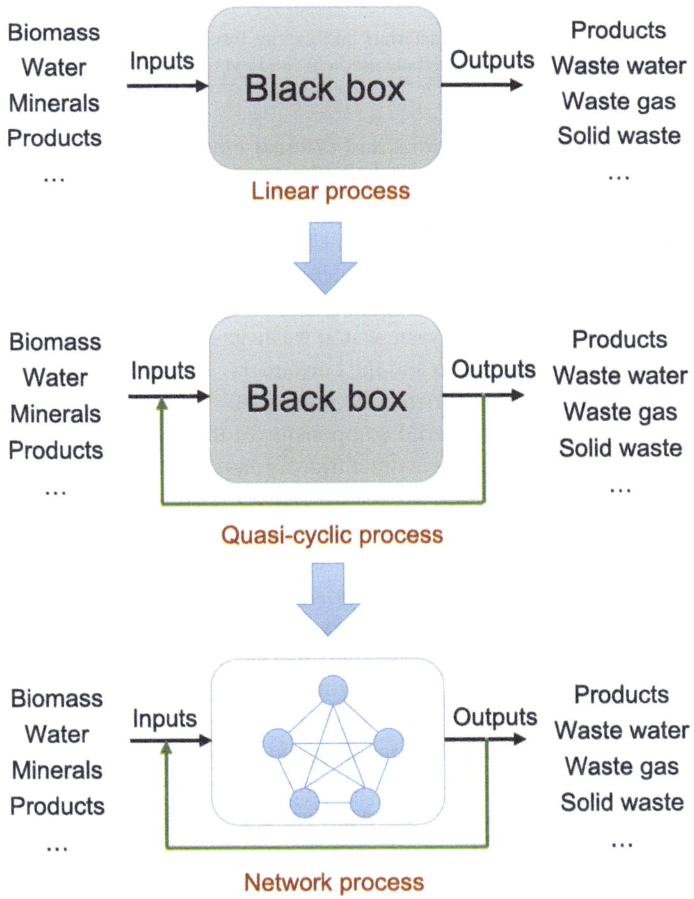

Fig. 1.4 Illustration of the evolution of perspectives for analysis of urban metabolic processes

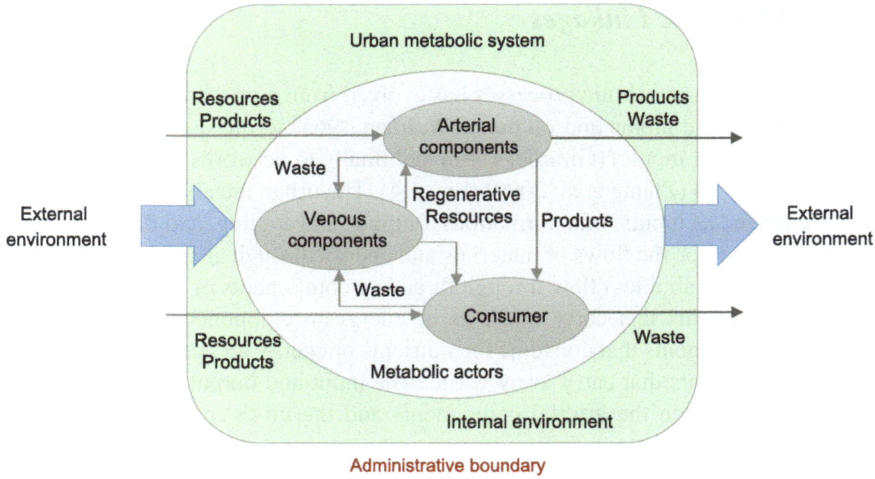

Fig. 1.5 Illustration of the flows of materials and energy through the urban metabolism (*Note* Products include final products, intermediate products, and byproducts)

environments (Fig. 1.5). The internal and external environments provide materials and energy such as biological substances, fossil fuels, metallic minerals and their (final or intermediate) products, non-metallic minerals and their (final or intermediate) products, and building materials, which act as nutrients that sustain operation of the components of the urban metabolism. These components export final and intermediate products and waste, including raw materials, semi-finished products, and finished products, as well as waste water, waste gas, and waste residues from the arterial components of the system and the consumers. Together, these constitute the input and output links for urban metabolic processes.

The exchanges among the arterial components of the system and between these components and the consumers constitute transformation links. The arterial components use many kinds of materials and energy from the environment, and turn them into products that meet human needs through mining, refining, processing, and manufacturing. The resource mining sector exploits various natural resources from the environment, and then the resource processing sector transforms the primary resources into partially processed materials that are consumed by the production sector. The raw materials are transformed into intermediate materials or primary products (semi-finished products or processed materials) through primary processing, and are then processed into products for human consumption. The recycling of resources and treatment of waste to render them harmless comprise the recycling sector. After treatment, the waste generated by venous components of the system and consumers can be turned into regenerative resources and become inputs for a subsequent production process.

1.4.5 Metabolic Chains

The primary chain of the urban metabolism starts with inputs of resources and some (final or intermediate) products from the internal or external environment, and then passes through processing, transformation, consumption, and output of products and waste by the urban system's components. The recovery chain transforms part of the waste into products that can be reused or consumed directly by other components of the system. The combination of the primary chain with the recovery chain constitutes a complete urban metabolic process (Fig. 1.6). The overall metabolic process can be decomposed into resource metabolism and waste metabolism based on whether the nutrients pass through the primary chain or the recovery chain. Resource metabolism and waste metabolism occur simultaneously. Although they appear to be independent (because they are not directly connected), they are inevitably connected because waste generation is an inevitable consequence of resource utilization.

Arterial components of the system and consumers play similar roles by connecting resource metabolism with waste metabolism. The venous components of the system play a critical role, as the system's ability to reduce waste generation by regenerating resources will significantly affect its overall metabolic characteristics, which is consistent with the role of the venous components in the catabolic phase. Division of metabolic processes into resource metabolism and waste metabolism allows a deeper analysis of the metabolic characteristics of each component. Resource metabolism is controlled by the demand for resources and by a component's resource utilization efficiency and economic contribution. Waste metabolism is linked to waste

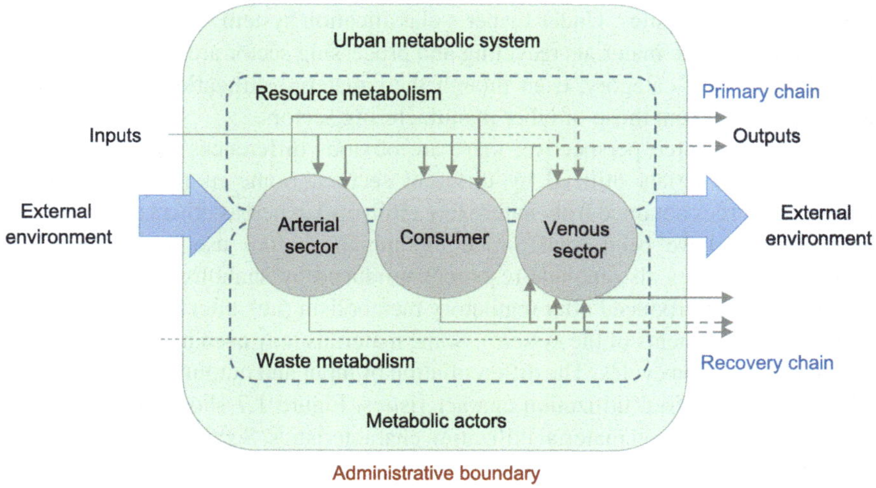

Fig. 1.6 Illustration of urban metabolic processes based on different metabolic objects (*Note* The horizontal dotted lines represent transmission and transformation pathways that exist in theory, but that do not currently exist in practice because these metabolic pathways have not yet been incorporated in a real-world urban metabolism)

generation, reuse, and final discharge, and tends to reflect the environmental impact characteristics of the system's components (Fig. 1.6).

1.4.6 Classification of the Metabolic Actors

Metabolic "actors" represent the components of the urban metabolism that participate in exchanges of materials or energy with other components (Tao 2003). Therefore, they include households, industries, animals, and artificial infrastructure, as well as natural components such as the atmosphere, bodies of water, and the soil. These metabolic actors can be classified based on the types of activities they perform, such as production (e.g., a manufacturing industry), or based on their functional group.

To allow a systematic analysis that facilitates comparisons between studies, it's necessary to define clear rules for classifying and subdividing metabolic actors. For example, all industries within a sector should have common characteristics, such as the types of material or energy they utilize and how they utilize these inputs. The division of components into sectors can use a pre-existing classification system, such as the 10 categories defined in the United Nations' *International Standard Industrial Classification of All Economic Activities* (https://unstats.un.org/unsd/classifications/Econ/ISIC.cshtml), the three industrial divisions proposed by Allan G. B. Fisher (1935), resource-intensive industries versus industries that consume relatively few resources, or industries at different stages of their evolution; this approach divides the actors into different groups based on their nature. These methods are widely used in China and Japan; however, the latter approach doesn't account for sector metabolic characteristics. Under Fisher's classification system, the waste resources sector and the waste materials recycling and processing sector are incorporated into the manufacturing category, even though their material utilization characteristics differ obviously from those of other manufacturing sectors.

From the resource perspective, there are obvious differences in the characteristics of the materials utilized by different sectors. Some of the resources are primary resources obtained from nature (e.g., iron ore), whereas others are secondary resources that have been transformed by humans before use (e.g., iron). The secondary resources also include resources produced by anabolism and the regenerative resources produced after regulatory metabolism (i.e., after being reduced by the venous components of the system). Some materials may even have gone through several regeneration cycles. The differentiation of input and output materials reflects differences in material utilization characteristics. Figure 1.7 shows the process of sector analysis based on material utilization characteristics. Sector I imports primary resources from both the internal environment and the external environment. Sector II uses both primary resources and secondary resources (intermediate products) provided by other industries. Sector III, on the other hand, mainly engages in the further processing of secondary resources. In addition, Sector IV is a venous component of the system that mainly takes wastes from other industries and turns it into

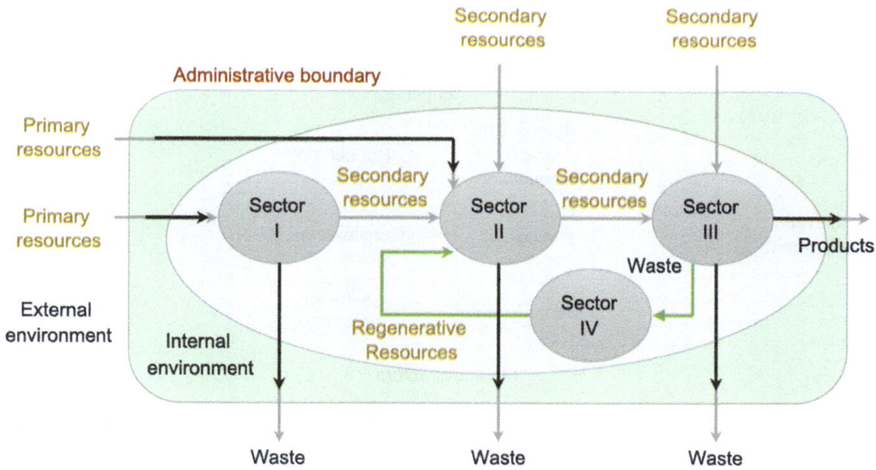

Fig. 1.7 Industry analysis based on differences in the material utilization characteristics of each sector (*Note* Black arrows represent flows to or from the internal environment, and grey arrows represent flows to or from the external environment)

regenerative resources for other industries to use. These characteristics can be used to classify industries in the four sectors.

1.4.7 Characteristics of the Metabolic Actors

Based on the differences in their material utilization characteristics, the arterial and venous components of the system and the consumers can be divided into the following sectors: agriculture, mining, manufacturing, material and energy conversion, construction, circular processing, and household consumption (Fig. 1.8). Urban arterial components include the mining and agriculture sectors, which introduce primary resources into the metabolic system; the material and energy conversion sector, which mainly uses primary resources; the manufacturing sector, which mainly uses secondary resources; and the construction sector, which mainly converts secondary resources into urban stock (e.g., buildings). The venous component of the system comprises the circular processing sector, which uses both regenerative resources (such as recycled paper) and wastes. In addition, it's possible to simply adopt the classifications used by statistical yearbooks, input–output tables, and other data sources to take advantage of the available data. Table 1.1 summarizes the characteristics of the seven metabolic actors shown in Fig. 1.8. In the following sections, we will discuss the metabolic actors in more detail.

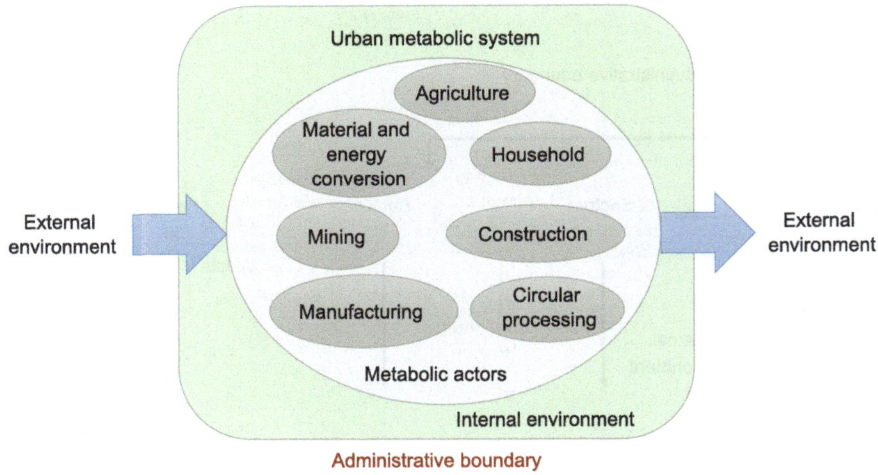

Fig. 1.8 Division of the urban metabolic system into the main metabolic actors

1.4.7.1 Mining Sector

The mining sector is the metabolic actor that introduces the local abiotic resources into the urban metabolic system. It is dominated by various kinds of mineral extraction industries that bridge the main metabolic actors with the environment. For resource-based cities, the scale and types of mining data are relatively comprehensive. However, to process this data accurately, it's necessary to consider the associated "storage transfer libraries" (Lu and Yue 2015). For example, cities dominated by energy exploitation and processing will receive inputs and outputs of many different energy materials, and will convert or transform the inputs of primary energy into secondary energy (e.g., transforming petroleum into gasoline). Therefore, it's necessary to create virtual energy storage and transfer libraries that process and aggregate data on these transformations. For cities that are not dominated by resource extraction, the mining sector's proportion of the total flows is small, and most resources are directly consumed by the key metabolic actors (processing and manufacturing) after they have been transformed and transported. The material utilization characteristics of the water production and supply industries are similar to those of the mining sector. Although water is generally regarded as a renewable resource, its actual regeneration process differs from that of plants and animals. Water must be purified and recycled in order to be reused. In nature, it therefore resembles a non-renewable resource that is directly obtained from the environment, and water production and supply can be classified as part of the mining sector, whereas water purification and treatment are classified as venous (recycling) components of the urban system.

Table 1.1 Subdivision of the metabolic actors in the urban system and description of their material use characteristics

Types		Typical industries	Material use characteristics	
Mining		Mining and washing of coal Extraction of petroleum and natural gas Mining and processing of ferrous metals ores Mining and processing of non-metal ores Production and supply of water	Non-renewable resources	Introduce primary resources into the system
Agriculture		Crop cultivation Fisheries Forestry Animal products	Renewable resources	Introduce primary resources into the system
Material and energy conversion		Production and supply of electricity and heat power	Produce energy	Mainly use primary resources
Manufacturing	Primary processing	Farm and farm byproduct processing Wood processing and wood, bamboo, rattan, and palm grass products Manufacture of raw chemicals and chemical products Cement manufacturing Smelting and processing of metals	Produce intermediate or manufactured products	Mainly use primary resources
	Processing and manufacturing	Food manufacturing Textile, clothing, footwear, and hat manufacturing Furniture manufacturing Reproduction in the printing and recording media Pharmaceuticals Plastic products Equipment manufacturing	Mainly use primary processing products	Mainly use secondary resources

(continued)

Table 1.1 (continued)

Types	Typical industries	Material use characteristics	
Construction	Housing and civil engineering construction	Mainly convert resources into urban stocks	Mainly use secondary resources

1.4.7.2 Agriculture Sector

Agriculture refers to crop cultivation, forestry, animal husbandry, fisheries, and other agricultural and related production industries. Two inappropriate methods have been used to deal with agriculture in the classification of metabolic actors. The first classifies agriculture as part of the city's internal environment, and not as a separate metabolic actor (Zhang et al. 2009). This approach regards agricultural products consumed by industries and households as inputs from the external environment. Although it simplifies the data processing and calculations, this approach is ineffective because it doesn't account for the agricultural imports that are required by most large cities (which cannot sustain their residents based only on agriculture within the city's boundaries). The second merges the agriculture and mining sectors. This approach assumes that the material utilization characteristics of the agriculture and mining sectors are similar, and that both primarily function to introduce primary resources into the metabolic system. However, the resources introduced by agriculture are mainly renewable resources, whereas the mining sector primarily consumes non-renewable mineral resources. To solve these problems, I have treated agriculture in this book as an independent metabolic actor that operates both in the city's internal environment and in its external environment.

1.4.7.3 Manufacturing Sector

The manufacturing sector can be divided into two categories based on their material utilization characteristics: the primary processing sector, which mainly uses primary resources, and the processing and manufacturing sector, which mainly uses secondary resources. The primary processing sector outputs intermediate products that are then consumed by the processing and manufacturing sector. The manufacturing sector is very diverse, and depending on the country, it may include as many as 31 categories. Among them, the recycling and processing of waste resources and waste materials and the processing and manufacturing of regenerative resources are excluded because they are venous components of the urban system, and therefore should not be included in the manufacturing sector.

1.4.7.4 Material and Energy Conversion Sector

The material and energy conversion sector converts primary energy into secondary energy (e.g., coal into electricity). It is similar to the primary processing sector in terms of its material utilization characteristics, since both are dominated by primary resource utilization. However, the outputs of the two sectors are quite different. The primary processing sector produces products in a physical form, whereas the material and energy conversion sector produces products in the form of energy. At the same time, the material and energy conversion sector resembles the mining sector because it transfers energy in the form of solid or liquid raw materials (e.g., coal, oil), and only some components of this sector (green energy) are renewable.

1.4.7.5 Construction Sector

The construction sector creates the municipal infrastructure (e.g., houses, roads), including key features such as water conservation facilities, but also maintains existing buildings or dismantles buildings to make room for other structures. As a result, it requires high material inputs and generates large waste outputs. The construction sector's material utilization characteristics are similar to those of the manufacturing sector at its input end, since both primary resources and secondary resources produced by the primary processing sector are the main inputs. However, its outputs differ greatly from those of the manufacturing sector. Most construction products are converted into urban stocks (i.e., permanent or semi-permanent structures such as buildings) rather than exported as products, and much of the waste cannot be decomposed and restored by the venous components of the urban system.

1.5 Urban Metabolic Characteristics

As urban organisms are large and complex systems, their metabolic processes show typical characteristics of living organisms such as growth and development, openness and dependence, and stability and robustness (Fig. 1.9).

1.5.1 Growth and Development

The growth and development of a city involves changes from disorder to order and from quantity to quality, gradually forming a complete organizational structure and functional layout. In the early phases of urban development, the city's ecological niche must expand to make room for more of the city, and the growth rate is slow. During this process, the environment is transformed and adapted to the needs of an urban organism. Once sufficient space is available, the city's growth increases rapidly

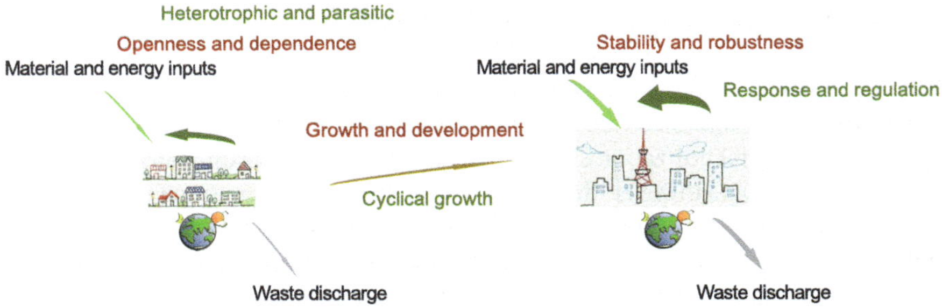

Fig. 1.9 Illustration of the metabolic characteristics of urban organisms

(potentially even exponentially). However, the organism's growth is restricted by the available space and the development conditions, and this restriction causes the growth rate to slow again. The alternation between rapid and slow growth produces a cyclical growth curve (Fig. 1.10). Unlike biological organisms, urban organisms can always expand to escape the size bottleneck, provided sufficient resources (including space) are available to sustain their growth. During this growth, they must constantly adjust their niche and change in response to changing development conditions, until new limiting factors constrain their growth (multi-stage logistics curve). So long as the growth does not exceed the environment's ability to sustain the growing organisms, the cyclical growth can represent a sustainable and healthy development process.

During the process of cyclical growth, urban organisms are constantly upgrading and renewing their stocks, but it is inevitable that age differentials will develop between parts of the city due to unequal rates of development in different parts of the organism. For example, Beijing's Tongzhou district is relatively young (18–30 years

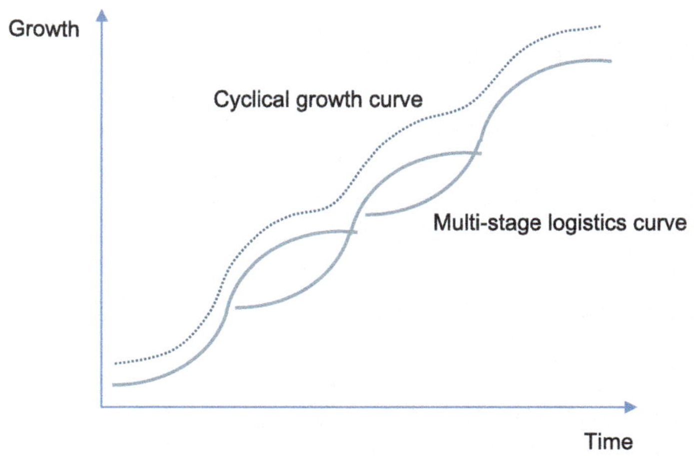

Fig. 1.10 The cyclical growth curve for an urban organism (*Note* The dotted line represents the overall trend)

old), and its spatial development and functional design have different characteristics from the land that lies inside Beijing's second ring road, which is significantly older (55–60 years old). Because of constraints such as existing residences in the latter area, it will be difficult for this area to change. In the context of metabolism, this leads to metabolic differences. For example, the older area will require more maintenance, whereas the new area is young enough that most of its parts don't yet require maintenance. Comprehensively considering the age distribution pattern of urban metabolic actors will reveal important differences between areas of the city, and will create a need for different solutions, such as maintaining the existing functions of the old areas and implementing newer and more efficient technologies in new areas, thereby maintaining each area's vitality and allowing it to continue to grow and develop.

The growth and development of urban organisms is reflected both by the expansion of the boundary between the city and its external environment and by improvement of the structures and functions inside the boundary. As communities grow, the markets that support consumption by the residents (e.g., food markets) eventually become too far from some residents; in response, new markets develop to serve the most distant residents. The process then repeats. In addition, new functional areas such as markets retain their original functions, but with differences related to any unique needs of the communities they serve.

The cyclical growth of urban organisms also reflects the local variations that arise as a city's boundary continuously expands, particularly in social cultures that promote innovations such as industrial upgrading and adjustment of some functions (e.g., changing from manufacturing only glass and ceramic products to manufacturing kitchen products). In the face of a changing market environment and changing resource or environmental conditions, the ecological niche occupied by an urban organism must be continuously extended (to provide sufficient resources), maintained (to ensure that the city can continue meeting the needs of its residents), and degraded (replaced) when necessary to make room for newer functional structures in response to the city's continuous development and any problems created by this evolution.

For example, consider a city based primarily on resource extraction, such as mining and processing of local mineral resources, harvesting of forests, and exploitation of other natural resources. As the city's scale expands, resource depletion becomes a bottleneck that slows growth and development, and this requires a transformation of the urban metabolism. One possibility is to seek management and technological innovations that improve the resource-use efficiency without requiring additional resources or a change in the industrial structure. Another possibility is to import more resources from the city's external environment to sustain that structure. Alternatively, the city could modify its industrial structure so that it is less dependent on resources by promoting the development of a services sector. Every urban organism must face these choices during their development (Fig. 1.11).

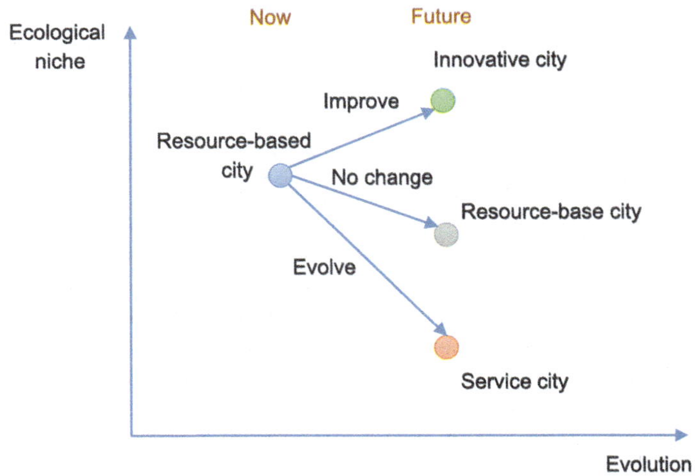

Fig. 1.11 Relationships between the size of a city's ecological niche, and the three main paths for its evolution (*Note* "Ecological niche" represents the position of a city within an ecosystem, describing both the range of conditions necessary for persistence of the city, and its ecological role in the ecosystem)

1.5.2 Openness and Dependency

Due to the large metabolic throughputs and losses by an urban organism, its growth requires a constant and growing supply of resources from its environment. However, at some point its internal environment cannot meet the demand, and a growing proportion of the nutrients that sustain the city must be obtained from its external environment. Simultaneously, the internal and external environments must also accept urban metabolites, such as waste. The result is increasing openness (i.e., increasing exchanges with the external environment). Some aspects of this evolution are linear. For example, the amount of food needed by the city's residents is proportional to their number, and at some point exceeds the amount of food that can be supplied by the internal environment. Similarly, as the scale of the producers (economic activities such as manufacturing, construction, and transportation) increases, this requires increasing supplies of materials and energy, and at some point, these must also be supplied by the external environment. Conversely, modern decomposition and recycling technologies are not yet very sophisticated, so most of the waste generated by urban production and living activities must be discharged into the environment for digestion and decomposition. This creates great pressure on the environment.

These problems are mainly due to the imperfect circulation mechanisms in an urban living system. Because the city cannot sustain itself (i.e., is analogous to a heterotrophic or parasitic organism rather than an autotrophic plant), it differs from a natural ecosystem, which depends on a balance among the producers that supply energy and materials for the ecosystem, consumers that exploit these resources, and decomposers that process their waste. Of course, the urban organism's characteristics of openness and dependence on its external environment offer both advantages and disadvantages, which lead to coexistence of risks and opportunities. Designers of cities must learn from natural ecosystems so they can transform these systems from open structures that cannot achieve self-sufficiency to closed systems or clusters of cities that are self-sufficient, thereby avoiding the risks (e.g., pollution, resource depletion) associated with an unbalanced and "leaky" system. Natural ecosystems achieve stability and robustness (see the next section for details) that cities cannot currently achieve. Urban metabolism research can help city planners and managers to achieve greater self-sufficiency by revealing excessive consumption or unbalanced flows and suggesting how those flows can be made more sustainable. In particular, by identifying the characteristics and flows of waste, these materials can be transformed from liabilities into assets that can be recycled by the urban organism.

1.5.3 Stability and Robustness

With continuous acceleration of urbanization and industrialization, the metabolic processes of urban organisms intensify. The intensity of material and energy flows in a large city is potentially much greater than that in a natural ecosystem, and this requires sophisticated logistics to maintain the urban metabolism in a relatively balanced and stable state. The greater metabolic intensity of urban organisms can only be sustained by high material inputs, high accumulation of stocks (e.g., people, capital, buildings, roads), and low outputs of goods and services. Although growing cities can provide more services, this also increases consumption of materials and energy. The contradiction between supply and demand increases continuously as the urban organism grows and develops. If this contradiction becomes large enough, the city's growth stagnates and the city may be unable to meet the needs of its residents. Thus, it becomes necessary to find ways to mitigate this contradiction.

The robustness of an urban organism reflects its ability to respond to the stress created by change with minimal disruption by developing ways to cope with and adapt to the stress. Other stresses include natural disasters, safety risks that result from unexpected phenomena, and social changes; each of these requires some form of preventative or mitigation measures, such as the creation of a disaster-response organization (e.g., fire and police departments) or the development of policies that mitigate or eliminate certain risks. These responses increase the city's stability and robustness. Since it isn't always possible to prevent or predict these stresses (e.g., prevent an annual hurricane season, predict earthquakes), urban organisms must also find ways to become more robust by adapting to the stresses; for example, construction regulations can require builders to create structures with greater resistance to wind, flooding, and earthquakes.

The stability and robustness of urban organisms depend mainly on their ability to regulate social and economic activities, and on their ability to protect and restore their natural ecological environment. Achieving this stability and robustness also requires city managers and planners to account for socioeconomic change during urbanization, since these changes also affect the natural ecological environment. These changes cannot be allowed to happen blindly, as this leads to metabolic disorders and even disasters; instead, the consequences of these changes must be predicted and included in planning and management.

As this chapter shows, the analogy between cities (urban organisms) and living organisms or ecosystems provides important insights. Like organisms and ecosystems, cities have a metabolism that comprises inputs of materials, energy, and information, and their metabolic activities produce outputs such as waste and manufactured goods. Understanding the flows of these resources between metabolic actors, which are analogous to biological systems such as consumption and digestion, provides clear insights into which flows are potentially the cause of metabolic disorders, and insights into how adjusting these flows can mitigate or prevent the metabolic disorders from becoming serious. In subsequent chapters, I will explore these concepts further and provide examples of how urban metabolism research reveals both problems and potential solutions.

References

Acuto M, Parnell S, Allen A E, et al (2018) Science and the future of cities: report on the global state of the urban science-policy interface (2018-12-1) [2022-1-1]. https://www.researchgate.net/publication/329717388_Science_and_the_Future_of_Cities

Barles S (2007) Urban metabolism and river systems: an historical perspective—Paris and the Seine, 1790–1970. Hydrol Earth Syst Sci 11(6):1757–1769

Bettencourt LMA (2013) The origins of scaling in cities. Science 340(6139):1438–1441

Bohle HG (1994) Metropolitan food systems in developing countries: the perspective of "Urban Metabolism." GeoJournal 34(3):245–251

Costa A (2008) General aspects of sustainable urban development (SUD). In: Clini C, Musu I, Gullino ML (eds) Sustainable development and environmental management. Springer, Dordrecht, The Netherlands, pp 365–380

Decker EH, Elliott S, Smith FA et al (2000) Energy and material flow through the urban ecosystem. Annu Rev Energy Env 25(1):685–740

Fischer-Kowalski M (1998) Society's metabolism: the intellectual history of materials flow analysis, Part I 1860–1970. J Ind Ecol 2(1):61–78

Fisher AGB (1935) The clash of progress and security. Macmillan, London, UK

Girardet H (1996) The Gaia Atlas of cities: new directions for sustainable urban living. Gaia Books Limited, London, UK

Golubiewski N (2012) Is there a metabolism of an urban ecosystem? An ecological critique. Ambio 41(7):751–764

Huang GH, Chen B, Qin XS (2006) Study of diagnosing, preventing and controlling illnesses of modern cities. Journal of Xiamen University of Technology 14(3):1–10 (in Chinese)

Jacobs J (2011) The death and life of great American cities. Random House, New York

Kennedy C, Cuddihy J, Engel-Yan J (2007) The changing metabolism of cities. J Ind Ecol 11(2):43–59

Kennedy C, Pincetl S, Bunje P (2012) Reply to "comment on 'The study of urban metabolism and its applications to urban planning and design' by Kennedy et al. (2011)". Environmental Pollution 167(1):186

Lu ZW, Yue Q (2015) The studies of industrial ecology. Science Press, Beijing

McDonald GW, Patterson MG (2007) Bridging the divide in urban sustainability: from human exemptionalism to the new ecological paradigm. Urban Ecosystems 10(2):169–192
Newman PWG (1999) Sustainability and cities: extending the metabolism model. Landsc Urban Plan 44(4):219–226
Odum HT (1996) Environmental accounting: energy and environmental decision making. Wiley, New York
Pincetl S, Chester M, Circella G et al (2014) Enabling future sustainability transitions: an urban metabolism approach to Los Angeles. J Ind Ecol 18(6):871–882
Tao ZP (2003) Eco-rucksack and eco-footprint. Economic Science Press, Beijing (in Chinese)
Warren-Rhodes K, Koenig A (2001) Escalating trends in the urban metabolism of Hong Kong: 1971–1997. Ambio 30(7):429–438
Wolman A (1965) The metabolism of cities. Sci Am 213(3):178–190
Zhang Y (2019) Urban metabolism. In: Fath BD (ed) Encyclopedia of ecology, 2nd edn. Elsevier, Oxford, UK, pp 441–451
Zhang Y, Yang ZF, Li W (2006) Analyses of urban ecosystem based on information entropy. Ecol Model 197(1):1–12
Zhang Y, Yang ZF, Yu XY (2009) Ecological network and emergy analysis of urban metabolic systems: model development, and a case study of four Chinese cities. Ecol Model 220(11):1431–1442
Zucaro A, Ripa M, Mellino S et al (2014) Urban resource use and environmental performance indicators: an application of decomposition analysis. Ecol Ind 47(4):16–25

Chapter 2
Progress in Urban Metabolism Research

2.1 The Significance of Urban Metabolism Research

2.1.1 Feasibility

As shown in Chapter 1, the flows of materials and energy through cities, and the associated growth and development, are analogous to those in an organism or an ecosystem, and this suggests that it is feasible to study urban metabolism using the same tools. Although information also flows within the system, these flows have not been well studied, and for the most part, they will not be discussed in this book. By examining these flows, it is possible to identify the key processes (e.g., ones that either constrain growth or represent inefficient resource use) and define the rules that govern growth and development. Understanding these rules enables urban planners and managers to adjust a city's operations to improve its vigor. A city's openness (ability to exchange materials and energy with its environment) and dependence on its external environment make it less efficient and less self-sufficient than a natural ecosystem or organism, and this can make it parasitic on its environment; however, by examining how these characteristics differ from those of a natural system, measures can be taken to improve the city's stability and robustness, thereby guiding the city towards more sustainable development.

For these reasons, urban metabolism is a promising new research perspective. This research begins with a summary of the processes that occur within the system and the links along which the flows occur, as well as the emissions that result from consumption activities. Urban metabolism research not only accounts for the inputs of raw materials and energy by the built environment (whether that environment comprises urban areas with dense groups of tall buildings or low-density rural settlements), but also pays attention to the transformation of raw materials, product synthesis, and waste-generation activities in which a city's metabolic actors participate, and further traces these products or waste backwards to identify their sources. This perspective emphasizes the relationships between resource consumption and environmental pollution, clarifies the sources, destinations, and circulation paths for materials and

Y. Zhang, *Urban Metabolism*,
https://doi.org/10.1007/978-981-19-9123-3_2

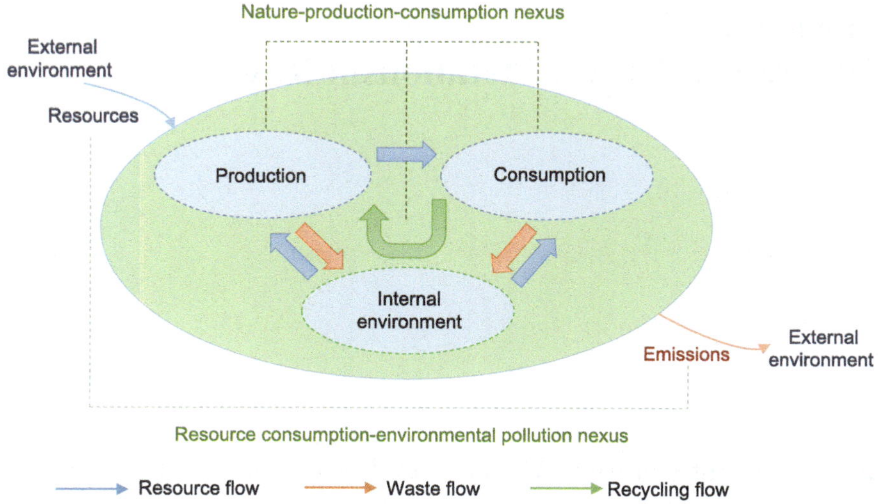

Fig. 2.1 A perspective for considering the flows within an urban metabolism

energy, and quantifies the scale and structure of these flows and their storage in the network that connects the metabolic actors. This perspective links nature, production, and consumption, analyzes the driving and response mechanisms of the metabolic actors, identifies the resource consumption and pollution emission that underlie urban metabolic disorders, and seeks measures to transform industries and their consumption to transform the city. To do so, it aims to improve the efficiency of the flows and the relationships between the sources and the actors that consume their outputs, thereby providing a more comprehensive picture of the city's functions and making it possible to improve their health (Fig. 2.1).

2.1.2 *Necessity*

Solving complex scientific and practical problems requires insights from researchers working in multiple disciplines and the integration of their different understandings of the problem to produce a more comprehensive solution. Urban metabolism research, therefore, incorporates insights from ecology, civil engineering, industrial engineering, network analysis, economics, and sociology. These different perspectives are required to fully understand the city's structure, the relationships among the city's components, and the relationships between these components and the city's internal and external environments. Because ecosystems are complex, their self-organizing and evolutionary processes are also important, particularly since many of the interactions are nonlinear. Such analyses must also account for key principles from physics, such as the conservation of energy and mass; for example, the inputs received by a metabolic actor can be transferred to another component of the urban system (e.g., the flows of products and embodied energy) or retained by the actor (i.e., they become part of its stock). Moreover, these processes, their structure, and their functional relationships change over time and vary spatially. Thus, urban metabolism research must find ways to identify the city's metabolic capacity from the perspectives of time, space, volume or mass or number, structure, and order (Wang and Ouyang 2012), and the interactions of these processes with thermodynamics (i.e., the conservation of energy).

Urban metabolism research attempts to account for as many metabolic actors as possible (e.g., families, enterprises, communities, the natural environment), multiple states of matter (solid, liquid, gas), and multiple elements (particularly carbon, nitrogen, and phosphorus) that are involved in the city's functioning (Fig. 2.2). Each metabolic actor functions either as a source of materials and energy, as a sink (a location where the materials and energy are temporarily stored), or as a generator or sink for pollutants and other metabolic waste, such as waste heat. Changes in the behavior of the metabolic actors will have a significant influence on the efficiency and sustainability of the city by affecting the supply of nutrients (inputs) and the generation of intermediate products and waste (outputs). Therefore, two important subjects of urban metabolism research are the regulation of the utilization scale for resources, and the establishment of feedback loops, for modifying the behavior of metabolic actors (i.e., their scale, structure, spatial layout) to improve their efficiency and sustainability.

2.1.3 Urgency

China'S diverse cities have developed different metabolic types as a result of differences in their dominant flows. This provides a natural laboratory for urban metabolism research. At the same time, serious ecological problems have emerged during China's rapid urbanization, and there is an urgent need to perform practical research to support the development of solutions to these problems. This combination has created an urgent need to develop basic theories and techniques to support case studies of urban metabolism. China's cities are divided into a range of administrative levels, and there are 600 cities of various sizes that are ranked as either prefecture-level (sub-provincial) or county-level cities. If these cities are grouped by metabolic scale (i.e., total energy consumption), the number in each size class forms a pyramid, with a smaller number of megacities at the top that have a large metabolic scale, and many small cities at the bottom that have a small metabolic scale. The need for solutions is particularly urgent in the megacities, but lessons learned from studying those cities will also have implications for the smaller cities, especially if those cities are growing.

In addition to the metabolic scale, there are also differences in the nutrient uptake and byproduct or waste generation structure, the metabolic efficiency (the production of outputs per unit input), and the effects among these cities. For example, metabolic rates in older cities are often lower than in young cities, and have become

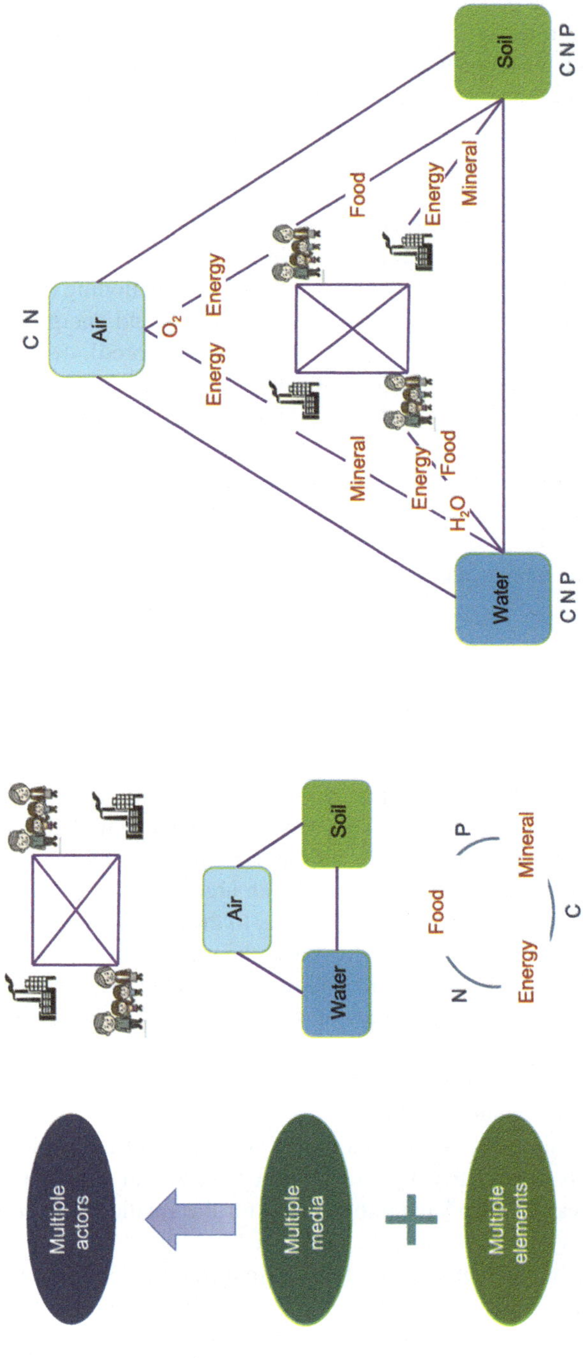

Fig. 2.2 Illustration of the multidimensional and complex nature of urban metabolism (*Note* C, N, and P represent carbon, nitrogen, and phosphorus, respectively)

relatively stable in mature cities. Some cities can be described as "lean" because they have a poor capacity to absorb inputs (possibly due to low cumulative stocks), whereas others have a high absorptive capacity (possibly due to high cumulative stocks), and can be described as "fat". Differences in the city's environment (e.g., the resource endowment and the environmental capacity to absorb pollution) can lead to differences among cities in their dietary structure and differences between cities in the distribution of ecological niches can create cities with different resource characteristics; for example, some cities are dominant consumers of resources, whereas others are dominant service providers. Healthy cities have relatively balanced inputs, outputs, and stocks, and have a metabolic efficiency that is suitable for their size. Urban metabolism is a theoretical and applied discipline that starts by identifying urban ecological and environmental problems based on the complex ecosystem theory, ecological thermodynamics, and systems ecology. It then seeks solutions to these problems by focusing on the relationships between a city and its environment, describing the laws that govern urban development and evolution, defining the internal mechanisms that control the metabolism, and providing support for improving the ecological environment and promoting sustainable development. To meet these urgent requirements, urban metabolism research must be deepened in terms of its research scale, subjects, and methods.

Because urgent problems can arise at multiple scales and over multiple time periods, the scale of urban metabolism research ranges from metabolic actors within a city to individual cities and the regions that the cities belong to; by integrating the results from these different scales, the research can be expanded to regional, national, continental, and even global scales. The focus of the research has evolved from studies of isolated natural and socioeconomic components of an urban system to complex hybrid natural and socioeconomic systems, and has deepened from simple descriptions of these systems to historical retrospectives and predictions of future states. Given the magnitude of global problems, such as climate change, and their impacts at regional and local scales, understanding the causes of these problems is increasingly urgent.

The research methods have also evolved from scientific reductionism (i.e., breaking a complex system into smaller parts that are easier to understand) to holistic, integrated research that accounts for the interactions among these parts. The research often combines top-down research methods (reductionism) with bottom-up research methods (integration) to provide different insights. The key difference between the top-down and bottom-up strategies is that the former decomposes the system into its component parts, whereas the latter synthesizes the overall effect of small structures into larger components. Top-down research starts from a general problem and decomposes it into smaller and simpler components that can be individually controlled, whereas bottom-up research starts with these controllable units and integrates them to construct an overall system.

This evolution has also led to the deepening of the research content through process-based analysis that reveals the structure, functions, and patterns of the components of an urban metabolism. The resulting improved understanding allows the simulation and the prediction of both the urban system's dynamic evolution

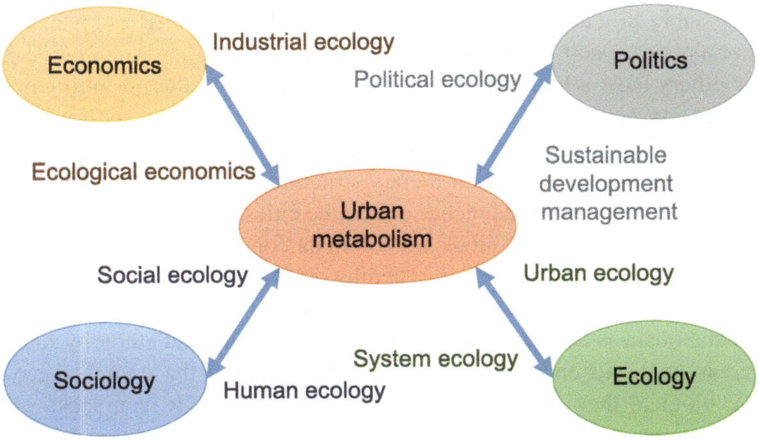

Fig. 2.3 Illustration of the multidisciplinary and intersectional perspectives of modern research on urban metabolism

over time and its spatial variation. This requires an improved understanding of the metabolic mechanisms and how the city responds to stress in order to permit the diagnosis of the city's health status and the regulation of the behavior of problematic metabolic actors. The deepening of the methodology is reflected in the shift from qualitative descriptions to the quantification of flows and quantitative simulation, while we simultaneously try to integrate multidisciplinary methods such as economics, sociology, politics, and ecology (Fig. 2.3).

The disciplines of applied economics, such as industrial ecology and ecological economics, can support urban metabolism research by accounting for the laws that govern the economic components of cities. The sub-disciplines of sociology, such as human ecology and social ecology, can support urban metabolism by providing insights into the mechanisms that influence social attitudes, culture, and social institutions related to material flow and energy flow processes. As branches of political science, the management of sustainable development and political ecology aim to set societal goals for urban development that will improve urban metabolism by choosing more sustainable development paths. The branches of ecology, such as systems ecology and urban ecology, provide support for understanding the environmental impacts of the other research disciplines.

2.2 CiteSpace Knowledge Mapping Analysis

Since Wolman (1965) initiated the field of urban metabolism research nearly 60 years ago, this field has evolved and become richer in terms of its scale and methodology, despite concerns over whether this biological metaphor is truly appropriate, and this research has become increasingly recognized and praised for its ability to support sustainable development. Some scholars have tried to summarize urban metabolism research from the perspectives of technical methods (Goldstein et al. 2013; Zhang 2013; Pincetl et al. 2012; Weisz and Steinberger 2010), development phases (Zhang et al. 2015; Kennedy et al. 2011), and contributions to sustainable development (Barles 2010); however, they did not synthesize these perspectives to provide an overview of urban metabolism. Therefore, it is urgently necessary to systematize the knowledge provided by this discipline so that we know where the field has come from, where it is now, and where it is going. To accomplish this, my research group used the CiteSpace literature analysis tool (Chen et al. 2015) to analyze the knowledge that this discipline has accumulated, with the goal of revealing both the research frontiers and the knowledge foundation that supports urban metabolism research (Wang et al. 2021). To do so, we examined the collaborative networks that have brought researchers together from around the world, the disciplines in which they are working, the key topics, the most influential (highly cited) authors and papers, and the structure, context, and evolution of this knowledge.

2.2.1 *The Number of Publications*

As of 29 April 2019, a search in the Web of Science Core Collection (https://clarivate.com/webofsciencegroup/solutions/web-of-science-core-collection/) using the phrase "urban metabolism" produced 1972 references (1970–2019). This decreased to 1069 papers after we excluded irrelevant results in medicine, botany, zoology, and microbiology (Fig. 2.4).

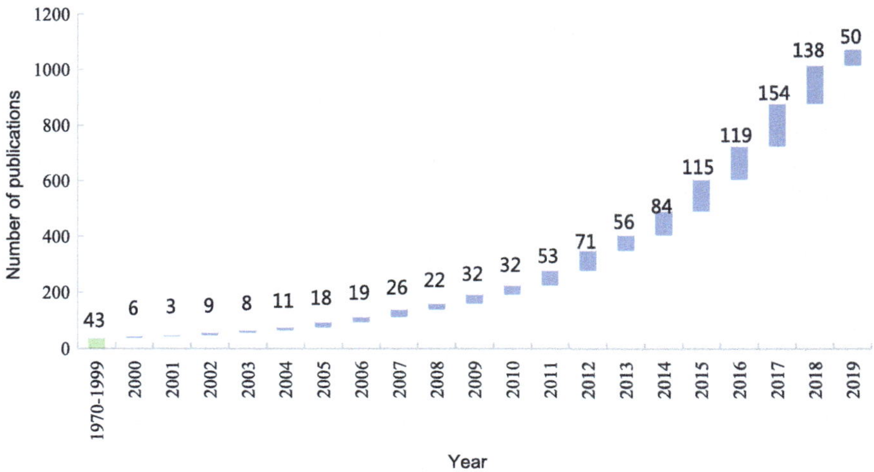

Fig. 2.4 Increase in the number of articles published in the field of urban metabolism research from 1970 to 2019

The data in Fig. 2.4 suggest that there were two distinct phases: before and after the year 2000. Before 2000, urban metabolism research developed relatively slowly, with an average of less than 1.5 papers per year (only 4% of the total number of publications published during the study period, with most of these publications in the 1990s). Since 2000, research in this field has increased exponentially. Since 2015, the annual number of publications has exceeded 100, with nearly half of the total publications published from 2015 to 2018. Because the data in 2019 covered only part of the year, we excluded this year from our analysis.

2.2.2 Collaborative Network Analysis

Collaborative network analysis focuses on the importance of cooperation among researchers from different institutions and countries, thereby forming a network of relationships. Since 1981, when urban metabolism research became an important topic in the United States, research in this field has spread to more than 30 countries (Fig. 2.5). The United States (frequency = 279) and China (frequency = 263) are the two countries with the largest frequency (number of publications), but with different institutional distributions. The number of institutions that performed urban metabolic research in the United States was relatively large, and included Arizona State University, Towson University, University of Queensland in America, University of Maryland, University of California at Los Angeles, University of Georgia, University of Michigan, and Yale University, each with fewer than 15 publications. In contrast, research in China has been concentrated at several major research institutions (Beijing Normal University, Chinese Academy of Sciences, and Tsinghua University), and the number of publications is larger than that in the United States. Beijing Normal University has been particularly prominent (frequency = 100), which is higher than the total for the third-ranked country (England, frequency = 81). The centrality of a particular country can be calculated to provide a quantitative measure of how many other countries are connected to it through collaborative research. Both the United States (centrality = 0.60) and China (centrality = 0.27) had high centrality as a result of close cooperation with researchers in the United Kingdom, Italy, Spain, Canada, Australia, The Netherlands, Austria, Germany, France, Japan, and Portugal.

More than 200 scholars from around the world have studied urban metabolism and published two or more papers. Canada's Christopher A. Kennedy (Sahely et al. 2003) began his research in 2003, and Zhang Yan and Yang Zhifeng of China's Beijing Normal University started their research in 2006, becoming the earliest researchers in China (Fig. 2.6). Of the 10 researchers who have published most frequently, 7 are from Beijing Normal University, with Zhang Yan being the most frequent (frequency = 45), followed by Chen Bin (frequency = 34) and Yang Zhifeng (frequency = 27), and these researchers have formed the largest China-U.S.-France collaborative research group, with Zhang Yan at its core. Scholars in this group from Beijing Normal University have cooperated closely with Chinese scholars such as Geng Yong from Shanghai Jiao Tong University and Zhang Tianzhu from Tsinghua University,

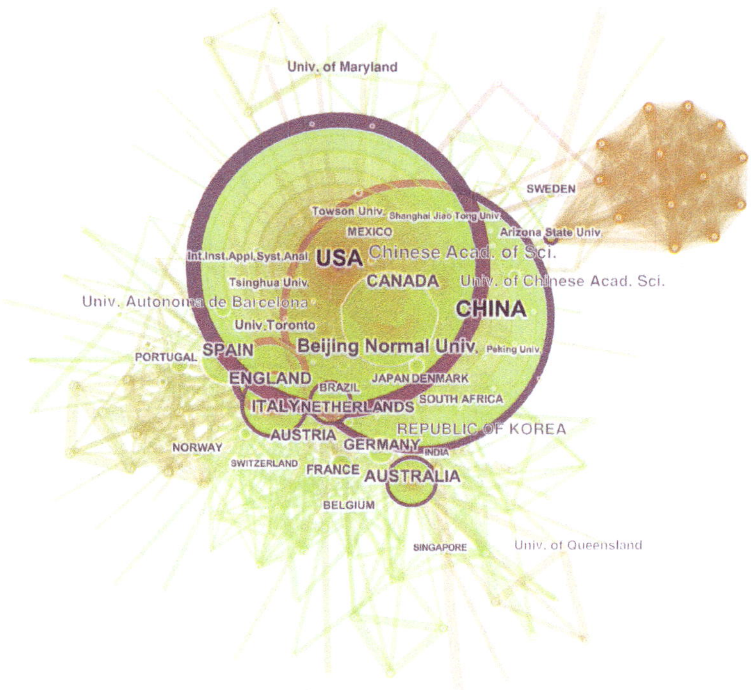

Fig. 2.5 Map of the network of cooperation among countries and institutions (Nodes with a frequency of ≥ 10 are named)

and with European and American scholars such as Brian D. Fath, Sergio Ulgiati, and Sabine Barles. From the perspective of cooperation intensity (a number ranging from 0 to 1, with higher values representing greater cooperation), the closest relationships were for Zhang Yan, who cooperated with Yang Zhifeng; Chen Shaoqing, who cooperated with Chen Bin and Zhu Feiyao; and Su Meirong, who cooperated with Liang Sai and Liu Gengyuan. Each had an intensity of 1.0. Christopher A. Kennedy from Canada's University of Toronto ranked 4th, with 20 publications, and his collaboration with Sabine Barles, Stephanie Pincetl, and other scholars formed the second-largest research group, with Halla R. Sahely, Shauna Dudding, and David N. Bristow closely surrounding Kennedy, each with an intensity of 1.0. In addition, the cooperation between Christopher A. Kennedy and Chen Bin facilitated the connection between these two research groups, so Kennedy and Chen became the nodes with the highest degree of network centrality (0.06).

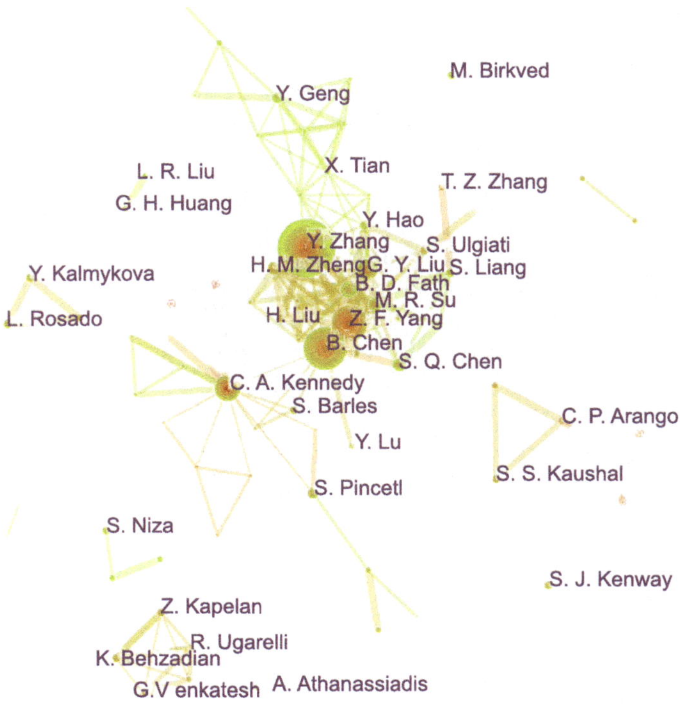

Fig. 2.6 Network map of collaborations among authors (Nodes with a frequency of ≥5 are named)

2.2.3 *Discipline Co-occurrence Analysis*

Discipline co-occurrence analysis aims to reveal the evolution of the mainstream disciplines in the research field, as well as the interdisciplinary characteristics and interconnections among these disciplines. The disciplines associated with urban metabolism research have evolved from science and technology to environmental ecology, then to geography and engineering, and finally to management (Fig. 2.7). Urban metabolism research first appeared in science and technology (frequency = 224, centrality = 0.17). In 1970, *Science* published an article on human settlements in which the author noted that the scale and quality of human settlements should be analyzed from a multidisciplinary and cross-disciplinary perspective, and that the interference of human activities with natural metabolism should be studied (Doxiadis 1970). This paper established a good foundation for urban metabolism research. Subsequently, the scale of human activities continued to expand, disrupting the original metabolic balance between cities and their environment, leading to an increasingly severe level of pollution. Coupled with the maturation of ecological research in the 1960s, this led to a growing interest in urban metabolism, with the field of environmental ecology beginning in 1977, including the sub-disciplines of environmental science and ecology, environmental studies, environmental sciences, ecology, urban studies, and water resources (Newcombe 1977). In 1987, geography focused mostly on the key elements of urban metabolism from the perspective of biogeochemical cycles, while interpreting the natural and spatial characteristics of urban metabolic processes from a multidisciplinary perspective (Bornkamm 1987). Since 1990, urban metabolism research has shifted to engineering applications, with disciplines such as engineering, environmental engineering, and energy and fuels emerging (Wright 1990). Since 1996, more emphasis has been placed on the purposes and ethical values of urban metabolism research, and a large number of studies on regional and urban planning, public administration, and green and sustainable science and technology have emerged, highlighting the significance and potential application of urban metabolism research (Baccini 1996). Of course, engineering research during this period applied urban metabolic practices, whereas geographic research (including multidisciplinary geosciences, geology, and geography) remained dominant in urban planning and design.

The urban metabolism discipline with the largest number of research articles was environmental ecology, where environmental science and ecology had the largest frequency and network centrality (frequency = 671, centrality = 0.56), with environmental science also having a large number of articles and centrality (frequency = 513, centrality = 0.22), suggesting that environmental ecology was the dominant discipline in the field of urban metabolism research (Fig. 2.7). Engineering (frequency = 317, centrality = 0.20) contained only half the number of articles, but with a similar centrality, whereas science and technology, the earliest discipline to emerge, remained relatively prominent (frequency = 224, centrality = 0.17). Some other disciplines had a publication volume similar to that of the science and technology disciplines, but with a difference of more than one order of magnitude in the degree

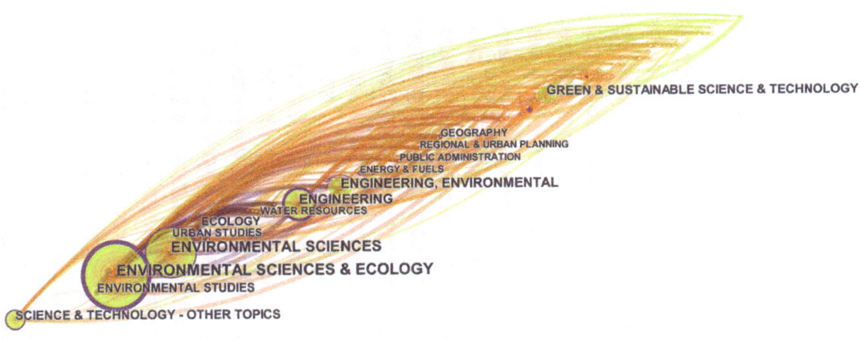

Fig. 2.7 Temporal evolution of the key disciplines in the field of urban metabolism research (*Note* Temporal evolution is from the early research at the left to the more recent research at the right)

of centrality, such as environmental engineering (frequency = 259, centrality = 0.08) and green and sustainable science and technology (frequency = 207, centrality = 0.04), which indicates that these disciplines were flourishing, though they were relatively independent and had no strong ties with other disciplines. In contrast, ecology (frequency = 117, centrality = 0.23) and urban studies (frequency = 81, centrality = 0.16) had a much lower number of publications but a high degree of centrality, suggesting that these disciplines focus on interdisciplinary integration. The disciplines in which the characteristic indicators (frequency and centrality) were most prominent determine the tone of urban metabolism research, but multidisciplinary cross- fertilization has been a constant feature of this research.

In the discipline correlation network (Fig. 2.8), the correlation between similar disciplines was strongest. For example, the connection intensity within the environmental sciences was 1.0 (the maximum value), with environmental science and ecology, environmental sciences, and environmental engineering belonging to environmental ecology, and with engineering and environmental engineering belonging to engineering. In addition, the correlations were all 1.0 between water resources, environmental engineering, ecology, and urban studies, between geography and ecology, and between geology and urban studies, which reflects the refinement of urban metabolism research in some disciplines and its interdisciplinary characteristics.

Fig. 2.8 Intensity mapping for correlations among the disciplines involved in urban metabolic research

2.3 Research Frontier Analysis

2.3.1 Timeline Analysis

Co-occurrence analysis of the literature indicates the keywords that occur together, which reveals both the research hotspots and their evolution (i.e., the emergence of new frontiers), and can clarify the development of a field of research. Figure 2.9 shows 179 urban metabolism research topics with at least three publications that cover five aspects of the research: research subjects, research contents, research methods, impact analysis (impact factors and effect analysis), and management objectives. Here, the research subject relates to a narrower category such as flows of carbon, whereas the research content relates to a broad category such as ecology or economics. There are obvious differences between these aspects in a given period, and their relative importance changed over time. From 1981 to 2009, the number of publications was relatively small, and the research content was emphasized. From 2010 to 2013, research methods (e.g., modeling) and management objectives became most important. From 2014 to 2015, the research content became important, particularly for process analysis, resource flows, and waste flows, and the research expanded to include different scales. From 2016 to 2018, nearly half of the papers were published, and the research content became increasingly important.

Keyword analysis focuses on the words researchers use to describe the subject of their research. Initially, relatively few keywords were used, with the proportion of the total keywords used between 1981 and 2018 smaller between 1981 and 2009 than in subsequent periods. The research content accounted for half of the total keywords

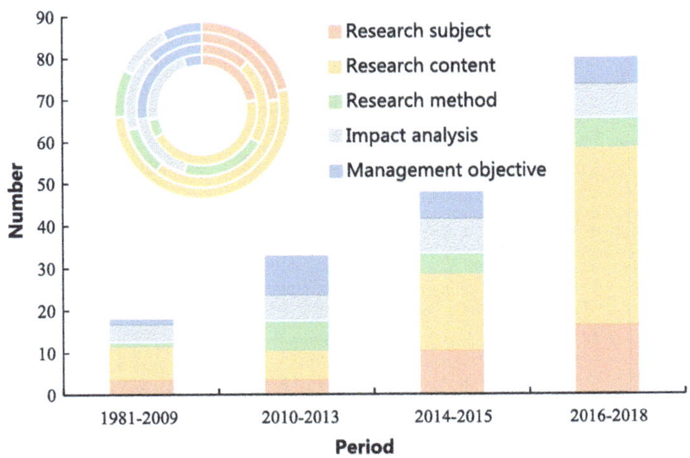

Fig. 2.9 Characteristics of urban metabolism keywords over different periods (*Note* The circular graph represents the proportions of the total accounted for by each category, with the inner ring representing 1981–2009 and the outer ring representing 2016–2018)

during this period, and research subjects and impact analysis each accounted for approximately 20% of the total (Fig. 2.10). In terms of the research content, the frequency (frequency = 218 in 2002) and centrality (centrality = 0.28) of urban metabolism were both high and this keyword was the most important node in the network. At the same time, the consumption and occupation of different materials and elements, which involved keywords such as material flows, construction materials, organic matter, organic carbon, and primary production, were prominent. Urban area (frequency = 107 in 2008) and urban system (frequency = 74 in 2007) were also frequent research subjects, but urban area had the highest centrality (0.31). In addition, keywords such as urban ecosystem and urban scale emerged. Since then, scholars began to pay attention to the analysis of impact factors, which had expanded from the earliest and macroscopic keyword (environmental impacts; frequency = 29 in 2005) to more specific keywords such as human activity (frequency = 19 in 2009) and ecosystem function (frequency = 3 in 2009). However, few of the keywords during this period related to management objectives. Sustainable development (frequency = 44) appeared as a management objective in 2006, and was the most important management objective during this period. It laid a foundation for the refinement and deepening of subsequent management objectives. There was only one keyword under research methods, substance flow analysis (i.e., copper; frequency = 3 in 2009), which focused on single elements.

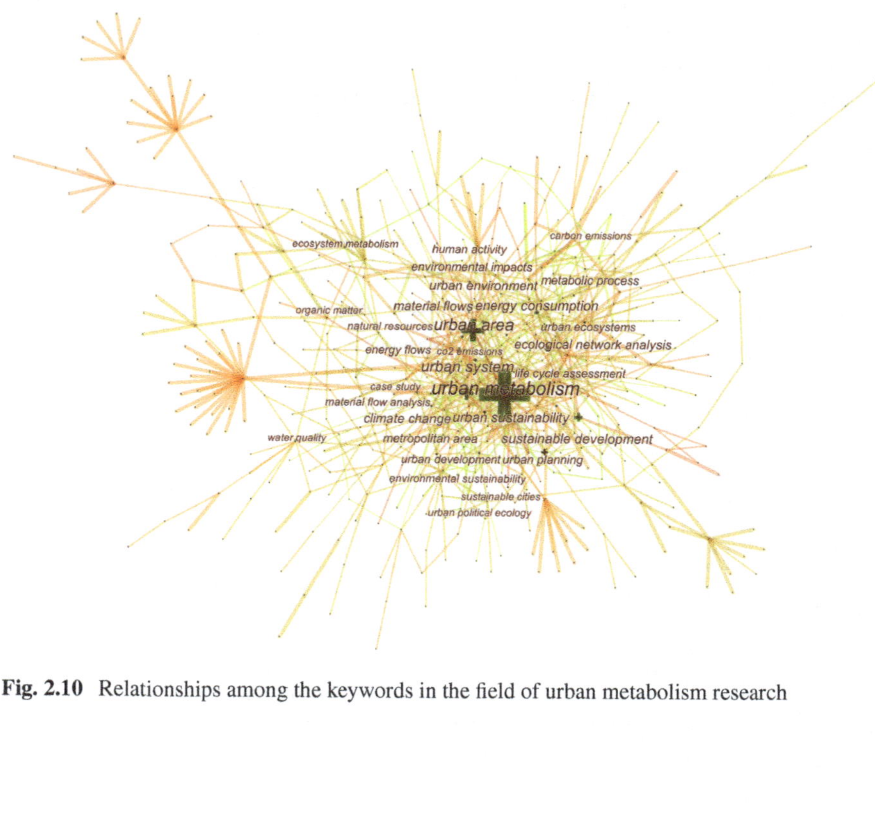

Fig. 2.10 Relationships among the keywords in the field of urban metabolism research

From 2010 to 2013, the number of keywords was twice the number of the previous period and a large number of keywords related to methods appeared. Ecological network analysis (frequency = 36, centrality = 0.24 in 2010) had the highest frequency and centrality, and became a mainstream method during this period. This method explores functional relationships within the urban network by constructing an ecological network model (frequency = 3 in 2010) and computing a network utility matrix (frequency = 3 in 2010), which reflects the transformation of urban metabolism from linear or cyclic models to a network model. Since then, life cycle assessment (frequency = 18, centrality = 0.06, in 2012), emergy analysis (frequency = 3 in 2009), and methodological framework (frequency = 3 in 2013) became important. (Here, emergy refers to embodied energy, which will be discussed briefly in Sect. 2.4.1 and in more detail in Chapter 4.) In addition, management objectives became the most common in this period. These objectives started from urban planning in 2010 and, in 2011, evolved to goals derived from sustainable development (such as urban sustainability, environmental sustainability, and sustainable urban development). Subsequently, more specific management objectives related to efficiency emerged in 2013, such as environmental performance and energy efficiency. At the same time, the research on impact analysis also expanded, including studies of social and economic influences, such as economic development and population growth, and studies of ecosystem services, factors that influence the ecological environment (such as climate change) and ecosystem services, and factors that influence consumption behavior, such as energy consumption and resource consumption. During this period, the number of research contents and subjects remained essentially the same as in previous periods; however, with a deepening of the research methods and impact analysis, the research contents and subjects changed. The study subjects shifted from urban areas (such as urban ecosystems and urban metabolic systems) to the environment and surrounding areas (such as the urban environment and urban surrounding areas), and the range of study scales expanded. Apart from the study of substance consumption (such as energy flows, metabolic flows, and natural resources), the research subjects also began to focus on waste (such as greenhouse gases). Meanwhile, more detailed urban metabolic processes (such as human metabolism and the effects of population density) emerged during this period, and ecological relationships were increasingly analyzed.

Of the 48 keywords that appeared from 2014 to 2015, research content was the largest category, accounting for nearly 40% of the total and focusing on process analysis (such as metabolic processes, urban metabolic processes, ecosystem metabolism, social metabolism, and energy metabolism), and on resource utilization and pollution emission (such as resource flows, gross domestic product, material resources, water quality, greenhouse gas emissions, carbon emissions, CO_2 emissions, and waste generation). Under the category of research content, findings related to metabolic processes (frequency = 26 and centrality = 0.09 in 2014) and case studies (frequency = 14 in 2015) became prominent, while studies such as literature reviews also emerged, and researchers began to pay attention to urban resilience (2015). During this period, research was carried out at different scales, accounting for 23% of the total keywords, which reflects the richness and development of research on urban metabolic processes. In addition to advancing the research at scales such as the urban metabolic system, urban space, and urban context in Beijing (China) and Oslo (Norway), the study scales expanded to include background scales such as the metropolitan area, which had a high frequency (23 in 2014). This area of research also began to include rural areas and subjects related to water, such as the urban water system, urban streams, urban watersheds, and urban water utilities. Impact analysis research continued to be an important part of the socioeconomic impacts category from the previous period, with rapid urbanization, economic growth, economic activity, urban growth, and final demand appearing. At the same time, water consumption and water supply also became research hotspots. During this period, material flow analysis (frequency = 18 in 2015), which was the traditional analysis method, remained the leading method, and became a mainstream direction for industrial ecology research. Model development and conceptual models also became hot topics. Compared with the previous period, management objectives became a stronger focus. Although the concepts of sustainable development (such as sustainable urban metabolism, sustainability challenges, and urban sustainability indicators) continued to be important, more emphasis was placed on urban political ecology and political ecology.

From 2016 to 2018, many keywords flourished, with research contents being abundant (accounting for 53% of the total number of keywords), among which in-depth research results such as process analysis, resource utilization, and pollution emission accounted for nearly 50%. The research contents associated with water and energy (such as water resources, water flows, fossil fuels, stream metabolism, and urban energy metabolism) were widespread, and the frequency and centrality of both water resources (frequency = 12 and centrality = 0.05 in 2016) and fossil fuels (frequency = 9 and centrality = 0.05 in 2016) were high. This expanded to include research on water-energy relationships, with keywords such as energy-water nexus and energy water. Meanwhile, the analysis of relationship types (such as mutualisms, competition, and carrying capacity) was also emphasized. During this period, more studies were conducted on the evaluation of footprints (such as the water footprint, carbon footprint, and ecological footprint), and the research also focused on metabolic characteristics (such as the urban metabolic rate). New research directions such as material stocks and international trade also emerged.

The research contents during this period not only focused on long-term comparative analysis, but also began to include scenario analysis to simulate the future. In addition, spatially explicit research topics, such as spatial distribution, spatial pattern, urban form, and spatial analysis, as well as analytical framework and conceptual framework, also received attention. The research subjects during this period were increasingly considered in the context of complex systems, with expansion of the details, targets, and background scales. For example, urban agglomerations in general and specific agglomerations, such as the Tianjin-Hebei region, appeared as the background scale; the detailed scale began to include the areas outside the city (such as the natural environment and the peri-urban area), residents (such as urban residents and urban population), and economic sectors (such as construction, urban infrastructure, different sectors, and urban energy systems). Further, keywords such as urban level, urban network, and Chinese cities, as well as specific cities, such as China's Hong Kong special administrative region and Syracuse (New York, USA), appeared as the target scale.

During this period, the bottom-up approach was more clearly defined and methods from other disciplines such as economics were introduced, for instance, input–output analysis and system dynamics. Based on in-depth research on sustainability assessment methods, the application of integrated approaches and multi-scale integration were emphasized. In addition, research on uncertainty increased, therefore, sensitivity analysis became a research hotspot. Based on the continuing emphasis on decision objectives (such as environmental benefits and main objectives), the focus expanded to include practical implementation of management objectives to support policy-makers, modeling of urban transformation, and urban design (such as intervention strategies, the circular economy, and urban design). Impact analysis research continued to focus on socioeconomic influences (such as economic benefits, final consumption, and socioeconomic development), on changes in natural conditions (such as environmental impacts, global warming, and environmental challenges), and on human-nature interactions (such as land-use change).

2.3.2 Cluster Analysis

Cluster analysis focuses on understanding the relationships among areas of research by examining the most closely related keywords. This analysis revealed 10 general categories, namely urban resilience (the keyword with the longest period), followed by metabolic system, energy use, urban water system, material flow, energy quantification, urban form, urban stream, carbon emission, and sustainable resource management (Fig. 2.11). Urban resilience was the cluster with the longest time span (from 2002 to 2018, i.e., 17 years). This cluster has expanded from its early research focuses (such as the urban system, urban ecosystem, urban scale, and peri-urban area) to management objectives and method frameworks. The purpose of research in this cluster is to buffer cities to allow them to cope with uncertainty and disturbances through reasonable preparations, so as to maintain normal operation of urban public security, the social order, and the economy. The development of this cluster shows that analysis of metabolic processes (such as urban metabolism, urban metabolism analysis, and urban energy metabolism) is an important perspective for the study of urban resilience. By paying attention to the changes of metabolic flows and actors, including construction materials, material flows, energy flows, and water quality, it explores the important social drivers behind urban resilience in depth, and provides an important basis for urban planning and development of a regulatory network for more sustainable cities. As the smallest cluster, sustainable resource management had a time span of only 3 years. This cluster emphasizes the strengthening of the management of raw materials and waste under the guidance of insights from industrial ecology, and takes the urban network as its form of organization, with the goal

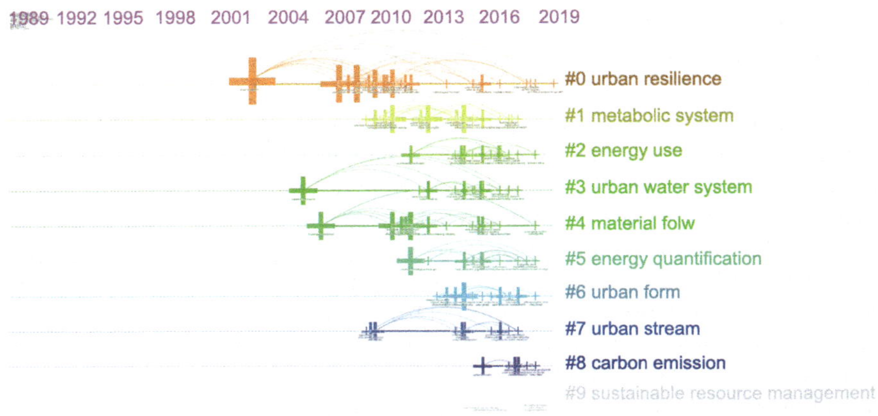

Fig. 2.11 Clustering and timeline distributions for keywords in urban metabolism research

of building an integrated framework and promoting the development of a circular economy (Fig. 2.11).

Metabolic system was the second-longest cluster and spanned 10 years (from 2009 to 2018). This cluster expanded from case studies that focused on research methods and impact analysis to include a range of scales. Methods such as substance flow analysis and ecological network analysis increased the resolution of descriptions of metabolic processes. In addition, using these methods, researchers also developed models, and studied ecological relationships and spatial patterns at different scales, with research that focused on a specific metabolic system (e.g., carbon), a specific city (such as Beijing), urban agglomerations, or a specific agglomeration (such as Tianjin-Hebei). Spatial analysis research was also reflected in the 7th cluster, i.e., urban form. This cluster had a time span of 5 years, and the research mainly focused on the specific spatial forms shown by cities, including the external geometrical form of the urban land, regional differentiation in the pattern of urban functions, and the spatial organization and appearance of urban architecture. The research focused on analyzing the factors that influence urban form and the refinement of conceptual models. This cluster also examined the ecological effects (such as resource consumption, water consumption, environmental performance, the water resource, carbon dioxide, and carbon footprint), which represented breakthroughs, and emphasized the purposes and ethical values of the research through conceptual models and urban design.

Material flow was the 5th cluster and spanned 12 years (from 2006 to 2018). It includes a rich diversity of management objectives and several different research subjects, and fully embodies the analysis of management targets oriented towards urban residents (such as sustainable urban development, urban development, economic benefits, main objectives, and energy efficiency), the environment (urban environment and natural environment), and energy (urban energy systems).

The clustering analysis included five sub-clusters related to water and energy research. Water-related clusters included the urban water system (cluster 4, 2005–2017) and urban streams (cluster 8, 2009–2017), which focus on water resource utilization and natural river ecosystems, respectively. Approximately one-third of the keywords in these two clusters related to research subjects such as the urban water system, urban water utilities, urban space, rural areas, and Chinese cities. Related studies focused on metabolic processes (such as the social metabolism, ecosystem metabolism, and stream metabolism), and used multi-scale integrated analysis with a bottom-up approach to examine problems such as carrying capacity and ecosystem respiration. These were often supported by disciplines such as the social sciences and urban political ecology, as well as by setting different scenarios and formulating different intervention strategies. Energy-related clusters included energy use (cluster 3, 2011–2018), energy quantification (cluster 6, 2011–2018), and carbon emission (cluster 9, 2015–2018). Terms in these clusters emphasized fewer management objectives (only environmental sustainability, urban sustainability, and socioeconomic development). Most studies related to these clusters sought to focus on changes of resource flows (such as fossil fuels, non-metallic minerals, and external resources) and waste flows (such as greenhouse gases, food wastes, demolition waste, and carbon emissions) and their impacts (such as environmental impacts and global warming) by means of material flow analysis, ecological footprints, metabolic rates, and an integrated approach.

2.3.3 *Burst Analysis*

"Bursts" represent periods when there is a sudden rapid increase in the citation frequency of a given keyword. Figure 2.12 shows the burst strength, which represents the increase in the number of citations compared with the number in the previous period, and the burst interval (the period during which the burst occurred) for the four dominant keywords that exhibited a burst, i.e., urban area (burst strength = 9.63), urban metabolism (burst strength = 9.44), urban planning (burst strength = 5.88), and urban development (burst strength = 4.04). The first three belong to the cluster of urban resilience, whereas urban development belongs to the material flow cluster. This indicates that within a specific time range, the subjects including urban area, urban metabolism, urban development, and urban planning were research hotspots and frontiers. From the perspective of burst duration, there were significant differences in the distribution of keywords. Urban metabolism appeared on a large scale from 1982 to 2008, which means that it has been a research hotspot for nearly 30 years. Subsequently, urban case studies became popular, and the keyword urban area experienced a burst from 2008 to 2013.

2.3.4 *Cluster Analysis for Co-cited References*

References that are co-cited in the same paper also represent clusters of related information. Co-citation analysis can, therefore, reflect the knowledge base of urban

Fig. 2.12 Burst degree (strength) and interval during which the burst occurred for the keywords with the four largest citation burst strengths in the urban metabolism field from 1970 to 2019

metabolism research and the relationships among its subjects. In the research on which this section was based (Wang et al. 2021), my research group found 40,226 references between 1956 and 2018 that were co-cited (i.e., that were cited together in at least one paper) by 1069 publications that we analyzed. Through keyword clustering, we identified nine clusters (Fig. 2.13) with four main aspects: research methods, management objectives, discipline foundations, and research subjects.

Four clusters were named after the corresponding research method: emergy, ecological network analysis, ecological footprint, and composite indicator. Emergy and ecological network analysis were the two largest clusters, with the longest time span (from 1965 to 2018, and from 1973 to 2017, respectively), whereas ecological footprint (from 1996 to 2013) and composite indicator (from 2011 to 2016) had the shortest time spans, and most of these publications belonged to the 1069 initial publications. Clusters of emergy and ecological network analysis accounted for 43% and 30% of the additional publications (which are the publications that remained after excluding the 1069 selected publications), respectively. Apart from the keywords used in the 1069 initial publications, such as emergy indicators, life cycle analysis, emergy analysis, and relationship and utility analysis, achievements related to emergy's basic parameters and ecological network methods and principles emerged. These included transformity, emergy synthesis, network environ analysis, ascendency, and information indexes.

The clusters named after the management objective were water efficiency, nutrient retention, and urban mining. Most of the keywords in the nutrient retention cluster were the same as those used in the 1069 selected publications, and emphasized the factors that affected the identification and management of interventions, such as land-use change, urbanization, global change, disturbance, and stream channel modification. As the third-largest cluster, water efficiency spanned the period from 1965 to 2016, of which 55% represented additional literature. The keywords in this cluster were consistent with those in the 1069 initial publications, including industrial ecology, material flow analysis, substance flow analysis, urban development, and sustainability, with the new keywords of urban ecology, territorial ecology, urban planning, resource management, and urban design. Publications in the urban mining cluster (2010–2016) were also mostly the 1069 initial publications. In addition, building stock, urban fabric, infrastructure, material flow analysis, material stock analysis, life cycle assessment, and dynamic modeling were added. These keywords reflect a management orientation and added basic models, such as recycling of construction materials, dynamic changes in the quantity of waste being produced, and flow modeling.

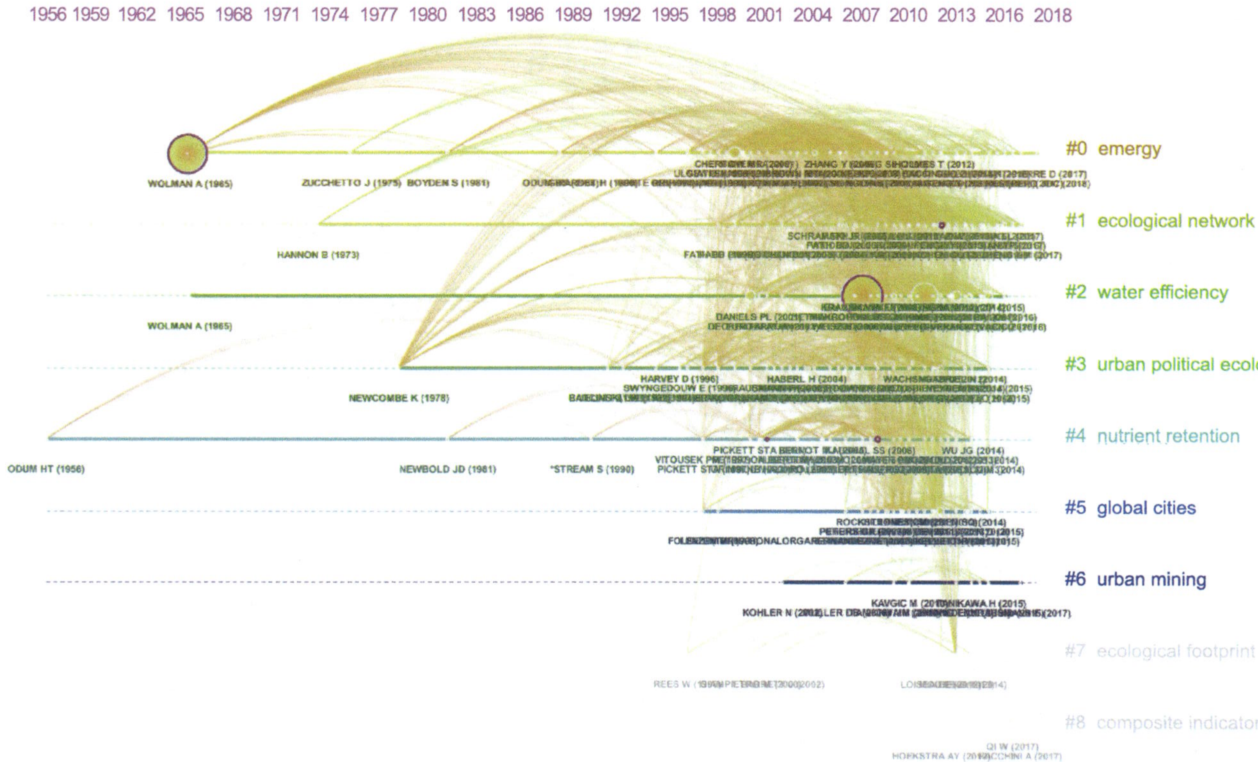

Fig. 2.13 Clusters of keywords in the 40,226 co-cited references (*Note* Red circles represent the publications that had the highest centrality. The small text, which could not be enlarged because of software limitations, represents author names and publication dates)

The cluster named after a discipline was urban political ecology, which covered the period from 1978 to 2015, with 13% of the additional literature. In addition to the keywords involved in the initial publications, such as industrial ecology, urban political ecology, industrial symbiosis, urban theory, and modes of social discipline, the knowledge bases of urban ecology theories and models (such as urban ecology theory and urban ecological models) were added.

The global cities cluster was named after the research subject, and covered the period from 1998 to 2015, and 20% of these publications were additional literature. The keywords (such as urban system, city, and cities) were related mostly to cities. Based on greenhouse gas emissions, carbon footprints, and urban energy uses, the keywords carbon cycle and complexity were added in this cluster. This cluster included input–output analysis, life cycle assessment, and other methods. In addition, thermodynamic methods were also added.

By comparing the co-occurrence and co-citation clusters, it appears that the latter had longer periods and more clusters. Although they do not have exactly the same cluster names, their clustering overlaps to a certain extent. Co-occurring clusters focused on metabolic materials related to water and energy (such as energy use, material flow, energy quantification, urban stream, and carbon emission) and study subjects (such as metabolic system and urban water system). Co-citation clustering focused more on research methods (such as emergy, ecological network analysis, ecological footprints, and composite indicators). In addition, the management objective clusters in both the co-occurrence and co-citation analyses had the same size, but the co-occurrence clusters focused on macroscopic management goals, such as sustainable resource management and urban resilience, whereas the co-citation clusters focused on innovative management concepts (such as water efficiency, nutrient retention, and urban mining), which more fully reflect methods that support different subjects in the field of urban metabolism research and the multidisciplinary characteristics of urban metabolism research.

2.3.5 Analysis of High-Frequency Co-cited Literature

The 30 most frequently cited references represent knowledge bases related to the development of this field of research, case studies, innovative models, and management innovations. Nearly 80% of these publications have inspired research frontiers for urban metabolism. Other frequently cited publications include discussions of social metabolism, material flow analysis, and emergy analysis, which also provide important support for the development of urban metabolism (Fig. 2.14).

Of the 30 most frequently cited references, 37% were reviews and perspective articles which comprehensively support the development of urban metabolism research from the perspectives of the name's connotation, interdisciplinary characteristics, technical frameworks, and universal laws. For example, the review article published by Kennedy et al. (2007) in the *Journal of Industrial Ecology* had the highest citation frequency (261), and moderate centrality (0.12). This study explored the evolution of

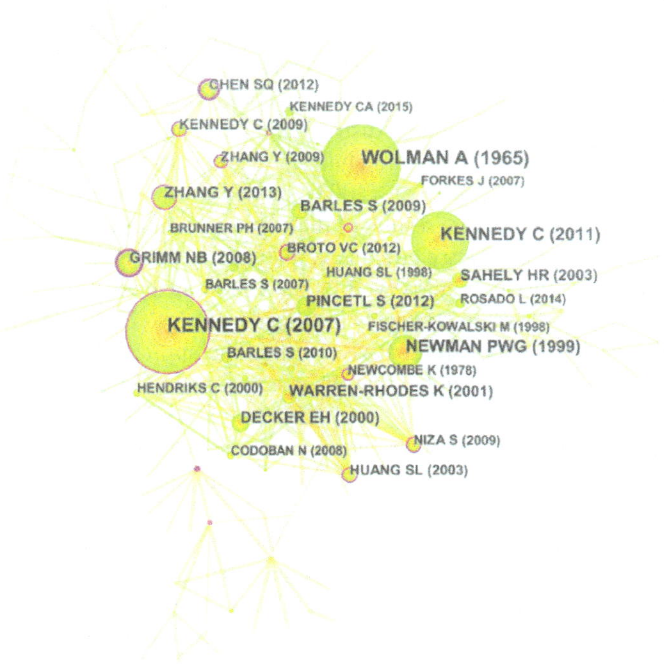

Fig. 2.14 Map of the 30 most frequently cited publications (co-citations) in urban metabolism research

the rules that govern urban metabolism based on data from eight metropolitan areas on five continents. The article published by Wolman (1965) in *Scientific American* was also frequently cited (261 times), but its centrality (0.04) was relatively low. In this paper, Wolman defined the connotation of urban metabolism for the first time, and emphasized the importance of viewing problems with the urban ecological environment from the perspective of urban metabolism. The opinion article published by Grimm et al. (2008) in *Science* had the highest centrality (0.39). In it, Grimm noted that achievements in global change and urban ecology research can provide cutting-edge support for urban metabolism research. Brunner (2007) published an opinion article in the *Journal of Industrial Ecology* (frequency = 216, centrality = 0.12), pointing out the important role of universal methods and laws for modeling urban metabolisms. Fischer-Kowalski (1998), Barles (2010), Kennedy et al. (2011), Broto et al. (2012), Pincetl et al. (2012), and Zhang (2013) systematically reviewed advances in urban metabolism evolution, material flow analysis, and multidisciplinary intersections. Zhang (2013) proposed a technical framework for the study of urban metabolism, and this paper had the third-highest centrality (0.14).

Case studies accounted for 30% of the most frequently cited studies. Newcombe et al. (1978) and Warren-Rhodes and Koenig (2001) conducted in-depth studies of Hong Kong, whereby Warren-Rhodes and Koenig updated Newcombe's groundbreaking research results and noted that such results play a significant role in assisting the formulation of urban management goals and action plans. Huang and Hsu (2003), S. Q. Chen and B. Chen (2012), and Kennedy et al. (2015, 2010) performed a case study of Taibei (China), of Vienna (Austria), and of 10 and 27 global cities (respectively), with special emphasis on the analysis of the causes leading to the current urban metabolic state. Sahely et al. (2003) studied the metabolism of the greater Toronto area and noted that the results of their study could provide important information to improve urban energy efficiency, material recycling, waste management, and infrastructure construction, as well as an important basis for the formulation and implementation of urban strategies in Canada. Barles (2007, 2009) carried out a case study for the metabolism of Paris, verified the applicability of material flow analysis to this research, and noted that the research results could assist technology research and development. Kennedy et al. (2010) analyzed the differences in urban greenhouse gas emissions and their causes, and because of the importance of this research for understanding global climate change, this made their study a hotspot with the highest centrality (0.21).

Research on modeling methods accounted for 30% of the frequently cited references. Huang (1998) and Zhang et al. (2009a, 2010a, b) introduced the emergy method and ecological network analysis into urban metabolism research, constructed the first urban energy flow diagram and ecological network model, and analyzed the evolution of metabolic processes in typical cities such as Beijing and Taibei. Newman (1999) proposed an extended, livability-oriented model of the metabolism of cities and tested it for Sydney, Australia; it was cited 131 times. Hendriks et al. (2000) and Codoban and Kennedy (2008) emphasized the important role of material flow analysis to support environmental decision-making and sustainable urban design, and performed case studies of Vienna and the Swiss lowlands, and of four typical communities, respectively. Niza et al. (2009) and Rosado et al. (2014) emphasized the importance of material flow analysis and applied it to a study of Lisbon. Rosado et al. (2014) constructed an urban flow-stock model based on material flow analysis.

Only one study of management was found. Forkes (2007) adopted a nitrogen-balance method to quickly assess the impact of urban waste management policies and plans on the urban nitrogen cycle and nitrogen metabolism, and identified opportunities for improving the waste nitrogen cycle, thus providing support for the strategic adjustment of urban waste management.

2.4 Development Stage of Urban Metabolism Research

Although the CiteSpace analysis provided a good overview of key aspects of urban metabolism research, it did not provide an integrated picture. To fully understand the history of this research, it was necessary to examine the changes over time. Based on the number of publications, urban metabolism research can be divided into three periods: initial proposal of the theory and case studies (from 1965 to 1980), slow growth during a period when the methodology had not yet been standardized (from 1981 to 2000), and a rising period that resulted from gradual standardization of the methods and application practices (from 2001 to the present).

2.4.1 Early Period (1965–1980)

Quantitative research on urban metabolism began in 1965, when Wolman built the first black-box input–output model of a city and performed a hypothetical case study of an American city with a population of 1 million. Case studies of real cities began in the early 1970s, with studies of Miami (Zucchetto 1975), Tokyo (Hanya and Ambe 1977), Brussels (Duvigneaud and Denayeyer-De Smet 1977), and Hong Kong (Newcombe et al. 1978). In addition to quantifying the human technological metabolism, the Tokyo, Brussels, and Hong Kong studies examined the natural components of urban metabolism, including the natural energy balance, plant biomass storage, and transformation processes. This was an important development because urban metabolism not only emphasizes changes in the city's natural environment caused by social and economic activities (i.e., urban ecology), but also identifies the natural environment as a key component of the urban ecosystem. Thus, it emphasizes coordination and symbiosis between the urban natural and human (social and economic) components of the system. This has led to much debate over whether the technological metabolism has been overemphasized (Kennedy et al. 2007) and whether the natural metabolism should be given equal importance to the technological metabolism (Golubiewski 2012).

During this period, the analysis of material and energy flows was most commonly used to study urban metabolism. Flows of water, materials, and nutrients were expressed in terms of their mass, whereas energy was mostly expressed in energy units such as joules (Baccini and Brunner 1991). The ecosystem theory of E. P. Odum (1953) and the heterotrophic characteristics of cities (Odum 1975) provided a theoretical foundation for urban metabolism research. Systems ecologist H. T. Odum (1971) studied the relationship between humans and the environment from an energy perspective, highlighting the importance of energy as a way of characterizing ecosystem metabolic processes, such as photosynthesis for organic matter production and respiration for consumption of the energy captured by photosynthesis. On this basis, H. T. Odum (1973) formally proposed the term "emergy" (embodied energy) that represented a way to convert flows of different, non-comparable materials, energy, and money, into a single unit of measurement (emjoules), thereby accounting for differences in their quality and emphasizing the integrity of all flows within the human economic system and between this system and its external environment. This approach provided a methodological basis for studying Miami's metabolic processes (Zucchetto 1975). However, this method faces some challenges, such as the need to eliminate double-counting and disagreement over the most appropriate conversion factors, which have prevented this approach from becoming a mainstream method in subsequent research. Some scholars continue to seek ways of improving emergy analysis, such as using an algebraic method to distinguish the emergy of products, associated products, and symbiotic products to avoid double-counting (Kennedy et al. 2011).

2.4.2 Slow Growth Period (1981–2000)

During this period, urban metabolism research expanded relatively slowly, and there were significant problems with non-uniformity (i.e., different ways to quantify flows) and difficulty standardizing the methodology (Daniels and Moore 2001). The theoretical foundation deepened through the addition of the concept of urban parasitism, in which the city depends on its internal and external environments for survival, but without providing any benefit to those environments (Odum 1989), and the concept of a hierarchical relationship among the system's components (Odum and Odum 1981). Odum conducted metabolic studies of Paris, France, in the 1850s using the emergy method (Odum 1983; Stanhill 1977).

In contrast with the emergy measurements of H. T. Odum, researchers who adopted material flow analysis focused on the resource stock and changes in this stock as a result of flows. For example, *Metabolism of the Anthroposphere* (Baccini and Brunner 1991) and *Regionaler Stoffhaushalt: Erfassung, Bewertung und Steuerung* (Baccini and Bader 1996) emphasized the need to account for materials such as nutrients, salt, materials, and water in the urban hydrological cycle. Under UNESCO's Man and the Biosphere Programme, researchers studied the urban metabolisms of Rome, Barcelona, and Hong Kong in the early 1970s (Celecia 2000; Boyden et al. 1981) (UNESCO: United Nations Educational, Scientific, and Cultural Organization). In these studies, researchers treated the cities as ecosystems, and defined the components of the urban system as resource consumers, producers, and waste decomposers, with the goal of more clearly describing the complex relationships between nature and humans that result from the flows of materials, energy, water, nutrients, and waste (Douglas 1983, 1981). The aim of this program was to promote multidisciplinary and integrated research (Bonnes et al. 2004), rather than studying specific issues, such as sustainable resource use, biodiversity conservation, and pollutant emissions in isolation. In 1993, the first international symposium on urban metabolism was held in Kobe, Japan, though researchers presented few results. Only one paper was published, which studied the urban food system in developing countries from the perspective of urban metabolism (Bohle 1994). Thus, the symposium did not have a strong impact on the field of urban metabolism research. Later, based on Wolman's pioneering research and under the influence of the 1992 Rio Summit on ecologically friendly development, Girardet (1996) focused on the relationship between urban metabolism and sustainable cities (Kennedy et al. 2007), which laid a foundation for the study of urban metabolism from the perspective of industrial ecology.

During this period, early urban metabolic models were developed, such as a black-box and subsystem model (a simplistic early subsystem model that nonetheless improved on previous black-box models) that considered risk and health impacts, a circular metabolism model designed to improve the sustainability of cities (Girardet 1996; Zhang et al. 2015), and a vitality metabolism model that accounted for social goals (Newman 1999). Akiyama noted that there are two ways to study the metabolic processes of cities. First, the black-box model extracts macroscopic indicators to represent the overall intensity and scale of urban activity by analyzing the inputs and outputs of substances, using indicators similar to those used to describe human health (such as weight, body temperature, and blood pressure). Second, subsystem models divide the black box into metabolic components based on detailed statistical data, allowing analysis of the flows between subsystems and the driving factors that control these flows (Zhang et al. 2015). The most appropriate research method can then be selected to support different research purposes. The black-box model can be used to compare the efficiency and sustainability of different cities when only a small amount of statistical data is available, whereas the subsystem model is helpful for diagnosing a city's problems when data is available at the subsystem level, thereby allowing managers to focus on the treatment of the most serious problems and answering a series of questions such as how to improve energy or material efficiency, how to improve the quality of life, and how to achieve sustainability.

Girardet (1996) proposed a circulation model for sustainable cities, based on the recognition that a linear model with no cycling loops fails to account for the continuous and circular flows within a real city. This is important because the acceleration of urban metabolism during growth and development of a city will generate a global crisis in a linear system, whereas a circular system can avoid or mitigate this problem by recycling some materials and energy (Kennedy et al. 2011). If urban consumers are replaced by transformers, a linear metabolism will be transformed into a cyclic metabolism (Girardet 2008). Newman (1999) proposed a vitality model that considers the goal of urban sustainability, noting that urban sustainability not only includes the reduction of metabolic flows (resource inputs and waste outputs) to reduce pressure on the environment, but also increases human vitality (demand for life services, health, and welfare). This model integrated socioeconomic and environmental factors. Newman's study of Sydney's metabolism based on the vitality model became an important part of the Australian State of the Environment report.

Most of the case studies during this period were conducted in the last decade of the twentieth century. They included a material flow analysis of Prague in the Czech Republic, of Yevler in Sweden, of the Swiss lowlands, and of Vienna carried out by the European Union (Hendriks et al. 2000; Baccini 1997), as well as studies of Taibei (Huang 1998), Sydney (Newman 1999; Newman and Kenworthy 1999), Brisbane (Mullins et al. 1999), 25 cities (Decker et al. 2000), and 5 coastal cities (Timmerman and White 1997). However, case studies of materials, energy, and pollutants have very different goals and results, so it was difficult to compare the metabolic conditions in different cities by continuing to apply such conceptual methods (Goldstein et al. 2013), and these differences also prevented the integration of research results and the assessment of their reliability (Rosado et al. 2014). These problems limited progress in this field to some extent. Therefore, it became urgent to formulate a unified accounting methodology for material and energy flows to clarify the basic data requirements. At the same, it became necessary for local governments to strengthen their capacity for monitoring and reporting.

2.4.3 *Rising Period (2001–Present)*

Since 2001, journals, conferences, projects, reports, and practical studies have been constantly emerging, and urban metabolism studies have become a research hotspot (Kennedy et al. 2011). For example, in 2007, the *Journal of Industrial Ecology* published a special issue on urban metabolism (Bai 2007). In 2008, the ConAccount conference emphasized that urban metabolism is an important tool for measuring the ecological state of cities (Havranek 2009). In 2011, the American Geophysical Union held a conference on urban metabolism characterization, simulation, and expansion of the subjects studied by researchers in this field. The main projects during this period (Table 2.1) included Sustainable Urban Metabolism for Europe (Schremmer and Stead 2009) and sustainaBle uRban plannIng Decision support accountinG for urban mEtabolism (BRIDGE) (Chrysoulakis 2008) by the European Union (EU) in 2008, Evolution of the Lisbon Metropolitan Area Metabolism (MEMO) Project in Portugal (Rosado et al. 2014), the Bangalore Urban Metabolism Project (BUMP) (Reddy 2013) in India, and the Public Interest Energy Research (PIER) by the California Energy Department (Pincetl et al. 2014). Research reports such as a metabolic analysis of ancient Caral in Peru (Fernández 2010) and a metabolic survey of megacities (Kennedy et al. 2015) were generated, which directly supported the feature recognition, planning, and design of different cities, as well as the development of management tools and standards. In addition, this research promoted the integration of the goals of stakeholders and researchers, and more fully reflected the value of applying urban metabolism research.

2.5 Historical Evolution of Urban Metabolism Research

2.5.1 *Accounting Evaluation Methods*

Material flow analysis was developed to analyze the data in the official EU statistical database in 2009 (Eurostat 2009), and has been widely used in studies of cities such as Lisbon (Rosado et al. 2014; Niza et al. 2009), Singapore (Schulz 2007), York (Barrett et al. 2002), Hamburg, Vienna, and Leipzig (Hammer and Giljum 2006), European cities (Schremmer and Stead 2009), and 8 metropolitan areas (Kennedy et al. 2007). In the field of industrial ecology, most scholars have applied substance flow analysis to study metal metabolism in studies of heavy metals in Stockholm (Lestel 2012; Hedbrant 2001), copper in Vienna (Europe), and Taibei (Asia) (Kral et al. 2014), urban nitrogen (Billen et al. 2009; Barles 2007), and carbon in Vienna (Chen and

Table 2.1 Major research projects on urban metabolism

Project	Funding agency	Time period	Purpose	Content	Goal	References
SUME (Sustainable Urban Metabolism for Europe)	EIJ	2000–2011	Design urban systems based on minimizing environmental damage	Starting with the built environment, analyze the influence of different urban forms on resource utilization	Evaluate the potential of spatial restructuring of urban buildings to significantly reduce resource and energy consumption	Schremmer and Stead (2009)
BRIDGE (sustainaBle uRban plannIng Decision support accountinG for urban mEtabolism)	EU	2008–2011	Integrate biophysical science systems to provide innovative planning strategies for urban planning and design in Europe	Account for energy, water, carbon, and pollutant flows, and consider the environmental and social impacts at all stages of planning and policy development, from problem identification and policy design to implementation and post-development assessment	Develop a decision-support system for sustainable urban management	Chrysoulakis (2008)

(continued)

Table 2.1 (continued)

Project	Funding agency	Time period	Purpose	Content	Goal	References
MEMO (Evolution of the Lisbon Metropolitan Area Metabolism)	Centre for Research on Innovation, Technology and Policy, Portugal	2013–2015	Analyze the changes of Lisbon's urban morphology and metabolic behavior characteristics during different historical periods	Evaluate the evolution of urban morphology by material-flow and substance-flow analysis	Identify the driving force of metabolic process transformation, simulate its influence and response mechanisms, and provide a guarantee for building a more sustainable urban environment	Rosado et al. (2014)
PIER (Public Interest Energy Research)	California Energy Department	2007–2012	Create a comprehensive community energy-use agreement and an energy baseline for California	Integrate life-cycle assessment with direct measurements, and indirect and supply chain energy consumption estimates, then determine the baseline of community development, assess the impact of decision, and analyze the spatial differentiation of energy use in combination with land-use information	Provide data for community energy type adjustments in California by developing an energy baseline, and support the development of energy conservation and energy- and land-use planning	Pincetl el al. (2014)

(continued)

Table 2.1 (continued)

Project	Funding agency	Time period	Purpose	Content	Goal	References
Ancient Caral urban metabolism	Holcim Seed Fund	2011	Find a possible transformation from non-fossil fuel cities to fossil fuel cities by reviewing the development history of ancient cities	Compare and analyze the resource consumption scale and intensity characteristics of an ancient city and a modern city (in Peru) based on a resource intensity index	Recommend efficient urban system development scenarios	Fernández (2010)
BUMP (Bangalore Urban Metabolism Project)	Indira Gandhi Institute for Development	2012	Understand the city's key processes and contribute to the development of effective resource policy	Analyze resource inflows, transfers, and transformations to measure the efficiency of resource utilization	Provide an effective framework for stock and flow accounting analysis	Reddy (2013)
Metabolic survey of megacities	Enel Foundation	2014	Identity differences in the biophysical characteristics of large cities	Collect energy, water, material, and waste flow data for 10–15 large cities to assess the impact of public infrastructure construction on urban metabolic processes	Sort out and summarize a development path for sustainable cities	Kennedy et al. (2015)

Chen 2012). Material flow analysis and substance flow analysis were both developed from the perspective of industrial ecology but are different method systems. They refer to the Euler method (materials) and the Lagrangian method of fluid mechanics (substances) to observe a given space (at a global, national, regional, or city scale) and track amounts of specific elements (Lu and Yue 2006). Substance flow analysis makes it easier to open the black box, track the flows of specific elements, and clearly show the reuse, recycling, and other paths along with the flows of elements; thus, the method can effectively evaluate the potential for developing more efficient flows of the element (Bringezu et al. 2009). Therefore, compared with material flow analysis, substance flow analysis can more easily open the black box, track the life cycle of a specific element, clearly show its reuse, recycling, and other paths, and effectively evaluate its development potential (Bringezu et al. 2009). Material flow analysis involves various types of biological and non- biological resources and their transformations, so it requires large amounts of data. However, this method can estimate the overall urban metabolic throughput, comprehensively reflect the overall intensity and scale of urban activities, and facilitate comparisons between cities.

Accounting for these flows provides a clearer picture of urban behavior, allows comparisons of the resource flow status in different cities, and provides insights into whether a city is sustainable and whether the environmental impacts it creates are acceptable. To perform this analysis, an effective indicator or system of indicators is required (Girardet 2014; Golubiewski 2012). Metabolic efficiency (i.e., the production of outputs per unit of input) can reflect the decoupling between resource use, environmental deterioration, and the city's social and economic growth during efforts to achieve sustainable development. The human appropriation of net primary productivity has been proposed when material flow and energy flow analysis is used to study a city's social metabolism (Haberl et al. 2007; Krausmann and Haberl 2002; Haberl 2001; Vitousek et al. 1986). It has been applied to 29 large cities in the basin ecosystem of the Baltic Sea (Folke et al. 1997), the food footprint in Paris (Billen et al. 2009), the spatial footprint of food consumption in Linköping, Sweden (Neset and Lohm 2005), and the occupancy of the city's internal natural environment in Vienna, Austria (Krausmann 2005). In addition, some scholars have used an ecological footprint index (i.e., the pressure per unit area that a city exerts on its environment) to represent the level of urban metabolism based on a matrix that combines resource consumption with land use (Sovacool and Brown 2010; Kenny and Gray 2009). This approach has been used to study Liverpool (Barrett and Scott 2001), London (Chambers et al. 2002), Cape Town (Gasson 2002), Cardiff (Collins et al. 2005), Paris (Chatzimpiros and Barles 2009), and 121 American cities (Jenerette et al. 2006).

2.5.2 *Model Simulation*

Model development during this period became more sophisticated, which is reflected in cyclic metabolism models (Beck et al. 2013; Shillington 2013; Beatley 2012; Gandy 2004), dynamic metabolism models (McDonald and Patterson 2007; Zhang et al. 2006a), and network models (Zhang et al. 2010a, b). By defining the processes that govern transfers of materials and energy between urban metabolic actors, the actors can be abstracted into nodes in a network, and the transfers between actors can be abstracted into paths between nodes; by doing so, the real world can be abstracted into networks. Network models go beyond black-box models by deeply examining the components of the urban metabolic system and analyzing the distribution of its structures and functions. Examples include studies of a transit network model (Samaniego and Moses 2008), a city's urban water metabolism (Zhang et al. 2010a), energy metabolism (Zhang et al. 2009b, 2010b), and material metabolism (Yang et al. 2014). Models can be static, and examine a single point in time, or dynamic, and examine changes over time. Scholars also improved existing static models to create dynamic simulation and prediction models, such as a model of the metabolic processes in the Toronto Metropolitan Area (Kennedy 2012), a dynamic material flow analysis of the built environment (Brattebø et al. 2009), and a dynamic model of the metabolism of urban water services (Venkatesh et al. 2014; Zeng et al. 2014). Although many model types were developed during this period, there have been few spatially explicit network models (Zhang et al. 2014a), which limits the real-world application of such models. Therefore, it is necessary to adopt a combined bottom-up and top-down approach to develop more refined urban models, and combine this with a big data approach to promote the adoption of these models in planning practices. For example, in the top-down approach, the system can be broken into smaller components that are targeted by specific policy and planning objectives; this allows the calculation of parameter values for these components that can then be used in simulation models that support policy development and urban planning.

Some scholars have emphasized the direct consumption or occupation of resources by cities, and have, therefore, focused on simulating the resource consumption embodied in upstream processes (Kennedy et al. 2011). This made it possible to introduce input–output analysis into the study of urban metabolism and the simulation of direct and indirect consumption. Here, "indirect" consumption refers to the consumption of energy or a material whose production earlier in a sequence of activities consumed energy. This approach has been applied in Suzhou (Liang and Zhang 2012), Lisbon (Rosado and Ferrão 2009) and Beijing (Zhang et al. 2014b). The difficulty of this method lies in how to transform the monetary inputs and outputs (the primary data source that is available to support such research) into physical flows and in how to distinguish the sources of imports and exports to account for differences in their technical levels (i.e., to account for differences in the quality of the materials embodied in a flow). This is information that cannot be obtained by examining the currency units of the economic data, which is the most commonly available data to support such analyses.

Ecological network analysis has become an important method to simulate direct and indirect effects of flows, their structures, and the city network's functional characteristics (Fath and Killian 2007). Such analyses have mainly focused on the combination of emergy analysis with material flow analysis. For example, researchers have used ecological network analysis to study the metabolic processes of multiple materials in four typical Chinese cities (Zhang et al. 2009b), and the metabolic processes related to water metabolism (Zhang et al. 2010a), energy use (Zhang et al. 2010b), and material consumption (Li et al. 2012) in Beijing, as well as to study the carbon metabolism in Austria (Chen and Chen 2012).

To understand the dynamic characteristics of metabolic networks, it is also necessary to understand the driving factors behind them. However, the research results have shown limited ability to explain the reasons for the differences or changes in different cities (Barles 2009; Sahely et al. 2003; Gasson 2002). For example, Warren-Rhodes and Koenig (2001) did not consider the changes of urban population size, income, and other socioeconomic factors when they studied Hong Kong's metabolism, so they failed to determine the reasons why Hong Kong transformed from a manufacturing center to a consumption center (Pincetl et al. 2012). Many scholars have also tried to associate the metabolic inflows and outflows with their driving factors, such as industrialization, urbanization, lifestyle changes, technological levels, and other social and economic factors that drive urban change (Li et al. 2019; Inostroza 2014; Barles 2009; Weisz et al. 2006; Liu et al. 2005; Douglas et al. 2002; Schandl and Schulz 2002). Researchers have also examined the impacts of the urban morphology and density, infrastructure, landscape, land use, and other urban structural characteristics on metabolic flows and stocks (Broto et al. 2012; Minx et al. 2011; Deilmann 2009; Krausmann et al. 2003). They have also paid attention to the impacts of urban natural factors on a city's metabolism. This included studies of the thermal environment of an urban heat island (Kennedy et al. 2007), the albedo of rooftops (Susca 2012), and urban forests (Manning 2008). Therefore, it is necessary to fully consider the social, economic, technological, and ecological factors that affect urban metabolic processes, and combine the most appropriate method with policy objectives, potential actions, and design schemes to demonstrate the feasibility and importance of urban metabolism research to support planning, management decision-making, and stakeholder consideration.

2.5.3 Application Research

Urban metabolism studies not only explain the human-nature interaction, but also provide support for public policies and actions (Barles 2009; Baker et al. 2007; Hashimoto and Moriguchi 2004). However, the existing studies on urban metabolism seldom support urban design and planning. Although these studies have provided indicators for quantifying the process of urban sustainable development, this research has produced few operational action plans; for example, one such study (Ferrão and Fernández 2013) provided recommendations for the reconstruction of New Orleans. Meaningful research has been carried out by the Federal Institute of Technology Zurich, Massachusetts Institute of Technology, University of Toronto, and Polytechnic University of Turin, Italy. Oswald et al. (2003) of the Federal Institute of Technology Zurich first proposed the concept of urban network design by combining morphological and physiological approaches to solve the problems with a core-periphery network model. In this approach, the urban network can be divided into core actors, which represent the origins of the network of connections with peripheral actors that interact mainly with the core actors rather than with other peripheral actors. I will discuss this approach in more detail in Chapter 5. Urban engineering students at the University of Toronto designed an enclosed circular infrastructure for the port community of Toronto based on their practical experience in green building design and energy substitution (Codoban and Kennedy 2008; Engel-Yan et al. 2005). Montrucchio (2012) at the Polytechnic University of Turin, Italy, used the metaphor of a single building as a pump to analyze the flows of water, energy, nutrients, and materials to support Swedish architectural design. In particular, Fernández and his research team at the School of Architecture, Massachusetts Institute of Technology successfully analyzed the reconstruction of New Orleans (Ferrão and Fernández 2013; Quinn 2007). This study was performed more than a year after hurricane Katrina, and combined analyses of the population and employment, housing demand, and growth priority data using material flow analysis to measure the material and energy inputs and outputs, the durability of residential types, construction costs, energy utilization, and waste production. On this basis, they developed a software tool to promote a more effective reconstruction of the city, development of its green space, and a balance among the needs of different stakeholders, which they then used to help redesign the city.

Although previous urban metabolism studies were rarely applied to support urban decision-making (Huang and Hsu 2003; Timmerman and White 1997; Wolman 1965), this research remains important for urban policy-makers who need to understand urban metabolic processes, including the availability of urban water, energy, materials, and nutrients, and the efficiency of their flows; whether or to what extent the surrounding resources are near depletion; and when and what measures will be taken to slow the development (Kennedy 2007). By combining strategic environmental assessment with urban metabolism research, planners can reveal the influence of strategies, projects, and planning on the flows, exchange, and transformation of energy, water, carbon, and pollutants, and the strength of this influence (González et al. 2013; Chrysoulakis et al. 2009). This will contribute to timely and more effective policy adjustment (González et al. 2013). Encouragingly, Pincetl et al. (2012) have attempted to match the specific energy and waste flows in Los Angeles County to people, land, and activities in the county. They matched flows with changes in land use and in social and demographic factors in an effort to understand who is using each type of energy, where they are using it, and what the use is intended to accomplish; their goal was to support more careful and effective decision-making.

2.5.4 Scales and Boundaries

2.5.4.1 Multidimensional Scaling

The openness of cities requires an expansion of research on urban metabolisms to a multidimensional spatial scale that includes global, national, regional, and urban scales, which have relatively low resolution, as well as to studies of communities, families, parks, and industries, at relatively high resolution (Agudelo-Vera et al. 2012; Kennedy and Hoornweg 2012; Pincetl et al. 2012). This research must not only examine the functions and roles of urban material and energy use and the future sustainability of these uses, but also the relationship between a given scale and the layers (regional, national, global) it is embedded in (Pincetl et al. 2012). From a study of the world's geochemistry in the mid-nineteenth century (Goldschmidt 1958), to research on anthropocentric metabolism in 1991 (Baccini and Brunner 1991) and social metabolism at a national scale (Kim and Barles 2012; Haberl et al. 2004), these studies have provided a background for current and future research at different scales.

Research on urban metabolic processes at multiple scales is becoming increasingly important because no city exists in isolation; all are embedded in larger systems (Fig. 2.15). Identifying the role of cities in global sustainable development requires an understanding of urban patterns and processes and their interaction with processes that occur at a larger scale. Similarly, it is also necessary to investigate data on parks, industries, communities, and households to support metabolic research on these different functional groups, which will, in turn, support management and control of the urban ecological space at a finer scale. Studies on a small scale can provide basic data for large-scale and high-precision studies. However, most urban metabolism studies have used coarse and highly aggregated data (Guo et al. 2014; Kennedy et al. 2011) due to a lack of higher-resolution data. This results in homogenizing the city being studied by not considering key differences between (for example) suburbs and inner cities, such as living arrangements (e.g., apartments versus single-family dwellings), service functions, and residential versus industrial functions (Kim and Barles 2012). Such oversimplification makes it hard to associate material and energy flows with specific locations or human activities (Pincetl et al. 2012). As a result, industrial and commercial water and energy utilization data cannot be mapped to specific land uses, and this makes it difficult to implement large-scale research results at the small scales where such results would typically be implemented (Rosado et al. 2014). Therefore, one reason why urban metabolism research has not been widely used to support urban planning and design is that the aggregated data that is available at an urban or regional scale adequately describe the diverse functions and spatial patterns within cities (Sahely et al. 2003).

The open nature of cities makes it difficult to obtain a complete picture of a city's metabolic characteristics by solely observing the city, as this ignores its interactions with its external environment. Thus, material and energy flows should also be measured at a regional scale. This will reveal urban problems that arise from causes outside the city, such as cross-border regional pollution transfers and the occupation of ecological resources or processes by upstream and downstream regions that affect resource utilization. Therefore, some efforts to solve resource utilization and waste treatment problems within a city's scope are ineffective because they attack the wrong problems. It is necessary to comprehensively consider resource supply and waste generation problems in the context of a higher level in the overall hierarchy. In addition, many urban metabolic problems are more prominent within a given city, but

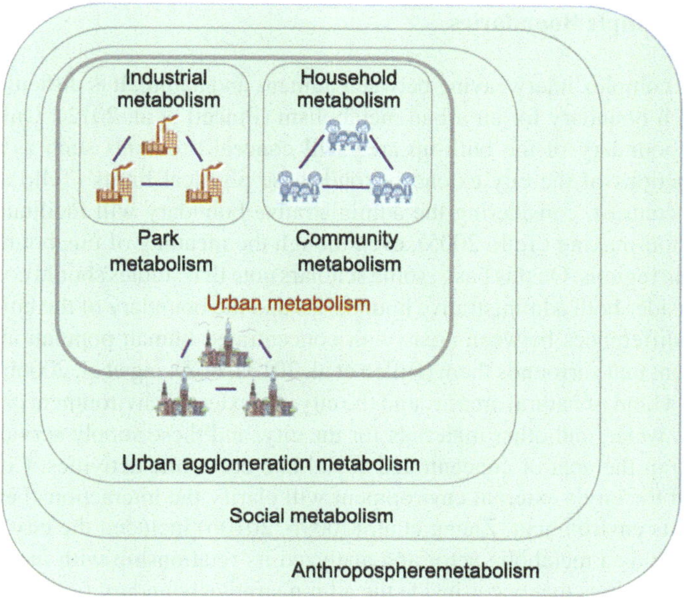

Fig. 2.15 Illustration of the multidimensional scales of urban metabolism research

appear less obvious when a study is performed at a regional scope. For example, in the Beijing-Tianjin-Hebei urban agglomeration, Beijing is predominantly a consumer, whereas Hebei is predominantly a producer (Zhang et al. 2016). Therefore, the use of resources generated in Hebei to sustain the economic activities in Beijing will lead to problems such as Beijing's large dependence on its external environment and the excessive occupation of ecological services that cannot be solved only within the scope of Beijing. However, if we view these relationships at the scale of the urban agglomeration, the problem may not be obvious. Because Beijing uses advanced technology to centralize its production, Hebei can also consume this resource-efficient production. At the same time, Beijing feeds Hebei through capital and technological investments to enhance Hebei's ecological services function, which is beneficial for the development of the urban agglomeration by ensuring high resource-utilization efficiency and enhancing the city's natural ecological service function. It is difficult to require every city to achieve economically, socially, and ecologically efficient coordination with its environment. Therefore, to achieve regional sustainable development, it is necessary to start from the overall perspective of the region, then fully consider the social, economic, technological, and resource factors, identify the different roles and positions of each city in regional sustainable development, and objectively diagnose its metabolic problems (Satterthwaite 1997).

2.5.4.2 Multiple Boundaries

Due to the complex interweaving between humans and nature, it is difficult to define the research boundary for an urban metabolism (Pincetl et al. 2012). Only considering the boundary of the built-up area will conceal problems such as when the actual functions of the city extend beyond these physical limits (Tello and Ostos 2012). In contrast, considering the administrative boundary will facilitate sustainable decision-making (Tello 2005), even though the meaning of this boundary may vary among regions. On this basis, some scholars note that studies should comprehensively consider both administrative boundaries and the boundary of the built-up area to reflect differences between areas with concentrated human populations and the environment that surrounds them (Billen et al. 2012a, b; Zhang et al. 2006b; Girardet 2004). As a kind of natural area around the city, the external environment can provide food, fuel, water, and other materials for the city, and these supply services distinguish it from the area of concentrated social and economic activities. Considering changes in the city's external environment will clarify the interactions between the city and this environment. Zhang et al. (2009b, 2010b) included the city's external environment as a metabolic actor and analyzed its relationship with the social and economic activities that occur inside the administrative boundary for Chinese cities.

An urban metabolism, therefore, involves multiple boundaries that result from different understandings of what a city represents. On the one hand, cities can be defined as large, densely populated settlements with a built-up area formed by non-agricultural industries and a non-agricultural population. On the other hand, they can be defined based on the administrative boundary, which may (as is common in China) include a considerable area of agriculture and a large agricultural population. The former boundary is redefined dynamically as the city expands, whereas the latter boundary is defined by humans to serve the needs of zoning management. The coexistence of these and other boundaries leads to different meanings of the research results, so these results require a careful consideration of the context.

Urban metabolism research based on the boundary of built-up areas can highlight the characteristics of the city's large metabolic scale, high intensity of activities, and severe dependence on its external environment. The advantages of research based on administrative boundaries is that data is often collected at this scale, which makes it easy to understand the system's boundaries and implement the conclusions obtained from the research. This is particularly true because the administrative boundaries define discrete decision-making units, making it more feasible to apply the research results. The difference between a city's built-up and administrative boundaries arises from the presence of non-urbanized areas, including rural areas responsible for production and consumption activities and natural areas (i.e., the ecological environment), which support urban metabolic processes and functions as metabolic actors that participate in the overall metabolism. Generally, the higher the urbanization level of a city, the closer the boundaries between its built-up area and its administrative boundary. Figure 2.16 shows the spatial relationship between the different boundaries.

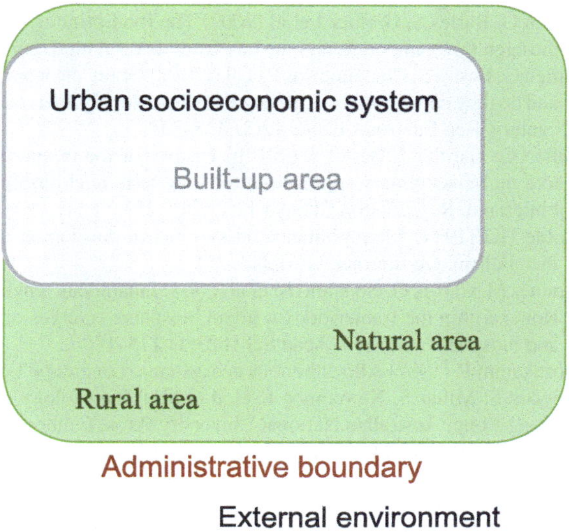

Fig. 2.16 Multiple boundaries for urban metabolic processes

References

Agudelo-Vera CM, Mels A, Keesman K et al (2012) The urban harvest approach as an aid for sustainable urban resource planning. J Ind Ecol 16(6):839–850

Baccini P (1996) Understanding regional metabolism for a sustainable development of urban systems. Environ Sci Pollut Res 3(2):108–111

Baccini P (1997) A city's metabolism: towards the sustainable development of urban systems. J Urban Technol 4(2):27–39

Baccini P, Bader HP (1996) Regionaler Stoffhaushalt: Erfassung, Bewertung und Steuerung. Spektrum Akademischer Verlag, Heidelberg

Baccini P, Brunner PH (1991) Metabolism of the anthroposphere. Springer Verlag, Berlin

Bai X (2007) Industrial ecology and the global impact of cities. J Ind Ecol 11(2):1–6

Baker L, Hartzheim P, Hobbie S et al (2007) Effect of consumption choices on fluxes of carbon, nitrogen and phosphorus through households. Urban Ecosyst 10(2):97–117

Barles S (2007) Feeding the city: food consumption and flow of nitrogen, Paris, 1801–1914. Sci Total Environ 375(1–3):48–58

Barles S (2009) Urban metabolism of Paris and its region. J Ind Ecol 13(6):898–913

Barles S (2010) Society, energy and materials: the contribution of urban metabolism studies to sustainable urban development issues. J Environ Planning Manage 53(4):439–455

Barrett J, Scott A (2001) An ecological footprint of Liverpool: a detailed examination of ecological sustainability (2001-2-1) [2022-7-28]. http://www.gdrc.org/uem/footprints/LiverpoolEFReport.PDF

Barrett J, Vallack H, Jones A et al (2002) A material flow analysis and ecological footprint of York. Stockholm Environment Institute, Stockholm, Sweden [2022-5-10]. https://www.researchgate.net/profile/Gary_Haq/publication/257494298_A_Material_Flow_Analysis_and_Ecological_Footprint_of_York/links/02e7e525555fbc95ab000000/A-Material-Flow-Analysis-and-Ecological-Footprint-of-York.pdf?origin=publication_detail

Beatley T (2012) Introduction: why study European cities. In: Beatley T (ed) Green cities of Europe: global lessons on green urbanism. Island Press, London, pp 1–28

Beck MB, Walker RV, Thompson M (2013) Smarter urban metabolism: earth systems re-engineering. ICE Proc: Eng Sustain 166(5):229–241

Billen G, Barles S, Garnier J et al (2009) The food-print of Paris: long-term reconstruction of the nitrogen flows imported into the city from its rural hinterland. Reg Environ Change 9(1):13–24
Billen G, Barles S, Chatzimpiros P et al (2012a) Grain, meat and vegetables to feed Paris: where did and do they come from? Localising Paris food supply areas from the eighteenth to the twenty-first century. Reg Environ Change 12(2):325–335
Billen G, Garnier J, Barles S (2012b) History of the urban environmental imprint: introduction to a multidisciplinary approach to the long-term relationships between Western cities and their hinterland. Reg Environ Change 12(2):249–253
Bohle HG (1994) Metropolitan food systems in developing countries: the perspective of urban metabolism. GeoJournal 34(3):245–251
Bonnes M, Carrus G, Bonaiuto M et al (2004) Inhabitants' environmental perceptions in the city of Rome within the framework for urban biosphere reserves of the UNESCO programme on man and biosphere. Ann N Y Acad Sci 1023(1):175–186
Bornkamm R (1987) Allochthonous ecosystems. Landscape Ecol 1(2):119–122
Boyden S, Millar S, Newcombe K et al (1981) The ecology of a city and its people: the case of Hong Kong. Australian National University Press, Canberra, Australia
Brattebø H, Bergsdal H, Sandberg NH et al (2009) Exploring built environment stock metabolism and sustainability using systems analysis approaches. Build Res Inf 37(5–6):569–582
Bringezu S, van de Sand I, Schütz H et al (2009) Analysing global resource use of national and regional economies across various levels. In: Stefan B, Raimund B (eds) Sustainable resource management: global trends, visions and policies. Greenleaf Publishing, Sheffield, UK, pp 10–51
Broto VC, Allen A, Rapoport E (2012) Interdisciplinary perspectives on urban metabolism. J Ind Ecol 16(6):851–861
Brunner PH (2007) Reshaping urban metabolism. J Ind Ecol 11(2):11–13
Celecia J (2000) UNESCO's Man and the Biosphere (MAB) programme and urban ecosystem research: a brief overview of the evolution and challenges of three-decade international experience. First Meeting of the Ad Hoc Working Group to Explore Applications of the Biosphere Reserve Concept to Urban Areas and Their Hinterlands, Paris
Chambers N, Heap R, Jenkin N et al (2002) City limits: a resource flow and ecological footprint analysis of Greater London [2022-7-28]. https://library.uniteddiversity.coop/Measuring_Progress_and_Eco_Footprinting/Ecological_Footprinting/City%20Limits%20London.pdf
Chatzimpiros P, Barles S (2009) Quantitative water footprint of meat consumption in historical perspective: first results for Paris (France), 1817 and 1906. In: Havránek M (ed) ConAccount 2008: urban metabolism, measuring the ecological city. Charles University Environment Center, Prague, Czech Republic, pp 215–239
Chen SQ, Chen B (2012) Network environ perspective for urban metabolism and carbon emissions: a case study of Vienna, Austria. Environ Sci Technol 46(8):4498–4506
Chen Y, Chen CM, Hu ZG et al (2015) Principles and applications of analyzing a citation space. Science Press, Beijing, China (in Chinese)
Chrysoulakis N (2008) Urban metabolism and resource optimisation in the urban fabric: the BRIDGE methodology. Proceedings of EnviroInfo 2008: Environmental Informatics and Industrial Ecology, Aachen, Germany
Chrysoulakis N, Vogt R, Young D et al. (2009) ICT for urban metabolism: the case of BRIDGE. In: Wohlgemuth K, Page B, Voigt K (eds) Proceedings of EnviroInfo 2009: environmental informatics and industrial environmental protection: concepts, methods and tools. Hochschulefür Technik und Wirtschaft, Berlin, Germany, pp 175–185
Codoban N, Kennedy CA (2008) Metabolism of neighborhoods. J Urban Plan Dev 134(1):21–31
Collins A, Flynn A, Netherwood A (2005) Reducing Cardiff's ecological footprint: a resource accounting tool for sustainable consumption. Cardiff University, Cardiff, UK
Daniels PL, Moore S (2001) Approaches for quantifying the metabolism of physical economies, Part I: methodological overview. J Ind Ecol 5(4):69–93
Decker EH, Elliott S, Smith FA et al (2000) Energy and material flow through the urban ecosystem. Annu Rev Energy Env 25(1):685–740

Deilmann C (2009) Urban metabolism and the surface of the city. In: Deilmann C (ed) Guiding principles for spatial development in Germany. Springer, Berlin, Germany, pp 1–16
Douglas I (1981) The city as ecosystem. Prog Phys Geogr 5(3):315–367
Douglas I (1983) The urban environment. Edward Arnold, London, UK
Douglas I, Hodgson R, Lawson N (2002) Industry, environment and health through 200 years in Manchester. Ecol Econ 41(2):235–255
Doxiadis CA (1970) The science of human settlements. Science 170(3956):393–404
Duvigneaud P, Denayeyer-De Smet S (1977) L'ecosystéme Urbain Bruxellois. Productivité en Belgique, Brussels, Belgium
Engel-Yan J, Kennedy C, Saiz S et al (2005) Toward sustainable neighbourhoods: the need to consider infrastructure interactions. Can J Civ Eng 32(1):45–57
Eurostat (2009) Economy wide material flow accounts: compilation guidelines for reporting to the 2009 Eurostat questionnaire. EUROSTAT, Luxembourg (2009-6-1) [2022-3-12]. https://unstats.un.org/unsd/envaccounting/ceea/archive/Framework/Eurostat%20MFA%20compilation%20guide_2009.pdf
Fath BD, Killian MC (2007) The relevance of ecological pyramids in community assemblages. Ecol Model 208(2):286–294
Fernández JE (2010) Urban metabolism of ancient Caral, Peru. [2021-10-12]. http://src.lafargeholcim-foundation.org/dnl/466e9f92-de70-4ed9-ac48-f72095668c53/F10_GreenWorkshop_Paper_FernandezJohn.pdf
Ferrão P, Fernández JE (2013) Sustainable urban metabolism. MIT Press, Boston, MA
Fischer-Kowalski M (1998) Society's metabolism: the intellectual history of materials flow analysis, Part I 1860–1970. J Ind Ecol 2(1):61–78
Folke C, Jansson A, Larsson J et al (1997) Ecosystem appropriation by cities. Ambio 26(3):167–172
Forkes J (2007) Nitrogen balance for the urban food metabolism of Toronto, Canada. Resour Conserv Recycl 52(1):74–94
Gandy M (2004) Rethinking urban metabolism: water, space and the modern city. City 8(3):363–379
Gasson B (2002) The ecological footprint of Cape Town: unsustainable resource use and planning implications. The National Conference of the South African Planning Institution, Durban, South Africa
Girardet H (1996) The Gaia Atlas of cities: new directions for sustainable urban living. Gaia Books Limited, London, UK
Girardet H (2004) The metabolism of cities. In: Heeler SMW, Beatley T (eds) The sustainable urban development reader. Routledge, London, UK, pp 125–132
Girardet H (2008) Cities people planet: urban development and climate change, 2nd edn. Wiley, Chichester, UK
Girardet H (2014) Creating regenerative cities. Routledge, New York, USA
Goldschmidt VM (1958) Geochemistry. Oxford University Press, London, UK
Goldstein B, Birkved M, Quitzau M et al (2013) Quantification of urban metabolism through coupling with the life cycle assessment framework: concept development and case study. Environ Res Lett 8(3):1–14
Golubiewski N (2012) Is there a metabolism of an urban ecosystem? An ecological critique. Ambio 41(7):751–764
González A, Donnelly A, Jones M et al (2013) Decision-support system for sustainable urban metabolism in Europe. Environ Impact Assess Rev 38(1):109–119
Grimm NB, Faeth SH, Golubiewski NE et al (2008) Global change and the ecology of cities. Science 319(5864):756–760
Guo Z, Hu D, Zhang FH et al (2014) An integrated material metabolism model for stocks of urban road system in Beijing, China. Sci Total Environ 470–471(1):883–894
Haberl H (2001) The energetic metabolism of societies, Part I: accounting concepts. J Ind Ecol 5(1):11–33

Haberl H, Erb KH, Krausmann F et al (2007) Quantifying and mapping the human appropriation of net primary production in Earth's terrestrial ecosystems. Proc Natl Acad Sci USA 104(31):12942–12947

Haberl H, Fischer-Kowalski M, Krausmann F et al (2004) Progress towards sustainability? What the conceptual framework of material and energy flow accounting (MEFA) can offer. Land Use Policy 21(3):199–213

Hammer M, Giljum S (2006) Materialflussanalysen der Regionen Hamburg, Wien Und Leipzig. Natur und kultur 7(2):62–78

Hanya T, Ambe Y (1977) A study on the metabolism of cities. In: Kaigi NG (ed) Science for a better environment. Science Council of Japan, Tokyo, Japan, pp 228–233

Hashimoto S, Moriguchi Y (2004) Proposal of six indicators of material cycles for describing society's metabolism: from the viewpoint of material flow analysis. Urban Ecosyst 40(3):185–200

Havranek M (2009) ConAccount 2008: urban metabolism, measuring the ecological city. Charles University Environment Centre, Prague, Czech Republic

Hedbrant J (2001) Stockhome: a spreadsheet model of urban heavy metal metabolism. Water Air Soil Pollut Focus 1(3–4):55–66

Hendriks C, Obernosterer R, Müller D et al (2000) Material flow analysis: a tool to support environmental policy decision making: case studies on the city of Vienna and the Swiss lowlands. Local Environ 5(3):311–328

Huang SL (1998) Urban ecosystems, energetic hierarchies, and ecological economics of Taibei metropolis. J Environ Manage 52(1):39–51

Huang SL, Hsu WL (2003) Materials flow analysis and energy evaluation of Taibei's urban construction. Landsc Urban Plan 63(2):61–74

Inostroza L (2014) Measuring urban ecosystem functions through 'Technomass': a novel indicator to assess urban metabolism. Ecol Ind 42(3):10–19

Jenerette GD, Marussich WA, Newell JP (2006) Linking ecological footprints with ecosystem valuation in the provisioning of urban freshwater. Ecol Econ 59(1):38–47

Kennedy CA (2007) Applying industrial ecology to design a sustainable built environment: the Toronto Port Lands challenge. Engineering Sustainability Conference, Pittsburgh, USA

Kennedy CA (2012) A mathematical description of urban metabolism. In: Weinstein MP, Turner E (eds) Sustainability science: the emerging paradigm and the urban environment. Springer, New York, USA, pp 275–291

Kennedy CA, Hoornweg D (2012) Mainstreaming urban metabolism. J Ind Ecol 16(6):780–782

Kennedy CA, Cuddihy J, Engel-Yan J (2007) The changing metabolism of cities. J Ind Ecol 11(2):43–59

Kennedy CA, Steinberger J, Gasson B et al (2010) Methodology for inventorying greenhouse gas emissions from global cities. Energy Policy 38(9):4828–4837

Kennedy CA, Pincetl S, Bunje P (2011) The study of urban metabolism and its applications to urban planning and design. Environ Pollut 159(8–9):1965–1973

Kennedy CA, Stewart I, Facchini A (2015) Energy and material flows of megacities. Proc Natl Acad Sci USA 112(19):5985–5990

Kenny T, Gray NF (2009) Comparative performance of six carbon footprint models for use in Ireland. Environ Impact Assess Rev 29(1):1–6

Kim E, Barles S (2012) The energy consumption of Paris and its supply areas from the eighteenth century to the present. Reg Environ Change 12(2):295–310

Kral U, Lin CY, Kellner K et al (2014) The copper balance of cities: exploratory insights into a European and an Asian city. J Ind Ecol 18(3):432–444

Krausmann F (2005) A city and its hinterland: the social metabolism of Vienna 1800–2000. Conference of the European Society for Environmental History, Florence, Italy

Krausmann F, Haberl H (2002) The process of industrialization from the perspective of energetic metabolism: socioeconomic energy flows in Austria 1830–1995. Ecol Econ 41(2):177–201

Krausmann F, Haberl H, Schulz N et al (2003) Land-use change and socioeconomic metabolism in Austria, Part I: driving forces of land-use change 1950–1995. Land Use Policy 20(1):1–20

Lestel L (2012) Non-ferrous metals (Pb, Cu, Zn) needs and city development: the Paris example (1815–2009). Reg Environ Change 12(2):311–323
Li SS, Zhang Y, Yang ZF et al (2012) Ecological relationship analysis of the urban metabolic system of Beijing, China. Environ Pollut 170(11):169–176
Li YX, Zhang Y, Yu XY (2019) Urban weight and its driving forces: a case study of Beijing. Sci Total Environ 658(6):590–601
Liang S, Zhang TZ (2012) Comparing urban solid waste recycling from the viewpoint of urban metabolism based on physical input-output model: a case of Suzhou in China. Waste Manage 32(1):220–225
Liu JR, Wang RS, Yang JX (2005) Metabolism and driving forces in Chinese urban household consumption. Popul Environ 26(4):325–341
Lu ZW, Yue Q (2006) Two methods of material flow analysis and their applications. Nonferrous Met Regen Util 3(2):27–28 (in Chinese)
Manning WJ (2008) Plants in urban ecosystems: essential role of urban forests in urban metabolism and succession toward sustainability. Int J Sust Dev World 15(4):362–370
McDonald GW, Patterson MG (2007) Bridging the divide in urban sustainability: from human exemptionalism to the new ecological paradigm. Urban Ecosyst 10(2):169–192
Minx J, Creutzig F, Medinger V et al (2011) Developing a pragmatic approach to assess urban metabolism in Europe: a report to the European Environment Agency. Technische Universität Berlin
Montrucchio V (2012) Systemic design approach applied to buildings: definition of a co-operative process. Int J Eng Sci 10(3):323–327
Mullins P, Natalier K, Smith P et al (1999) Cities and consumption spaces. Urban Aff Rev 35(1):44–71
Neset TSS, Lohm U (2005) Spatial imprint of food consumption: a historical analysis for Sweden, 1870–2000. Hum Ecol 33(4):565–580
Newcombe K (1977) Nutrient flow in a major urban settlement, Hong Kong. Hum Ecol 5(3):179–208
Newcombe K, Kalma J, Aston A (1978) The metabolism of a city: the case of Hong Kong. Ambio 7(1):3–15
Newman PWG (1999) Sustainability and cities: extending the metabolism model. Landsc Urban Plan 44(4):219–226
Newman PWG, Kenworthy J (1999) Sustainability and cities: overcoming automobile dependence. Island Press, Washington, DC
Niza S, Rosado L, Ferrão P (2009) Urban metabolism: methodological advances in urban material flow accounting based on the Lisbon case study. J Ind Ecol 13(3):384–405
Odum EP (1953) Fundamentals of ecology. Saunders, Philadelphia, USA
Odum HT (1971) Environment, power, and society. Wiley Interscience, New York, USA
Odum HT (1973) Energy, ecology and economics. Ambio 2(6):220–227
Odum EP (1975) Ecology: the link between the natural and social sciences. Holt, Rinehart and Winston, New York, USA
Odum HT (1983) Systems ecology: an introduction. Wiley Interscience, New York, USA
Odum EP (1989) Ecology and our endangered life-support systems. Sinauer, Sunderland, USA
Odum HT, Odum EC (1981) Energy basis for man and nature. McGraw-Hill, New York, USA
Oswald F, Baccini P, Michaeli M (2003) Netzstadt: designing the urban. Birkhäuser Basel, Basel, Switzerland
Pincetl S, Bunje P, Holmes T (2012) An expanded urban metabolism method: towards a systems approach for assessing the urban energy processes and causes. Landsc Urban Plan 107(3):193–202
Pincetl S, Chester M, Circella G et al (2014) Enabling future sustainability transitions: an urban metabolism approach to Los Angeles. J Ind Ecol 18(6):871–882
Quinn D (2007) Urban metabolism: ecologically sensitive construction for a sustainable New Orleans. MIT Press, Cambridge, USA
Reddy BS (2013) Metabolism of Mumbai—expectations, impasse and the need for a new beginning [2021-12-1]. http://www.igidr.ac.in/pdf/publication/WP-2013-002.pdf

Rosado L, Ferrão P (2009) Measuring the embodied energy in household goods: application to the Lisbon city. In: Havránek M (ed) ConAccount 2008: urban metabolism, measuring the ecological city. Charles University Environment Center, Prague, Czech Republic, pp 159–181
Rosado L, Niza S, Ferrão P (2014) A material flow accounting case study of the Lisbon metropolitan area using the Urban Metabolism Analyst model. J Ind Ecol 18(1):84–101
Sahely HR, Dudding S, Kennedy CA (2003) Estimating the urban metabolism of Canadian cities: Greater Toronto Area case study. Can J Civ Eng 30(2):468–483
Samaniego H, Moses ME (2008) Cities as organisms: allometric scaling of urban road networks. J Transp Land Use 1(1):21–39
Satterthwaite D (1997) Sustainable cities or cities that contribute to sustainable development. Urban Study 34(10):1667–1691
Schandl H, Schulz NB (2002) Changes in United Kingdom's natural relations in terms of society's metabolism and land use from 1850 to the present day. Ecol Econ 41(2):203–221
Schremmer C, Stead D (2009) Sustainable urban metabolism for Europe (SUME). Proceedings of the World Bank's Fifth Urban Research Symposium, Marseille, France
Schulz NB (2007) The direct material inputs into Singapore's development. J Ind Ecol 11(2):117–131
Shillington LJ (2013) Right to food, right to household urban agriculture, and socionatural metabolism in Managua, Nicaragua. Geoforum 44(1):103–111
Sovacool BK, Brown MA (2010) Twelve metropolitan carbon footprints: a preliminary comparative global assessment. Energy Policy 38(9):4856–4869
Stanhill G (1977) An urban agro-ecosystem: the example of nineteenth-century Paris. AgroEcosystems 3(3):269–284
Susca T (2012) Multiscale approach to life cycle assessment evaluation of the effect of an increase in New York city's rooftop albedo on human health. J Ind Ecol 16(6):951–962
Tello E (2005) Changing course? Principles and tools for local sustainability. In: Marshall T (ed) Transforming Barcelona. Routledge, London, UK, pp 225–250
Tello E, Ostos JR (2012) Water consumption in Barcelona and its regional environmental imprint: a long-term history (1717–2008). Reg Environ Change 12(2):347–361
Timmerman PR, White R (1997) Megahydropolis: coastal cities in the context of global environmental change. Glob Environ Chang 7(3):205–234
Venkatesh G, Sægrov S, Brattebø H (2014) Dynamic metabolism modelling of urban water services: demonstrating effectiveness as a decision-support tool for Oslo, Norway. Water Res 61(17):19–33
Vitousek PM, Ehrlich PR, Ehrlich AH et al (1986) Human appropriation of the products of photosynthesis. Biosci 36(6):368–373
Wang RS, Ouyang ZY (2012) Social-economic-natural complex ecosystem and sustainability. Bull Chin Acad Sci 27(3):337–345 (in Chinese)
Wang XJ, Zhang Y, Zhang J et al (2021) Progress in urban metabolism research and hotspot analysis based on CiteSpace analysis. J Clean Prod 281(2):125224
Warren-Rhodes K, Koenig A (2001) Escalating trends in the urban metabolism of Hong Kong: 1971–1997. Ambio 30(7):429–438
Weisz H, Steinberger JK (2010) Reducing energy and material flows in cities. Curr Opin Environ Sustain 2(3):185–192
Weisz H, Krausmann F, Amann C et al (2006) The physical economy of the European Union: cross-country comparison and determinants of material consumption. Ecol Econ 58(4):676–698
Wolman A (1965) The metabolism of cities. Sci Am 213(3):178–190
Wright DH (1990) Human impacts on energy-flow through natural ecosystem, and implications for species endangerment. Ambio 19(4):189–194
Yang ZF, Zhang Y, Li SS et al (2014) Characterizing urban metabolic systems with an ecological hierarchy method, Beijing, China. Landsc Urban Plan 121(1):19–23
Zeng WH, Wu B, Chai Y (2014) Dynamic simulation of urban water metabolism under water environmental carrying capacity restrictions. Front Environ Sci Eng 10(1):114–128

Zhang Y (2013) Urban metabolism: a review of research methodologies. Environ Pollut 178(7):463–473

Zhang Y, Yang ZF, Yu XY (2006a) Measurement and evaluation of interactions in complex urban ecosystem. Ecol Model 196(1):77–89

Zhang Y, Yang ZF, Li W (2006b) Analyses of urban ecosystem based on information entropy. Ecol Model 197(1):1–12

Zhang Y, Yang ZF, Yu XY (2009a) Evaluation of urban metabolism based on energy synthesis: a case study for Beijing (China). Ecol Model 220(13):1690–1696

Zhang Y, Yang ZF, Yu XY (2009b) Ecological network and emergy analysis of urban metabolic systems: model development, and a case study of four Chinese cities. Ecol Model 220(11):1431–1442

Zhang Y, Yang ZF, Fath BD (2010a) Ecological network analysis of an urban water metabolic system: model development, and a case study for Beijing. Sci Total Environ 408(20):4702–4711

Zhang Y, Yang ZF, Fath BD et al (2010b) Ecological network analysis of an urban energy metabolic system: model development, and a case study of four Chinese cities. Ecol Model 221(16):1865–1879

Zhang Y, Xia LL, Xiang WN (2014a) Analyzing spatial patterns of urban carbon metabolism: a case study in Beijing, China. Landsc Urban Plan 130(5):184–200

Zhang Y, Zheng HM, Fath BD et al (2014b) Ecological network analysis of an urban metabolic system based on input-output tables: model development and case study for Beijing. Sci Total Environ 468–469(1):642–653

Zhang Y, Yang ZF, Yu XY (2015) Urban metabolism: a review of current knowledge and directions for future study. Environ Sci Technol 49(19):11247–11263

Zhang Y, Zheng HM, Yang ZF et al (2016) Urban energy flow processes in the Beijing-Tianjin-Hebei (Jing-Jin-Ji) urban agglomeration: combining multi-regional input-output tables with ecological network analysis. J Clean Prod 114(3):243–256

Zucchetto J (1975) Energy-economic theory and mathematical models for combining the systems of man and nature, case study: the urban region of Miami, Florida. Ecol Model 1(4):241–268

Chapter 3
Theory, Paradigms, and Technical Methods for Urban Metabolism

3.1 Composite Ecosystem Theory

The composite ecosystem theory describes a city as a kind of hybrid ecosystem that includes natural, social, and economic components. Although it is dominated by human behavior, it's supported by the natural environment and by resource flows within the city, as well as between the city and its external environment (Wang 2000; Ma and Wang 1984). The natural subsystem is comprised of ecological elements and their interactions; the economic subsystem is comprised of human activities such as production, circulation, and consumption; and the social subsystem includes cultural concepts such as the state (i.e., government). These three subsystems interact in ways that mutually reinforce each other or contradict each other on different time scales (e.g., intra-generation, inter-generation), spatial scales (e.g., region, watershed), quantity scales (e.g., materials, energy), and structural scales (e.g., industrial structure, resource structure), as well as at different levels of order (entropy change, coordinated symbiosis). These differences lead to different forms of coupling among the subsystems; this, in turn, determines the direction the ecosystem evolves and the pathways that develop within the composite ecosystem (Wang and Ouyang 2012). In the following sections, I will describe the three subsystems, combining the social and economic subsystems because of their tight integration.

3.1.1 Natural Subsystem

The natural subsystem is comprised of the environment, on which the city depends for survival. It can be described based on the Chinese five elements (metal, wood, water, fire, and soil) and the relationships among them. In the ecosystem model of a city, these five elements correspond to mineral resources, biological resources, water resources, energy resources (fossil fuel), and land resources. The relationships among

Y. Zhang, *Urban Metabolism*,
https://doi.org/10.1007/978-981-19-9123-3_3

the five elements and the constraints each one creates for the others are analogous to natural processes.

Let's take wood as the starting point to explain the circular relationships among the five elements (Fig. 3.1). In this metaphor, wood represents natural biological resources in general, including plants, animals, and microorganisms. All of these components require energy and, by analogy, have combustible characteristics (i.e., they consume and generate energy). Thus, fire is both hidden within wood and separate from it. Wood produces fire, which refers to natural energy; natural energy includes solar energy, wind, geothermal energy, and tidal energy, whereas mineral energy includes fossil fuels such as coal, oil, and natural gas, as well as radioactive fuels such as uranium. Because combustion produces ash, which becomes part of the soil, fire provides inputs to the soil. Most metallic minerals (i.e., minerals that contain metal chemical elements) are buried in the soil, which suggests that soil produces metal. Because metal mines are often near rivers, minerals are related to water and can be crafted into water collection tools. Because water is essential for life and a key component of all living organisms, it is an essential input for wood. This completes the circle by bringing us back to our starting point—wood. At the same time, the five elements also counteract or neutralize each other. For example, excessive metal content in the soil will reduce plant productivity (wood), whereas excessive plant growth can deplete soil water.

The interface between socioeconomic activities and the natural world occurs through the soil. Soil refers to earth, land, and the various minerals (building materials, metal, and non-metallic minerals) it contains, as well as to the topography,

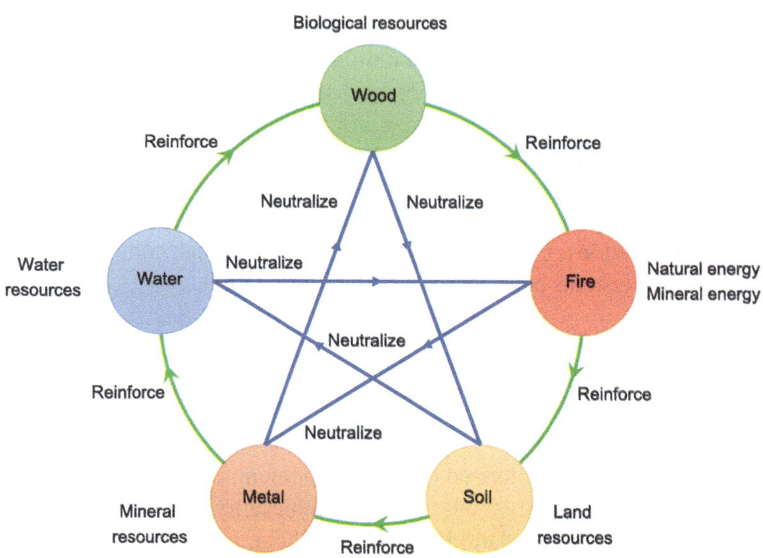

Fig. 3.1 Interpretation of the relationships among the components of a city's natural subsystems based on the Chinese five elements model (metal, wood, water, fire, and soil) (*Note* Wood represents living components of the ecosystem and fire represents the energy components)

geology, landforms, landscapes, and locations of various features of the ecosystem. Humans rely on these ecosystem components for food, fiber, and support for the development of social and economic activities. Soil is, therefore, the foundation of human survival.

3.1.2 Socioeconomic Subsystem

The core of the urban socioeconomic subsystem is people. This subsystem carries out production and consumption activities through the combination of different human behaviors. The production activities ultimately aim to serve top consumers; that is, all consumption activities are designed to meet human needs. These activities can be described based on their scale, nature (e.g., generating energy vs. materials), density (spatial concentration), and patterns (e.g., sequence, spatial distribution), and taken together, they define the city's current state and direction of evolution. The socioeconomic subsystem also integrates human knowledge, culture, perceptions, and organizational systems to develop a knowledge system (including philosophy, science, and technology) and a cultural system (interpretation and inheritance of ethics, beliefs, and context) to form a conceptual model of how urban memory is preserved and spread and how it can transform urban development. The result is a social system that includes organizations, regulations, and policies that shape the spirit and soul of the city (Fig. 3.2).

Based on the characteristics of the utilization processes for materials and energy, the economic subsystem is composed of sectors that are responsible for the production, conversion, and circulation of materials and energy, synthesis of products, purification of waste, and control (regulation) of these processes. The system's materials and energy conversion components convert natural materials and energy into the nutrients needed by other components of urban metabolism. As a result, it is located at the start or in the middle of most economic chains. The outputs from these components are transmitted through the system via its materials and energy circulation component, which provides nutrients for other components that are responsible for the production of intermediate or final products. The production of these products generates waste, including waste from human consumption of nutrients or use of products, and the waste becomes inputs for the purification component. The purified waste can be restored to nature or can re-enter the material conversion and synthesis components for reuse (i.e., recycling). Some waste cannot be purified or recycled, so they are released into the environment without treatment.

The material and energy control component regulates the material flow and energy flow processes. By constructing an organizational structure, residents of a city perform an organizational role by developing and implementing plans, policies, and regulations. Together, these controls regulate production and consumption behaviors and the associated flows of energy and materials.

Cities function like complex organisms or ecosystems, with producers, consumers, and recyclers that are continuously engaging in metabolic processes (Haughton and

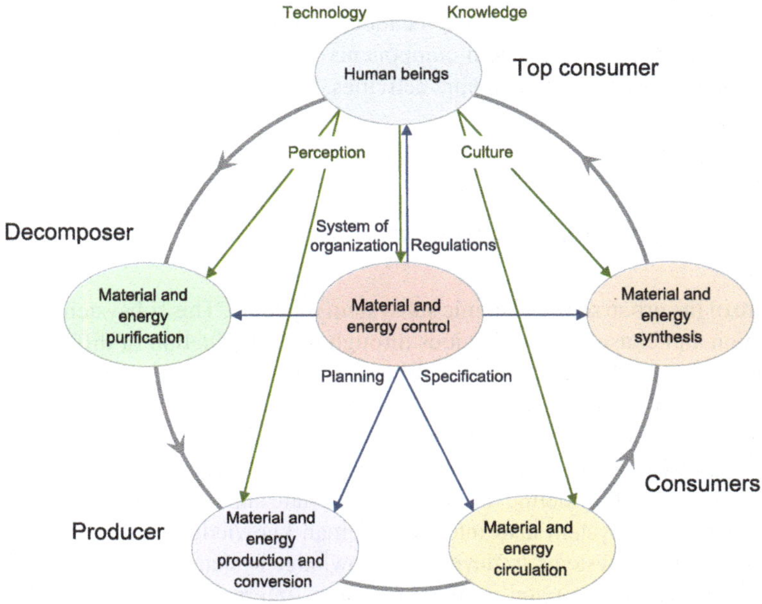

Fig. 3.2 Illustration of a city's socioeconomic subsystem based on material flow and energy flow processes

Hunter 1994; Tjallingii 1993). These processes can lead to system optimization by improving the flows, recycling, and regeneration of materials and energy (Newman 1999; Huang et al. 1998).

3.1.3 Structural Features

The urban hybrid ecosystem is composed of both natural and socioeconomic subsystems that are hierarchical, heterogeneous, and diverse. This variation forms a structure similar to that of a natural ecosystem. The urban ecosystem can be divided into producers, consumers, and decomposers based on each component's role in the urban structure.

Broadly speaking, the natural subsystem plays the role of a primary producer, while the socioeconomic subsystem consumes the nutrients from this producer to play the roles of consumer (which takes in these nutrients) and producer (which transforms these nutrients into intermediate or final products or waste). Note that here, I am distinguishing between intermediate products, which will undergo further transformation before their ultimate use, and final products, which are ready for that final use. Studying the socioeconomic subsystem can identify multiple metabolic agents that play the roles of producer, consumer, and decomposer, and the insights provided by such studies can reveal ways to refine or optimize urban metabolism. The extraction and agriculture sectors, as well as others like material and energy conversion, provide cities with natural biological products (e.g., agricultural, forestry, animal husbandry, and fisheries products) and mineral products (industrial raw materials and fossil fuel energy), and, therefore, fulfill roles similar to those of primary producers in an ecosystem. The material and energy circulation and synthesis components of the urban system transfer their primary products to consumers that will use them to produce intermediate or final products. Although these consumers produce various products, their consumption depends on their ability to receive the necessary nutrient inputs from the material and energy transformation components.

The city's natural subsystem performs ecological service functions such as climate regulation, pollution purification, water production and conservation, spatial isolation of components of the urban system, and recycling waste produced by metabolic activities. Most of the waste generated in a city must be processed so they can be decomposed into less toxic substances. This processing also ensures that the waste amount can be reduced by recycling them as inputs for other metabolic activities. Enterprises and environmental protection infrastructure, such as sewage treatment plants, that are involved in various aspects of waste treatment and recycling should be classified as purification or decomposer sectors (Zhang et al. 2006a). Similarly, ecological restoration and construction activities to provide green space, clean water and air purification services needed by organisms that maintain these services should also be categorized as decomposers (Fig. 3.3).

3.1.4 Balance Between Pressure and Support

The division of the urban ecosystem into the structures described in the previous section fully reflects the interactive relationship between the socioeconomic subsystem and the natural subsystem, revealing both the pressures the urban system exerts on its support systems and the support provided by those systems. Urban socioeconomic activities generate pressure through resource utilization and waste emission, whereas the ecosystem's natural resource endowments and environmental capacity to receive discharged waste support the city's social and economic activities (Zhang et al. 2006a). The interactions within the socioeconomic subsystem integrate decomposition and regeneration. Decomposition produces waste products that are then regenerated and detoxified or converted into inputs for subsequent metabolic processes (Patten and Costanza 1997).

Rapid economic growth inevitably results in high consumption of resources and proportionally high emission of waste (such as pollution), and these two processes create tremendous pressure on the city's natural subsystem. If we define the pressure on the ecosystem created by socioeconomic development as P, then we can compare this with the natural ecosystem's capacity to support this pressure (S). As P increases, S must increase to counteract that pressure. Together, the two forces generate a dynamic equilibrium. When $S \geq P$, the urban system is balanced or has excess

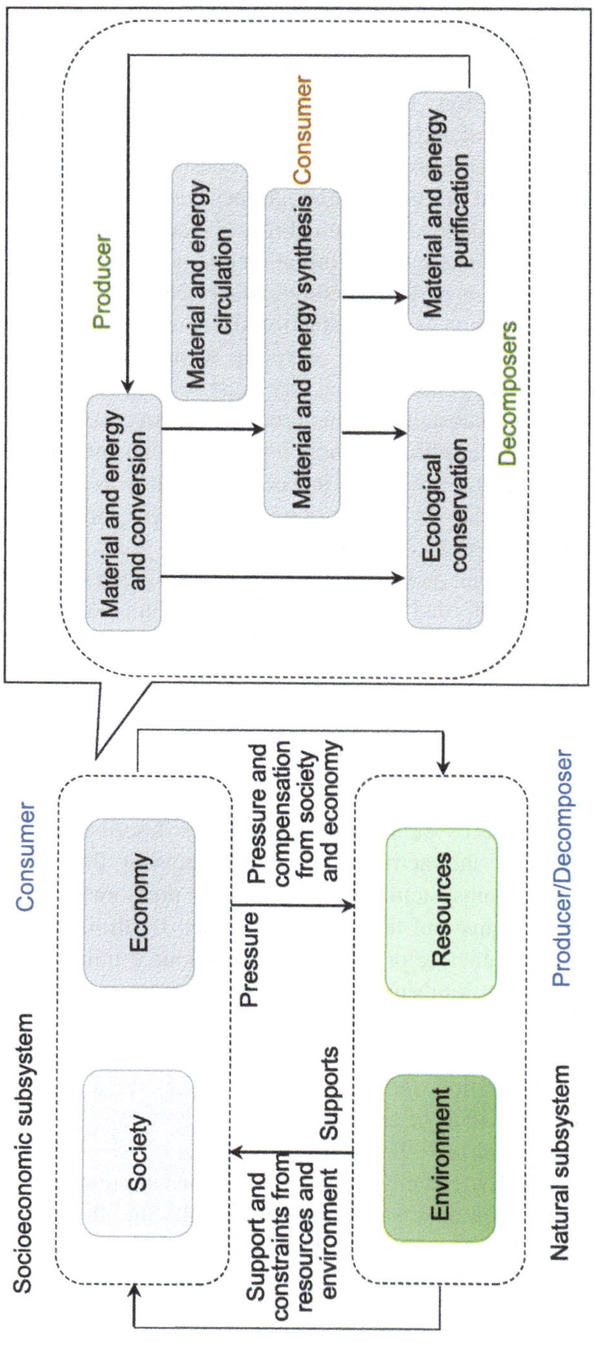

Fig. 3.3 Illustration of the structure of the hybrid natural and socioeconomic urban ecosystem

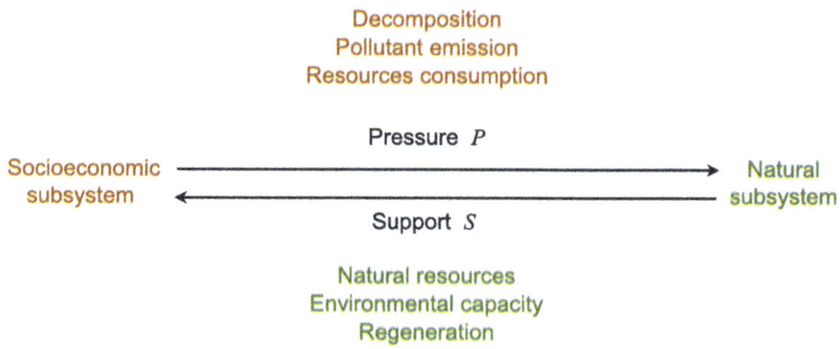

Fig. 3.4 Illustration of the relationship between the pressure exerted by the urban metabolic system on its natural environment (P) and the environment's capacity to support that pressure (S)

capacity, and the system has achieved a harmonious and balanced state. When $P > S$, the urban system is unbalanced and places more pressure on the natural environment than it can sustain, leading to a metabolic disorder. The urban system's decomposition subsystem must then have its decomposition and recycling capacity increased (e.g., through technological innovation) or the pressure must be reduced (e.g., through more efficient consumption or less-polluting processes). Because both P and S change dynamically, the dynamic balance between them is constantly shifting to a new point (Fig. 3.4).

3.2 Thermodynamics Theory

Physics teaches us that energy and mass are conserved, but that both can change forms. In this section, I will discuss the implications of this conservation for urban metabolism.

3.2.1 Vitality Metabolism

The first law of thermodynamics states that energy is conserved. Similarly, although matter can change forms (e.g., from solid to liquid), its mass is conserved. This means that, if we can identify and quantify all of the flows of energy and materials through a system, the energy and mass flows will balance (i.e., neither energy nor mass will be lost). Thus, when a city consumes resources, the difference between what it immobilizes within the city as stock and what it consumed to produce that stock will be released as waste (i.e., waste heat, pollutants) or as intermediate products that are transferred to another metabolic actor. To relieve the pressure these processes place on the urban ecosystem, it's necessary to find ways to reduce resource consumption and waste generation. This can be achieved through the development of more efficient

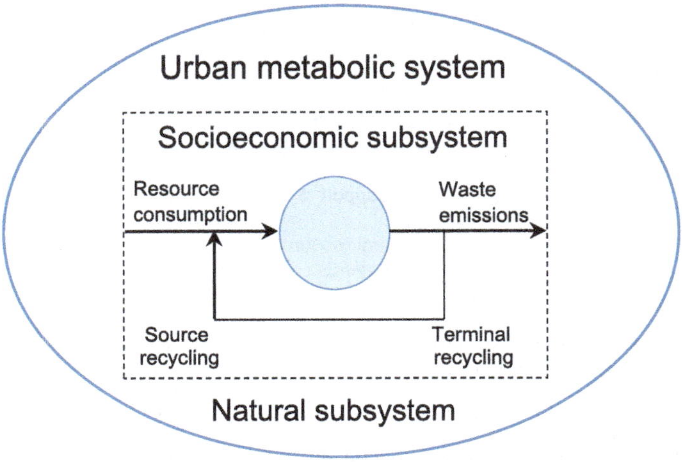

Fig. 3.5 Illustration of circular flows and a city's vitality metabolism

processes (e.g., technological innovation) and recycling to recover as much of the waste as possible.

The waste generated by these activities should be recycled as much as possible to reduce both resource consumption and waste generation (Girardet 2004; Jordan and Vaas 2000). This approach is the basis for creating what is called a "circular society", in which materials and energy flow continuously through the urban system, with waste from one component becoming resources for another component. Achieving this state creates metabolic vitality (Fig. 3.5).

The vitality of an urban system refers to the quality of its organizational structure, functional relationships, and interactions with its environment. Vitality occurs when these factors create a balanced and orderly state for the urban metabolic processes, thereby strengthening both social and economic development, conserving resources, supporting the natural environment, and improving ecological conservation. A vital system is able to efficiently transport the nutrients that nourish the city to where they are needed and process all or most of the waste. A vital urban system is self-renewing and continuously improving, and its resilience, coordination, and sustainability are continuously improving.

The vitality of an urban system can be described from two perspectives: efficiency and scale. Efficiency focuses on optimization of the city's internal processes so that they consume the smallest amount of a resource for each unit of output, whereas scale focuses on the size (mass, area, or volume) of the system. The two are inseparable since increasing efficiency allows the city to grow (increase scale) without unacceptably increasing pressure on its environment. Thus, maintaining both at acceptable levels is essential for the promotion of social and economic vitality as well as ecological and environmental health. Constantly making efforts to improve efficiency and to keep the city's scale within the environment's carrying capacity is necessary to rejuvenate the city and prevent it from developing the metabolic disorders associated with aging. The spatial and temporal scales of a city's processes also determine its vitality. When a city stops growing and evolving, it may cease to exist or be replaced by younger and more vigorous cities. One way to promote efficiency and allow an expansion of a city's scale is to examine the city's urban metabolic processes, which requires a diagnosis of important metabolic disorders (e.g., an imbalance between supply and demand) and, in turn, allows the original urban pattern to renew itself, improving its vigor.

This model of urban vitality analyzes the resource consumption problems that lie behind waste discharge, explores the recycling potential of the urban metabolic system, and provides a way to diagnose problematic processes to allow optimization and regulation of these processes. The characteristic indicators used in this model relate to metabolic flows (e.g., resource consumption and waste discharge) and vitality indicators (e.g., circulation). Circulation indexes can be used to quantify these flows (Zhang et al. 2006b). They can be divided into two types: source and terminal circulation indexes (Zhang et al. 2006b). Source indexes characterize the vitality of urban life through the city's use of waste as a resource and reflect the recycling and regeneration of these materials by the city's socioeconomic subsystem. In contrast, terminal indexes characterize urban metabolic flows that focus on reducing harm to the city's natural subsystem.

3.2.2 Entropy

The second law of thermodynamics states that the entropy (disorder) of any system increases over time unless energy is applied to temporarily reverse this process, although not all processes can be reversed. For example, some of the energy consumed by a metabolic activity is always lost (usually as waste heat). Although some of that heat can be recovered, there will always be some loss. The metabolic processes related to urban resource utilization and waste discharge comply with this law, as urban organisms are self-organized dissipative systems (i.e., systems that are not in equilibrium with their environment). As a dissipative system, urban organisms must obtain materials and energy from their external environment and continuously output products and waste to maintain a stable and orderly state (Huang 2002; Boyden et al. 1981); as a result, cities are open systems. As such, the city generates a flow of entropy through its interactions with external forces. If the entropy flow $\mathrm{d}S_\mathrm{e}$ is greater than the entropy production ($\mathrm{d}S_\mathrm{i}$), the entropy in the system decreases; that is, $\mathrm{d}S = \mathrm{d}S_\mathrm{e} + \mathrm{d}S_\mathrm{i} < 0$, where $\mathrm{d}S$ is called the entropy change (relative entropy reduction), which characterizes the system's direction of evolution (Haken 1988; Weber et al. 1988). Entropy production $\mathrm{d}S_\mathrm{i}$ and entropy flow $\mathrm{d}S_\mathrm{e}$ can be positive (an increase in entropy, which means greater disorder), negative (a decrease in entropy, which means greater order), or zero. The generation and transmission of negative entropy (i.e., $\mathrm{d}S < 0$) by the city's natural subsystem can be regarded as an offset for the positive entropy (i.e., $\mathrm{d}S > 0$) by the socioeconomic subsystem, thereby maintaining an entropy balance between the two subsystems. However, if the natural subsystem generates insufficient negative entropy, the city needs to rely on its external environment to supply this entropy. A positive value of dS_i indicates that the city's restoration metabolic function has improved (ecological restoration) and the urban organism is showing an orderly development trend.

Based on an analysis of the structure and function of the urban ecosystem, it's possible to determine the entropy production and flows and thereby calculate the entropy change; which, in turn, reveals opportunities and measures to improve the regulation of urban metabolic functions (Weber et al. 1988). The interaction between the city's socioeconomic and natural subsystems, such as the interaction between its pressure on the environment and the environment's support for that pressure, is embodied in the input and output processes that lead to flows of materials and energy, which together comprise the entropy flow dS_e of the urban organism. Entropy production comes from within the socioeconomic subsystems through processes such as waste discharge and environmental degradation, which lead to increasing entropy (positive entropy production). The increase in urban decomposer functions (e.g., construction of municipal waste treatment infrastructure and ecological restoration investments) is the inverse of the waste discharge process and, therefore, restores order to the system by reducing its entropy (i.e., its generation of negative entropy) to offset and overcome the positive entropy production (waste discharge) that forms the entropy generation dS_i by urban organisms (Svirezhev 2000). Based on the entropy generation dS_i and entropy flow dS_e, the entropy change dS of urban organisms can be calculated.

The entropy flow dS_e represents the material and energy exchanges between the urban socioeconomic subsystem and the natural subsystem and can, therefore, represent the coordination between the two subsystems. Entropy production dS_i comes from the balance between the deterioration of environmental quality of the urban social and economic subsystems and ecological restoration, and this balance can characterize the vitality of an urban organism and the direction of its evolution.

The entropy flow generated by the support received from the natural subsystem and the socioeconomic subsystem represents a supporting input of entropy and the entropy flow generated by the pressure of the social and economic subsystems on the natural subsystem represents the pressure output entropy. The materials and energy absorbed by the metabolic activities or urban organisms are partly transformed into stocks, such as the city's environmental protection infrastructure and ecological restoration projects, generating reduced metabolic entropy (i.e., negative entropy production). Pollutants are discharged into the atmosphere, water, and soil either directly or after treatment to create metabolic entropy (i.e., positive entropy production). This reflects the integration of decomposition and regeneration processes in the socioeconomic subsystem (Fig. 3.6).

If $dS_2 = dS_1 + dS_e^2 – dS_e^1 – dS_i^1 + dS_i^2 < 0$ and $dS_2 < dS_1$, then the urban organism is in a state of dynamic development. Conversely, if $dS_2 = dS_1 + dS_e^2 – dS_e^1 – dS_i^1 + dS_i^2 < 0$ and $dS_2 > dS_1$, then the urban organism is in a degrading state. Finally, if $dS_2 = dS_1 + dS_e^2 – dS_e^1 – dS_i^1 + dS_i^2 > 0$, then the urban organism is in a further degrading state.

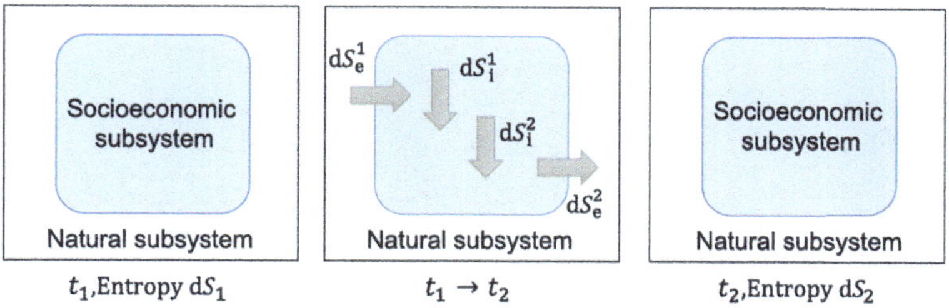

Fig. 3.6 Illustration of the entropy changes over time for an urban organism (*Note* dS_1 and dS_2 represent the entropy at times t_1 and t_2, respectively, and $\Delta S = dS_2 - dS_1$ represents the change in entropy. dS_e^1 represents the flow of support entropy received from the natural subsystem (support input entropy). dS_e^2 represents the flow of pressure entropy from the socioeconomic subsystem to the natural subsystem (pressure output entropy). dS_i^1 represents the entropy production (reduced metabolic entropy) formed by ecological restoration and construction of environmental protection infrastructure. dS_i^2 represents the entropy production (decomposition metabolic entropy) caused by environmental pollution)

3.3 System Ecology Theory

3.3.1 Integration of Holism and Reductionism

Scientists work from two complementary perspectives. First, reductionism attempts to break a complex system into its component parts so that details of each component can be understood in isolation. Second, holism attempts to integrate knowledge of those components into a single overall understanding of the system. This difference is analogous to the difference between walking into a forest and examining individual trees versus attempting to understand the forest as a single functioning ecosystem. Neither perspective is sufficient on its own; they must be combined to fully understand the forest (Miao 2005).

The reductionist way of thinking has the advantage of permitting detailed top-down decomposition of the system, but also has the disadvantage of failing to provide a comprehensive bottom-up diagnosis of the overall state of the system. Excessive reductionism loses track of the original associations among components of the system and their functions. Although this permits a sharper focus on the individual components, it produces a blurry overall image (Wang 2008). Holism emphasizes the functions and attributes of the system as a whole and often reveals universal general laws that affect the whole system, but it lacks the necessary focus to understand how these laws arise from each of the system's many components. This makes it difficult to form a rigorous logical description of the system that is both intuitive and generalizable.

Systems theory attempts to overcome these problems by organically integrating the two perspectives by considering the relationship between the whole and its component parts. As a result, it combines the benefits of the two approaches in a way that also mitigates their weaknesses. Systems theory emphasizes that the components of a system cannot be separated from the whole and cannot exist in isolation.

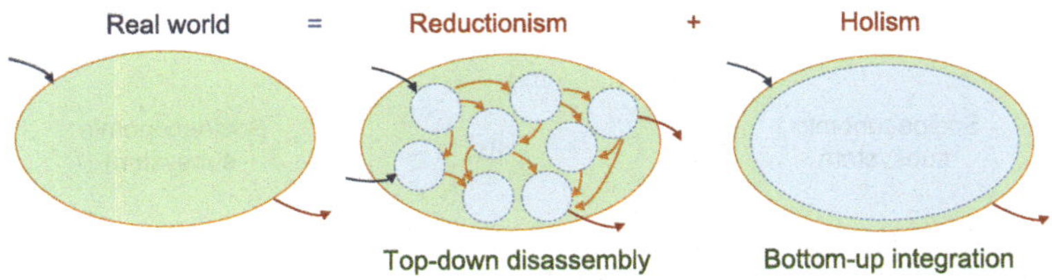

Fig. 3.7 Illustration of the relationship between reductionism and holism

Isolated components no longer function in the same way they would as mere parts of the overall system. In effect, "the whole is greater than the sum of its parts", because each component can add to the effects of other components (i.e., synergy).

Systems ecology combines systems theory and ecology. The theory emphasizes the need to treat the subject of the research, cities, as an integrated system. Figure 3.7 illustrates the relationship between reductionism and holism. Analyzing the ecological functions of the system permits in-depth analysis of the interactions among its subsystems and clarifies the rules that govern the operation of the system as a whole. The thought process behind systems ecology balances concrete detailed knowledge with abstraction. For example, when you see a city from a distance, you are likely to focus on its overall appearance (an abstraction), and not profoundly understand what you are seeing. To understand what you're seeing, it's necessary to step into the city and walk along its streets and inside its homes and factories to understand its composition and the interactions among its components; at that point, you will perceive certain specific aspects of the city and the connections among them (i.e., concrete details). When you leave the city, you are more able to regard it as a whole and, by combining the details and their complex connections, form a comprehensive and specific understanding of the city.

3.3.2 Urban Metabolism Research Based on Systems Ecology

To diagnose urban metabolic disorders, it's helpful to combine the approaches of traditional Chinese medicine with those of Western medicine, which offer different advantages. Traditional Chinese medicine is holistic and looks at connections between parts of the body and among its various systems (Wang 2007). In contrast, Western medicine is more reductionist and tends to examine the body as a series of isolated parts. As in the case of systems theory, combining the two provides a much better understanding of the urban organism's body, its disorders, and how to cure those disorders.

The modern study of urban metabolism is based on the theory of systems ecology; that is, it recognizes the need to understand an ecosystem's individual systems, while also examines the ecosystem as a whole to understand how those systems combine to produce the overall behavior. One insight provided by traditional Chinese medicine relates to the forces that are described by the concepts of *yin* and *yang*. Simplistically, these represent seemingly opposite forces that are, in reality, complementary and connected in complicated ways. Another insight relates to the Chinese concept of *qi*, which basically can be described as flows of vital forces and the need to balance these flows. The relationships of these concepts with the characteristics of urban metabolism are clear, as is their ability to reveal imbalances that lead to metabolic disorders.

Urban organisms, which are analogous to living organisms such as the human body, operate through flows of materials and energy between and within the body's systems. Each organ, tissue, and cell plays a unique role in the overall functioning of the organism. Each must be healthy, function efficiently, and be in balance with other components of the system for the overall organism to be healthy. If any component is unhealthy, inefficient, or imbalanced, the overall organism will suffer. In the analogy between a city and an organism, organs represent large urban components such as sectors involved in the extraction of natural resources; tissues represent more finely divided components of an organ, such as the agriculture sector; and cells represent the smallest functional units, such as individual farms. Urban metabolic disorders can result from the malfunction of an organ, tissue, or cell, or an imbalance between them (Shi and Chen 2016).

Studying urban metabolism not only emphasizes the components of living bodies (organs, tissues, and cells) but also emphasizes the interactions among these parts. Therefore, the construction of an urban metabolic model requires a consideration of the extraction of raw materials that serve as inputs (nutrients) for the system and a consideration of their destinations (e.g., a specific manufacturing industry). The same is true for the processes that regulate them, such as integration of their operations, changes of the spatial arrangement of different components, and governance mechanisms (Wang et al. 2006). Urban metabolism research starts with the collection of original data for each of these factors and develops linear or nonlinear metabolic models that provide the best description of these factors and help to reveal the optimal solution for urban health. These models rely on indicator systems that allow researchers to quantify key aspects of the system. The models can then be used to optimize urban metabolism to mitigate or eliminate metabolic disorders.

To support this approach, it's necessary to begin by dividing urban organisms into components for which data is available. For systems as large and complex as a city, this may require compiling massive amounts of data, particularly if the analysis will be dynamic (i.e., covers a period of time instead of providing a static snapshot of one point in time). It may be necessary to describe the system's initial condition, define variables and parameters that must be quantified, define the driving forces and the response variables they cause to change, and describe other characteristics of the system. Researchers should avoid the assumption that simple, linear causality can describe the relationships among the system's components since some relationships are nonlinear. Next, it's necessary to identify indicators that will reveal inefficiencies, imbalances, and other disorders. It's also necessary to identify the most important materials, such as raw materials and water, and the most important flows of materials and energy, as well as the interactions among sectors that result from these flows.

Producers and consumers must be identified. The sensitivity of each process to small changes in initial conditions should be explored to explain the associations between initiating factors or driving forces and the laws that govern how processes change in response to them. The hierarchical structure of the system should also be explored. One of the key challenges involves how to consolidate the massive quantities of data required by this analysis into a form that makes it possible to understand individual processes and how they interact and combine to shape the system's overall behavior. The result is an ecological network model that can simulate the influence of the many abovementioned factors on the urban organism's metabolic processes at different scales or hierarchical levels and their responses to interactions among components of the system. The models can then guide urban planning and design.

3.4 Research Paradigms

Urban metabolism research has accumulated rich and unique results in terms of its theoretical foundations, research frameworks, and technical methods. These results have strongly promoted the field's development. Urban metabolism relies on an analogy that compares cities to living organisms and ecosystems, applying the metabolic principles and methods that have been developed by ecological researchers to plan and design cities in ways that will alleviate their ecological and environmental problems. Urban metabolism is still a relatively new discipline. Although it originated from biology, it has developed with inputs from multiple disciplines, leading to the transformation and expansion of its research paradigms. These paradigms now include aspects related to the metabolism of the city's natural subsystem, socioeconomic subsystem, and integrated (hybrid) natural and socioeconomic system. The interactions among these different paradigms, and how they complement each other, are promoting further evolution of urban metabolism research.

3.4.1 The Relationship Among the Three Research Paradigms

The abovementioned three paradigms differ in their maturity, modeling methods, and temporal and spatial complexity. Fusion and integration of the three paradigms have benefited from adopting the perspective of cities as essential and highly influential parts of the biosphere and from interdisciplinary research supported by multiple disciplines. Wolman (1965) provided the first important example of an urban metabolism research paradigm. In the nearly 60 years since this paper, technical methods and case studies of this approach have been continuously enriched and improved. However, it was not until the 2000s that urban metabolism research formed a trinity that

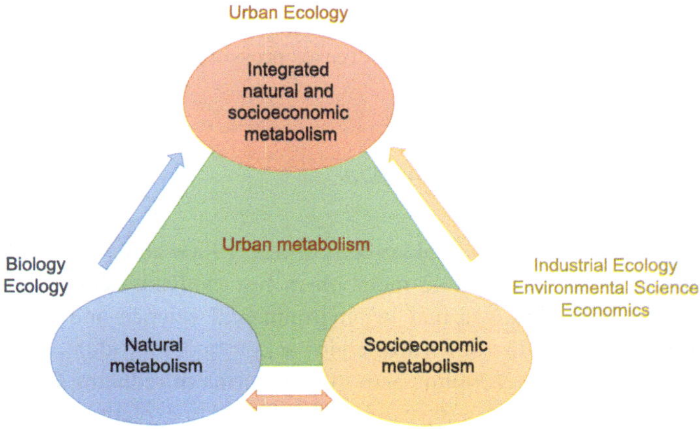

Fig. 3.8 Illustration of the three research paradigms that together define the field of urban metabolism research

comprised the different aspects of natural metabolism, socioeconomic metabolism, and a hybrid of the two types of metabolism (Fig. 3.8).

3.4.2 Natural Metabolism

Since the 1930s, the continuous development of biological research and ecological research have supported the emergence of the natural urban metabolism research paradigm. Researchers in these fields have used their techniques to focus on urban, suburban, and mosaic villages that exist within a matrix of different types of natural and artificial ecosystems, as well as analyzing the laws that govern the evolution of urban ecological space. These laws govern a city's animals, plants, and natural communities, which have parallels in the processes that occur in natural habitats such as forests and wetlands. The main difference is that the driving factors and their interactions are far more intense in urban areas and cities are, thus, unlikely to be as self-sufficient as natural ecosystems. In addition, researchers have been likely to regard the associated processes from the perspective of external conditions (i.e., by treating the city as something separate from its natural components) and from the perspective of the driving forces that initiate and regulate metabolic flows (Liang and Zhang 2011). As a result, they have analyzed the city's interference with its natural environment and the pressure it places on that environment through its effects on species diversity, natural processes, and the ecosystem's structural and functional characteristics. This has led to consideration of the pressure created by the urban system on natural metabolism and the responses to this stress, including adaptations and factors that interfere with natural processes, such as the establishment of impermeable surfaces that prevent water infiltration, occupation of natural spaces by infrastructure such as roads and buildings, greatly increased population density, and large-scale economic activity that increases pressure on the natural metabolism.

3.4.3 Socioeconomic Metabolism

With the development of industrial ecology in the 1980s and 1990s, the research paradigm of urban socioeconomic metabolism has evolved. Most researchers in this field come from a background in environmental science and environmental economics. Starting from their own academic background, they identified a group of shared subjects (i.e., socioeconomic activities) in terms of reducing costs and mitigating environmental pollution. Researchers have tried to explain or modify a city's socioeconomic behaviors from the perspective of ecology. With the transformation of the research paradigm from natural metabolism to socioeconomic metabolism, the subject of the research also shifted from the natural components of the city to the human-created components.

Early research started with the perspective that the natural environment provides fuel for the fire of socioeconomic development, leading to clarification of the physical boundary between the urban built-up area and its environment. On this basis, it was possible to establish two "red lines"—one for ecological protection that defined the upper limit of resource utilization and another for environmental quality that defined the external constraints imposed on urban activities.

As the research evolved, researchers began to define socioeconomic activities as the metabolism itself, not as the driving factors and external conditions that were key parts of the natural metabolism research paradigm. Through detailed analysis of the scale, structure, pattern, processes, interactive relationships, and dynamic evolution of the socioeconomic system, researchers further analyzed the complexity of mechanisms that initiated and regulated socioeconomic activities (Dai and Wang 2018).

In contrast to the natural metabolism paradigm, the socioeconomic metabolism paradigm treats the natural environment as an external condition that constrains a city's development and focuses on the material and energy flows caused by socioeconomic activities that are intended to improve human health and well-being. This approach emphasizes the urban socioeconomic subsystem as the research subject without fully considering the problems this system creates for the natural metabolism. Researchers mostly use material flow analysis to study the relationships between the city and its environment. The socioeconomic metabolism research paradigm focuses on the artificial components of the city (industrial and commercial areas, residential areas, roads, and other infrastructure) by examining the scale, structure, and heterogeneity of the socioeconomic activities that create and modify these components of the city. These depend on diverse economic behaviors and social systems. Multiple types of industrial activities and human cultural activities combine to define the urban matrix and external driving factors in the natural metabolism research paradigm (McPhearson et al. 2016).

In the context of global urbanization, the socioeconomic metabolism research paradigm fully incorporates and integrates the man-made characteristics of cities, which permits comparisons between cities so that insights from one city can guide development in another city.

3.4.4 *Integrated (Hybrid) Natural and Socioeconomic Metabolism Paradigm*

Although the natural metabolism and socioeconomic metabolism paradigms have been developing in parallel, they have increasingly overlapped as this field of research has grown. Although the natural metabolism paradigm treats socioeconomic activities as driving factors, it also seeks to understand the state of these driving factors and how they are developing. Although the socioeconomic metabolism paradigm regards the environment as a constraint, it also seeks to analyze the composition, support capacity, structure, functions, and recovery capacity of the ecological environment (Pickett et al. 2016). The two paradigms, therefore, continue to merge as they evolve. This coevolution has gradually led to the emergence of an integrated, or hybrid, natural and socioeconomic metabolism that is providing new insights for urban planners, decision-makers, and the public (Fig. 3.8). This evolution is similar to how reductionism and holism have gradually merged to produce a richer and more comprehensive understanding of how cities function, as I noted earlier in the chapter.

The integrated paradigm evolved in the 1980s and 1990s. Researchers who work under this paradigm regard cities as complex ecosystems in which the ecological environment is as important as socioeconomic activities and the two, taken together, form an organic whole. This paradigm also pays more attention to comprehensive questions such as the degree of human well-being, the livability of the urban environment, and biological and ecological health and abundance. This research paradigm links ecology with urban processes and can provide important support for the restoration, management, and sustainable development of urban ecosystems (Pickett et al. 2016).

The integrated research paradigm not only focuses on natural and semi-natural ecological spaces, such as urban parks, green spaces, and water bodies but also characterizes the "texture" formed where these natural components interact with urban socioeconomic activities. To support this research, researchers conduct a comprehensive survey of all components of a city to determine the scale and area covered by components such as trees, shrubs, grasslands, crops, bare land, built-up land with an impermeable surface, water bodies, and buildings. Dynamic characteristics such as comparisons between components and cities and analysis of the interactions among components more fully demonstrate the heterogeneity of natural patches within the urban matrix and provide a powerful tool for answering core questions about the interaction between the urban socioeconomic structure and natural ecology.

The integrated research paradigm will not replace the natural metabolism and socioeconomic metabolism paradigms but will, instead, benefit from progress by researchers who are using those paradigms and from results obtained using the paradigms (e.g., their unique insights into urban metabolism). The goal of the integrated paradigm is to comprehensively and systematically analyze all urban metabolic processes and provide a holistic understanding of the city. This will be essential to identify problems, diagnose their causes, and find solutions. The emergence of this hybrid approach reflects the maturity of urban metabolism research. Comparison of the results of the three paradigms provides additional insights and is being accompanied by a shift from a single-discipline research perspective to one that is integrative and multi-disciplinary. The cooperation among researchers from different disciplines is making urban metabolism research more innovative, while also revealing ways to improve environmental integrity, social equity, and economic efficiency. The result is a better ability to propose comprehensive, science-based solutions for planners, decision-makers, and the public.

3.5 Technical Framework

As in any other field of research, urban metabolism has followed a progression from theory (providing a research framework and asking questions) to method development (finding ways to use theory to guide research), and finally to application (identifying and solving problems). This evolution has led to the development of a four-part technical framework that starts with process analysis, continues with an accounting analysis to quantify flow quantities, then an evaluation of the meaning of the accounting results, and concludes with the development of simulation models that help urban managers optimally regulate urban metabolism. This approach provides important support for ensuring urban ecological security (Zhang 2013).

Process analysis is the theoretical cornerstone and starting point for urban metabolism research, as it provides the basis for subsequent research. The accounting evaluation and simulation models are important technical methods to objectively describe and analyze problems, as well as provide decision-making support for practical application of the theory to help planners and managers optimize regulation of urban metabolism. Figure 3.9 illustrates the overall process.

Process analysis is a bridge that connects the real world with scientific research. Using this tool, researchers can interpret the real world in ways that are conducive to current research and can provide theoretical support for subsequent investigations. Process analysis has evolved from a focus on linear models (largely based on input–output analyses) to circular models that more accurately capture cyclical flows (e.g., treating outputs from one component as inputs for other components, recycling waste as inputs for other processes), and then to a network model that shows the connections among components of urban metabolism.

Accounting evaluation is an important technical means to quantify metabolic processes and the associated flows to make the analysis less subjective. It can also

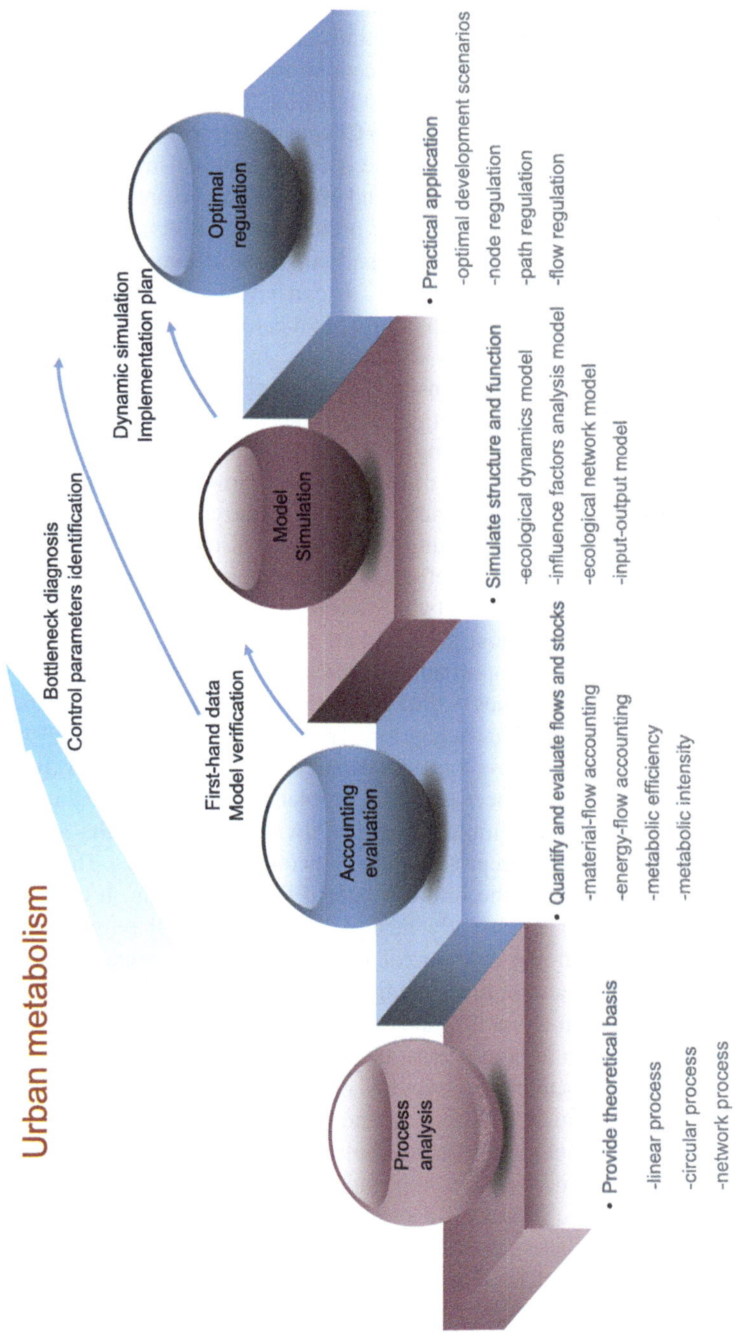

Fig. 3.9 Illustration of the technical framework for urban metabolism research

reveal metabolic bottlenecks. Accounting can provide first-hand data for the development of simulation models, a historical reference that can be used for model verification, and policy control parameters that can optimize regulation of urban metabolism. In addition to the traditional material flow accounting methods (element flow and batch flow analyses), energy accounting methods (exergy and emergy) have begun to appear. I will discuss emergy in more detail in Chapter 4, but will not discuss exergy because this has played a relatively minor role in urban metabolism research.

All accounting methods have advantages and offsetting disadvantages. The material flow accounting method is based on the principle of conservation of mass. It measures the consumption, transformation, and emission of materials and energy by urban metabolic processes in mass units. It can vividly reflect the weight of materials involved in the urban organism's life, but it cannot account for the quality of these materials and energy or their social effects. In contrast, the energy flow accounting method can provide these insights. The energy flow accounting method fully considers the efficiency and quality of various substances based on the energy and information they contain, converting various material flows, energy flows, and information flows into a solar emergy or exergy value. However, one drawback of this approach is that it depends on accurately calculating the conversion rate between material and energy flows. The accounting methods have also evolved to account for the scale of metabolic activities, their efficiency, their intensity, and their impact on metabolism. This is usually supported by the development of a system of evaluation indexes that can be used to diagnose the health of urban metabolism.

Simulation models are an important step because they apply theoretical and accounting knowledge to seek clues to the rules that govern a metabolism's behavior and to let researchers test various scenarios. The key is to build a model based on process analysis, parameterize the model based on accounting, and dynamically simulate metabolism to detect potential problems that could explain or lead to metabolic disorders. Model simulation can be divided into two parts—external representation of the system and representation of its internal processes. The external representation can help predict the future health of the metabolism. It can answer questions such as whether a currently healthy metabolism will remain healthy. The internal representation can help reveal whether the components of an urban organism's organs and tissues are functioning in a coordinated and balanced manner and whether the metabolism's structure and functioning are harmonious. The external representation of the urban organism may assume that the organism is healthy, but there is no guarantee that its internal tissues and organs are not disordered. Therefore, it's necessary to use ecological dynamics methods to simulate the long-term evolution of the internal components of the metabolism. This can be done using factor decomposition, ecological network analysis, input–output analysis, spatially explicit analysis using a geographic information system, and other methods that permit the development of a model that simulates the metabolism's structure, its internal and external processes, and its functioning, to explore its dynamic evolution and the laws that govern that evolution. Simulation models can support the development of an implementation plan for optimal regulation of urban metabolism and, as a result, can help to regulate metabolic actors, the paths that connect them, as well as the flows along those paths, from a network perspective.

Therefore, the modern technical framework for urban metabolism research links process analysis, accounting evaluation, simulation models, and optimization (regulation) of the metabolism while also considering the interactions among the city's metabolic actors. Through systematic analysis of the status of the urban organism's metabolism and diagnosis of its disorders, it becomes possible to develop strategies for cures. The modern theoretical and technical frameworks provide important support for a science-based regulation of urban metabolism to improve management of the urban ecosystem and promote the development of a healthier urban metabolism.

References

Boyden S, Millar S, Newcombe K et al (1981) The ecology of a city and its people: the case of Hong Kong. Australian National University Press, Canberra, Australia

Dai TJ, Wang WJ (2018) The characteristics and trends of socioeconomic metabolism in China. J Ind Ecol 22(5):1228–1240

Girardet H (2004) Cities people planet: livable cities for a sustainable world. John Wiley & Sons Ltd., Chichester, UK

Haken H (1988) Information and self-organization. Springer, New York, USA

Haughton G, Hunter C (1994) Sustainable cities, regional policy and development. Jessica Kingsley, London, UK

Huang CH (2002) Is the structure and function of the urban ecosystem a replica of the natural ecosystem? China Popul Resour Environ 12(3):134–136 (in Chinese)

Huang S, Wong J, Chen T (1998) A framework of indicator system for measuring Taibei's urban sustainability. Landsc Urban Plan 42(1):15–27

Jordan SJ, Vaas PA (2000) An index of ecosystem integrity for Northern Chesapeake Bay. Environ Sci Policy 3(S1):559–588

Liang S, Zhang TZ (2011) Urban metabolism in China achieving dematerialization and decarbonization in Suzhou. J Ind Ecol 15(3):420–434

Ma SJ, Wang RS (1984) Socio-economic-natural complex ecosystem. Acta Ecol Sin 14(1):1–9 (in Chinese)

McPhearson T, Pickett STA, Grimm NB et al (2016) Advancing urban ecology toward a science of cities. Bioscience 66(3):198–212

Miao DS (2005) On system thinking (3): combination of holistic thinking and analytical thinking. J Syst Dialectics 13(1):1–5 (in Chinese)

Newman PWG (1999) Sustainability and cities: extending the metabolism model. Landsc Urban Plan 44(4):219–226

Patten BC, Costanza R (1997) Logical interrelations between four sustainability parameters: stability, continuation, longevity, and health. Ecosyst Health 3(3):136–142

Pickett STA, Cadenasso ML, Childers DL et al (2016) Evolution and future of urban ecological science: ecology in, of, and for the city. Ecosyst Health Sustain 2(7):e01229

Shi L, Chen WQ (2016) Retrospect and prospect of the development of China's industrial ecology. Acta Ecol Sin 36(22):7158–7167 (in Chinese)

Svirezhev YM (2000) Thermodynamics and ecology. Ecol Model 132(1–2):11–22

Tjallingii SP (1993) Ecopolis: strategies for ecologically sound urban development. Backhuys Publishers, Leiden, The Netherlands

Wang RS (2000) Frontier research progress of urban ecology in the transition period. Acta Ecol Sin 20(5):830–840 (in Chinese)
Wang RS, Ouyang ZY (2012) Socio-economic-natural complex ecosystem and sustainable development. Bull Chin Acad Sci 27(3):337–345 (in Chinese)
Wang X (2008) The turning point of business management thinking: from system theory to reduction theory. IT Manager World 475(S1):58–60 (in Chinese)
Wang YY (2007) The relation between system theory and reduction theory in traditional Chinese medicine research. World Science and Technology 9(1):70–73 (in Chinese)
Wang YY, Zhang QM, Zhang ZB (2006) Extraction of syndrome elements and targets. J Shandong Univ Tradit Chin Med 30(1):6–7 (in Chinese)
Weber BH, Depew DJ, Smith JD (1988) Entropy, information and evolution: new perspectives on physical and biological evolution. MIT Press, Cambridge, MA
Wolman A (1965) The metabolism of cities. Sci Am 213(3):178–190
Zhang Y (2013) Urban metabolism: a review of research methodologies. Environ Pollut 178(7):463–473
Zhang Y, Yang ZF, Li W (2006a) Analyses of urban ecosystem based on information entropy. Ecol Model 197(1):1–12
Zhang Y, Yang ZF, Yu XY (2006b) Measurement and evaluation of interactions in complex urban ecosystem. Ecol Model 196(1):77–89

Part II
Methods

Based on the technical and theoretical framework for urban metabolism research presented in Chapt. 3, this section introduces the systems of methods used by urban metabolism researchers from the perspectives of accounting evaluation (Chap. 4), simulation models (Chap. 5), and regulation and optimization of an urban metabolism (Chap. 6). Urban metabolism research can provide information to support more effective efforts to improve a city's energy efficiency, material circulation, waste management, and infrastructure construction. The main accounting methods are material flow analysis (and the related method of substance flow analysis) and energy flow ("emergy") accounting. Based on these methods, I will describe a comprehensive index system that can be used to evaluate the evolution of an urban system, the relationships among its components, and the metabolic level of an urban organism. In Chap. 5, I describe how to construct a metabolic network model using the material and energy accounting methods introduced in Chap. 4, with the network nodes identified and refined by means of input-output analysis. I then propose a set of technical methods that can be used to simulate the structure and function of an urban organism. Next, I construct a factor-decomposition model that simulates the system's internal characteristics and their relationships with external driving factors, discuss the key factors that drive the urban metabolic processes, and analyze the relationships among metabolic actors and their spatial distribution. In Chap. 6, I conclude by simulating the dynamic evolution of the metabolic actors and based on the results, propose optimization scenarios and regulation schemes.

Chapter 4
Accounting Evaluation of Urban Metabolism

As I noted in Chapter 3, an important component of the urban metabolism research tools is quantification of the flows and stocks of materials. In this chapter, I will describe the key points of the main accounting methodologies.

4.1 Material Flow Analysis

Early urban metabolism research mostly used material flow accounting to track the flows of materials and energy, including the input, storage, transformation, and output of materials. This approach has been used in the design and management of many cities (e.g., Hendriks et al. 2000). Urban development consumes (via flows) and accumulates (via stocks) large amounts of resources and energy. The flows that occur during urban development, and the resulting accumulation of stocks are two crucial factors in material flow analysis, whether the analysis focuses on one or multiple materials. The scale of flow and material accumulation represents the throughput and metabolic scale of urban metabolism. Stocks represent the accumulations of a given material in a specific state, such as the building or road stock. In material flow analysis, stocks are defined in mass units, and therefore reflect the weight of these materials in an urban metabolism. There is a close relationship between flows and stocks: small stocks result when the required materials flow slowly, whereas large stocks accumulate when this flow is rapid.

Both flows and stocks can represent urban weight, which refers to the mass of the throughput and stocks in the urban material metabolism, and reflects the resource inputs and waste outputs required to maintain normal operation of the city. The weight can vividly reflect a city's development characteristics. For example, a mature city's stock weight remains relatively stable, so the flow weight is the dominant factor that characterizes its resource consumption and the resulting environmental effects. In contrast, the stock of young cities tends to grow rapidly as the cities expand, so the metabolic rate (the rate of resource consumption) is therefore high. The metabolic

Y. Zhang, *Urban Metabolism*,
https://doi.org/10.1007/978-981-19-9123-3_4

rate of a mature city tends to be low, since the city is no longer growing rapidly, but it can also be high if the metabolism is inefficient or if there is a large stock that must be maintained by input flows. Conversely, a younger city often grows fast, so its metabolic rate tends to be high even if it has a relatively low stock; the small amount of stock that must be maintained leads to a low input requirement, but the rapid addition of new stock requires a higher flow. In most recent research, the accounting system for flows and stocks tends to treat the socioeconomic subsystem as its subject, and treats the natural components of the urban system and the city's external regions as the environmental components (Guo et al. 2020; Liu et al. 2020).

Quantifying urban weight lets researchers track materials as they pass through the urban system. Based on their improved understanding of the system, they are then able to refine their model of urban resource consumption and determine areas where an intervention or regulations are required to improve urban resource-utilization efficiency or reduce the environmental impact (Swilling et al. 2018). Although research on a single material such as water lets researchers focus on the impact of that specific material on urban development, the tradeoff is that this fails to account for other materials that may flow and accumulate along with the focal material. Nonetheless, the knowledge gained from this form of analysis can reveal problems related to a key material and help urban managers develop ways to increase the efficiency of its utilization in order to reduce the environmental impact. In contrast, multi-material research can focus on synthesizing the flows of a greater proportion of the total material flows during urban development, thereby helping decision-makers more comprehensively understand the problems faced by cities and formulate policies that are more likely to achieve sustainable development.

4.1.1 Flow Accounting

In the study on which this section was based (Wang et al. 2020), my research group established an overall framework for urban material flow analysis by referring to the national-scale framework formulated by the European Union (Eurostat 2018). It was also based on the availability of urban statistical data for China, my primary research subject, since China's government has gathered the large quantities of data required to quantitatively describe the material flows that occur during urban metabolic activities. This approach can reveal the hidden resource and environmental problems created by these activities. The material flow accounting framework combines top-down analysis with bottom-up analysis. First, the city is divided (from the top down) into components, which allows tracking of the flows of materials (input, storage, transformation, and output) between these components. When statistics can't be obtained directly from official statistical data, various estimation and conversion methods can be used. For example, when the accounting is based on mass units but only economic data are available for a material, the economic data can be converted into mass data by dividing the total economic value by the cost per unit (thereby obtaining the number of units), then multiplying that number by the mean mass per unit. The flow data can then be integrated (from the bottom up) to obtain one or more quantitative indicators of the scale of these flows, which together indicate the size of the overall urban metabolic throughput. The framework and measurement methods used by urban material flow analysis can provide a template for the calculation of urban material flows anywhere in the world, providing the necessary data are available. Using a consistent framework also makes the accounting results more regular and comparable between studies, which both improves the quality of the accounting and facilitates comparisons between cities to see whether insights from one city can guide the management of others.

4.1.1.1 Accounting Items

The materials whose flows are quantified include biomass, metallic minerals, non-metallic minerals, industrial products, fossil fuels, and various types of byproducts, pollutants, waste, and balance items. Table 4.1 describes the specific contents of each type of material. The processes that cause flows through the system include inputs (i.e., local exploitation, transfers-in from regions outside the study city), outputs (i.e., transfers-out into regions outside the study city), hidden flows, and balancing items.

O_2, CO_2, and H_2O consumed or produced by respiration and photosynthesis, or O_2 consumed and water vapor produced by fuel combustion, are also incorporated into the framework as balancing items that ensure conservation of mass. Due to the large mass of balancing items and the few factors that influence them, there has been little research on their value. As a result, these materials are generally not included in calculations of the final material inputs and outputs, or in indexes that summarize these flows. In contrast, the amount of water input into the socioeconomic system is far greater than the total amount of dry matter. If this amount was included within the material flow analysis framework, it would outweigh the inputs, outputs, and consumption of other materials. In addition, clear statistics are available on the water resource supply, consumption, and discharge in government statistical databases, so it is relatively easy to analyze the flows of water in the socioeconomic system separately from other materials. Therefore, I have not included these water resources in the material flow analysis framework described in this chapter.

During mining or harvesting of primary raw materials, a large amount of materials must be transported and processed, and this will damage the environment. The portions of these materials that do not directly enter the socioeconomic system are called "hidden flows." There are 3 main kinds of hidden flow. First, there are the tailings left after selecting and washing the mineral ore, since these materials are excavated with the minerals. The hidden mining flow for these materials can be obtained by multiplying the quantity of the material mined with a hidden-flow coefficient that accounts for the content of tailings in the primary material. Second, there is disturbance of the soil layer, such as soil erosion during and after agricultural production, and excavation of soil and rock to produce construction materials. The hidden flow resulting from agricultural harvests can be obtained by calculating the crop straw production coefficient (as a proportion of the crop that is harvested) and the

Table 4.1 Items included in the urban material flow accounting framework

Category	Items	Scope
Material inputs	Biomass	Production of crops, forestry, animal husbandry, and fisheries
	Metallic minerals	Iron ore and non-ferrous metallic minerals
	Non-metallic minerals	Industrial non-metallic minerals and building materials
	Fossil fuels	Raw coal, crude oil, natural gas, gasoline, kerosene, diesel, fuel oil, and liquefied petroleum gas
	Transfers-in and imports	Raw materials, finished products, semi-finished products and other products imported from outside the city
Material outputs	Air pollutants	CO_2, SO_2, NO_x, smoke, and dust
	Water pollutants	Chemical oxygen demand, ammonia nitrogen, petroleum, volatile phenolic compounds, etc
	Solid waste	Industrial solid waste, construction waste, urban household waste, etc
	Dissipative materials	Unused inorganic fertilizer, pesticides, and agricultural plastic film
	Transfers-out and exports	Raw materials, finished products, semi-finished products, and other products
Balancing items	Inputs	Oxygen, CO_2 (photosynthesis)
	Outputs	Water vapor, O_2 (photosynthesis), CO_2 (respiration)
Hidden flows	Hidden flows within the city	Unused materials in processes such as energy and mineral extraction, engineering excavation, and biomass harvesting
	Transfers-in and hidden import flows	Unused materials in processes such as energy and mineral extraction, engineering excavation, and biomass harvesting

utilization ratio (the proportion of the total biomass not left behind at the harvest site), and then multiplied by the crop harvest (Li 2014). Third, the residues of agricultural and forestry crops, such as straw and branches, must be considered, since they are harvested together but not used. Government statistics do not include these residues, so they are generally estimated based on conversion coefficients from research literature. Most of the excavation hidden flow (the soil and rock that remains after creation of the excavation) in the Construction component of the metabolic system is based on the completed area of urban residences and not on the calculation of the actual material utilization (Bao 2010).

Hidden flows can be divided into inner and outer flows. Inner hidden flows can be assigned to the urban components or sectors that participate in the urban material flows, or can be quantified in the flows to regions surrounding the city through transfers-out and exports of materials. They therefore impact both the domestic environment outside the city and the foreign environment. In contrast, the outer hidden flows represent transfers-in from the city's external environment, and include materials that are consumed within the city's boundaries, even though they primarily impact the external environment (Li et al. 2019). Although hidden flows are not directly counted as material flows in cities, they significantly affect the consumption of resources and the impact of this consumption on the environment. Therefore, I have included these hidden flows in the accounting framework described in this section.

4.1.1.2 Accounting Subjects

Since the total amount and structure of the flows of materials that enter and leave a city are inevitably affected by the material flow processes within the city, it's not appropriate to treat a city as if it were a black box; instead, it's necessary to divide it into different metabolic actors that function inside the box. According to the availability of data and the actual differences in material utilization of different parts within a city, I have divided the urban socioeconomic system into eight metabolic components: Agriculture, Mining, Manufacturing, Recycling, Household Consumption, Construction, Transportation, and Energy Transformation (Fig. 4.1).

Let's start by considering the Agriculture component. Biomass input to the Agriculture component includes crops and their byproducts (e.g., straw) harvested directly from the land, and forage grass for livestock. Industrial products are the products required by Agriculture, such as inorganic fertilizer, livestock fodder, pesticides, and plastic film mulches. Fossil fuels are the energy materials consumed by agricultural production activities. The input data that must be calculated can be divided into two categories before they are included in the accounting: materials such as wood that must be converted from a volume to a mass using a conversion coefficient, and materials such as straw, forage grass, and artificial feed that must be calculated separately for each crop or animal. Biomass leaves the Agriculture component in the form of harvested crops and livestock products. The main air pollutant is CO_2, generated by fuel consumption and burning of straw. In future research, the CH_4 emitted by cattle, rice paddies, and reservoirs should be added to the accounting, as these flows are large enough to be significant. The water pollutants represent the chemical oxygen demand and ammonia nitrogen in discharged agricultural water. Dissipative materials are pesticides, inorganic fertilizer, and plastic mulches that are not fully consumed or not used at all by crops and are thus dissipated into the environment. Among the biomass component flows, the main materials that must be converted to mass units

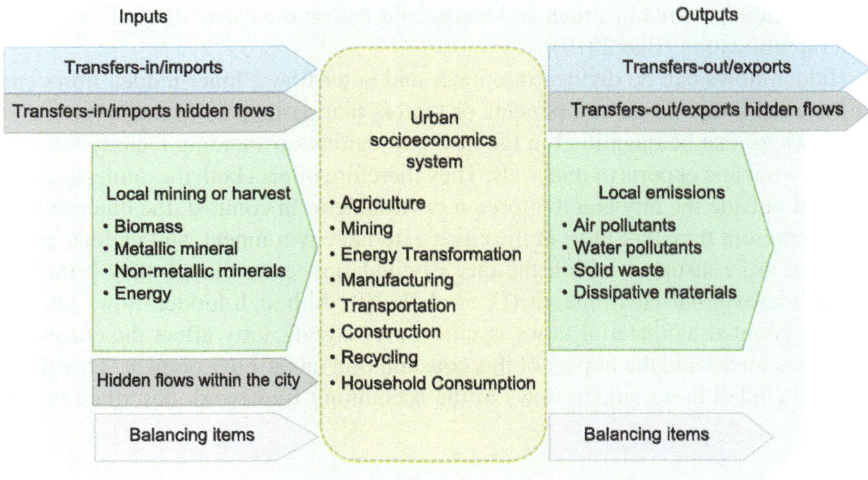

Fig. 4.1 An accounting framework for the material flows among the different urban metabolic actors (components of the system) (*Note* Hidden flows are allocated to different actors based on the flows of minerals, fuels, biomass resources, and industrial products that each hidden flow is associated with. There is no inter-component transfer of materials for the balancing items; in the framework used in this chapter, these balancing transfers occur only between sectors and the city's internal environment, and these flows are mainly derived from organisms and fossil fuels, so there are inputs and outputs for every sector except Recycling and Energy Transformation. The balances and hidden flows are not detailed in the diagram)

are the dissipative materials and air pollutants. Some cities do not record statistics on the water pollutants discharged by Agriculture; for these cities, the quantity can be estimated based on the ratio of the cultivated land area in the focal city to the cultivated area in neighboring cities that do record these pollutants; the ratio can then be multiplied by the pollutant discharge of the latter cities.

Next, let's consider the Mining component. The input materials are the locally mined and imported metallic and non-metallic minerals, as well as both the fossil fuels that are locally mined and those obtained from the Energy Transformation component and consumed by the Mining component. Except for the fossil fuels consumed by the Mining component, other inputs flow through this component and are transferred to other components or exported to regions outside the city. The data for flows of minerals and fossil fuels into and out of the Mining component in some cities can be directly obtained from statistical data; for cities that do not provide these data, the value can be estimated based on the ratio of total values for the focal city to the total values for the Mining sector in neighboring cities that do provide this data; the ratio can then be multiplied by the flows in the latter cities.

Inputs to the Manufacturing component are raw materials or semi-finished products required for the production of industrial manufactured products, including metals and non-metallic minerals, biomass, and industrial products imported from local and external sources, as well as fossil fuels that provide energy for industrial production.

Outputs are semi-finished or final products, as well as some waste generated by the production processes.

The inputs for the Transportation component are mainly vehicles and the fossil fuels they consume. The outputs are the air pollutants produced by fuel combustion. The number of all kinds of vehicles can be directly obtained from the city's statistical yearbooks. The quantity of vehicles that represent inputs into the Transportation component can be calculated based on the weight coefficient (i.e., the weight per unit) for each kind of vehicle. The consumption of fossil fuels is calculated by multiplying the number of vehicles by a fuel-consumption coefficient based on the number of passengers or weight of freight transported by that type of vehicle. The air pollutants emitted by each type of vehicle are calculated according to the energy consumed multiplied by the pollution emission factor for each mode of transport.

The inputs to the Construction component are building materials (including non-metallic minerals, wood, and industrial products) and fossil fuels. The materials at the output end are mainly air pollutants and solid waste generated by construction activities. The building materials at the input end are converted into stocks based on a coefficient that is expressed as a value per unit floor space under construction, and the pollutants at the output end are similarly converted based on a coefficient per unit of the completed floor space.

The inputs of the Recycling component are pollutants and waste from the Manufacturing and Household Consumption components. The materials at the input end of the Recycling component can be obtained from the pollutant production by the other components. The materials at the output end of the Recycling component comprise the emission of pollutants to the environment at levels that meet government standards, their removal rate, and the comprehensive utilization amount (for treated materials supplied to the Manufacturing component). These data can be directly obtained from the relevant government statistical yearbooks.

Finally, the input end of the Household Consumption component comprises biomass and industrial (manufactured) products such as food, clothing, and the durable goods needed by residents of the city, as well as fossil fuels such as liquefied petroleum gas and natural gas to provide energy. These values can be directly obtained from the city's statistical yearbooks. It is worth noting that the consumption of food in some cities is recorded in economic terms (i.e., in RMB), and these values must be translated into quantities (mass values) based on the mass per unit of the material after obtaining the recorded local prices. In some cities, the mass of clothing cannot be obtained directly, so the average annual consumption of clothing must be converted into mass values based on the population (i.e., the per capita amount). Durable goods in the statistical yearbooks are often counted as the number per 100 households, and the new quantity consumed in each year (i.e., the flow rather than the stock) must be calculated as the difference in numbers between consecutive years. Then, according to the number of households and the mass per unit of durable goods, the total mass of new durable goods can be calculated and used to represent the flow to the Household Consumption component. The output materials are CO_2 generated by fuel combustion, chemical oxygen demand, and ammonia nitrogen in wastewater, as well as household garbage and feces. Except for CO_2 directly discharged into the environment, all other pollutants are sent to the Recycling component. CO_2 is calculated based on the use of fossil fuels and respiration (oxygen consumption and CO_2 release) by livestock and humans. Other pollutant data can be obtained from the city's statistical yearbooks. For some cities that have no pollution source data, values from nearby cities with a similar development level can be used, with total values calculated based on per capita values (Fig. 4.2).

The main sources of raw data used in such analyses are government statistical yearbooks, economic yearbooks, or economic statistical yearbooks for a city. Statistical yearbooks for specific industries, such as energy statistics yearbooks and transportation statistics yearbooks, are also important sources, as are environmental statistics yearbooks. The data in these yearbooks are usually collected by different government departments. Statistical data are not available for most of a city's imports and exports. In some cities, the proportions of imports and exports can be estimated for some types of materials based on the actual situation, and then the estimated mass of these materials can be obtained using conversion coefficients. For example, the export proportion of fossil fuels can be obtained by investigating the sales direction (to components inside or outside the city) of the products of the main energy extraction enterprises in the city. Most of the other materials are calculated mainly through the balance between the import, production, consumption, export, and stock changes of the materials (where the stock change = local production + import − consumption – exports). Obtaining more and better data for these flows will be an important challenge for future researchers.

According to the material supply relationships among the various components in Fig. 4.2, we can assume the materials that are consumed locally come first from local production or extraction of such materials. When the local supply of this material is insufficient, the required material is provided by imports. When local production or exploitation is insufficient, we can assume that the allocation priority for the local material supply is in the following order: Household Consumption, Agriculture, Mining, Energy Transformation, Manufacturing, Construction, Transportation, and Recycling. That is, the demand created by Household Consumption is satisfied first, then if any of the material has not been allocated to Household Consumption, it is then allocated to Agriculture, and so on. Any unsatisfied demand is provided by imports. When the amount of a material produced or extracted locally is greater than the local demand, the surplus material is exported, except for materials consumed by changes in stocks such as those in the Construction component.

4.1.1.3 Accounting Indicators

Based on the material flow analysis framework described in the previous section, my research group developed an index system to facilitate comparisons between cities, and longitudinal comparisons using a time series for a city. We defined the following scale indexes: the direct material input (DMI), total material input (TMI), total material requirement (TMR), direct material consumption (DMC), domestic processed output (DPO), direct material output (DMO), and total material output

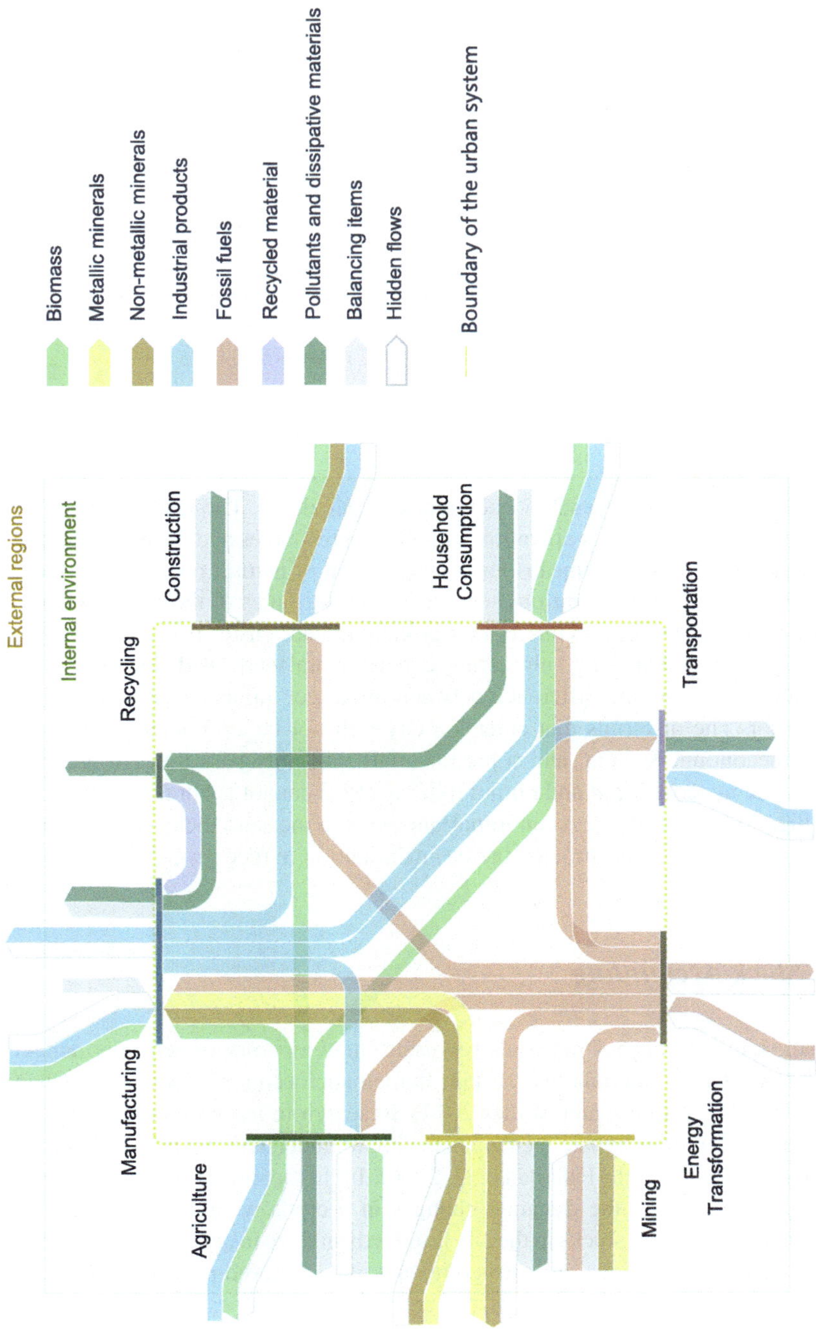

Fig. 4.2 Illustration of the flow processes in an urban material metabolism

Table 4.2 Scale indexes for input, consumption, and outputs

Category	Index	Meaning	Formula
Input	DMI	Direct material input	DMI = local mining + transfers-in + imports
	TMI	Total material input	TMI = DMI + hidden flows from within the city
	TMR	Total material requirement	TMR = TMI + hidden flows from outside the city
Consumption	DMC	Direct material consumption	DMC = DMI – transfers-out – exports
Output	DPO	Domestic processed output	DPO = pollutants and waste
	DMO	Direct material output	DMO = DPO + transfers-out + exports
	TMO	Total material output	TMO = DMO + hidden flows within the city

(TMO). Table 4.2 summarizes how each index is calculated. The smaller the city's scale index, the less impact its operation will have on resources and the environment. For example, a smaller DMI means less consumption of natural resources and less external support required for various socioeconomic activities of the city; similarly, a smaller DPO means the city emits less pollutants and waste. It is worth noting that DMC is not obtained by subtracting exports from total DMI, but is instead calculated by summing the difference between the direct inputs and the exports of each material. The materials that enter the city will not be completely consumed by its socioeconomic system, and some materials will only pass through the city. Therefore, the material input and output indexes alone cannot accurately reflect the city's utilization of materials. As an important index to measure the consumption of urban materials, DMC has attracted increasing attention in recent years.

4.1.2 Stock Accounting

The urban stocks of various materials represent the temporary or semi-permanent accumulation of the materials that sustain human socioeconomic activities (Chen and Graedel 2015; Pauliuk and Müller 2014). Research to improve accounting for these materials and their dynamic changes is vital to understanding how to improve an urban organism's health (Fishman et al. 2016). In this section, I will use material flow analysis to analyze the dynamic changes in a city's weight (i.e., the actual gravitational weight of all stocks in the city), and will analyze in detail the distribution of 3 types of stock: buildings, infrastructure, and manufactured products. Based on my research on Chinese cities (Fig. 4.3), this accounting comprises 11 sub-categories, and a total of 54 accounting items in these sub-categories. Studies of other cities may

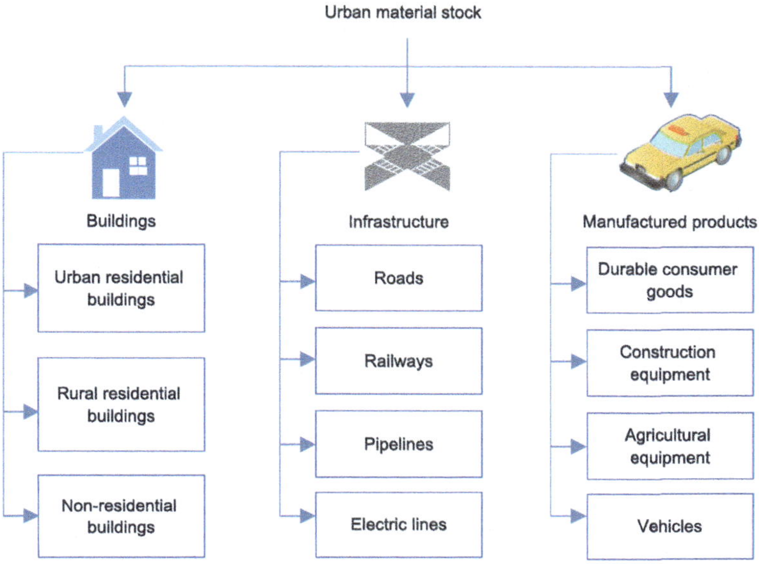

Fig. 4.3 Illustration of the accounting framework for the urban metabolic stocks

require different categories and accounting items due to differences in available data or in the metabolic characteristics of a city.

Under the 11 stock sub-categories, we defined 54 accounting items (Table 4.3). Due to data limitations, we could not subdivide the 3 types of buildings examined in this paper into finer-grained accounting items to provide more resolution, but the infrastructure could be subdivided into 3 road items, 2 railway items, 2 pipeline items, and 8 electric power line items. The manufactured products could be divided into 9 types of durable consumer goods, 11 types of construction equipment, 11 types of farm machinery, and 5 vehicle types. We obtained the total in-use stock by converting the stock of each accounting item into a weight (Mt).

For building and infrastructure stock (except electric power lines), I multiplied the original size data (I), such as the area of roads, by the material-use intensity (MI; the weight of the material per unit), to obtain the material stock (MS) for material i, as shown in Eq. (4.1). The stock of electric power lines is the product of the cross-sectional area of the wires, their density (mass per unit volume), and their length. For stocks of manufactured products (MS), I multiplied the quantity of the product (Q, the number of items) for product i by the product's common unit weight (the material quantity, MQ), as shown in Eq. (4.2).

$$\mathrm{MS}_i = I_i \times \mathrm{MI}_i \tag{4.1}$$

$$\mathrm{MS}_i = Q_i \times \mathrm{MQ}_i \tag{4.2}$$

Table 4.3 Subtypes and accounting items used to quantify the in-use stock

Category	Subtype	Items
Buildings	Urban residential buildings	—
	Rural residential buildings	—
	Non-residential buildings	—
Infrastructure	Roads	Expressway; ordinary highway; urban road
	Railways	Aboveground railway; subway
	Pipelines	Sewage pipe; water pipe
	Electric power lines	500-kV aboveground line; 220-kV aboveground line; 110-kV aboveground line; 35-kV aboveground line; 220-kV buried cable; 110-kV buried cable; 35-kV buried cable; 10-kV buried cable
Manufactured products	Durable consumer goods	Shower water heater; washing machine; color TV set; refrigerator; camera; air conditioning; computer; mobile telephone; household car
	Construction equipment	Shovel; bulldozer; scoop-tram (front-end loader); tracked excavator; wheeled excavator; onboard crane (loader); tower crane; loader; concrete mixer; air compressor; pile driver
	Farm machinery	Large or medium tractor; small tractor; tractor-operated machine; mobile sprayer; powered rice-transplanting machine; combine harvester; powered (driven) thresher; grain thrower; rice noodle processing machine; mechanical milker; feed grinder
	Vehicles	Truck; commercial passenger car; private car; subway vehicle; public transit vehicle (bus or trolley)

where MS represents the material stocks in the built environment, I represents the built size, MI represents the material intensity (the quantity of a material stocked per in-use unit), Q represents the quantity of various products, MQ_i represents the unit weight of the product, and i represents the stock type.

For my research in China, the raw data were collected mainly from statistical yearbooks such as the Urban Statistical Yearbook, China Construction Industry Statistical Yearbook, China Traffic Yearbook, and China Urban Construction Statistical Yearbook. Cities in other countries will have comparable information sources. The MI parameter values (i.e., the coefficients used to convert raw data into weight values) mostly came from previous studies of cities in China or Europe.

4.2 Substance Flow Analysis

Whereas material flow analysis focuses on materials that are composed of multiple substances, substance flow analysis focuses on the substances contained in the materials (e.g., the material CO_2 contains the substances carbon and oxygen). This is particularly useful for tracking flows of elements such as carbon and nitrogen that are the key factors in certain biogeochemical cycles. Substance flow analysis can also be used in analyses of metabolic processes, but provides a much narrower focus. Because carbon and nitrogen are involved in many metabolic processes, such as a city's natural metabolism and socioeconomic metabolism, natural subsystems and socioeconomic subsystems are generally given equal importance in the accounting.

4.2.1 Carbon Accounting

Urban carbon accounting involves carbon emission, absorption, and transfer between metabolic actors. Emission accounting includes energy consumption, industrial processes, waste disposal, and other anthropogenic activities, as well as biological (human and livestock) processes such as respiration, digestion, and soil respiration. Carbon absorption accounting mainly includes vegetation absorption, dry and wet atmospheric deposition, and other processes. I also analyzed the carbon transfers related to the processes of raw material extraction, processing, consumption, and use. Therefore, carbon accounting includes eight aspects: biomass, industrial products, food, non-metallic minerals, fossil fuels, renewable resources, pollutants and dissipative substances, and biological carbon fixation and atmospheric deposition, which mainly reflect carbon consumption that forms urban stock such as buildings and durable consumer goods (Table 4.4).

Based on the abovementioned carbon metabolic processes, I identified a total of 18 metabolic actors. Four of them are natural metabolic actors, namely the Atmosphere, Forest, Grassland, and Surface Water. With reference to relevant information such as the city's urban statistical yearbook, I divided the main socioeconomic metabolic actors into 12 industrial sectors: Crop Cultivation, Animal Husbandry, Fisheries, Manufacturing, Mining, Production and Supply of Electric Power and Heat, Energy Transformation, Construction, Transportation, Wholesale and Retail (which includes accommodation), Other Services and Waste Disposal, and 2 human metabolic actors (rural and urban households) (Fig. 4.4). Among these components, Crop Cultivation combines natural and socioeconomic attributes because it includes farmland, but from the perspective of defining actors, it is dominated by production activities, so I classified it with the socioeconomic actors.

The actors involved in the city's carbon metabolism have relatively obvious upstream and downstream distributions and hierarchical relationships; that is, the flow directions are generally clear. There are 4 main carbon-transfer chains. Crop Cultivation, Forest, Grassland, and Surface Water absorb CO_2 from the atmosphere

Table 4.4 Items included in urban carbon flow accounting

Category	Scope
Biomass	Straw, wood, fuelwood, biomass industrial raw materials, CO_2 dissolved in water
Industrial products	Plastics, inorganic fertilizer, livestock feed, paper products, agricultural film, grease, furniture
Food	Grain, dried fresh fruits, vegetables, meat, eggs, milk, aquatic products, industrial processed food
Non-metallic minerals	Limestone, sand, and rocks
Fossil fuels	Raw coal, crude oil, natural gas, and secondary energy
Renewable resources	Waste plastics, waste paper, livestock and poultry manure, biogas sludge, human manure
Pollutants and dissipative substances	Fossil energy combustion, biomass combustion, emission from ruminant animals, animal respiration, soil respiration, fertilizer decomposition, methane emission from paddy fields, residual feed, carbon from soil and water loss, solid waste, wastewater, sewage sludge, and treated feces
Biological carbon fixation and atmospheric deposition	Grassland carbon fixation, forest carbon fixation, cultivated land carbon fixation, atmospheric dry and wet deposition

and convert it into organic matter, and some carbon is transmitted to humans and livestock through the food chain; the rest is transmitted to Manufacturing for further processing, and is finally utilized or consumed by humans and livestock. Fossil fuels and minerals are introduced into the urban metabolic processes through extraction from the internal environment. Some of these substances enter the Energy Transformation sector for further processing, and some directly enter a terminal consumption sector (i.e., any sector where they are consumed without undergoing additional processing). The external environment also provides semi-finished products such as fuel oil and naphtha, raw materials such as wood, and industrial products such as paper and plastic. Some of them enter the Manufacturing or Energy Transformation sectors, whereas others enter a terminal consumption sector, where they serve multiple socioeconomic actors. The waste and byproducts produced by metabolic actors through their consumption of carbon-containing substances are released directly into the natural environment or are processed (e.g., treated, recycled) before that release. These take the form of CO_2 emissions into the atmosphere, or discharge of solid waste and wastewater into the waste disposal sector. Treated waste, along with some untreated waste, is finally discharged into bodies of water or the internal environment.

The data I have used in this analysis in my Chinese research mainly came from a city's statistical yearbook, the *China Energy Statistics Yearbook*, the *China Rural Statistical Yearbook*, the *China Environmental Statistics Yearbook*, and the *China*

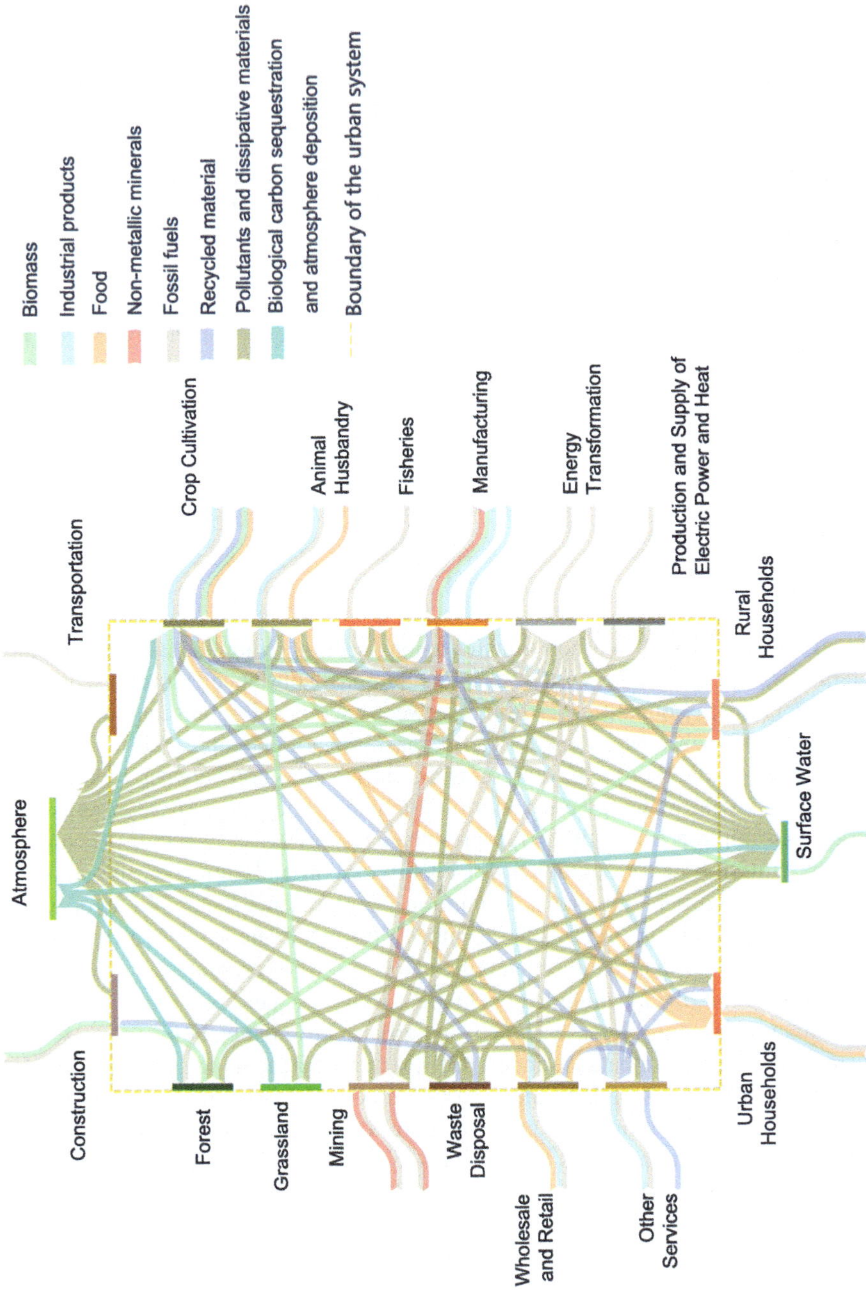

Fig. 4.4 Illustration of the carbon flows related to urban carbon metabolism processes

Plastics Industry Yearbook. Comparable sources of information should exist for most cities elsewhere in the world. Because most of the data I collected were based on urban inputs and outputs, which were treated as a black box by researchers, they cannot be matched directly to the process paths. Therefore, it was necessary to define principles for distributing the flows represented by these data among the paths. For the food distribution, I used the conservation of mass principle to calculate the flows. I assumed that local production was first allocated to meet local consumption needs, with the remainder becoming outputs (exports) to the external environment. When the local production was insufficient to meet the local demand, the missing amount was provided by inputs (imports) from the city's external environment. Specific accounting items are shown in Table 4.4.

Empirically derived coefficients for the carbon content of various materials can be used to convert the masses of those materials into the corresponding masses in carbon emission and absorption flows [Eqs. (4.3)–(4.5)]:

$$\mathrm{CT}_i = \mathrm{Mt} \times \mathrm{TF} \tag{4.3}$$

$$\mathrm{CE}_j = \mathrm{Me} \times \mathrm{EF} \tag{4.4}$$

$$\mathrm{CS}_k = A_k \times \mathrm{SF} \tag{4.5}$$

where CT_i represents the carbon transfer among sectors (components of the urban system) for material category i that contains the focal substance, Mt represents the substance or energy quantity transferred between sectors, TF represents the carbon coefficient in the substance or energy (i.e., a transfer factor that represents the proportion of carbon in the food), CE_j represents the carbon emission between sectors and the atmosphere due to energy consumption for energy category j, Me represents the energy consumption (mass of energy-producing materials) by a sector, EF represents the emission factor coefficient for energy carbon emission (i.e., the amount of carbon emitted per unit fuel consumption), k represents metabolic actor, CS_k represents carbon absorption by photosynthesis for metabolic actor k, A represents the area covered by that natural metabolic actor, and SF represents the quantity of carbon absorption (the stocking factor) by photosynthesis per unit area.

To quantify the problems arising from urban carbon metabolic processes, I constructed a carbon imbalance index (CII), which represents the ratio of carbon emission to carbon absorption, and a carbon external dependence index (CEDI), which represents the ratio of external resources consumed to internal resources consumed. CII mainly accounts for the paths in the urban carbon metabolic system related to the atmosphere. As densely populated areas, cities mostly have actors with large carbon emissions, and have fewer actors with a potential for carbon fixation, leading to a small potential. Thus, carbon emission is generally far greater than carbon absorption, resulting in a serious imbalance of the carbon metabolism. The larger the CII, the more serious the carbon imbalance. CII is calculated as follows:

$$\mathrm{CII} = \sum e_i / \sum s_i \tag{4.6}$$

where e_i is the carbon emission by carbon metabolic actor i and s_i is the carbon fixation (absorption) by that actor.

CEDI mainly accounts for the paths related to the external and internal environments. As highly open ecosystems, cities frequently exchange large amounts of carbon with their external environment. Because very few cities can meet their own carbon needs, the processes in an urban carbon metabolism generally depend strongly on inputs from the external environment. Therefore, the ratio of inputs from the external environment to inputs from the internal environment reveals the degree of dependence of the urban metabolism on external resources. In view of the different metabolic pressures produced by the use of renewable and nonrenewable external resources, it is possible to assign different weights to these resources. For the sake of simplicity, I assumed that renewable resources had a weight of 0.5 and non-renewable resources had a weight of 1.0. In future research, these weights could be determined empirically for a given city or adjusted to account for changes such as strong implementation of solar power to reduce consumption of fossil fuels. The greater the CEDI, the greater the dependence on the external environment. The formula for CEDI is as follows:

$$\mathrm{CEDI} = (0.5 \sum Z_{i,\mathrm{r}} + \sum Z_{i,\mathrm{n}} - \sum 0.5 y_{i,\mathrm{r}} - \sum y_{i,\mathrm{n}}) / W_i \tag{4.7}$$

where $Z_{i,\mathrm{r}}$ represents the quantity of carbon in renewable resources (r) that is input to actor i from the external environment, $Z_{i,\mathrm{n}}$ represents the quantity of carbon in non-renewable resources (n) that is input to actor i from the external environment, $y_{i,\mathrm{r}}$ represents the quantity of carbon in renewable resources (r) that is output to actor i from the external environment, $y_{i,\mathrm{n}}$ represents the quantity of carbon in non-renewable resources (n) that is output to actor i from the external environment, and W_i is the quantity of carbon that is supplied by the internal environment from actor i.

4.2.2 Nitrogen Accounting

Urban nitrogen accounting involves biological nitrogen fixation, industrial nitrogen fixation, atmospheric deposition, and nitrogen release. Urban biological nitrogen fixation refers to the amount of organic nitrogen stored in organisms, including crops, livestock, poultry products, and other animal products. Industrial nitrogen fixation includes the production of feed, inorganic fertilizer, and industrial chemicals. Atmospheric deposition refers to nitrogen deposited into water and onto land in rain or snow or in dust fall. Urban releases of nitrogen include NO_x emissions into the atmosphere, discharge of nitrogen-containing pollutants in sewage, runoff loss or volatilization after agricultural nitrogen application, nitrogen release from floodwaters and wetlands, and denitrification in sewage treatment plants (Duh et al. 2008). Consumption of all nitrogen-containing substances in a city depends not only on local production, but also on large transfers or imports that increase the mass of nitrogen.

Based on the nitrogen metabolic flow processes, I defined 15 sectors that are involved in these processes: Household Consumption, Industry, Animal Husbandry, Crop Cultivation, Fisheries, Forestry, Services, Construction, Transportation, Sewage Treatment, and natural components of the urban system (Surface Water, Atmosphere, Forest, Grassland, Farmland). This division covers both the natural and socioeconomic subsystems. Household Consumption refers to the rural and urban residents who consume food and other products. Industry includes extractive, processing, and manufacturing industries, and its production starts with minerals and agricultural products as raw materials. Animal Husbandry is a component of the system that obtains meat, eggs, milk, and other animal-based foods by raising pigs, cattle, sheep, poultry, and other animals. Crop Cultivation produces agricultural products such as food crops (including both grains and vegetables), cash crops, forage crops, pastures, and crop straws and other straws. Fisheries includes aquatic products, but excludes the water released from aquaculture facilities. Forestry cultivates and protects forest resources to obtain wood and other forest products, and uses the natural characteristics of forest trees to play a protective role (e.g., to reduce wind erosion). Services include wholesale, retail, accommodation, and catering (i.e., food services such as restaurants). Construction specializes in civil engineering, house construction, and equipment installation, as well as engineering surveys and design work. Transportation undertakes transportation functions both for people and materials. Sewage Treatment centrally treats industrial wastewater and household sewage to separate solid pollutants from the water and reduce organic pollutants and eutrophic substances (mainly nitrogen and phosphorus compounds) in water. Surface Water is the general term for dynamic and static water (i.e., rivers and lakes) on the land surface, and is also known as "land water"; my analysis does not include subsurface water (e.g., aquifers) due to a lack of data. Atmosphere refers to the thin layer of air near the ground, at a height of between 10 and 100 m. Forest refers to the land covered by natural forests of trees and dense shrubs, secondary forests, and artificial forests. Grassland is land that mainly grows herbs and sparse shrubs and that is suitable for the development of animal husbandry. Farmland is the land dedicated to growing crops.

The first 10 actors belong to the system's socioeconomic components that represent the major sources of anthropogenic nitrogen. The remaining 5 actors belong to the system's natural components. Most of them receive waste (here, nitrogen-containing waste) produced by the socioeconomic actors, but some may also participate in the natural metabolic processes. Figure 4.5 summarizes these components of the overall urban nitrogen metabolism.

In addition, I subdivided the Agriculture component into 3 subcomponents (Crop Cultivation, Animal Husbandry, and Fisheries) that reflect differences in their nitrogen recycling activities, and in human consumption of different types of nitrogen-containing food. Urban energy consumption is also an important source of

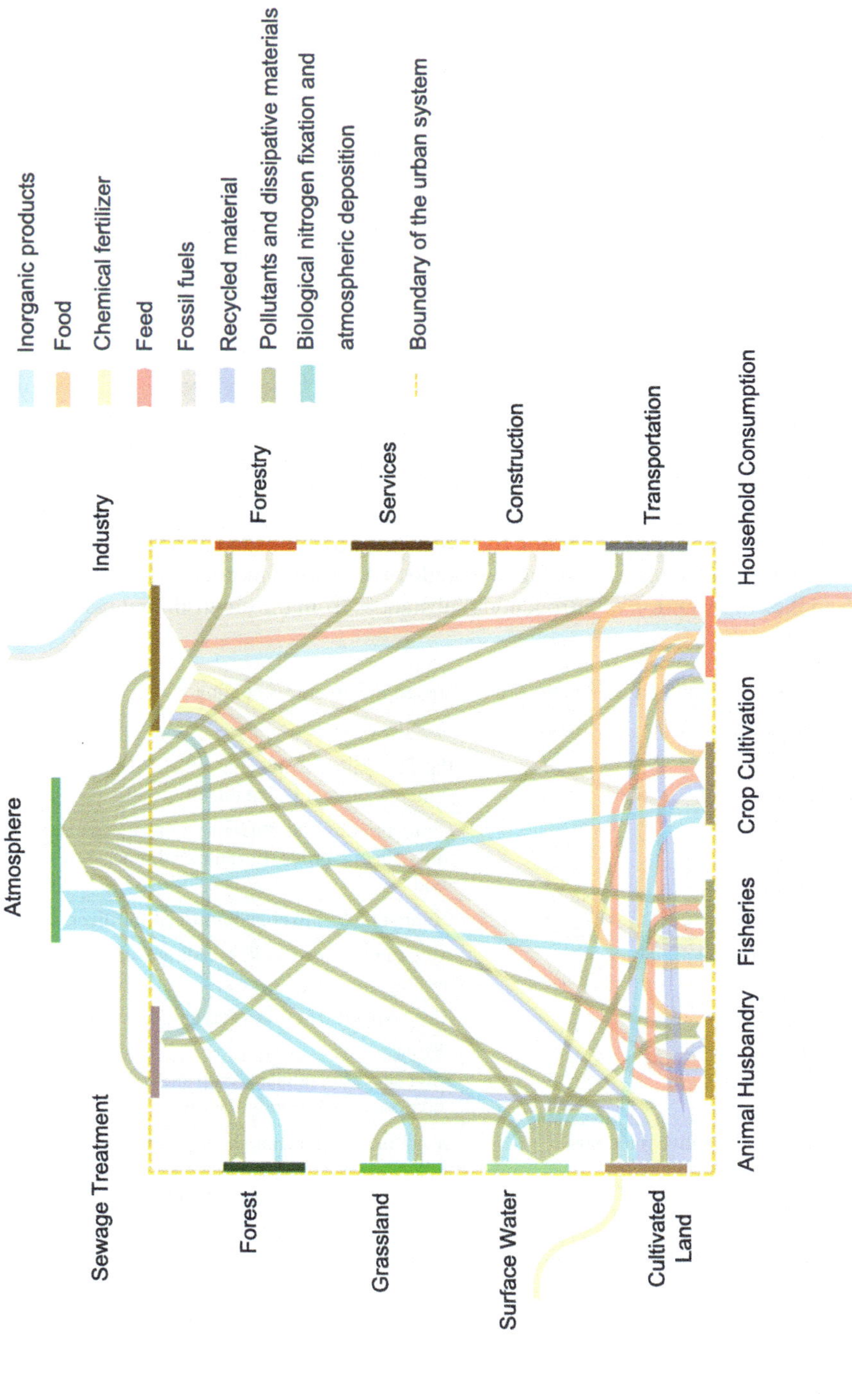

Fig. 4.5 Illustration of the nitrogen flows in an urban nitrogen metabolism

nitrogen (i.e., in the form of NO_x emitted by the combustion of fossil fuels), so the main energy consumption sectors should be treated as separate subcomponents. I defined Sewage Treatment as a separate node to highlight the environmental impact of this activity (primarily the removal of ammonia by waste treatment plants). I subdivided the natural components of the urban ecosystem into 5 subcomponents to reflect differences in their transfer, governing circumstances, and accumulation of nitrogen. I also subdivided terrestrial ecosystems into Grassland, Forest, and Farmland to highlight their different biological nitrogen absorption abilities. The accounting items are divided into 8 aspects (shown at the right side of Fig. 4.5): chemical products, food, inorganic fertilizer, feed (animal fodder), fossil fuels, recycled materials, pollutants and dissipative materials, and biological nitrogen fixation and atmospheric deposition (Table 4.5).

The city's input of reactive nitrogen (Q_N) is an important indicator that represents the level of urban nitrogen accumulation. (Here, "reactive" refers to forms of nitrogen other than N_2 gas, which is relatively non-reactive.) This indicator includes inputs of nitrogenous substances from the external environment, the consumption of nitrogenous substances inside a city's administrative boundary, and the amounts of natural nitrogen fixation and atmospheric deposition. Natural reactive nitrogen includes biological nitrogen fixation by forest and grassland, and atmospheric deposition; anthropogenic reactive nitrogen includes agricultural biological nitrogen fixation and the consumption of nitrogenous substances from the environment inside the

Table 4.5 Accounting items used in the urban nitrogen metabolic accounting

Accounting items	Specific contents
Chemical products	Plastic, synthetic rubber, synthetic detergent, synthetic ammonia, chemicals
Food	Grain, dried fruits, vegetables, meat, eggs and milk, aquatic products, industrial processed food
Inorganic fertilizer	Compound fertilizer and nitrogen fertilizer
Feed (animal fodder)	Ammoniated feed, bait, breeding feed, fish protein
Fossil fuels	Raw coal, crude oil, natural gas, and secondary energy
Recycled materials	Kitchen waste, manure, crop straw, solid waste compost, sewage sludge reuse
Pollutants and dissipative materials	Fossil fuel combustion, household wastewater, industrial waste water, loss of animal manure, feed residues, runoff loss, denitrification, volatilization, soil leaching
Biological nitrogen fixation and atmospheric deposition	Seeds, crop nitrogen fixation, forest land nitrogen fixation, grassland nitrogen fixation, water nitrogen, dry and wet sedimentation

Note Fig. 4.5 illustrates the flows of these accounting items among the system's components urban administrative boundary and in the external environment. The consumption of nitrogenous substances includes human food, livestock and aquaculture feed, fossil fuel combustion to provide energy, inorganic fertilizer, and chemicals. The consumption of nitrogenous materials is mainly calculated by multiplying the consumption (a mass or number of units) by the nitrogen coefficient, which represents the nitrogen content of the nitrogen-containing material. The consumption data used in my Chinese research were obtained from China's National Bureau of Statistics and urban statistical yearbooks. Comparable information sources should exist for cities elsewhere in the world. The formula for Q_N is as follows:

$$Q_N = BNF_f + BNF_g + BNF_a + D + C_f + C_{feed} + C_e + C_{fert} + C_c \quad (4.8)$$

where BNF represents biological nitrogen fixation by forest (f), grassland (g), and agriculture (a), including nitrogen fixation by seeds and crops; D represents atmospheric deposition; and C represents the N consumption by food (f), livestock feed (feed), energy (e), inorganic fertilizer (fert), and chemical products (c). I converted all amounts into a total mass of nitrogen N.

4.3 Emergy Analysis

"Emergy" refers to embodied energy, which represents the energy consumed by the production of a substance or material, and accounts for differences in the quality of different forms of energy. Odum (1971) originated emergy theory, and introduced 2 key terms: emergy itself, and transformity. Emergy is the total energy of one kind that is required, both directly and indirectly, to create a resource, product, or service. Its units are solar emjoules (seJ), based on the recognition that the sun (through photosynthesis) is the ultimate energy source for life on Earth. Transformity is the emergy input per unit of available energy output. For example, if 10,000 seJ are required to generate 1 J of wood, then the solar transformity of that wood is 10,000 seJ/J. Transformity is used to characterize energy quality. The greater the transformity, the more solar energy is needed to produce the product.

The emergy analysis method uses transformity values to convert all materials and energy that flow through a system into a consistent system of units (i.e., solar emergy). This overcomes the problem that different logistics, energy flows, information flows, and currency flows cannot be measured uniformly without conversion to a consistent system of measurement, and transforming these different measurements into seJ permits a unified evaluation of ecological and economic value (Odum 1988). Emergy accounting links the socioeconomic system with the ecosystem in which it's embedded (Hall et al. 1986). Solar emergy can be calculated by multiplying the mass or energy by their corresponding solar emergy transformity (Odum 1996), and the product's emergy (M, in seJ) can be expressed as:

$$M = \tau E \quad (4.9)$$

where M is the emergy, τ is the emergy transformity, and E is the available energy of a product, which is expressed in J (for energy flows), g (for mass flows), or a currency value for monetary flows. These different units are converted into seJ using the corresponding transformity values.

Figure 4.6 shows the major system components and emergy flows that drive an urban metabolism, and reflects the general evaluation of urban socioeconomic development and environmental quality. To sustain the functions of the urban metabolic system, the basic emergy flows represented by several indexes must be ensured. These indexes are renewable resources emergy (R), non-renewable resources emergy (N), transfers-in and imported emergy (purchased inputs imported from the external environment, IMP), transfers-out and exported emergy (exported products, EXP), and total emergy use (U).

In the research on which this section is based (Zhang et al. 2009), R mainly includes emergy from sunlight, rain, wind, and Earth cycles. Since these energies are all composite products from the same process, my research group selected the maximum renewable flow (sunlight, wind energy, rain potential energy, rain chemical energy, Earth's rotational energy, and river potential energy) to collectively indicate the renewable sources emergy received by the system, thereby minimizing the risk of double-counting. We also defined an emergy value for N, which equals the sum of the emergy of dispersed rural sources, such as soil erosion, and local resources for concentrated use, such as fuels that can be obtained from inside the system, iron ore, limestone, sand, and related construction materials.

We also accounted for fuels from outside the system, goods, and services from outside the system (transfers-in and imports, IMP) and produced goods and services that are exported from the system (transfers-out and exports, EXP). The sum of all these flows equals the total emergy use (U), whose magnitude is a measure of the system's real wealth, including both resources and products. It also indicates the magnitude of the resource consumption required to sustain socioeconomic development. At the same time, biological forms of renewable production (the direct and indirect production of free environmental resources) are not accounted for in the total emergy use; these production processes include local renewable resources such as agricultural production, forestry production, and fish (to some extent) that are used to meet the city's production needs and support household use. Because these items are supported by free environmental flows (e.g., sunlight for plant growth), they are not included in the total emergy use to avoid double-counting. Table 4.6 summarizes the remaining items.

Urban development is centered around economic activities, with the ecological environment treated as the city's support system. Therefore, if the goal is to perform an emergy analysis for a city, it's necessary to create emergy indexes that evaluate the metabolic processes related to economic development, resource utilization, pollutant emission, and other relevant indicators reflecting the characteristics of efficiency, pressure, intensity, and environmental load. These indicators can comprehensively represent the operation of urban metabolic processes. Based on the quantification of emergy flows, a series of indexes and their ratios can be defined to evaluate the urban system (Brown and Ulgiati 1997): the emergy self-support ratio (ESR), the emergy

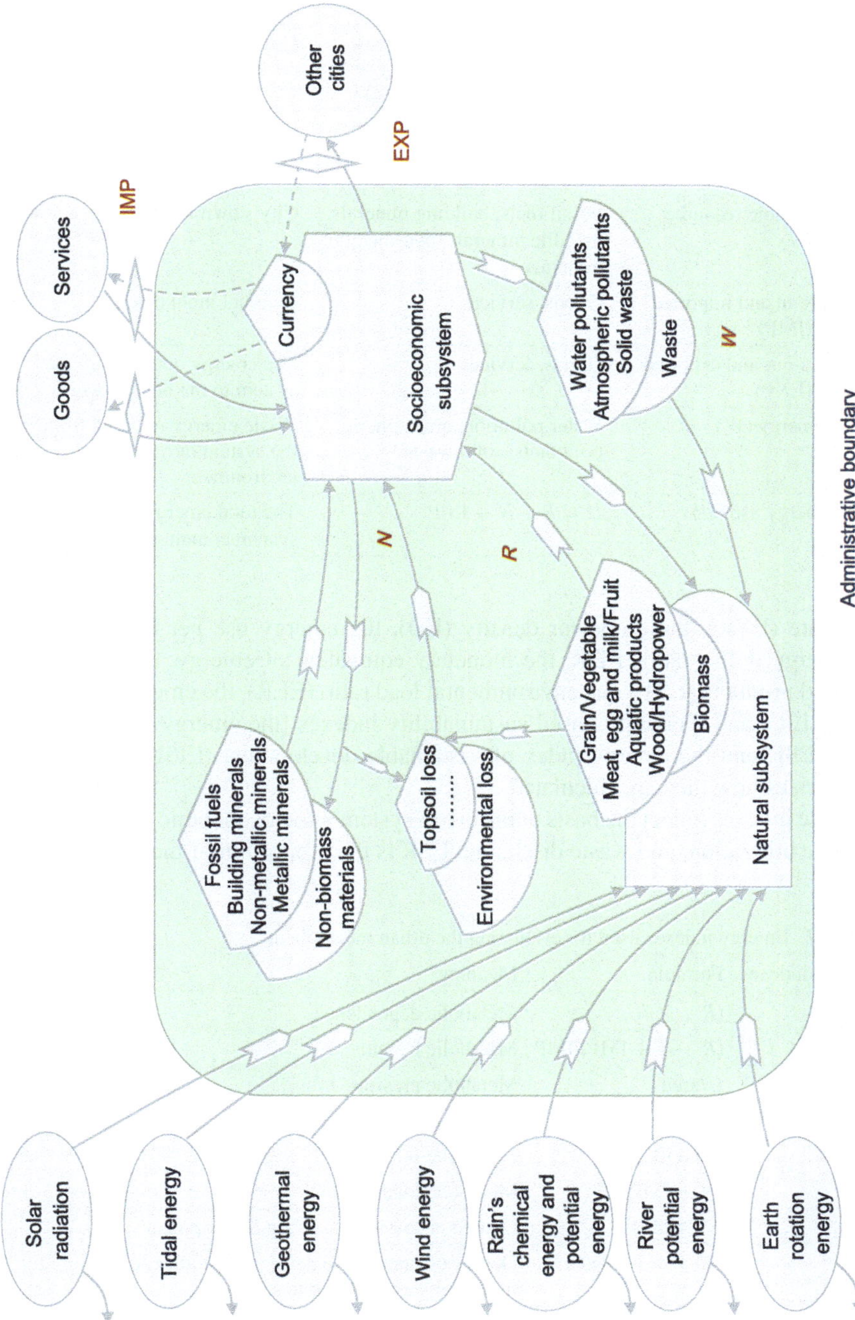

Fig. 4.6 Illustration of the main emergy flows in an urban metabolism (*Note: W* represents waste emergy)

Table 4.6 Accounting items for an urban metabolism based on emergy analysis

Items	Scope	Meaning
Renewable resources emergy (*R*)	Solar radiation, tidal energy, geothermal energy, wind energy, rainwater potential energy and chemical energy, river potential energy, Earth's rotational energy	City's own wealth
Non-renewable resources emergy (*N*)	Fossil fuels, building minerals, metallic minerals, non-metallic minerals	City's own wealth
Transfers-in and imported emergy (IMP)	Goods, services	External input emergy
Transfers-out and exported emergy (EXP)	Goods, services	The emergy output from the system to the outside world
Waste emergy (*W*)	Water pollution, atmospheric pollutants, solid waste	Waste emergy released from the system into the environment
Total emergy use (*U*)	$U = R + N + \text{IMP}$	The total emergy that the system contains

yield rate (EYR), the empower density (ED), the emergy use per capita (EPC), the emergy dollar ratio (EDR, the monetary equivalent of emergy, in emdollars), exported products (EXP), the environmental load ratio (ELR), the emergy exchange ratio (EER), and 2 emergy-based sustainability indexes [the emergy sustainability index (ESI) and the emergy index of sustainable development (EISD)]. Table 4.7 summarizes how they are calculated.

These indexes reflect the basis of the urban system's socioeconomic development, resource utilization, and waste discharge. ESR is the proportion of the total emergy

Table 4.7 Emergy indexes used for evaluating the urban metabolism

Emergy indexes	Formula	Meaning
ESR	$(R + N)/U$	Metabolic dependence
EYR	$(R + N + \text{IMP})/\text{IMP}$	Metabolic benefits
ED	U/area	Metabolic pressure
EPC	U/P	Metabolic intensity per capita
EDR	U/GDP	Metabolic intensity
ELR	$(N + \text{IMP})/R$	Metabolic loading
ESI	EYR/ELR	Degree of sustainable urban development
EISD	(EYR × EER)/ELR	Degree of sustainable development after accounting for the emergy exchange rate

Note ESI was developed by Brown and Ulgiati (1997); EISD was created by Lu et al. (2002); GDP, gross domestic product (in U.S. dollars); *P*, population driving a system that is derived from internal emergy inputs; a large ratio suggests a low level of dependence on the external environment and increased sustainability. EYR is the ratio of total emergy to imported emergy. This ratio measures the ability of a process to exploit local resources and make them available; in other words, it measures the use of local resources to generate emergy.

ED is the ratio of the total emergy driving a system to the relevant area; a large ratio suggests a high level of metabolic pressure. Because emergy measures the energy that goes into creating a product, it also measures the real wealth the product contributes to the economy and can be used to justify its production. EPC is the ratio of the total emergy driving a system to the population; this per-capita emergy can be used as a measure of the potential mean standard of living of the population, since a large ratio suggests a high level of metabolic intensity for each person. EDR is the emergy amount that a given amount of money buys (i.e., the ratio of emergy to money), with money measured using the GDP. EDR is thus a measure of the money's ability to purchase emergy. A higher value shows that the system's economy purchases a higher quantity of emergy flows in exchange for its investment in emergy-producing activities. ELR is the ratio of emergy imported plus nonrenewable internal emergy to renewable environmental resources emergy; a large ratio suggests a high metabolic load on the environment.

ESI (the emergy sustainability index) is the ratio of the emergy yield rate (EYR) to the environmental load ratio (ELR). It assesses the level of sustainable development based on emergy theory, but doesn't consider the impact of scientific and technological progress on waste recycling. It also assumes that the emergy yield rate has a positive effect, and is beneficial to humans (i.e., that the output of pollutants has negative consequences). The higher the EYR, the more likely it is that the urban metabolism will not benefit humanity and will not be conducive to sustainable development. EISD (the emergy index of sustainable development) is proportional to the system's sustainable development capability. The higher the EISD value, the higher the socioeconomic benefits per unit of environmental pressure, and the greater the potential for sustainable development (Lu et al. 2002). Some scholars introduced the emergy exchange ratio (EER) to improve on EISD, pointing out that under the same output conditions, the transaction process is also affected by financial markets, a city's culture, ethical considerations, and other factors (Li et al. 2006; Lu et al. 2002). They proposed that EER provides a better indicator of the city's benefits from exchanging emergy with its external environment, and therefore reflects different effects of urban development.

4.4 Measuring the system's Evolution

As urbanization progresses, discovering how to ensure the health, order, and vigor of the urban ecosystem becomes an important research topic. Based on ecological thermodynamics theory and the theory of complex integrated natural and socioeconomic systems (Chapter 3), my research group developed an indicator system and a model based on information entropy (Zhang et al. 2006a) that can measure the various flows within an urban ecosystem. By calculating the indicator values at different times, it becomes possible to quantify the city's evolution and its directions of change. The model for harmonious development is based on 4 types of information entropy, including the sustaining inputs, imposed outputs, destructive (catabolic) metabolism, and regenerative (anabolic) metabolism. The model can be used to present the degree of order (total entropy), health (entropy flows), and vigor (entropy production) of the urban ecosystem, thereby reflecting the urban ecosystem's state of evolution at any point in time.

4.4.1 Measurement Index System

Because urban ecosystems are so complex, the system's state and changes cannot be described using only a few indexes. Instead, a system composed of many indexes should be developed to describe the numerous aspects of the system's development. First, the theory of dissipative structures (Prigogine and Lefever 1973) suggests that a portion of the energy flowing through a system is used to maintain its equilibrium, while entropy outside the system increases. On this basis, the entropy flow and production should be determined. Based on these results, a second level of indexes can be developed. Second, the interactions between the socioeconomic subsystem and the city's natural ecological environment must be considered. These interactions primarily embody human reproduction, the pressures imposed by socioeconomic activity on the ecological environment, and the ability of the environment and natural resources to support these socioeconomic activities. Third, the rate of pollution production within the urban ecosystem and the ability of the ecological environment to purify this pollution must be evaluated. In developing the indicator system based on this framework, the environmental pollution indicators are not classified as imposed output indicators because waste can be considered analogous to a misplaced resource that should be utilized by the system to achieve a recycling (circular) economy. In effect, it is the internal factors of the socioeconomic subsystem and their interaction with the regenerative metabolism that determines entropy production in the urban ecosystem, and this interaction provides information to support decision-makers and the public.

To construct the indicator system, my research group followed 3 principles: representativeness (i.e., the ability to describe key features of the system), hierarchy (nesting of indicators), and operability (feasibility based on the available data). Based on the data available for my research in China, the system includes 2 secondary levels for entropy flows and entropy production, and 4 tertiary indicators (support inputs, imposed outputs, destructive metabolism, regenerative metabolism), and 4 fourth-level indicators, for a total of 40 indicators (Table 4.8). Similar sources of information should be available for cities elsewhere in the world to support the use of similar indicators.

Table 4.8 Indicator system for measuring urban metabolic evolution

Type of indicators	Specific indicators
Sustaining inputs	Area planted in grain crops or the total crop yield (hm^2 or t), area or yield of vegetables (hm^2 or t), area planted to produce fruits or the total fruit yield (hm^2 or t), total area of forest or output of major forest products (hm^2 or t), livestock production (t), area of saltwater and freshwater aquaculture or total aquatic products for cities close to a body of water (hm^2 or t), mineral exploitation (t), amount of solar energy utilized (kJ), and investment in culture and education (RMB)
Imposed outputs	Natural population growth rate (‰), urban population density (persons/km^2), per capita annual expenditure for consumption by urban residents (RMB), per capita annual expenditure for consumption by rural residents (RMB), electricity consumption unrelated to living (kWh), electricity consumption by urban and rural residents (kWh), the Engel coefficient (the food consumption ratio, %), urban economic density (RMB/km^2, equal to the ratio of GDP to land area), energy consumption per 10^4 RMB of GDP (t coal equivalent [tce] per 10^4 RMB; in China, "coal equivalent" is often used as the measurement unit for energy, and coal is considered to provide 29,271 kJ per kg), and water consumption per 10^4 RMB of GDP (t/10^4 RMB)
Destructive metabolism	Industrial wastewater discharge (t), total wastewater discharge (t), chemical oxygen demand in industrial wastewater (t), industrial sulfur dioxide emissions (t), industrial soot emissions (t), industrial dust emissions (t), industrial waste gas emissions (t), and industrial solid waste produced or discharged (t)
Regenerative metabolism	Investment in environmental protection (10^4 RMB), percentage of sewage released after treatment (%), rate of innocuous disposal of solid waste or garbage (%), proportion of industrial wastewater that meets discharge standards (%), rate of water reused by industry (%), proportion of industrial solid waste that is utilized (%), rate of treatment and utilization of industrial solid waste (%), rate of treatment of excrement and urine (%), rate of automobile exhaust emissions (%), disposal rate for hazardous solid waste (%), rate of removal of sulfur dioxide from industrial waste gases (%), vegetation restoration in developed areas (% of total area), vegetation cover in developed areas (% of total area), rate of vegetation restoration (%), rate of plantation growth (%), and rate of region closed for afforestation (%)

This indicator system provides a synoptic list suitable for appraising the urban ecosystem's direction of evolution. Researchers who are studying different cities or regions should accept or replace certain indicators (i.e., the ones that are not representative of their unique conditions or for which data are not available) in order to design a feasible indicator system that will meet their specific needs. The indicator system can describe the urban ecosystem's main problems from economic, environmental, and social perspectives, and can link the coordinated development of the city with the problem of achieving a cyclic (circular) metabolism; however, it is important to integrate these indicators in such a way that decision-makers and the public can analyze changes in the degree of sustainable urban development. The urban organism evolves under the dual influence of external disturbances and internal responses in the context of the theory of dissipative structures (Zhang et al. 2006a). Therefore, the information entropy method can reveal the causes of the urban ecosystem's evolution and, if necessary, allow urban managers to propose countermeasures to undesirable changes.

4.4.2 *Information Entropy Index*

Shannon (1948) introduced the concept of entropy into information theory, and named the resulting theory "information entropy." Entropy expresses the degree of disorder in a system, and provides a measure of the urban ecosystem's state, evolution, and sustainability. In an uncertain system, if X is a random variable such that $X = \{x_1, x_2,\ldots, x_n\}$ for $n \geq 2$, we can use this variable to represent the system's status, and the corresponding probability of each value of x_n is $P = \{p_1, p_2,\ldots, p_n\}$, for $0 \leq p_i \leq 1$ and $i = 1$ to n and $\sum p_i = 1$, then the information entropy can be described as:

$$S = -\sum p_i(\ln p_i) \tag{4.10}$$

According to information entropy theory, multidimensional information can be quantified and integrated through the summation in Eq. (4.10) (Shannon 1948). The annual and indicator-based information entropy can be obtained by calculating the information entropy of the overall urban ecosystem. The calculation of entropy flows and production mainly uses the annual information entropy because information is rarely available for shorter time periods. The annual information entropy is calculated from annual statistics, and the indicator-based information entropy is calculated from indicator values in the focal year. In the research on which this section is based (Zhang et al. 2006a), my research group calculated 4 types of information entropy (for sustaining inputs, imposed outputs, destructive metabolism, and regenerative metabolism) using annual information entropy to analyze the degrees of order and complexity of the system. In the indicator-based information entropy, indicator weights are calculated based on information entropy, and the appraisal scores for the urban ecosystem are calculated by standardizing each indicator value and then multiplying the value by an appropriate weight; standardization is achieved by dividing each value by the maximum value, thereby producing values that range from 0 to 1. The resulting indicator scores can be applied to identify the degrees of vigor, order, development, and health of the urban ecosystem.

For a system with $i = 1$ to n indicators and with $j = 1$ to m appraisals (e.g., each year for which the value of each indicator is evaluated), we can assess the change in entropy flow (dS). The reliability and freedom from bias of the assessment is improved by standardizing each indicator (as described in the previous paragraph) to eliminate the effects of its dimensions (units of measurement) on the assessment results. This method transforms the raw data into standardized values between 0 and 1 for each index. The standardized values (q_{ij}) are then used to calculate dS. dS is calculated using dS_e^1, , which represents the flow of support entropy received from the natural subsystem (i.e., the input support entropy); dS_e^2, , which represents the flow of pressure entropy from the socioeconomic subsystem (i.e., the output pressure entropy); dS_i^1, , which represents the entropy production (metabolic entropy reduction) that results from ecological restoration and construction of environmental protection infrastructure; and dS_i^2, , which represents the entropy production (decomposition metabolic entropy) caused by environmental pollution. The resulting indicator-based information entropy is calculated as follows:

$$\mathrm{dS} = -\frac{1}{\mathrm{lnm}} \sum_{\mathrm{i}=1}^{\mathrm{n}} \frac{\mathrm{q_{ij}}}{\mathrm{q_j}} \ln \frac{\mathrm{q_{ij}}}{\mathrm{q_j}} \tag{4.11}$$

where $q_j = \sum_{i=1}^{n} q_{ij}$

$$Q_i = \frac{1 - E_i}{n - e_e} \tag{4.12}$$

where $\sum_{i=1}^{n} Q_i = 1(0 \leq \mathrm{Q_i} \leq 1)$, $E_i = -\frac{1}{\ln m} \sum_{j=1}^{m} \frac{q_{ij}}{q_i} \ln \frac{q_{ij}}{q_i}$, $q_i = \sum_{j=1}^{m} q_{ij} (i = 1, 2, ..., n; j = 1, 2, ..., m)$, and $e_e = -\sum_{i=1}^{n} \sum_{j=1}^{m} \frac{q_{ij}}{q_i} \ln \frac{q_{ij}}{q_i}$.

The weight of each indicator (Q_i) is determined using the indicator-based information entropy and combining the standardized value to obtain a comprehensive score (G) that integrates the weighted values for all indicators: $G = \sum Q_i q_{ij}$. The evaluation years can then be sorted according to the size of G to rank them in terms of their relative sustainability. The larger the G, the greater the sustainability and thus, the better the health of the urban superorganism in that year (Miyano 2001; Yelshin 1996; Renyi 1961).

4.4.3 Harmonious Development Model

In the research on which this section is based (Zhang et al. 2006b), my research group developed an evaluation score based on information entropy to determine the degrees of development and harmony of the urban system. Sustainable development of cities requires harmony (a stable entropy flow), vigor (low entropy production), and a combination of order and health (negative entropy change). The model of harmonious development can then be applied to measure the degrees of development and harmony. In essence, the model measures the changes from 2 perspectives: entropy flow between human activities and the natural environment, and the extents of dissipative and reductive metabolism. The goal is to seek a dynamic equilibrium among dS_e^1, dS_e^2, dS_i^1, and dS_i^2 (Zhang et al. 2006a).

For the harmonious development model to be useful in decision support, it must transform the entropy values into more intuitive indexes, because information entropy only determines the direction of the system's evolution and does not reflect its level of sustainable development. The model in Fig. 4.7 reflects the possible trends for urban development and harmony. The development curves A, B, and C are determined by referring to the production possibility curve (Zhang and Yang 2007). Based on the pressure a city exerts on its environment (PE) and its relationship with the environment's capacity to support that pressure (SC), we can define a production possibility curve. This represents all possible combinations of PE and SC under a given set of conditions. The curve represents the projection of the socioeconomic development function on the Cartesian coordinate system, and denotes the maximum degree of development the city can achieve under the given conditions with the amount of resource input and the level of technology used by the city's metabolic processes. The curves in Fig. 4.7 reflect the interplay between PE and SC. In the new knowledge economy and the new recycling economy, the economy will grow fast while the carrying capacity of the natural resources increases slightly due to improved resource utilization ratios and a higher degree of resource substitution (i.e., greater use of recycled materials, leading to decreased use of natural resources). At the same time, the economy and the environment change at different rates, thereby affecting the degree of harmony in different ways, but as long as resource substitution occurs, this will result in a gradual increase in harmony.

In the middle panel of Fig. 4.7, the area between the axes is divided into 4 equal regions (I to IV) by 3 curves: $y = y_0 - x^3$, where $y_0 = 1.00, 0.75$, and 0.50 (i.e., represents half the GDP of a value of 1.0), and represents a criterion for classification of

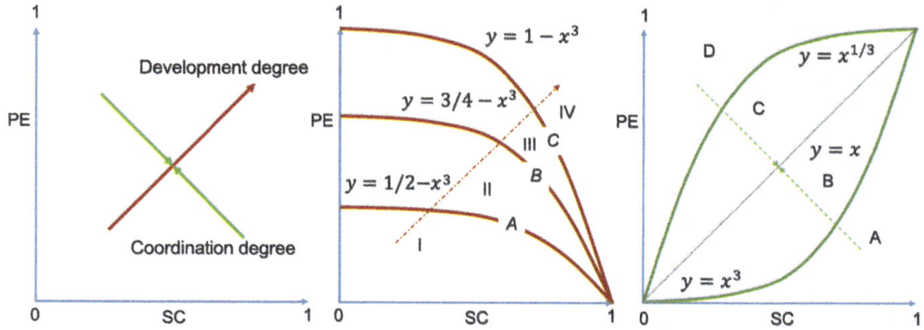

Fig. 4.7 Relationships between the pressure on the environment generated by socioeconomic activities (PE) and the environment's support capacity (SC) (*Note* Curves *A*, *B*, and *C* represent 3 production possibility curves, which define 4 regions (I to IV) that have different ranges of combinations for PE and SC)

the degree of development. Dividing the degree of development into 4 levels permits standardization of results and facilitates comparisons: the development may be rudimentary (I), elementary (II), intermediate (III), or advanced (IV). It is clear that both limited resources and a lack of advanced technology can restrict economic activities. Under the conditions of fixed investments in technology and limited resources, and for certain values of natural ecosystem support (carrying) capacity and pressure from socioeconomic activities, these curves represent the maximum degree of urban socioeconomic development. The metabolic subsystems of the urban ecosystem have similar conditions, including an interplay between destruction and regeneration. In the right panel of Fig. 4.7, the area between the axes is divided into 4 equal partitions by 3 curves ($y = x$, $y = x^3$, and $y = x^{1/3}$) that reflect different degrees of harmony with the natural subsystem: strong socioeconomic pressure (area A), moderate socioeconomic pressure (area B), moderate ecological carrying capacity (area C), and strong ecological carrying capacity (area D). If we define scores for sustaining inputs (x), imposed outputs (y), destruction (x^*), and regeneration (y^*), we can calculate the degrees of development (a and a^*) and harmony (b and b^*). For example, substituting x and y into $y = a–x^3$ predicts the degree of development a. In contrast, substituting x and y into $y = x^c$ to obtain the degree of quasi-harmony c, then substituting c into $b = 1/c$ (for $c > 1$ or $b = c$ for $c \leq 1$), calculates the degree of harmony (b).

The degrees of urban development, harmony, and quasi-harmony are indexes that reflect the development and harmony of the metabolic system. The degree of system development increases as the system grows. The degree of harmony reflects the balance between socioeconomic development and ecological environmental protection during the process of urban development.

4.5 Measuring Interactions Between the Natural and Socioeconomic Systems

In the previous sections, I developed a system of indicators and a model for evaluating an urban ecosystem to determine the degrees of development and harmony within the system. The indicators and model can be used to identify potential methods to improve the health and vitality of the urban organism, and provide a basis for ecological management decisions (Zhang et al. 2006b).

4.5.1 Measurement Index System

It's important to note that interaction between humans and their environment may change over time as conditions alter, or as a city passes through different evolutionary stages (Rosser 1995; Costanza et al. 1993; Holling 1987). For example, Odum (1983) suggests that "mutualism seems to replace parasitism as ecosystems evolve toward maturity, and it seems to be especially important when some aspect of the environment is limiting". In such cases, Odum suggests that mutualism can lead to the emergence of more resilient properties within a system. Such properties are important, because without ecosystem resilience, discontinuous changes in ecosystem functions may result in a sudden loss of biological productivity and a reduced capacity to support human life (Arrow et al. 1995). There are interactions in the form of competition, mutualism, and the gradual development of order among the components of a complex ecosystem (Hao and Qin 2003; Wang 2003). Though scholars have deeply discussed the mechanisms that govern the evolution of complex ecosystems, there has been insufficient research on the combination of development, harmony, and regeneration of the urban ecosystem to achieve sustainable development (Hao and Qin 2003; Wang 2003). In addition, there have been no quantitative schemes that describe these interactions (Zhang et al. 2006b).

Based on the interactions between the socioeconomic and natural subsystems, my research group developed an indicator system for measuring urban metabolic interactions (Zhang et al. 2003a). The system includes 3 types of indicators: socioeconomic development pressure on the environment (PE), the support (carrying) capacity of the natural environment (SC), and the regeneration ability of the ecosystem (RA). We developed 20 secondary indicators for these indicators (Table 4.9). We then used these indicators to evaluate the sustainability of urban development from 3 perspectives (i.e., the degrees of development, harmony, and recycling) using factor analysis and the analytic hierarchy process to define a sustainability index, as described in the next section.

4.5.2 Sustainability Index

Factor analysis focuses on the main factors responsible for the value of a sustainability index and the contributions of various indicators to the values of these factors (Zhang et al. 2006b). As a result, it can effectively analyze the many relationships among the components of a complex urban ecosystem (Zhang et al. 2003b). Based on the values of many indicators, factor analysis can confirm the numbers and types of the main factors required to fully model the urban system, and can account for at least 70% of the original information content using the main factors (Zhang et al. 2003b). Factor analysis minimizes the number of main factors in order to simplify the model, and the main factors that are retained are independent, have clear implications, and are easy to analyze. This approach also reveals primary factors with contrasting effects, avoids distortion of the message from the original indicators, and reveals the essential phenomena involved in the model of the ecosystem. For these reasons, the method is objective, accurate, and believable, and is thus one of the better operational methods for assessing interactions. The main purpose of factor analysis is to group all variables with similar factor loads under a common factor. When the initial factor cannot explain a sufficiently high proportion of variance in the predicted response variable, and therefore does not explain the response variable sufficiently, the matrix of load

Table 4.9 Indicator system for measuring the interactions in an urban metabolism

Type	Indicators
Socioeconomic development pressure on the environment (PE)	Coverage of the land by forestry activities, proportions of natural reserves, per capita public green areas in developed areas, coverage rate of forests in developed areas
Support (carrying) capacity of the natural environment (SC)	Energy consumption per 10^4 RMB of GDP (including "clean" energy), population density, GDP density, water consumption per 10^4 RMB of GDP, index of environmental pollution [refers to the dimensionless value obtained by treating environmental parameters according to certain principles and methods (i.e., the rate of pollutant monitoring concentration value to environmental standard value) to reflect the degree of environmental pollution]
Regeneration ability of the ecosystem (RA)	Proportion of industrial effluent that meets the local discharge standards, proportion of motor vehicle exhaust that meets the local discharge standards, proportion of smoke and dust eliminated from waste gases, proportion of treated industrial waste gases, proportion of treated industrial solid waste, and proportion of treated dangerous waste materials

factors for the initial factor's indicators can be rotated and the analysis repeated; the result is generally a more reasonable explanation of the response variable. To perform the analysis, it is necessary to collect statistical data on the socioeconomic pressure and natural support system indicators, identify the main factors, and calculate the contribution of each indicator to the main factors. On this basis, a model of the degree of harmonious development can be identified, a quantitative sustainability index can be obtained, and the state of urban development can then be comprehensively analyzed and evaluated using the sustainability index.

Based on the regenerative metabolism indicators (RA), we applied the analytic hierarchy process to evaluate the city's recycling degree, and developed a 3-level model (Zhang et al. 2006b). The recycling degree was defined in level 1. In level 2, we developed weights for indicators of the degrees of source recycling and terminal recycling. We then defined concrete indicators in level 3. The indicators include the resource utilization rate of the socioeconomic subsystems, the level of waste recycling, and the level of science and technology (Table 4.9). Using the analytic hierarchy process, we can calculate the weight of regenerative metabolism indicators, and then integrate with the indicator value to obtain the recycling degree (r). Based on the degrees of development, harmony, and recycling (Table 4.10), we proposed a sustainable development index (SDI) for an urban organism that can be used to assess the level of urban sustainable development. This is defined as follows:

Table 4.10 Measurement standards for the interaction analysis

Development degree (a)	Quasi-harmony degree (c)	Harmony degree (b)	Recycling degree (r)
Rudimentary $0 < a \leq 0.5$	Strong socioeconomic pressure $c \leq 1/3$	Rudimentary $0.75 < b \leq 1$	Rudimentary $0 < r < 0.25$
Elementary $0.5 < a \leq 0.75$	Moderate socioeconomic pressure $1/3 < c \leq 1$	Elementary $0.5 < b \leq 0.75$	Elementary $0.25 < r \leq 0.5$
Intermediate $0.75 < a \leq 1$	Moderate ecological carrying capacity $1 < c \leq 3$	Intermediate $0.25 < b \leq 0.5$	Intermediate $0.5 < r \leq 0.75$
Advanced $1 < a$	Strong ecological carrying capacity $3 < c$	Advanced $0 < b \leq 0.25$	Advanced $0.75 < r \leq 1$

Note The variables *a*, *b*, and *c*, are defined in Sect. 4.3 of this chapter

$$\text{SDI} = 0.4a + 0.3b + 0.3r \tag{4.13}$$

where, a represents development degree, b stands for harmony degree, and r means recycling degree, as can be seen in Table 4.10.

References

Arrow K, Bolin B, Costanza R et al (1995) Economic growth, carrying capacity, and the environment. Ecol Econ 15(2):91–95

Bao ZM (2010) Material flow analysis (MFA) of environmental-economic system of Dalian. Dalian University of Technology, Dalian (in Chinese)

Brown MT, Ulgiati S (1997) Emergy-based indices and ratios to evaluate sustainability: monitoring economies and technology toward environmentally sound innovation. Ecol Eng 9(1–2):51–69

Chen WQ, Graedel TE (2015) In-use product stocks link manufactured capital to natural capital. Proc Natl Acad Sci USA 112(20):6265–6270

Costanza R, Wainger L, Folke C et al (1993) Modeling complex ecological economic systems: toward an evolutionary, dynamic understanding of people and nature. Bioscience 43(8):545–555

Duh JD, Shandas V, Chang H et al (2008) Rates of urbanisation and the resiliency of air and water quality. Sci Total Environ 400(1–3):238–256

Eurostat (2018) Economy-wide material flow accounts handbook (2018 edition). (2018-6-1)[2022-1-2]. https://ec.europa.eu/eurostat/documents/3859598/9117556/KS-GQ-18-006-EN-N.pdf/b621b8ce-2792-47ff-9d10-067d2b8aac4b?t=1537260841000

Fishman T, Schandl H, Tanikawa H (2016) Stochastic analysis and forecasts of the patterns of speed, acceleration, and levels of material stock accumulation in society. Environ Sci Technol 50(7):3729–3737

Guo J, Fishman T, Wang Y et al (2020) Urban development and sustainability challenges chronicled by a century of construction material flows and stocks in Tiexi, China. J Ind Ecol 25(1):162–175

Hall CAS, Cleveland CJ, Kauffmann R (1986) Energy and resource quality: the ecology of the economic process. John Wiley and Sons, New York, USA
Hao X, Qin SS (2003) Relations between the complexity of compound ecosystem and sustainable development. J Syst Dialectics 11(4):23–26 (in Chinese)
Hendriks C, Obernosterer R, Müller D, et al (2000) Material flow analysis: a tool to support environment policy decision making. Case-studies on the city of Vienna and the Swiss lowlands. Local Environ 5(3):311–328.
Holling CS (1987) Simplifying the complex: the paradigms of ecological function and structure. Eur J Oper Res 30(2):139–146
Li G (2014) Analysis of material flows for sustainable development of agriculture in China. J Northwest Agric & for Univ (social Science Edition) 14(4):55–60 (in Chinese)
Li JP, Chen FP, Wang ZS (2006) The emergy synthesis and sustainability analysis of city's environment and economy. Acta Ecol Sin 26(2):439–447 (in Chinese)
Li YX, Zhang Y, Hao Y et al (2019) Exploring the processes in an urban material metabolism and interactions among sectors: an experimental study of Beijing, China. Ecol Indic 99(2):214–224
Liu J, Wang M, Zhang C et al (2020) Material flows and in-use stocks of durable goods in Chinese urban household sector. Resour Conserv Recycl 158(24):104758
Lu HF, Lan SF, Peng SL (2002) Studies on emergy indices for evaluating system sustainable development property. China Environ Sci 22(4):380–384 (in Chinese)
Miyano H (2001) Identification model based on the maximum information entropy principle. J Math Psychol 45(1):27–42
Odum EO (1983) Basic ecology. Saunders, New York, USA
Odum HT (1971) Environment, power, and society. Wiley Interscience, New York, USA
Odum HT (1988) Self-organization, transformity, and information. Science 242(4882):1132–1139
Odum HT (1996) Environmental accounting: energy and environmental decision making. Wiley, New York, USA
Pauliuk S, Müller DB (2014) The role of in-use stocks in the social metabolism and in climate change mitigation. Glob Environ Chang 24(1):132–142
Prigogine I, Lefever R (1973) Theory of dissipative structures//Haken H. Synergetics. Springer Vieweg Verlag, Berlin, Germany, pp 124–135.
Renyi A (1961) On measures of information and entropy//Neyman J. In: Proceedings of the Fourth Berkeley Symposium on Mathematics, Statistics and Probability. University of California Press, Berkeley, USA, pp 547–561
Rosser JB (1995) Systemic crises in hierarchical ecological economies. Land Econ 71(2):163–172
Shannon CE (1948) A mathematical theory of communication. Bell Syst Tech J 27(3):379–423
Swilling M, Hajer M, Baynes T, et al (2018) The weight of cities: resource requirements of future urbanization. United Nations Environment Programme, Nairobi, Kenya (2018-1-31)[2022-4-12]. https://www.resourcepanel.org/reports/weight-cities
Wang RS (2003) Approach of industry ecology for the development of recycling economy. Ind Environ 49:48–52
Wang XJ, Li YX, Liu NY et al (2020) An urban material flow analysis framework and measurement method from the perspective of urban metabolism. J Clean Prod 257(6):120564
Yelshin A (1996) On the possibility of using information entropy as a quantitative description of porous media structural characteristics. J Membr Sci 117(1–2):279–289
Zhang KM, Wen ZG, Du B et al (2003a) Evaluation and index system of eco-city. Chemical Industry Press, Beijing, China (in Chinese)
Zhang Y, Shang JC, Yu XY (2003b) Sustainable utilization of water resources in Jilin Province. Adv Water Sci 14(4):489–493 (in Chinese)
Zhang Y, Yang ZF (2007) Eco-efficiency of urban material metabolism: a case study in Shenzhen, China. Acta Ecol Sin 27(8):3124–3131 (in Chinese)
Zhang Y, Yang ZF, Li W (2006a) Analyses of urban ecosystem based on information entropy. Ecol Model 197(1):1–12

Zhang Y, Yang ZF, Yu XY (2006b) Measurement and evaluation of interactions in complex urban ecosystem. Ecol Model 196(1):77–89

Zhang Y, Yang ZF, Yu XY (2009) Evaluation of urban metabolism based on emergy synthesis: a case study for Beijing (China). Ecol Model 220(13–14):1690–1696

Chapter 5
Network Models to Simulate Urban Metabolism

As noted in previous chapters, the metabolism of an organism or ecosystem operates through a network of connections between the organism's or ecosystem's metabolic components. In this chapter, I describe how to construct network models and use them to quantify, track, and simulate the flows through an urban metabolic system. The basic units of an ecological network are "nodes" and "paths". Nodes perform a specified function (i.e., they are metabolic actors), and thus serve as the functional units of the ecosystem, whereas paths serve as transmission channels for materials, energy, and money between nodes. The models were created at different times and for different purposes, and were based on different datasets; the urban metabolic components thus differ among the models. This variation demonstrates the flexibility of the modeling approach and the different levels of detail that can be used for model development. Furthermore, the categories used for flow tracking were generally based on the categories defined in Chinese statistical yearbooks and databases, thus the category names and contents may differ from those in statistical datasets from other countries.

5.1 Network Models Based on Physical Metabolism

5.1.1 Urban Water Metabolic Network Models

5.1.1.1 Conceptual Model

A conceptual model of urban water metabolism processes can be developed based on how water flows through an urban system. This model has clear links among the system's three key trophic levels: the local ecological environment; the terminal consumption sectors; and the wastewater (sewage) regeneration sector. The model complies with the flows of water and water-containing resources among these levels.

Y. Zhang, *Urban Metabolism*,
https://doi.org/10.1007/978-981-19-9123-3_5

However, it mainly reflects the utilization of freshwater, recycled water, and rainwater, as well as the reuse of water and discharge of wastewater. The fresh water of the local ecological environment can be used for industry, agriculture, and household. But sometimes the water must be sourced from the external environment (e.g., ecological flows to maintain water bodies). The external environment contains the adjacent areas located upstream of the study area within the same watershed and other areas where water resources are transferred to the study area as a result of large hydrological projects (Fig. 5.1).

Under the constraint of a limited water supply, which is common in developing countries and in areas such as northern China that are experiencing a water shortage, city managers must consider deeply how to make the best use of fresh water and reuse wastewater. The wastewater recycling system receives the industrial wastewater and household sewage and then uses part of the treated wastewater to recharge the ecological environment, irrigate municipal green space, wash streets, as well as for agricultural irrigation and industrial utilization. Another way to recharge the ecological environment is by collecting rainwater, which can also be used as supplemental water for industry, agriculture, and household. In addition, in order to address the problem of high-water consumption, the industrial sector may reuse much of its water. In the process of water use, there is wastewater discharged from industry, agriculture, and domestic sectors (Zhang et al. 2010a).

5.1.1.2 Network Model

Different components of the urban system can be defined as producers, consumers, and reducers based on the trophic levels of natural ecosystems. We can also identify the water flows among the components of the system. In this model, the internal environment and the artificial rainwater harvesting system work as the producers; industrial, agricultural, and household sectors serve as consumers; while the wastewater recycling system plays the role of reducers.

The role of each component may change at different times because of their complicated chain of relationships. Take the internal environment as an example, firstly it plays the role of producer but meanwhile, it also needs water resources to maintain its operation and absorb (reduce) wastewater from the urban system. The wastewater recycling subsystem has the same condition. It can support the operation of the urban system by purifying the wastewater (i.e., like reducer) as well as providing regenerated (recycled) water (i.e., like producer). These changes in the roles of components lead to a (network) reticular (network) system structure rather than a linear structure. When it comes to metabolism, we will automatically link it to the biological concept. However, we may also use it as an analogy for the urban system because the urban water metabolic system can also acquire and process resources as well as produce waste. From this perspective, cities can be seen as "urban superorganisms" which demonstrate metabolic processes. To make the huge flows of matter and energy easy to understand, we can use the trophic levels of natural ecosystems to make an analogy. Then we can divide the urban water metabolic system into different roles

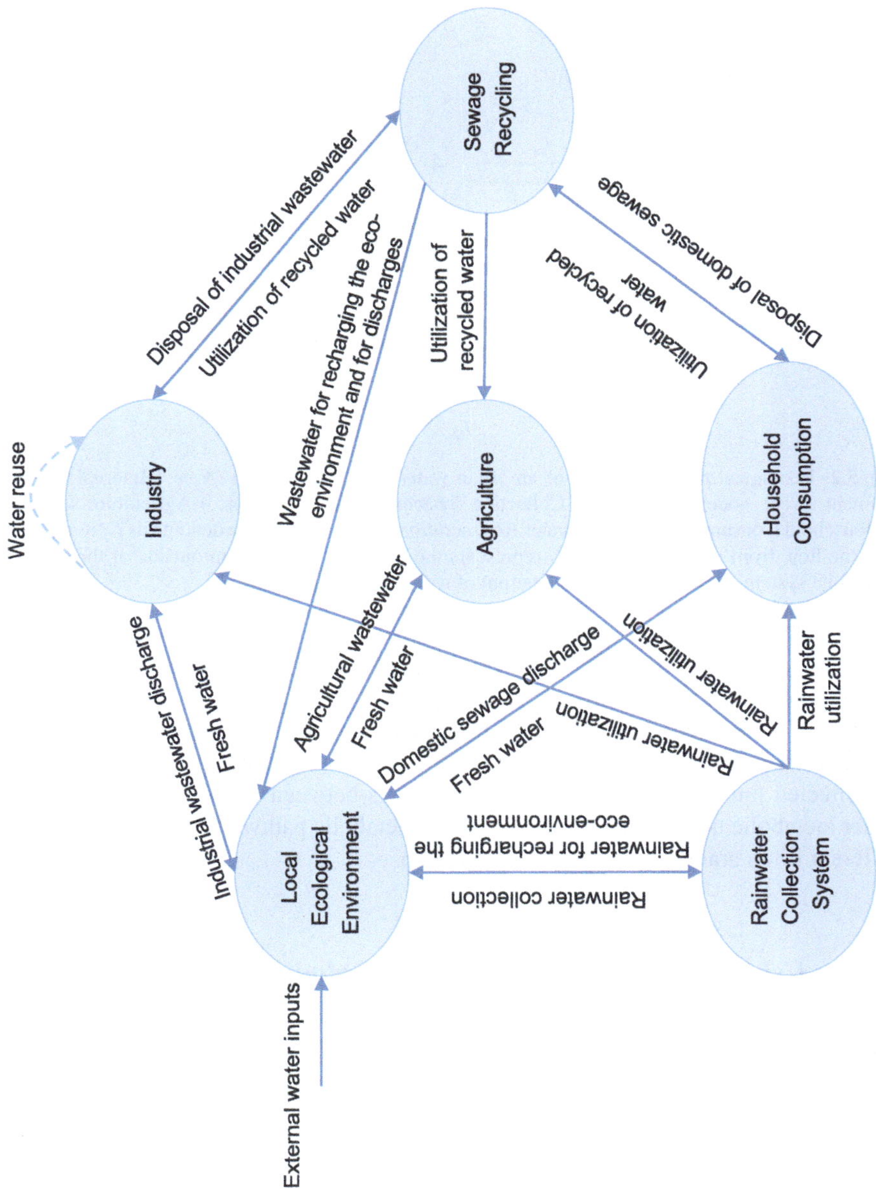

Fig. 5.1 Conceptual model of water flows within urban water metabolism

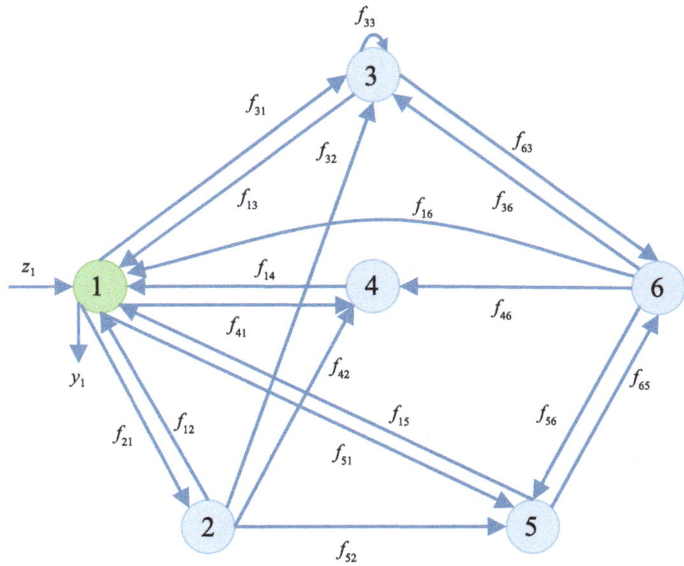

Fig. 5.2 Ecological network model of an urban water metabolic system (*Note* 1-Internal Environment (green node); 2-Rainwater Collection System; 3-Industry sector; 4-Agriculture sector; 5-Household Consumption; 6-Wastewater Regeneration System. Flows are denoted as f_{ij} to represent the flow from node j to node i, z_i represents the flow into node i from outside of the water metabolic system, and y_1 represents the output of water from the system)

like producers, consumers, and reducers, and also identify the water flows among the components of the system.

We can build an ecological network model based on the analysis of the process of urban water metabolism (Fig. 5.2). In an ecological network model, we always use the directed lines to represent the flows/pathways between different nodes. In our water metabolic model (Fig. 5.2) there are 18 metabolic pathways that represent the different flows among the six nodes (Table 5.1).

5.1.2 *Urban Energy Metabolic Network Models*

We can develop similar models for energy flows following the same approach to the conceptual and network models for water flows in the previous section.

5.1.2.1 Conceptual Model

By abstracting and summarizing the traditional energy utilization systems, the same approach used for water flows can be used to identify the key links in urban energy

Table 5.1 Nodes and paths in the urban water metabolic network model

	1	2	3	4	5	6	z
1		f_{12}	f_{13}	f_{14}	f_{15}	f_{16}	z_1
2	f_{21}						
3	f_{31}	f_{32}	f_{33}			f_{36}	
4	f_{41}	f_{42}				f_{46}	
5	f_{51}	f_{52}				f_{56}	
6			f_{63}		f_{65}		
y	y_1						

Note Flows are from the components in the first row to the components in the first column. Based on these node definitions, we can include the following flows in the model: f_{12}-rainwater for recharging the Internal Environment; f_{13}-wastewater discharged into the Internal Environment by the Industrial sector; f_{14}-wastewater discharged into the Internal Environment by the Agricultural sector; f_{15}-sewage discharged into the Internal Environment by the Household Consumption sector; f_{16}-recycled water used to recharge the Internal Environment and discharged wastewater; f_{21}-rainwater collection from the Internal Environment; f_{31}-fresh water utilized by the Industrial sector; f_{32}-rainwater utilized by the Industrial sector; f_{33}-recycled flows within the Industrial sector; f_{36}-recycled water utilized by the Industrial sector; f_{41}-fresh water utilized by the Agricultural sector; f_{42}-rainwater utilized by the Agricultural sector; f_{46}-recycled water utilized by the Agricultural sector; f_{51}-fresh water utilized by the Household Consumption sector; f_{52}-rainwater utilized by the Household Consumption sector; f_{56}-recycled water utilized by the Household Consumption sector; f_{63}-wastewater disposal by the Industrial sector; f_{65}-sewage disposal by the Household Consumption sector; y_1-flows from the Internal Environment to the external environment; and z_i-the flow into node i from outside of the water metabolic system

metabolic processes, thereby permitting the development of a conceptual model of the role of energy in these processes (Fig. 5.3). The conceptual model of urban energy metabolic processes reflects the basic relationships involved in exploitation and utilization of urban energy. In this model, there are links among three key trophic levels that are analogous to those of a natural ecosystem: the energy exploitation sector functions as an energy producer; the energy transformation sector functions as a primary consumer of this energy; and the terminal consumption sectors (i.e., other industries and households) function as secondary consumers.

The energy exploitation sector plays an important role in urban energy metabolic processes. It produces primary energy source which supplies both the transformation and terminal consumption sectors as well as the external areas of the system. The energy transformation sector, including oil refining, power generation, and co-generation, can use locally produced primary energy or energy imported (input) from outside the system to produce energy for secondary consumers, and a portion of this production is exported (output) to areas out of the system. The terminal consumption sectors, or in other words, the energy-using industries and the urban households, feed on primary and secondary energy from internal and external sources. The recycling processes associated with the recovery of by-product resources are another important part of the urban energy metabolism. These processes include recovering energy

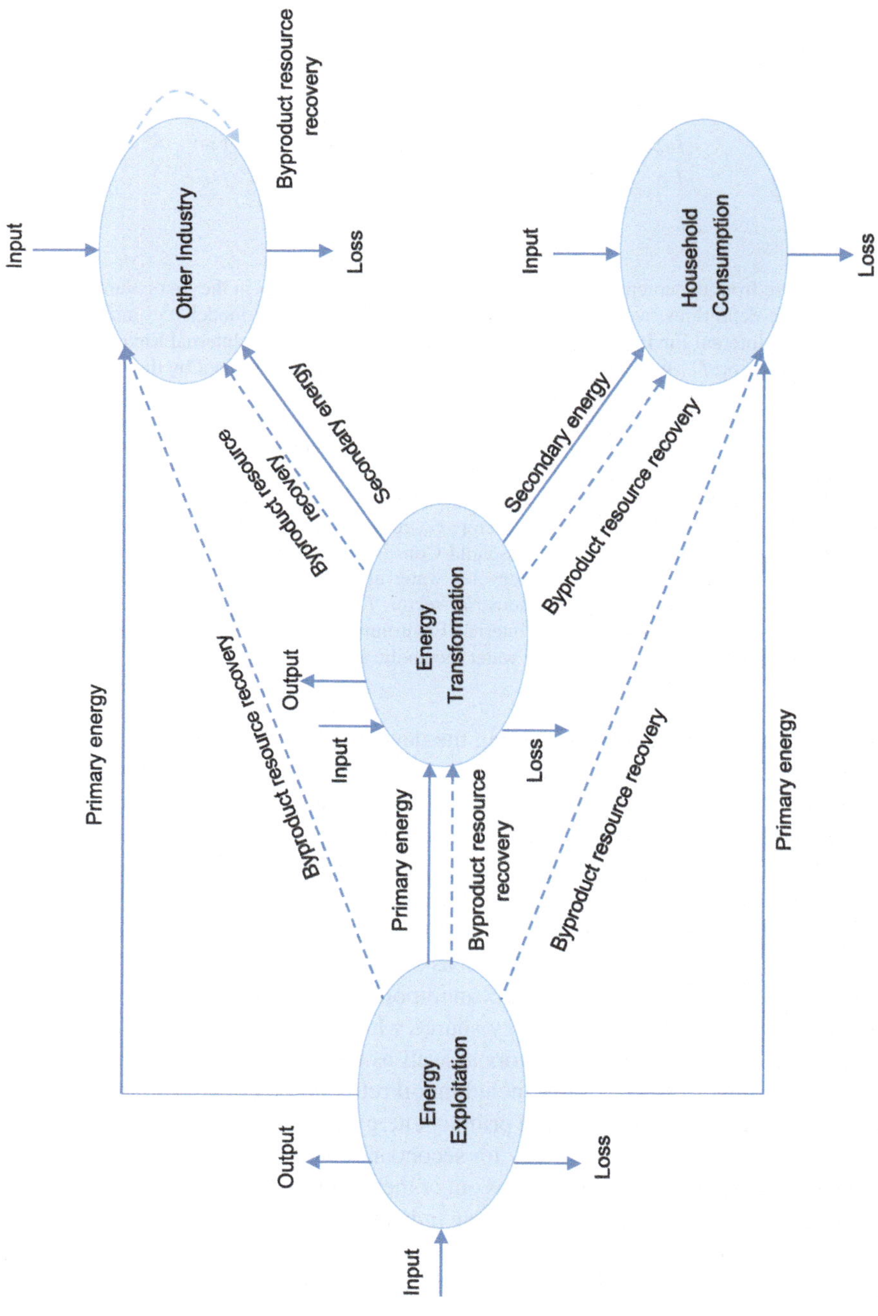

Fig. 5.3 Conceptual model of urban energy metabolic processes

from primary and secondary energy generation processes and from industrial and household processes.

5.1.2.2 Network Model

Through the analysis of urban energy metabolic processes in this model, we find that these sectors are connected by a network consisting of 11 flows which describe the different pathways of metabolism among five nodes (Fig. 5.4). We build the ecological network model of the urban energy metabolic system by using directed lines connecting the nodes in the network to plot all the energy flows among the components according to their energy flow direction. The pairing between ecological and social functions is not precise, but it is assumed that the energy exploitation, energy conversion, industrial and household consumers, and restoration components of the urban system perform similar ecological roles. In specific, the Energy Exploitation sector corresponds to the primary producers of the ecosystem, the Energy Transformation sector stands for the primary consumers, the Household Consumption and Other Industrial sectors are secondary consumers, and the Energy Recovery sector acts as a reducer. The metabolic actors and paths are summarized in Table 5.2.

By making additional analysis of the urban energy metabolic processes in the conceptual model, a more detailed ecological network model of the system energy

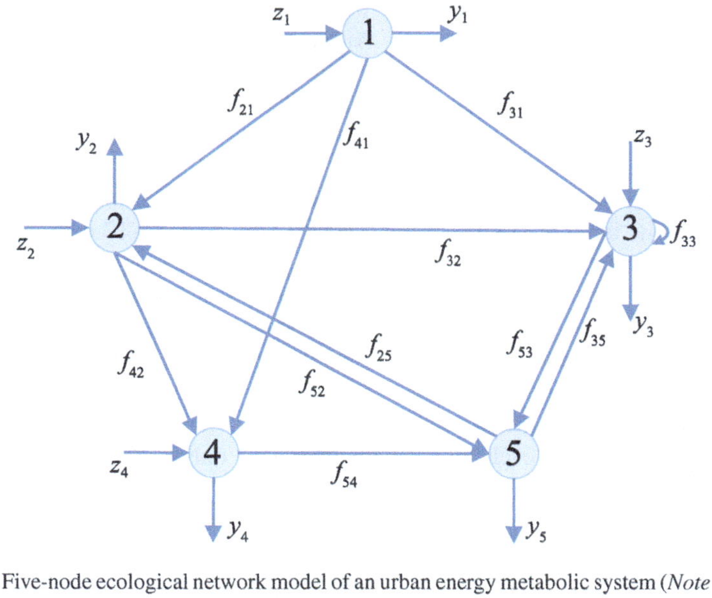

Fig. 5.4 Five-node ecological network model of an urban energy metabolic system (*Note* 1-Energy Exploitation sector; 2-Energy Transformation sector; 3-Other Production sector; 4-Household Consumption sector; 5-Energy Recovery sector. f_{ij} represents the flow from node j to node i, z_i represents the flow into node i from outside of the energy metabolic system, and y_i represents the boundary outflows from node i)

Table 5.2 Nodes and paths in the urban energy metabolic network model

	1	2	3	4	5	z
1						z_1
2	f_{21}				f_{25}	z_2
3	f_{31}	f_{32}	f_{33}		f_{35}	z_3
4	f_{41}	f_{42}				z_4
5		f_{52}	f_{53}		f_{54}	
y	y_1	y_2				

Note Flows are from the components in the top row to the components in the first column. f_{21}, f_{31}, and f_{41} represent the primary energy used in the Energy Transformation sector, Other production sectors, and Household Consumption sector, respectively; f_{32} and f_{42} are the secondary energy used in the Other Production sector and Household Consumption sector, respectively; f_{52}, f_{53}, and f_{54} are the energy recovered in the Energy Transformation sector, Other Production sector, and Household Consumption sector, respectively; f_{33} is the energy recovered from Other Production sector; f_{25} and f_{35} are the recovered energy used in the Energy Transformation sector and Other Production sector, respectively; z_1, z_2, z_3, and z_4 are the energy supplied by the external regions to the Energy Exploitation sector, Energy Transformation sector, Other Productive sector, and Household Consumption sector, respectively; and y_1 and y_2 are the energy exported to the external regions by the Energy Exploitation sector and the Energy Transformation sector, respectively

flow can be formulated. The model components reflect the fundamental connections among urban energy exploitation, transformation, consumption, and recovery. The new model classifies the urban metabolic system into 17 components (sectors): ① Energy Exploitation; ② Coal-fired Power; ③ Heat Supply; ④ Washed Coal; ⑤ Coking; ⑥ Oil Refining; ⑦ Gas Generation; ⑧ Coal Products; ⑨ Agriculture; ⑩ Industry; ⑪Construction; ⑫Transportation; ⑬Wholesale, Retail, Accommodation, and Catering; ⑭Household Consumption; ⑮Other Services; ⑯Recovery; and ⑰Energy Stocks.

Based on the structure of natural ecological systems, we can use the roles of the components in an urban metabolic system to divide these 17 components into producers, consumers, and reducers or decomposers, the latter of which recover energy lost from other system components. Through the analogy of a natural ecological system, the reducers in an urban metabolic system can also be considered to be producers because of their role in recovering energy and returning it to the system. Alternatively, they can be considered to be final consumers. Section 5.4.2 discusses the problem of how to classify reducer components of an urban system in more detail. Consumers at all levels must obtain energy not only from producers, but also from reducers and energy stocks. The recovery sector and energy stocks are thus both considered to be producers (Zhang et al. 2010b). Based on their ecological trophic levels, the energy flows can be described by directional lines that connect the nodes in the network, resulting in an ecological network model for the urban energy metabolism system. In the model in Fig. 5.5, 45 metabolic pathways can be defined that reflect the exchanges of flows among these 17 nodes.

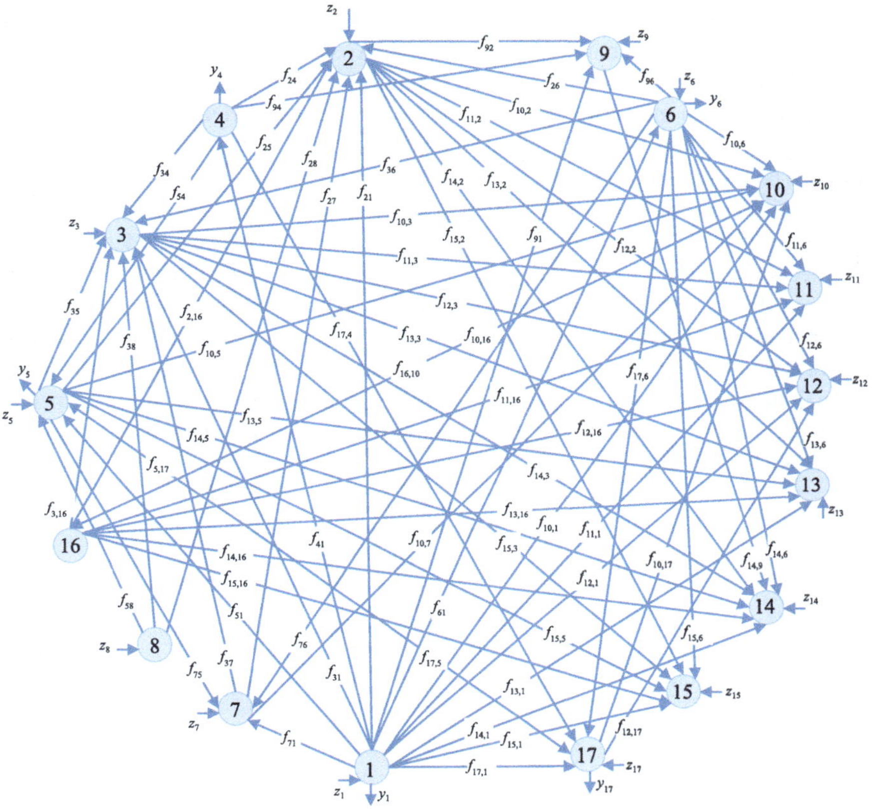

Fig. 5.5 Illustration of a 17-node ecological network model of an urban energy metabolic system (*Note* 1-Energy Exploitation; 2-Coal-fired Power; 3-Heat Supply; 4-Washed Coal; 5-Coking; 6-Oil Refining; 7-Gas Generation; 8-Coal Products; 9-Agriculture; 10-Industry; 11-Construction; 12-Transportation; 13-Wholesale, Retail, Accommodation, and Catering; 14-Household Consumption; 15-Other Services; 16-Recycling; 17-Energy Stocks. The energy types are as follows: raw coal, cleaned coal, other washed coal, mould coal (i.e., compressed coal powder), coke, coke oven gas, other gas, crude oil, gasoline, kerosene, diesel oil, fuel oil, liquefied petroleum gas, refinery gas, natural gas, other petroleum products, other coking products, heat, electricity, and other energy (all measured in tonnes of coal equivalent). f_{ij} represents the flow from node j to node i, z_i represents the flow into node i from outside of the energy metabolic system, and y_i represents the boundary outflows from node i)

5.1.3 Urban Carbon and Nitrogen Metabolic Network Models

We can also use the methods described in the previous sections for individual chemical elements. For example, we can develop a network model of urban carbon and nitrogen metabolism by tracing the carbon or nitrogen flows in food, energy, and other materials in a city, as well as the processes of carbon or nitrogen emission and fixation. As in water and energy models, the main concept is that the natural and socioeconomic components of an urban system can be abstracted as nodes in a network, and that the exchanges among these nodes can be treated as paths. Based on the available data and a rational classification of the key transfer processes in urban carbon or nitrogen metabolism processes, we can define models with 18 nodes for carbon (Fig. 5.6) and 15 nodes for nitrogen (Fig. 5.7) in an urban metabolic system. In the carbon metabolic network, there are 4 nodes related to natural components, 14 nodes related to socioeconomic sectors of the system, and 85 carbon transfer paths among these nodes. In the nitrogen metabolic network, there are 5 nodes related to natural components, 10 nodes related to socioeconomic sectors of the system, which represent the major sectors that generate anthropogenic nitrogen, and 51 nitrogen transfer paths among these nodes. By combining the nodes and paths, we can develop network models of urban carbon and nitrogen metabolic processes. Taking the urban internal environment as a node allows us to consider the pressure and influence of metabolic actors on the city's internal environment, as listed in Table 5.3 (Li et al. 2018) and Table 5.4 (Zhang et al. 2016a).

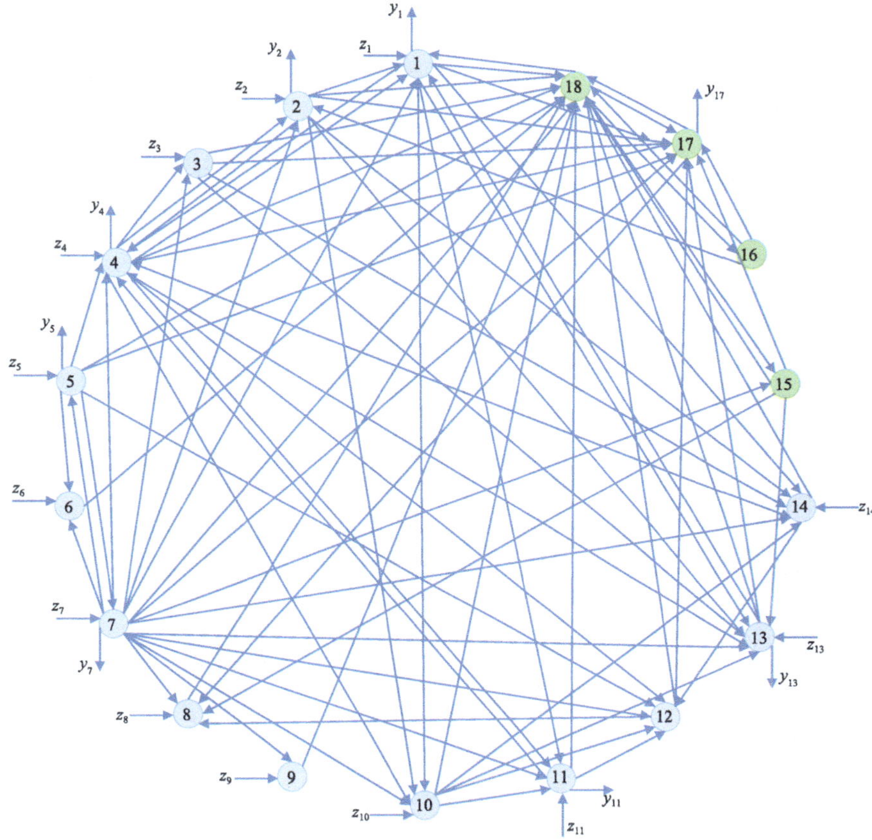

Fig. 5.6 Ecological network model of an urban carbon metabolic system (*Note* 1-Agriculture; 2-Animal Husbandry; 3-Fisheries; 4-Manufacturing; 5-Mining; 6-Production and Supply of Electric Power and Heat; 7-Energy Transformation; 8-Construction; 9-Transportation; 10-Retail and Catering; 11-Other Services; 12-Waste Disposal; 13-Rural households; 14-Urban Households; 15-Forest; 16-Grassland; 17-Surface Water; 18-Atmosphere. f_{ij} represents the flow from node *j* to node *i*, z_i represents the flow into node *i* from outside of the carbon metabolic system, and y_i represents the boundary outflows from node *i*. Green nodes are natural metabolic actors and blue nodes are the socioeconomic metabolic actors)

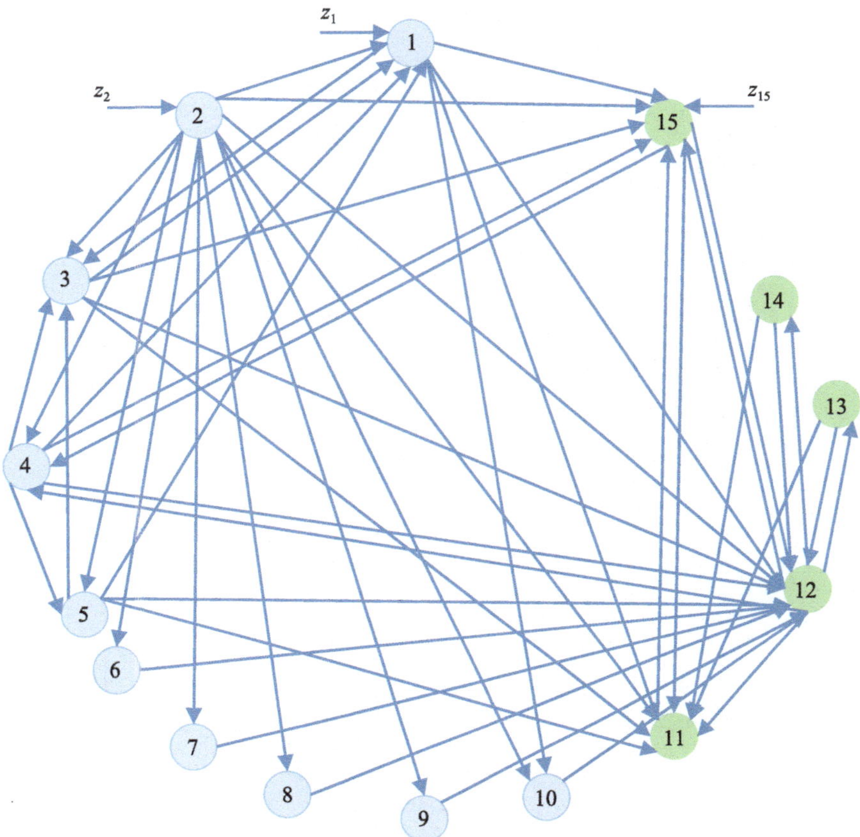

Fig. 5.7 Ecological network model of an urban nitrogen metabolic system (*Note* 1-Household Consumption; 2-Industry; 3-Animal Husbandry; 4-Crop Cultivation; 5-Aquaculture; 6-Forestry; 7-Services; 8-Construction; 9-Transportation; 10-Sewage Treatment; 11-Surface Water; 12-Atmosphere; 13-Forest; 14-Grassland; 15-Farmland. f_{ij} represents the flow from node j to node i, z_i represents the flow into node i from outside the nitrogen metabolic system, and y_i represents the boundary outflows from node i. Green nodes are natural metabolic actors, and blue nodes are the socioeconomic metabolic actors)

Table 5.3 Nodes and paths in the urban carbon metabolic network model

	1	2	3	4	5	6	7	8	9	10	11	12	13	14	15	16	17	18	z
1		f_{12}		f_{14}			f_{17}					$f_{1,12}$	$f_{1,13}$					$f_{1,18}$	z_1
2				f_{24}			f_{27}									$f_{2,16}$			z_2
3				f_{34}			f_{37}												z_3
4	f_{41}	f_{42}			f_{45}		f_{47}				$f_{4,11}$		$f_{4,13}$	$f_{4,14}$					z_4
5							f_{57}												z_5
6					f_{65}		f_{67}												z_6
7					f_{75}														z_7
8							f_{87}					$f_{8,12}$			$f_{8,15}$		$f_{8,17}$		z_8
9							f_{97}												z_9
10	$f_{10,1}$	$f_{10,2}$		$f_{10,4}$			$f_{10,7}$												z_{10}
11	$f_{11,1}$			$f_{11,4}$			$f_{11,7}$			$f_{11,10}$									z_{11}
12				$f_{12,4}$	$f_{12,5}$		$f_{12,7}$			$f_{12,10}$	$f_{12,11}$			$f_{12,14}$					
13	$f_{13,1}$	$f_{13,2}$	$f_{13,3}$	$f_{13,4}$			$f_{13,7}$			$f_{13,10}$					$f_{13,15}$				z_{13}
14	$f_{14,1}$	$f_{14,2}$	$f_{14,3}$	$f_{14,4}$			$f_{14,7}$			$f_{14,10}$									z_{14}
15							$f_{15,7}$											$f_{15,18}$	

(continued)

Table 5.3 (continued)

	1	2	3	4	5	6	7	8	9	10	11	12	13	14	15	16	17	18	z
16																		$f_{16,18}$	
17	$f_{17,1}$	$f_{17,2}$	$f_{17,3}$	$f_{17,4}$	$f_{17,5}$		$f_{17,7}$					$f_{17,12}$	$f_{17,13}$		$f_{17,15}$	$f_{17,16}$		$f_{17,18}$	
18	$f_{18,1}$	$f_{18,2}$	$f_{18,3}$	$f_{18,4}$	$f_{18,5}$	$f_{18,6}$	$f_{18,7}$	$f_{18,8}$	$f_{18,9}$	$f_{18,10}$	$f_{18,11}$	$f_{18,12}$	$f_{18,13}$	$f_{18,14}$	$f_{18,15}$	$f_{18,16}$	$f_{18,17}$		
y	y_1	y_2		y_4	y_5		y_7				y_{11}		y_{13}				y_{17}		

Note Flows are from the components in the first row to the components in the first column of the table. f_{41}-paper straw, fodder straw, and industrial biomass feedstock; $f_{10,1}$-grain, fresh and dried fruits, and vegetables; $f_{11,1}$-waste plastics; $f_{13,1}$-grain, fresh and dried fruits, and vegetables, and energy straw; $f_{14,1}$-grain, fresh and dried fruits, and vegetables; $f_{17,1}$-soil erosion; $f_{18,1}$-fuel combustion, soil respiration, fertilizer decomposition, and rice field emissions; f_{12}, $f_{17,2}$-animal manure; f_{42}-livestock and poultry products; $f_{10,2}$, $f_{13,2}$, $f_{14,2}$-meat, eggs, and dairy products, respectively; $f_{18,2}$-fuel combustion, animal ruminant respiration, and animal waste disposal; $f_{13,3}$, $f_{14,3}$-fish products; $f_{17,3}$-fish manure and residual feed; $f_{18,3}$-fuel combustion and respiration; f_{14}-fertilizers, plastics, and sludge; f_{24}, f_{34}-feedstuffs; $f_{10,4}$, $f_{13,4}$, $f_{14,4}$-tobacco, alcohol and grease, paper products, and plastics, respectively; $f_{11,4}$-paper products; $f_{12,4}$, $f_{12,5}$, $f_{12,7}$, $f_{12,10}$-solid waste and wastewater; $f_{17,4}$, $f_{17,5}$, $f_{17,12}$-wastewater; $f_{18,4}$-fuel combustion and industrial processes; f_{45}-other coal washing, coal waste, cement production, steel smelting, and limestone for refining; f_{65}-raw coal and other washed coal; f_{75}-cleaned coal; $f_{18,5}$, $f_{18,6}$, $f_{18,7}$, $f_{18,8}$, $f_{18,9}$, $f_{18,10}$, $f_{18,11}$-fuel combustion; f_{17}, f_{27}, f_{37}, f_{47}, f_{57}, f_{57}, f_{67}, f_{87}, f_{97}, $f_{10,7}$, $f_{11,7}$, $f_{13,7}$, $f_{14,7}$, $f_{15,7}$-secondary energy; $f_{17,7}$-carbon in wastewater; $f_{11,10}$-recycled plastics; $f_{13,10}$, $f_{14,10}$-food; $f_{4,11}$, $f_{4,13}$, $f_{4,14}$-waste paper and waste plastics; $f_{12,11}$-waste paper; $f_{1,12}$-manure, sludge; $f_{8,12}$-sludge; $f_{18,12}$-solid waste, wastewater, sludge, manure disposal; $f_{1,13}$-manure, digestate; $f_{17,13}$-feces; $f_{18,13}$-fuel and biomass combustion, respiration, feces disposal; $f_{12,14}$-manure, household waste, household wastewater; $f_{18,14}$-fuel combustion and respiration; $f_{8,15}$-wood; $f_{13,15}$-fuelwood; $f_{17,15}$-soil erosion; $f_{18,15}$-fuel combustion, soil respiration; $f_{2,16}$-pasture; $f_{17,16}$-soil erosion; $f_{18,16}$-soil respiration; $f_{8,17}$-irrigation; $f_{18,17}$-CH_4 emissions from surface water; $f_{1,18}$-crop carbon sequestration, atmospheric wet and dry deposition; $f_{15,18}$, $f_{16,18}$, $f_{17,18}$-carbon fixation, and atmospheric wet and dry deposition; z_1-primary and secondary energy, fertilizer; z_2-primary and secondary energy, feed; z_3-primary and secondary energy; z_4-primary and secondary energy, energy raw materials, limestone and chert, virgin paper, paper wood and straw, industrial food raw materials; z_5-primary and secondary energy, limestone; z_6-primary and secondary energy; z_7-primary and secondary energy; z_8-primary and secondary energy, wood; z_9-primary and secondary energy; z_{10}-primary and secondary energy, plastics, food; z_{11}, z_{13}, z_{14}-primary and secondary energy, food, paper, plastics, sugar fats and oils, furniture; y_1-agricultural products, straw, waste plastics; y_2-meat, eggs and milk; y_4-industrial products such as oils and fats, agricultural films; y_5-primary and secondary energy, building materials such as limestone; y_7-secondary energy; y_{11}-waste paper, waste plastics; y_{13}-waste plastics, waste paper, and household waste; y_{17}-carbon in water

Table 5.4 Nodes and paths in the urban nitrogen metabolic network model

	1	2	3	4	5	6	7	8	9	10	11	12	13	14	15	z
1		f_{12}	f_{13}	f_{14}	f_{15}											z_1
2																z_2
3	f_{31}	f_{32}		f_{34}	f_{35}											z_3
4		f_{42}										$f_{4,12}$			$f_{4,15}$	
5		f_{52}		f_{54}												
6		f_{62}														
7		f_{72}														
8		f_{82}														
9		f_{92}														
10	$f_{10,1}$	$f_{10,2}$														
11	$f_{11,1}$	$f_{11,2}$	$f_{11,3}$		$f_{11,5}$					$f_{11,10}$		$f_{11,12}$	$f_{11,13}$	$f_{11,14}$	$f_{11,15}$	
12	$f_{12,1}$	$f_{12,2}$	$f_{12,3}$	$f_{12,4}$	$f_{12,5}$	$f_{12,6}$	$f_{12,7}$	$f_{12,8}$	$f_{12,9}$	$f_{12,10}$			$f_{12,13}$	$f_{12,14}$	$f_{12,15}$	
13												$f_{13,12}$				
14	$f_{14,1}$											$f_{14,12}$				
15	$f_{15,1}$	$f_{15,2}$	$f_{15,3}$	$f_{15,4}$						$f_{15,10}$	$f_{15,11}$	$f_{15,12}$				z_{15}

Note Flows are from the components in the first row to the components in the first column. f_{31}-food waste and household sewage; $f_{10,1}$-treated household sewage; $f_{11,1}$-untreated household sewage; $f_{12,1}$, $f_{12,2}$, $f_{12,3}$, $f_{12,4}$, $f_{12,5}$, $f_{12,6}$, $f_{12,7}$, $f_{12,8}$, $f_{12,9}$-fuel combustion; $f_{14,1}$-pet waste; $f_{15,1}$-household solid waste composting and human and pet waste; f_{12}-chemical products, energy, and pet feed; f_{32}-ammonized feed and energy; f_{42}, f_{62}, f_{72}, f_{82}, f_{92}-energy; f_{52}-fertilizer and energy; $f_{10,2}$-treated industrial wastewater; $f_{11,2}$-untreated industrial wastewater; $f_{15,2}$-industrial solid compost and inorganic fertilizer; f_{13}-meat products; $f_{11,3}$-loss of livestock manure; $f_{15,3}$ - livestock manure; f_{14}-agricultural products; f_{34}-agricultural feed; f_{54}-bait; $f_{15,4}$-crop residues; f_{15}-fish; f_{35}-fish protein; $f_{11,5}$-feed residuals; $f_{11,10}$, $f_{12,10}$-ammonia emissions from sewage treatment plants; $f_{15,10}$-sludge return to fields; $f_{15,11}$-irrigation water; $f_{4,12}$-seeds and agricultural nitrogen fixation; $f_{11,12}$, $f_{15,12}$-sedimentation; $f_{13,12}$-sedimentation and nitrogen fixation in woodland; $f_{14,12}$-sedimentation and nitrogen fixation in grassland; $f_{11,13}$, $f_{11,14}$, $f_{11,15}$-runoff; $f_{12,13}$-denitrification; $f_{12,14}$, $f_{12,15}$-denitrification and volatilization; $f_{4,15}$-nutrients drawn from soil; z_1-transfers of chemical products, food and pet feed; z_2-transfers of ammonia; z_3-transfers of feed; z_{15}-transfers of fertilizer

5.1.4 Urban Material Metabolic Network Models

An analysis that fully accounts for all components of an external environment (e.g., other cities, regions, or countries) becomes very difficult to handle because of its complexity. Thus, to simplify the analysis, it is possible to include all of the flows outside of the urban boundary in the national-scale model by treating the external flows as if they flow to a single external environment. In analyses at the urban scale, previous studies generally did not include areas outside of the urban boundary as separate nodes, but rather took the internal environment as a node. This simplification allows the model to fully consider the pressure of the metabolic actors on the internal environment. With urban development and the improved quantity and quality of data and statistics, it becomes possible to disassemble the urban material metabolism network model into more socioeconomic metabolic actors. This allows a more refined simulation of the socioeconomic sectors involved in the metabolic processes and of the environment as nodes and the material flows among them as pathways.

Based on the differences in the material utilization characteristics and data availability, we can divide an urban metabolic system into its internal environment and seven internal components: the Agriculture, Mining, Industry (manufacturing), Recycling, Household Consumption, Construction, and Energy Transformation sectors. The inclusion of the internal environment increases the model to eight components. However, a significant improvement of the urban transportation sector typically occurs during socioeconomic development, which makes it necessary to consider the role of this sector. the Transportation sector is therefore included to create a final network model with nine nodes (Fig. 5.8).

We can also define the imports and exports as the inflows from the external environment to the metabolic system and the outflows from the metabolic system to the external environment, respectively. Using these components as nodes and the material flows between them as the paths between nodes, we can construct a network model that consists of the nine nodes and 30 paths among them (Table 5.5). Among these, the Internal Environment comprises the natural subsystems within the urban administrative boundaries; Agriculture includes agriculture, forestry, animal husbandry, and fisheries; mining includes the extraction of non-living resources (e.g., coal, petroleum) and their processing, as well as storage associated with the exchange of energy within the city and between the city and its external environment; Industry includes all primary and secondary processing and manufacturing; Energy Transformation comprises the production and supply of electricity and heat, including storage for the conversion and exchange of energy within the city; Construction is the housing and civil engineering construction industry, which also includes the demolition of existing buildings; Recycling is pollutant treatment, waste processing, and recycling; household refers to consumption resulting from the living activities of the resident population; and Transportation is the delivery or transport of goods or passengers to their destination, including land, water, shipping, and pipeline transportation (Li et al. 2019a, b, 2012; Zhang et al. 2013).

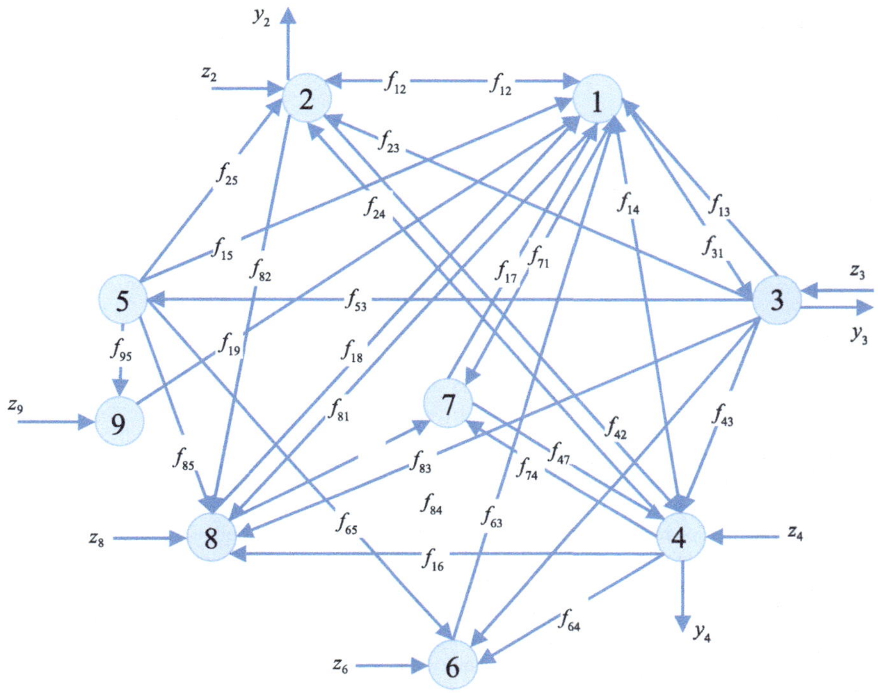

Fig. 5.8 Ecological network model of an urban material metabolic system (*Note* 1-Internal Environment; 2-Agriculture; 3-Mining; 4-Industry; 5-Energy Transformation; 6-Construction; 7-Recycling; 8-Household Consumption; 9-Transportation. f_{ij} denotes the flow from node j to node i, z_i denotes the external flow of matter into node i, and y_i denotes the flow from node i to the external environment)

Cities exist within the context of a larger society (i.e., nation) that surrounds them. This results in exchanges between cities and their surrounding regions at a national scale. This section shows how the urban-scale models previously discussed in this chapter can be modified to function at larger scales. The internal and external natural environment of social metabolic system is comparable to the environment inside and outside the organic body. The formation of metabolism within an organism is due to the exchange of substances in its internal environment. The formation of the external metabolism of an organism is due to the exchange of substances with external environment (Fig. 5.9). It is necessary to consider both of them when studying the holistic metabolism of an organism, as they are in competition. The exchange of substances and energy between metabolic participants in the internal environment constitutes internal metabolism. However, the metabolic activity of a system cannot be sustained by its internal resources alone; external metabolism is also necessary, i.e., the substance and energy exchange between social metabolic actors and their external environment. Both internal and external metabolic processes are important in studies dealing with the substance and energy exchange between

Table 5.5 Inter-node flows in the urban material metabolism network model

	1	2	3	4	5	6	7	8	9	z
1		f_{12}	f_{13}	f_{14}	f_{15}	f_{16}	f_{17}	f_{18}	f_{19}	
2	f_{21}		f_{23}	f_{24}	f_{25}					z_2
3	f_{31}									z_3
4		f_{42}	f_{43}				f_{47}			z_4
5			f_{53}							
6			f_{63}	f_{64}	f_{65}					z_6
7	f_{71}			f_{74}				f_{78}		
8	f_{81}	f_{82}	f_{83}	f_{84}	f_{85}					z_8
9					f_{95}					z_9
y		y_2	y_3	y_4						

Note Flows are from the components in the top row to the components in the first column. f_{12}-agricultural pollutants and hidden agricultural flows; f_{13}-pollutants from the Mining sector and its hidden flows; f_{14}-wastewater, gas, and sludge from Industry; f_{15}-construction residues and solid waste; f_{16}-construction residues and garbage; f_{17}-wastewater, gas, and sludge from Recycling; f_{18}-waste from Household Consumption; f_{19}-pollutant emissions from Transportion; f_{21}-harvesting biomass and its hidden flows; f_{23}-primary energy consumption from Agriculture; f_{24}-chemicals, plastics, and fertilizers, etc. consumed in Agriculture; f_{25}-consumption of secondary energy for Agriculture; f_{31}-minerals extracted from Mining and its hidden flows; f_{42}-agricultural products consumed by industry; f_{43}-minerals and primary energy consumed by industry; f_{47}-renewable resources used in Industry; f_{53}-transformation of primary energy; f_{63}-minerals and primary energy consumed by Construction; f_{64}-processed and manufactured products consumed by Construction; f_{65}-secondary energy consumed by Construction; f_{71}-oxygen consumption for pollutant recycling, etc.; f_{74}-Industry waste that must be treated or discarded; f_{78}-recycled household waste; f_{81}-oxygen consumed by residents breathing; f_{82}-agricultural products consumed by residents; f_{83}-primary energy consumed by residents; f_{84}-processed and manufactured products for Household Consumption; f_{85}-secondary energy for Household Consumption; f_{95}-secondary energy for Transportation consumption; z_2-imports of Agriculture; z_3-imports of Mining; z_4-imports of Industry; z_6-imports of building materials; z_8-imports of consumer goods; z_9-imports of Transportation; y_2-exports of agricultural products; y_3-exports of minerals and primary energy; y_4-exports of industrial products

metabolic actors and their environment, and they have different effects on social metabolic systems. Therefore, in order to reach more comprehensive conclusions, it is sensible to study the environment and its interactions by considering the internal and external environment independently.

Metabolic actors in a system can be classified according to their shared metabolic features. For instance, societal metabolic actors can be clustered into primary industries, advanced industries, recycling industries, and the household sector. Primary industries supply inputs like valuable raw materials to other metabolic players to support their operations. These inputs are used by advanced industrial and household consumption sectors to manufacture goods and services, and the waste (including pollutants) generated by the industrial and household consumption sectors is handed over to the recycling industry for processing and recovery. The exploitation sector and agricultural sector are the two most important sectors of primary industries. They

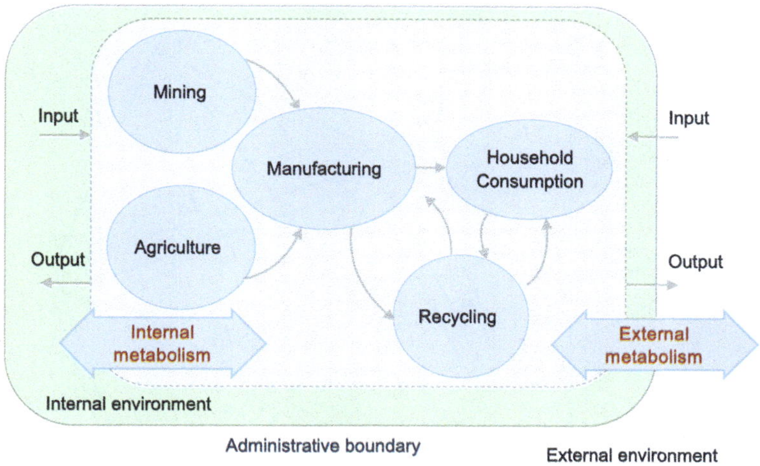

Fig. 5.9 Illustration of the metabolic processes in a societal metabolic system

convert natural resources into raw materials which are later used by the advanced industrial and household consumption sectors. Raw materials are converted into primary (semi-finished) products for immediate consumption (e.g. water, food) or processed by advanced industries (e.g. manufacturing) into more elaborate products that are fit for human to consume. All waste and energy generated during resource exploitation, product processing and manufacturing as well as household consumption can be recovered to enable extra production and consumption activities. In this section, the exploitation sector involves some activities associated with the refining of raw materials, therefore some refined minerals may be served as inputs to other sectors. The manufacturing sector consists of refining raw materials and conventional manufacturing procedures. The metabolic processes of a social system can be generalized to inputs, transformations, cycles, and outputs based on such depiction of the participants which are shown in Fig. 5.9.

The establishment of an ecological network model for the system becomes available after the analysis of the metabolic processes of a social system. For illustration, an integrated model such as the one shown in Fig. 5.10 can be developed based on the descriptions of the conceptual model. For this model, we can divide the social metabolic system into seven parts (environments and sectors): the Internal Environment, the Agriculture sector, the Exploitation sector, the Manufacturing sector, the Household Consumption sector (i.e., citizen domestic life), the Recycling sector, and the External Environment (Zhang et al. 2012).

The model shown in Fig. 5.10 has 26 metabolic pathways defined to represent the flows among the seven constituents of the network. Even though the internal and external environments do not participate in the metabolism, they still need to be treated like nodes in the system in order to allow analysis of the variation in how metabolic participants interact with these two environments. Due to the fact that a specific network component is likely to play several roles, it is essential to identify the

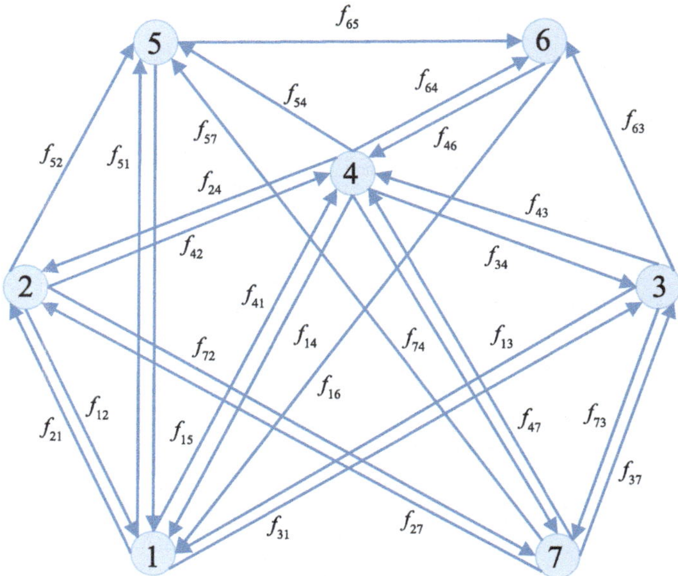

Fig. 5.10 Model of the ecological flows in a societal metabolic system (*Note* 1-Internal Environment; 2-Agriculture sector; 3-Exploitation sector; 4-Manufacturing sector; 5-Household Consumption sector; 6-Recycling sector; 7-External Environment. f_{ij} represents the flow from node j to node i, z_i represents the flow into node i from outside of the material metabolic system, and y_i represents the boundary outflows from node i)

different metabolic flows in each direction in the process of building an ecological network model of a social metabolic system. For such analysis, it is necessary to compute the values of each metabolic flow such as shown in Table 5.6.

5.1.5 Urban Emergy Metabolic Network Models

It is essential to unify the measurements of various units that are used to describe the different flows (i.e., material, energy, and monetary flows) in urban systems, so that they can be integrated within the same analysis. We use the emergy value analysis method in order to tackle this problem (Zhang et al. 2009). Compared with competing methods, the emergy analysis has its own advantage in offering a common foundation for estimating all these flows (material, energy and money) in the same measurement unit: the solar emjoule (seJ). A transformation factor named transformity is used to convert the parameter values into seJ from their original units (e.g., J for energy, t for mass, $ for money). Urban metabolic system consists of both metabolic constituents and internal and external environments. Among them, the internal environment refers to the natural areas within the administrative boundary of the city. Metabolic components are the independent basic material flows in the production and living sectors

Table 5.6 Nodes and paths in a national-scale material metabolic network model

	1	2	3	4	5	6	7
1		f_{12}	f_{13}	f_{14}	f_{15}	f_{16}	
2	f_{21}			f_{24}			f_{27}
3	f_{31}			f_{34}			f_{37}
4	f_{41}	f_{42}	f_{43}			f_{46}	f_{47}
5	f_{51}	f_{52}		f_{54}			f_{57}
6			f_{63}	f_{64}	f_{65}		
7		f_{72}	f_{73}	f_{74}			

Note Flows are from the components in the top row to the components in the first column. f_{21}-input of local agricultural production and forage and oxygen consumed by livestock; f_{31}-local non-renewable resources and mining, and unused household extraction; f_{41}-oxygen consumed by fuel combustion; f_{51}-oxygen consumed by citizens; f_{12}-agricultural indirect flows, including fertilizer and pesticide loss, and the gases emitted by burning agricultural waste and livestock; f_{42}-local agricultural products that must be manufactured for citizens and cannot be consumed directly; f_{52}-local agricultural products that can be consumed directly by citizens and non-commercial energy consumption; f_{72}-export of agricultural products; f_{13}-mining, unused household extraction, and untreated mining waste; f_{43}-local non-renewable resources used in the manufacturing sector; f_{63}-mining waste; f_{73}-export of mining products; f_{14}-untreated waste, gases emitted by combustion, and unused household extraction by the manufacturing sector; f_{24}-agricultural production resources, equipment, and energy consumption; f_{34}-mining equipment and energy consumption; f_{54}-food and living goods and energy consumption by citizens; f_{64}-industrial waste and energy consumption; f_{74}-export of manufactured products; f_{15}-gases produced by people and untreated waste; f_{65}-household waste produced by citizens; f_{16}-final emissions after treatment in the recycling sector; f_{46}-recycling materials; f_{27}-import of production materials and livestock; f_{37}-import of mining products; f_{47}-import of industrial products; f_{57}-import of food and living goods

of the urban metabolic system, including the agricultural, industrial, and household consumption sectors (Zhang et al. 2009). Urban systems are open-ended (i.e., its flows of materials to and from metabolic actors outside of its boundaries) that is, the limited urban space is not sufficient to support the metabolic activities of the city alone and needs the help of the external urban environment. The external environment constantly delivers the "nutrients" to the urban system to maintain its metabolism, and at the same time, it receives the metabolic and other wastes from the city. The external environment consists of all areas outside the administrative boundaries of the city where materials and energy are exchanged with the city. We can divide the urban metabolic system into five parts (the Internal Environment, External Environment, and the Agricultural, Industrial, and Household Consumption sectors) (Fig. 5.11) for the purpose of investigating the influence and the reliance of urban metabolic activities on the internal and external environments.

By analyzing the components of an urban metabolic system, it is possible to determine the direction of flows through the system and define its metabolic pathways. The multiple roles played by the system components create complicated functions for these components, which means that the flow direction among the components is not a linear chain, but rather a network. For this reason, we can construct

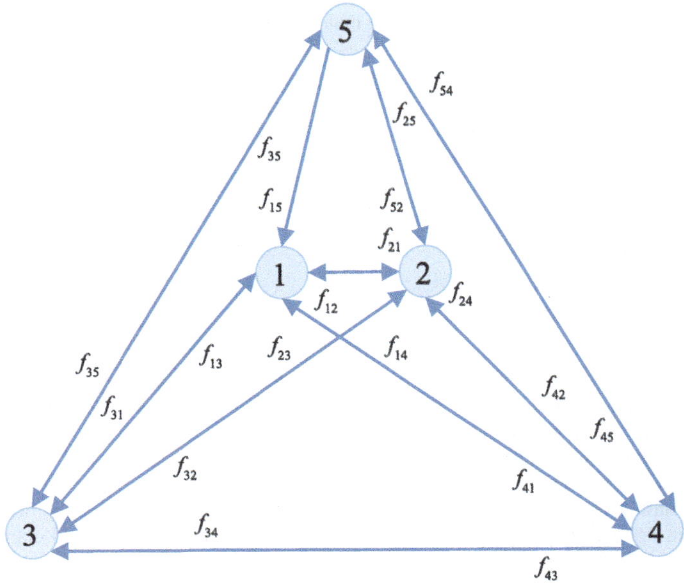

Fig. 5.11 Ecological network model of an urban energy system (*Note* 1-Internal Environment; 2-External Environment; 3-Agriculture; 4-Industry; 5-Household Consumption. f_{ij} represents the flow from node j to node i)

an ecological network model for the urban metabolic system. Ecological networks resemble biological networks, and describe the structure of the flows of materials, energy, and currency between different system components. Using the five metabolic actors shown in Fig. 5.11, we can develop an ecological network model for an urban metabolic system (Zhang et al. 2009). In this model, we can define 20 metabolic pathways that reflect the exchange of flows among these five nodes (Table 5.7).

5.2 Spatially Explicit Models Based on Land Use and Cover Change

5.2.1 Principles for Developing Spatially Explicit Carbon Metabolic Network Models

A highly abstract form of analysis has been thus far presented, in which the metabolic actors have no physical location. However, these actors have specific locations in the real world, and it is necessary to account for the spatial aspects of their distribution. In practice, accounting for specific locations at a relatively fine resolution is computationally infeasible because of the large number of locations in a large city. We

Table 5.7 Inter-node flows of emergy in an urban energy metabolic network model

	1	2	3	4	5
1		f_{12}	f_{13}	f_{14}	f_{15}
2	f_{21}		f_{23}	f_{24}	f_{25}
3	f_{31}	f_{32}		f_{34}	f_{35}
4	f_{41}	f_{42}	f_{43}		f_{45}
5	f_{51}	f_{52}	f_{53}	f_{54}	

Note Flows are from the components in the first row to the components in the first column. f_{21} represents the trans-boundary transfer of pollutants; f_{31} and f_{41} represent the resources provided by the Internal Environment for Agriculture and Industry, respectively; f_{51} represents the service function that the Internal Environment provides for household life; f_{12}, represents the total resource inputs from the External Environment; f_{13}, f_{14}, and f_{15} represent the pollutants discharged by Agriculture, Industry, and Household Consumption, respectively; f_{32}, f_{42}, and f_{52} represent the resource inputs from the External Environment for Agriculture, Industry, and Household Consumption, respectively; f_{23}, f_{24}, and f_{25} represent the outputs of agricultural products, industrial products, and labor services, respectively; f_{43} and f_{53} represent the agricultural raw materials consumed for industrial production and household consumption, respectively; f_{34} and f_{35} represent the industrial products and labor services consumed by agricultural production, respectively; f_{54} represents the industrial products consumed by household consumption; and f_{45} represents the labor services used for industrial production

can instead consider land use and cover types as a proxy for geographic locations, in which they represent the type of ecosystem in which the urban metabolism is located. Carbon provides a good example of such models because of its importance in an urban metabolism and the importance of how an area's land use or cover type influences its ability to receive or transfer carbon.

Urban development has dramatically changed the land surface and caused dramatic changes in land use and cover types (Grimm et al. 2008; Alberti and Marzluff 2004). The resulting complex relationships and their spatial structure can be abstracted as a network, which represents the urban carbon metabolic spatial network based on changes in land use or cover type over time. These changes can be recorded in a land-transfer matrix, which shows the area of a given land use or cover type that changed to (or was transferred to) a different land use or cover type during a given period. If that land use or cover type has a specific storage or metabolism (i.e., transformation) of a material or element, such as carbon, its storage or metabolic capacity can be determined. This can then be multiplied by the area of the change to determine the amount of carbon transfer (i.e., flux) that results from this change. When land is exchanged between metabolic actors, the absorption or release of carbon will also change. The metabolic capacity of the original land use or cover type is then replaced by that of the new type, and the type that received this land accordingly experiences an increase of total metabolic capacity. The change of carbon metabolic capacity by each actor will affect the carbon exchange between the actor and its environment (for carbon, typically the atmosphere), which can be quantified as an increase of the release capacity or absorption capacity of carbon, which represent the output flow y and input flow z. The difference of the carbon metabolic density (ΔCMD) between natural and socioeconomic actors should be considered in the construction of an urban carbon metabolic network model, because this determines the flows through a city's carbon metabolism and transfers among its components, f_{ij} (Fig. 5.12) (Zhang et al. 2016b, 2014a).

A city is a complex network of both socioeconomic and natural activities with a wide range of material and energy exchanges among the metabolic actors. These complex relationships and their spatial structures can be abstracted into a spatially explicit network model. A spatial model of a network's carbon flows that links the system's artificial and natural components can help planners achieve more environmentally sustainable urban spatial patterns by controlling key socioeconomic and ecological processes. Providing support for these activities is fundamental to

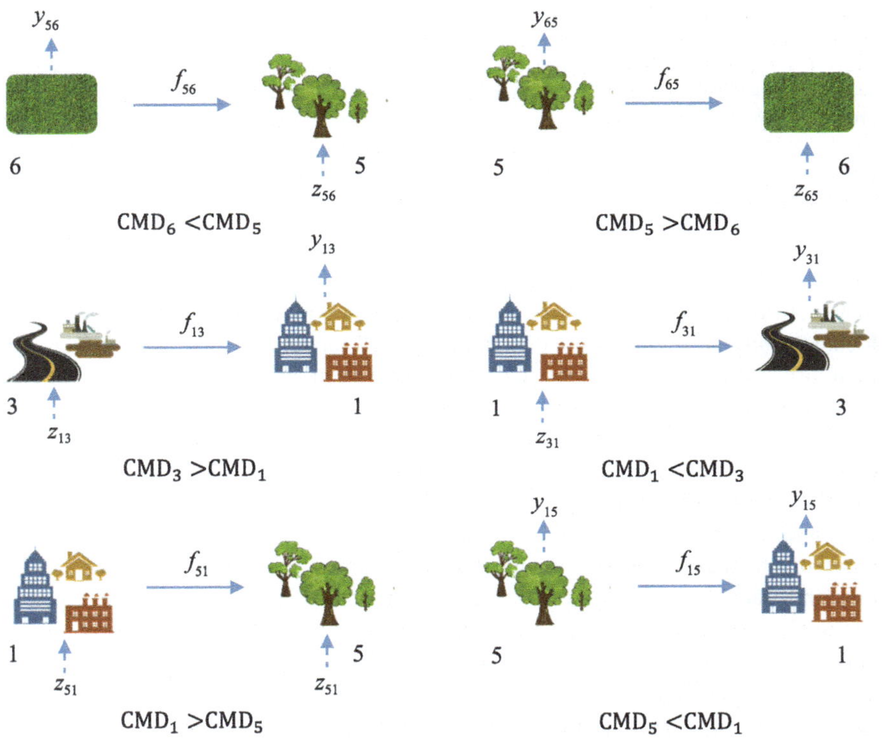

Fig. 5.12 Examples of urban carbon metabolic processes that result from the flows of energy and materials among the nodes of the system (*Note* 1-Urban Land (urban built-up area); 3-Transportation and Industrial Land; 5-Forest; 6-Grassland. f_{ij}, carbon flow from component j to component i; z_{ij}, input flow from component j to component i; y_{ij}, output flow from component j to component i. CMD_i, carbon metabolic density of component i, which is expressed as the change in carbon stock per unit time and per unit area for each component. The first line shows changes between the natural components of the system, the second line shows changes between the socioeconomic components, and the third line shows changes between the natural and socioeconomic components)

the construction of spatially explicit network models, because an improved understanding of the urban metabolism can help planners take measures to adjust the urban spatial patterns by changing the energy consumption and energy flows among the system components. Previous carbon network research only modeled the exchanges of carbon between the atmosphere and biosphere from the perspective of carbon emissions and sequestration (i.e., by only considering the overall flows), but this makes it difficult to understand the network's spatial structure. In addition, models based on network environ analysis, which reflects the mathematical proximity of nodes in a network rather than their spatial distribution, lack a spatially explicit component and can therefore not determine the directions and locations of urban evolution or reveal possible adjustment mechanisms. A landscape ecology network, which reflects the evolutionary changes between land use and cover types, can be used to identify possible paths among system components, but can only be used to describe the natural changes that result from ecosystem succession. It is therefore necessary to develop a model that accounts for both the paths and flows along those paths in a spatially explicit manner. To meet this need, a spatial model of a network's carbon metabolism can be developed based on the increases and decreases of carbon stocks along the links among the urban components in response to changes in land use or cover type.

Urban development dramatically changes the land surface and land use (Grimm et al. 2008; Alberti and Marzluff 2004), resulting in frequent carbon transitions as land use or cover type changes. Therefore, a city-wide carbon transition must be based on changes in land use or cover type within a network. To illustrate how this works, we can develop a model that describes the changes in carbon metabolism capacity, carbon transitions, and transfers of land among land use and cover types. Meanwhile, the development of a spatially explicit model can be supported by the use of a land-use transfer matrix. Changes between any two components may result in changes in carbon metabolism capacity (including carbon uptake or release capacity) in the affected land. When converting one landscape type to another, this is analogous to the output of that area of land, which would reduce the carbon metabolism capacity of the original land type. Conversely, that output of land is analogous to an input (i.e., increase in area) for a different land type. The carbon metabolic capacity then changes by an amount equal to the change in carbon metabolic density of the land that receives the input, multiplied by the area of land received as an input (Zhang et al. 2016b). For example, if 10 hm^2 of forest changes to grassland, the carbon absorption or release by those 10 hm^2 of forest is accounted for using the metabolic capacity of the new grassland (Zhang et al. 2016b, 2014a).

Landscape components and the external environment (here, the atmosphere) exchange the carbon metabolic capacity in form of input flows (z) and output flows (y) as a result of changes in land use or cover type. The carbon stock change per unit time and per unit area for each component [$kg/(m^2 \cdot a)$] is called carbon metabolic density (CMD), and we use the equivalent carbon flow along each pathway from component i to component j (i.e., f_{ij}) generated by the difference in carbon metabolic density (ΔCMD) between the two components involved in the exchange to express the carbon flux along each path in the network. Flows in the network can be classified into beneficial flows (absorption) that alleviate carbon metabolic imbalances and harmful flows that exacerbate systemic imbalances (release) based on the division of carbon metabolic capacity (carbon release and carbon absorption). It is appropriate to consider both processes with a spatial model of carbon metabolism in urban networks which connects the natural and artificial components of the system together and also represents the interactions between different components (Fig. 5.12).

The difference in metabolic density between any two components of the system results in a flow between components. To illustrate, the carbon that a given area of grassland absorb is usually less than the same area of forest, while a given area of urban land will release less carbon than the same area of transportation and industrial land. However, the relative area of each land type determines the net absorption (i.e., beneficial flow) or release (i.e., harmful flow, which represents carbon loss); the calculation of the total capacity of carbon absorption or release requires multiplying this area by the metabolic density per unit area. Therefore, based on the calculations, specific changes in land use or cover type may produce beneficial or harmful flows. Flow can occur along three path types: between natural system components; between artificial components; as well as between natural and artificial components.

These flows can be either beneficial or harmful. Figure 5.12 demonstrates the beneficial flow generated by the carbon transition between the two natural components. A reduction in grassland area (i.e., output) leads to a decrease in the total carbon absorption capacity of that land use type, which in turn leads to carbon release (y_{56}) and a reduction in carbon stocks. When the output of land changes into forest (i.e., an input), the aggregate carbon absorption capacity will increase, which in turn leads to carbon absorption (z_{56}), and the carbon stock also increases (i.e., a net increase of benefits). The reverse processes occur when the area of grassland has a net increase (z_{65}) while the area of forest has a decrease (y_{65}). The middle row in Fig. 5.12 is an example of a carbon transition between two manual system components. The change from transportation and industrial land to urban land (outputs of transportation and industrial land and inputs of urban land) leads to carbon absorption (z_{13}) and a net increase in benefits. On the contrary, there is a carbon release (y_{13}) when urban land is transformed into transportation and industrial land, leading to a net loss of benefits.

The last row in Fig. 5.12 represents the changes between natural and artificial system components. The capacity for carbon metabolism of these two components are different, which leads to a different process compared to the first two exchange types. When urban land changes to forest (i.e., an output of urban land and input of forest land), the carbon release capacity decreases, resulting in carbon absorption (z_{15}) as the forest area increases (i.e., an input of forest), thereby increasing the carbon absorption capacity, resulting in net carbon absorption (z_{15}) and an increase in benefits (Zhang et al. 2016b). On the contrary, when the forest becomes urban land, the capacity for carbon release will increase, leading to a loss of benefits.

5.2.2 Spatially Explicit Models of an Urban Carbon Metabolic Network

Almost all human activities require land as a fundamental carrier. Urban land, rural land, and transportation and industrial land are the primary types of artificial land use and therefore are the major carriers of carbon emissions. Compared to this, the natural components of urban systems, such as forests, grasslands and surface waters, have a significant contribution to carbon sequestration. All of the associated socioeconomic and natural processes represent the carbon shift in the biosphere as well as show the carbon exchange due to land use and cover changes. Based on the land transfer matrix described in the previous section, it is possible to consider carbon transition processes as an important part of urban metabolism, and to develop a spatially explicit network model of carbon metabolism that links the natural and artificial components based on the land's carbon metabolic ability (Xia et al. 2017). The spatial network developed in this study includes the processes of carbon metabolism generated by socioeconomic and natural activities, with a focus on carbon transitions between and within the system's natural and artificial components. Potential carbon transfers within the biosphere are also covered by the spatial network simultaneously. The model's quantification is on the basis of socioeconomic metabolic processes, natural metabolic processes, and amounts of carbon transferred to the biosphere from both types of processes. The carbon metabolism generated by socioeconomic activities is a part of socioeconomic metabolic processes, which mainly comes from urban land, transportation and industrial land, cultivated land, and rural land. The accounting items range from the energy consumption in industry and by life to human respiration, and include the carbon emissions from the livestock breeding industry, fertilizer applications, and mechanical cultivation (Xia et al. 2018). Natural metabolic processes mainly involve the carbon absorption from the natural components of cities, such as forest, grassland, surface water, and cultivated land (Fig. 5.13).

The quantification of the flows through the urban network is based on accounting for the carbon metabolic density of each land use or cover type. The carbon metabolic density equals the carbon metabolic rate per unit area. The flows through the spatial network are quantified by calculating the difference in carbon metabolic density between the output and input ends of a component. Previous research provided spatially explicit digital land-use data at a scale of 1:100,000 for China from 1980 to 2020 (Xu et al. 2014; Liu et al. 2009). Land-use data from consecutive years was overlaid using geographical information system software such as ArcMap or ArcGIS to create a land-use transition matrix. The carbon metabolic density (CMD) includes the carbon absorption metabolic density (CMD_A) and carbon emission metabolic density (CMD_E) for each component. The carbon metabolic difference (ΔCMD) between two components for which the land use changed is based on the carbon flow accounting results. The total carbon flow f_{ij} along each path between two nodes can be obtained by multiplying the area that was transferred between two land use types by the corresponding ΔCMD value, $f_{ij} = \Delta CMD \times \Delta A$, where ΔA represents the land area change. A flow matrix for a spatial model of the urban network's carbon

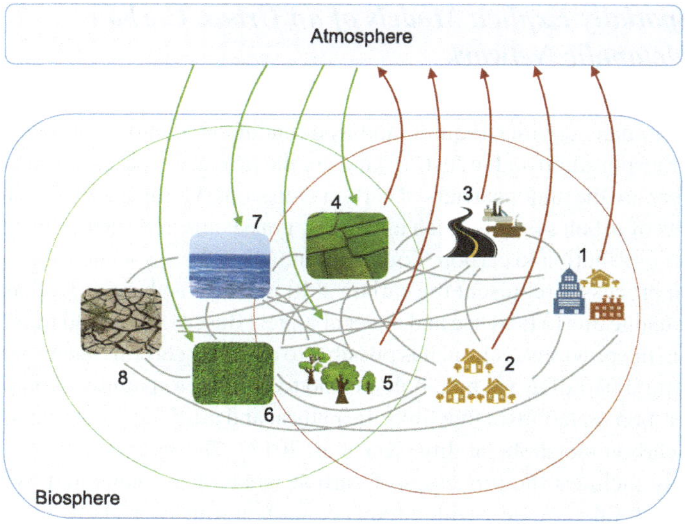

Fig. 5.13 Illustration of the urban carbon metabolism model (*Note* 1-Urban Land; 2-Rural Land; 3-Transportation and Industrial Land; 4-Cultivated Land; 5-Forest; 6-Grassland; 7-Surface Water; 8-Bare Land)

metabolism can then be built based on the carbon flows (f_{ij}), and is called the direct flow matrix ($\boldsymbol{F}$). Data on metabolic densities can be obtained from the literature, government statistical reports, and databases, or from a combination of multiple sources (e.g., empirical research data on carbon flows combined with government statistics on the area of each land use).

A spatial model of the carbon metabolism of a network can be developed that contains both natural and artificial components as the nodes and focuses on the carbon fluxes along flow paths between these nodes. Figure 5.14 illustrates the connections among the model's eight components: Urban Land, Rural Land, Transportation and Industrial Land, Cultivated Land, Forest, Grassland, Surface Water, and Bare land. Here, Bare Land refers to unused land with no vegetation cover and no sign of human activities (e.g., bare exposed rock), and is therefore assumed to have no significant carbon flow. Each component is associated with the entire network through several state variables (Xia et al. 2016; Zhang et al. 2016b).

Furthermore, we can subdivide the natural components and cultivated land into several sub-components, resulting in the 18-node model shown in Fig. 5.15 (Xia et al. 2018). A similar subdivision can be performed for any of the city's other components if sufficient data are available. The new model demonstrates the complex interrelationships among the sub-components of a city's natural components and cultivated land that results from the city's complex spatial pattern.

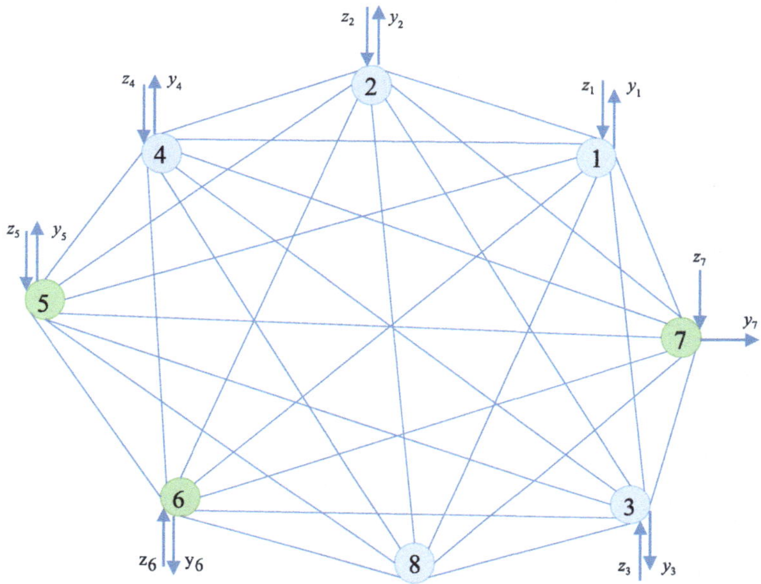

Fig. 5.14 Illustration of a spatially explicit eight-node model of an urban carbon metabolism network (*Note* 1-Urban Land; 2-Rural Land; 3-Transportation and Industrial Land; 4-Cultivated Land; 5-Forest; 6-Grassland; 7-Surface Water; 8-Bare Land. z_i represents the input flow to node i, and y_i represents the output flow from node i)

5.3 Network Models Based on Input–Output Tables

Input–output (IO) tables represent datasets that show the inputs of materials or energy, and their outputs, for each component of an urban metabolism. As a result, they represent a fundamental dataset for research on the flows of materials and energy through an urban system. However, because much of the statistical data that governments collect in these tables is recorded in monetary terms, it becomes necessary to convert this monetary data into flows of mass or energy when the goal is to track these flows through an urban network.

5.3.1 Development of an Input–Output Table

To convert the economic data into physical data, it is necessary to use an embodied ecological element intensity factor that transforms the data in monetary input–output tables into physical input–output tables (Chen 2011); the same principle can be applied to energy. In this context, "ecological elements" represents both chemical elements such as carbon and nitrogen, and substances such as water, which are essential components of the metabolic flow of urban ecosystems. This coefficient

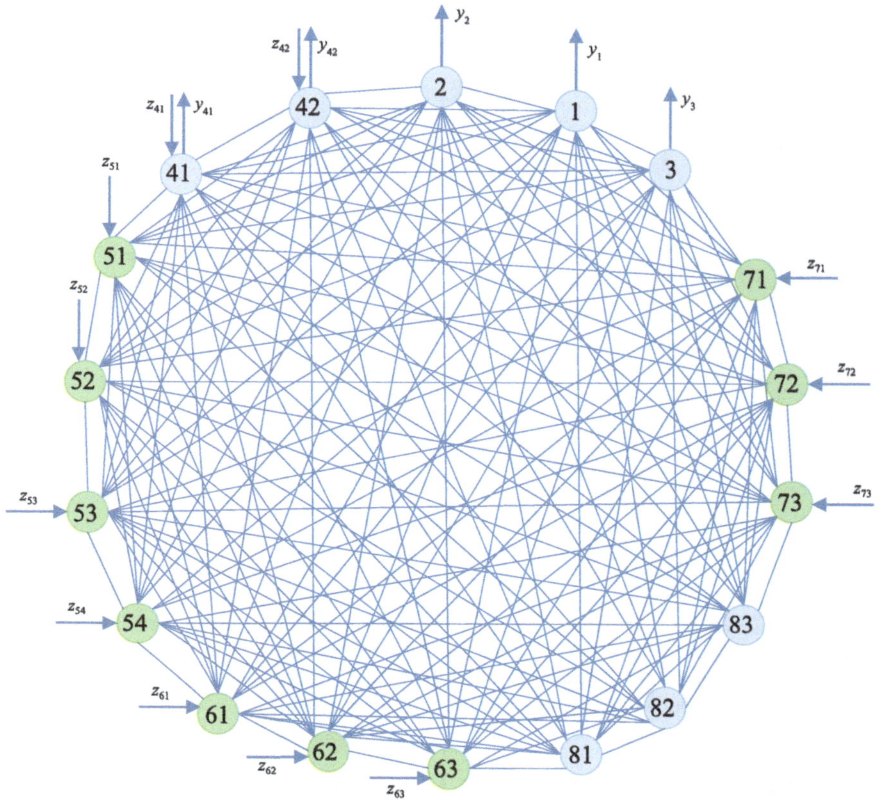

Fig. 5.15 Spatially explicit 18-node model of an urban carbon metabolism network (*Note* 1-Urban Land; 2-Rural Area; 3-Transportation and Industrial Land; 41-Paddy Land; 42-Dry Cultivated Land; 51-Forest; 52-Shrubland; 53-Sparse Woodland; 54-Other Woodland; 61-High-coverage Grassland (>50% vegetation cover); 62-Medium-coverage Grassland (20–50% vegetation cover); 63-Low-coverage Grassland (5–20% vegetation cover); 71-River; 72-Reservoir; 73-Wetland; 81-Sand; 82-Barren Earth; 83-Bare Exposed Rock. z_{ij} represents the input flow from *j* to *i*, and y_{ij} represents the output flow from *j* to *i*)

expresses the number of ecological elements per unit of monetary value implied in the exchange of that element within the system, which enables us to evaluate the intensity of ecological element consumption for each sector or component of the system. Sectors or components of urban metabolic processes can not only consume ecological elements directly, but also can use them indirectly, this is due to the fact that the production and consumption of intermediate products that are subsequently consumed by a different component of the system. This conversion factor, when multiplied by the economic value of a flow, results in the integral flow of ecological elements implied in the entire life cycle of a given product within the system, from creation, transport and recycling to reuse (this is similar to the emergy analysis described in Sect. 4.3 of Chapter 4). The consumption of ecological elements to maintain the operation of the system involves the consumption of primary ecological elements (e.g., raw materials) as well as the implied flows of ecological elements. The implied flows can be determined by multiplying the embodied ecological element intensity factor (i.e., the flow per unit of money) by the corresponding economic flow, thereby generating physical input–output tables from the monetary tables (Zhang et al. 2014c).

We can begin with the development of an input matrix for the initial ecological elements which will be used as the key elements of a monetary input–output table. Then, we can construct input–output tables that integrate the monetary and physical flows of these inputs, which are the basis for all subsequent flows among sectors and components. In our study period, each year's order of the matrix (i.e., its dimensions) for the economic input–output table (i.e., the number of rows and columns in the matrix) is $n \times n$, where n stands for the number of sectors in the system. The input matrix of ecological elements is in the order of $m \times n$, where m shows the number of types of ecological elements (including both resources and waste). We can divide the input–output table into a value module (including monetary flows) and a physical module (including the flows of ecological elements) without changing the structure of the initial table. Based on this data, physical-monetary input–output tables that can identify socioeconomic and environmental flows in urban metabolism can be constructed. Although this can be confusing, input–output tables use a different definition of flow directions from the notation we have used to describe all other flows: ij always represents a flow from i to j (Fig. 5.16).

The value of the flows into and out of sector i and the specific processes are described in Fig. 5.17. Based on these flows, it is possible to establish the ecological element equation for sector i. In addition to the direct ecological elements consumed by sector i, the sector also indirectly consumes ecological elements that are embodied in intermediate products that sector i receives from other sectors. Therefore, when a sector exchanges a value flow with other sectors, these value flows embody ecological element flows (Lenzen 1998).

Taking a regional sector i as an example, it is possible to create an equilibrium equation for sector i based on its flow of money. In addition to direct material consumption, sector i also consumes materials indirectly when it consumes intermediate products produced by other sectors, so that the exchange of value flows with other sectors implies the movement of substances (Lenzen 1998). The economic equation for sector i (i.e., monetary values) is

$$\sum_{j=1}^{n} x_{ji} + d_{\mathrm{M},i} + d_{\mathrm{E},i} + w_i = \sum_{j=1}^{n} x_{ij} + y_{\mathrm{M},i} + y_{\mathrm{L},i} + y_{\mathrm{C},i} + y_{\mathrm{E},i} \tag{5.1}$$

The ecological element equilibrium equation for sector i is then established as follows:

$$z_{ki} + \sum_{j=1}^{n} \varepsilon_{kj} x_{ji} + \varepsilon_{\mathrm{M},ki} d_{\mathrm{M},i} + \varepsilon_{\mathrm{E},ki} d_{\mathrm{E},i} = \sum_{j=1}^{n} \varepsilon_{ki} x_{ij} + \varepsilon_{ki} y_{\mathrm{M},i}$$

Input \ Output		Intermediate use				Final use			
		Sector 1	Sector 2	…	Sector *n*	Consumption	Capital formation	Outputto external regions in the same country	Exports
Intermediate input	Sector 1	x_{11}	x_{12}	…	x_{1n}	$y_{1,L}$	$y_{1,C}$	$y_{1,E}$	$y_{1,M}$
	Sector 2	x_{21}	x_{22}	…	x_{2n}	$y_{2,L}$	$y_{2,C}$	$y_{2,E}$	$y_{2,M}$
	⋮	⋮	⋮	⋮	⋮	⋮	⋮	⋮	⋮
	Sector *n*	x_{n1}	x_{n2}	…	x_{nn}	$y_{n,L}$	$y_{n,C}$	$y_{n,E}$	$y_{n,M}$
Input from external regions in the same country		$d_{E,1}$	$d_{E,2}$	…	$d_{E,n}$	$d_{E,L}$			
Imports		$d_{M,1}$	$d_{M,2}$	…	$d_{M,n}$	$d_{M,L}$			
Non-industrial input		w_1	w_2	…	w_n				
Initial resources inputs	Category 1	z_{11}	z_{12}	…	z_{1n}				
	Category 2	z_{21}	z_{22}	…	z_{2n}				
	⋮	⋮	⋮	⋮	⋮				
	Category *m*	z_{m1}	z_{m2}	…	z_{mn}				

Value module

Physical module

Fig. 5.16 The basic form of a monetary-physical input–output table (*Note* x_{ij} represents the value flows from sector i to sector j; $d_{E,i}$ represents the value flow from the external environment (parts of China outside the system being studied) into sector i; $d_{M,i}$ represents the value flow (import) from the foreign environment into sector i; $y_{i,C}$ represents the total capital formation in the system itself; $y_{i,E}$ represents the value flow from sector i to the external environment (parts of China outside the system being studied); $y_{i,L}$ represents the final consumption value flow for sector i within the system itself; $y_{i,M}$ represents the value flow (export) from sector i to sectors outside of China; w_i represents the non-industrial input value flow from labor and government services into sector i; and z_{ki} represents the flow of resource element k ($1 \leq k \leq m$) into sector i. If z is the waste emitted by sector i, the subscript is ik)

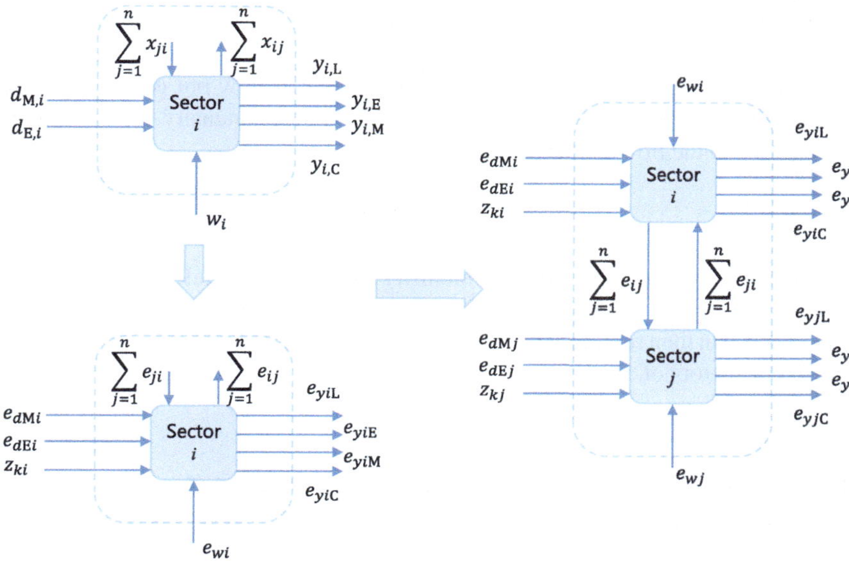

Fig. 5.17 The accounting relationships for the flow values and element flows among the sectors (*Note* The "d" variables refer to the variables defined in Fig. 5.16. The ecological elements embodied in the value flows are e_{ji} for x_{ji}, e_{ij} for x_{ij}, e_{dEi} for $d_{E,i}$, e_{dMi} for $d_{M,i}$, e_{yiC} for y_{yiC}, e_{yiL} for $y_{i,L}$, e_{yiE} for $y_{i,E}$, e_{yiM} for $y_{i,M}$, and e_{wi} for w_i. The w_i parameter in the input–output diagram is an added value that represents a weight used to adjust z_{ki}, which represents the value of the initial ecological element input into the system, so that the implied ecological element e_{wi} of w_i equals 0 (Duchin 2009))

$$+ \varepsilon_{ki} y_{L,i} + \varepsilon_{ki} y_{C,i} + \varepsilon_{ki} y_{E,i} \quad (5.2)$$

where z_{ki} represents the flows of resource or pollutant element k into sector i.

Equation (5.2) can be simplified as follows:

$$z_{ki} + \sum_{j=1}^{n} \varepsilon_{kj} x_{ji} = \varepsilon_{ki} \left(\sum_{j=1}^{n} x_{ij} + y_{L,i} + y_{C,i} + y_{E,i} + y_{M,i} - d_{M,i} - d_{E,i} \right) \quad (5.3)$$

that is

$$z_{ki} + \sum_{j=1}^{n} \varepsilon_{kj} x_{ji} = \varepsilon_{ki} X_i \quad (5.4)$$

Let us assume that $h_{ji} = x_{ji}$, $\boldsymbol{H} = [h_{ij}]_{n\times n}$, $\boldsymbol{Z} = [z_{ki}]_{m\times n}$, $\boldsymbol{\varepsilon} = [\varepsilon_{ki}]_{m\times n}$, $g_{ji} = X_i$ (for $i = j$) or $g_{ji} = 0$ (for $i \neq j$), and $\boldsymbol{G} = [g_{ki}]_{m\times n}$ for an urban metabolic system that includes n sectors and m types of ecological elements. Then, Eq. (5.4) can be presented in matrix form in terms of the embodied ecological element intensity matrix ($\boldsymbol{\varepsilon}$):

$$\boldsymbol{Z} + \boldsymbol{\varepsilon H} = \boldsymbol{\varepsilon G} \tag{5.5}$$

where $\boldsymbol{Z}$ represents the resources and waste that enter or flow out of a component of the metabolism, $\boldsymbol{\varepsilon}$ represents the embodied ecological intensity matrix, and $\boldsymbol{H}$ represents the value flow matrix. Then

$$\boldsymbol{\varepsilon} = \boldsymbol{Z}[\boldsymbol{G} - \boldsymbol{H}]^{-1} \tag{5.6}$$

The embodied ecological element intensity matrix can be calculated by multiplying the sector value by the corresponding intensity coefficient. The material input–output matrix can then represent the relationships for material use due to flows among the different sectors of an economy.

5.3.2 Compilation of the Input–Output Table Based on the Material Consumption Intensity Coefficient

The intensity coefficient for an economy, city, or sector can be estimated as a country's, city's, or sector's resource consumption and pollutant emission divided by its total output. The result can be used to convert the value-based data (economic flows) into material data (e.g., CO_2 flows), which represent the flow along each path in the network. In China, much of the data required for such an analysis can be obtained from multi-regional input–output tables and the corresponding price data. In this way, it is possible to calculate material flows and exchanges between sectors or regions that result from transfers of goods and services. Taking sector j as an example, the intermediate and final use of its production and consumption are added together to obtain the sector's total output, which is then multiplied by the sector's material consumption intensity. This calculation transforms the monetary input–output table into a material table. Here, input–output tables use a different definition of flow directions, as ij represents a flow from i to j. The formula is

$$f_{ij} = x_{ij} \times m_j \tag{5.7}$$

$$m_j = c_j / x_j \tag{5.8}$$

where f_{ij} represents the material transfer from sector i to sector j, x_{ij} represents the capital flow between sector i and sector j, m_j represents the material transfer coefficient for sector j, c_j represents the material consumption or pollutant emission by sector j, and x_j represents the total output of sector j.

5.3.3 Analogy Between Trophic Levels and Metabolic Network Models

The urban metabolism in physical input–output tables can be divided into production (industrial) and consumption components. The environment consists of both the natural environment within city's administrative boundaries (in China, this covers large rural areas surrounding the built-up urban core), as well as the economic entities and external areas located outside the administrative boundaries. It is obvious that the environment offers the needed support to the socioeconomic system, therefore the study of urban metabolic systems should not only focus on the exchange of materials and energy among the industrial sectors within the system and between the industrial and consumer sectors, but also take into account the inputs from and outputs to the environment.

In the original research that was the basis for this section (Zhang et al. 2014b), the only consumption sector was household consumption (which, for the sake of simplicity, included government consumption). This stands for the final consumption component in the physical input–output table. The definition of urban metabolic network model is based on the division of the system's components and the relationships among the components in the physical input–output table. In this model, the nodes stand for different components and the directional line between two nodes indicates the material exchange between the nodes. By using input–output method, it is possible to define the inputs and outputs for the exchanges with the environment and the exchanges of ecological elements among the system's components.

The insights provided by the models described earlier in this chapter can be enhanced by comparing with the trophic levels in a food web or food chain of natural ecosystem. Building on this analogy, we can use a bottom-up approach to cluster the components of the urban metabolic system into groups of the same nutrient level (i.e., producers, consumers, and reducers). We can define the functions of components based on this division, which in turn, help determine their positions in the ecological hierarchy. Producers in the urban metabolic system refer to the components that can use the basic elements of natural environment (raw materials such as water and minerals) to generate primary products. Agriculture and mining are among these producers. In comparison, consumers refer to components that can make use of primary products to generate advanced products. Primary manufacturing, advanced manufacturing, and material and energy transformation are among these. Consumers also include components that directly consume terminal products, tertiary industries that do not produce products (e.g., service industries), but that instead utilize the products produced by producers, and the construction sector, which transforms resources into capital stock (Zhang et al. 2014c). According to this description, we can divide the city's components into eight sectors: agriculture, mining, primary manufacturing, advanced manufacturing, material and energy transformation, construction, services, and household consumption (Table 5.8). Note that the categories in this table are based on the names of the categories in China's statistical databases; other countries may use different names and categories.

Table 5.8 The eight metabolic actors and their composition

Nodes (metabolic actors)	Node composition	Specific materials
Agriculture	Agriculture, Forestry, Animal Husbandry, Fisheries	Agricultural products, forest products, animal husbandry products, fisheries products
Mining	Coal mining and washing	–
	Petroleum and gas mining	–
	Metal mining	Ferrous and non-ferrous metals
	Non-metallic materials and other mining	Non-metallic mineral extraction products, mining support services and other mining products
Primary Manufacturing	Food processing and tobacco	Grain products, feed products, vegetable oil products, sugar, and sugar products, slaughtering and meat processing products, aquatic products, vegetables, fruits, nuts and other processed non-staple foods, convenience foods, dairy products, condiments, fermented products, other foods, alcohol and alcohol beverages and refined tea products, tobacco products
	Textiles	Cotton, inorganic fiber textiles and dyeing and finishing products, wool textiles and dyeing and finishing products, hemp, silk textiles and processing products, knitting, or crocheting and their products, textile finished products, textile clothing and apparel, leather, fur, feathers and their products, shoes
	Wood processing and furnishing	Wood products and wood, bamboo, rattan, palm and grass products, furniture
	Papermaking, printing, stationery, etc	Paper and paper products, reproductions of printed and recorded media, cultural, educational, artistic, sporting, and recreational supplies
Advanced Manufacturing	Petroleum refining, coking, etc	Refined petroleum and nuclear fuel processing products, coking products

(continued)

Table 5.8 (continued)

Nodes (metabolic actors)	Node composition	Specific materials
	Chemical industry	Basic chemical raw materials, fertilizers, pesticides, paints, inks, pigments, and similar products, synthetic materials, special chemical products and explosives, pyrotechnics and fireworks, daily chemical products, pharmaceutical products, chemical fiber products, rubber products, plastic products
	Non-metallic products	Cement, lime and gypsum, cement and similar articles, bricks, tiles, stone and other building materials, glass and glass products, ceramic products, refractory products, graphite, and other non-metallic mineral products
	Metallurgic products	Steel, iron and its castings, steel rolling products, ferroalloy products, nonferrous metals and their alloys and castings, nonferrous metal calendaring products
Advanced Manufacturing	General and specialist machinery	Boilers and power generation equipment, metal processing machinery, material handling equipment, pumps, valves, compressors and similar machinery, cultural and office machinery, other general equipment
	Specialized equipment manufacturing	Special equipment for mining, metallurgy, construction, the chemical industry, timber harvesting, special equipment for non-metal processing, special machinery for agriculture, forestry, animal husbandry, and fisheries, other special equipment
	Transport equipment	Complete vehicles, auto parts and accessories, railway and urban rail transport equipment, ships and related installations, other transport equipment

(continued)

Table 5.8 (continued)

Nodes (metabolic actors)	Node composition	Specific materials
	Electrical equipment	Electrical machinery, power transmission and distribution and control equipment, wires, cables, optical cables and electrical equipment, batteries, household appliances, other electrical machinery and equipment
	Communications equipment, computers, and other electronic equipment manufacturing	Computers, communications equipment, radio and television equipment and radar and ancillary equipment, audio-visual equipment, electronic components, other electronic equipment
	Instruments and meters	–
	Other manufacturing	–
	Scrap waste	Waste resources and waste materials recovered and processed products
	Metal products, machinery and equipment repair services	
Material and Energy Transformation	Electricity and hot water production and supply	
	Gas production and supply	–
	Water production and supply	–
Construction	–	Housing construction, civil engineering construction, building installation, architectural ornamentation, and other building services
Services	Wholesale and retailing	–
	Transport and storage	Railway transport, road transport, water transport, air transport, pipeline transport, loading, unloading, and transportation agents, warehousing, postal services
	Hotels and restaurants	Accommodation, catering
	Information transmission, software, and information technology services	Telecommunications and other information transmission services, software and information technology services, money, finance, and other financial services, capital market services, insurance
Services	Financial services	–

(continued)

Table 5.8 (continued)

Nodes (metabolic actors)	Node composition	Specific materials
	Real estate	–
	Leasing and commercial services	–
	Scientific research	Research and development, professional and technological services, science and technology promotion and application services
	Water, environment, and public facilities management	Water management, ecological protection and environmental management, public facilities management
	Other services	Household services and other services
	Education, health, and social work	–
	Culture, sports, and entertainment	News and publishing, radio, television, film and video recording production, culture and the arts, sports, entertainment
	Public administration, social security, and social organizations	–
Household Consumption	Household consumption and government consumption	–

Note Categories are based on the accounting system used by the National Bureau of Statistics of China

The ecological inputs sub-table in the physical and monetary input–output table represents the resources provided or the pollutants absorbed by the environment. For example, the initial input resources for a sector in the ecological elements obtained from the natural ecosystem are resources. In general, there are four resource types: biological resources (e.g., crops, plants, agricultural products), energy (mining), non-energy minerals (metals and non-metals), and water. In China, the current data allows for a further breakdown of these four categories into 20 particular materials. Moreover, the natural ecosystem can absorb six types of waste (wastewater, sulfur dioxide, smoke, dust, solid waste, and carbon dioxide). The material balance equation (which ensures conservation of mass in the calculations) allows calculation of the material flows among sectors, thereby providing further detail in an urban or regional metabolic network model (Zhang et al. 2014c).

5.3.4 Compilation of the Input–Output Table Based on the Energy Consumption Intensity Coefficient

In the ecological input sub-table of the physical and monetary input–output table, I have only considered energy. Using this approach (described in Sect. 5.3.1), direct energy flow coefficients for all sectors can be obtained. The economic data can then be multiplied by the corresponding energy coefficient to convert the monetary data into the corresponding energy units. These flows represent the energy exchanges among pairs of sectors (i.e., with no intermediates, so the path length is 1); accounting for flows with a path length > 1 can be done using ecological network analysis (described in Sect. 5.4). These flows are the basis for developing a network model of the system, which can range in scale from a single city to a region or group of regions. To demonstrate this approach to the analysis, we can use the results of a previous study (Zhang et al. 2016c), in which my research group divided China's 30 provinces into seven regions based on the administrative divisions defined by the Chinese government: Northeastern, Northern, Eastern, Central, Southern, Southwestern, and Northwestern.

Using China's multi-regional input–output table, a network model of energy metabolism can be constructed that reflects the energy exchanges among the seven regions (Shi and Zhang 2012; Li 2010). The network model was constructed based on the assumption that the technology level used to produce imported products is equal to the technology level used to produce household products; that is, the imported and household products are not distinguished based on their energy contents (however, differences in energy contents could be quantified in future research to support a study of ways to reduce energy consumption in inter-regional or international trade.) Using each of China's seven regions as nodes in the network and energy transfer between regions as paths between the nodes, the regional energy metabolism network model shown in Fig. 5.18 can be constructed (Zhang et al. 2016c, 2015a).

The same approach can be used for smaller regions, such as the group of cities in an urban agglomeration such as the Beijing-Tianjin-Hebei urban agglomeration. To demonstrate this, we can create an energy flow network for the agglomeration based on a multi-regional input–output table that includes the two cities and Hebei Province. The data from this table can be used to establish a steady-state network model (i.e., a model in which in which the inputs for each city or sector are balanced by the outputs) for the relationships among the three regions and their external environment (Fig. 5.19). Each of the cities functions as a node in the model, and the flows among the nodes follow the paths shown at the left of Fig. 5.19 (Zheng et al. 2017). We can also establish a steady-state model that describes the relationships among the five sectors within each region (the right side of Fig. 5.19). In this model, the sectors in each region function as nodes, and the energy flows among sectors follow paths that identify the roles of the sectors in each region (Zheng et al. 2018, 2017; Zhang et al. 2016d). In the original research, we simplified the model so that we did not include transfers between a sector in one city and one or more sectors in another city; this

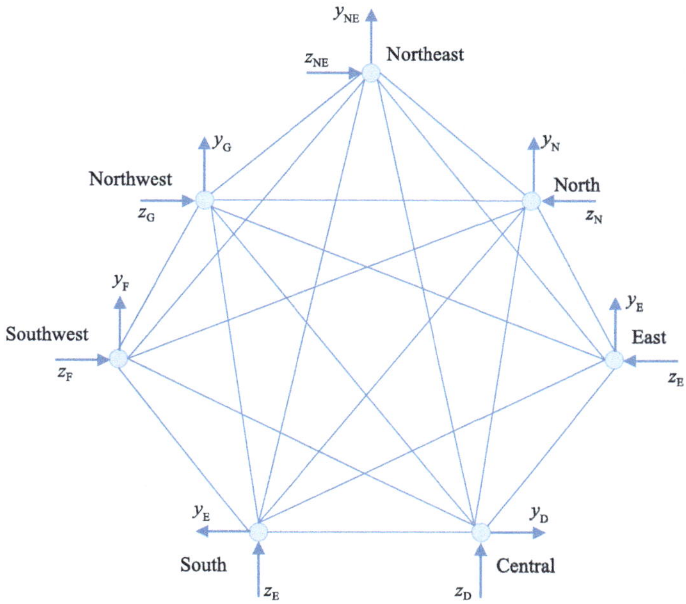

Fig. 5.18 Model of the latent energy metabolic network for the seven regions of China (*Note* Northeast includes Liaoning, Jilin, and Heilongjiang; North includes Beijing, Tianjin, Hebei, Inner Mongolia, and Shanxi; East includes Shanghai, Zhejiang, Jiangsu, Shandong, Anhui, and Fujian; Central includes Henan, Hubei, Hunan, and Jiangxi; South includes Guangxi, Guangdong, and Hainan; Southwest includes Sichuan, Chongqing, Guizhou, and Yunnan; and Northwest includes Xinjiang, Qinghai, Gansu, Ningxia, and Shaanxi. In the model, z and y represent input and output flows, respectively, for these regions)

could be done in future research to produce a more sophisticated model if data on these flows becomes available.

I have also chosen the Beijing-Tianjin-Hebei urban agglomeration as a case study to establish a framework that included all key cities in the agglomeration across a range of scales. We can increase the scale of the network model illustrated in Fig. 5.20 by progressively adding more nodes as the model's resolution increases. For example, at the left of the figure, the initial model deals with only the three regions. At the right of the figure, the model adds 11 key cities in Hebei Province to show the resulting greater complexity of the 13-node network. The paths among these nodes represent the energy flows embodied in the exchanges of resources between pairs of nodes. The advantage of this approach is that it provides insights at different network scales, ranging from 3 to 13 nodes; on the one hand, this provides a high-level overview for only three regions, but on the other hand, breaking Hebei into 11 sub-regions (here, cities) provides insights into what flows drive the relationships between Hebei and the other two regions. This reveals the functional ecological attributes of the system components and identifies the nodes with the highest energy consumption or flows. By examining the system across multiple scales, different perspectives are obtained

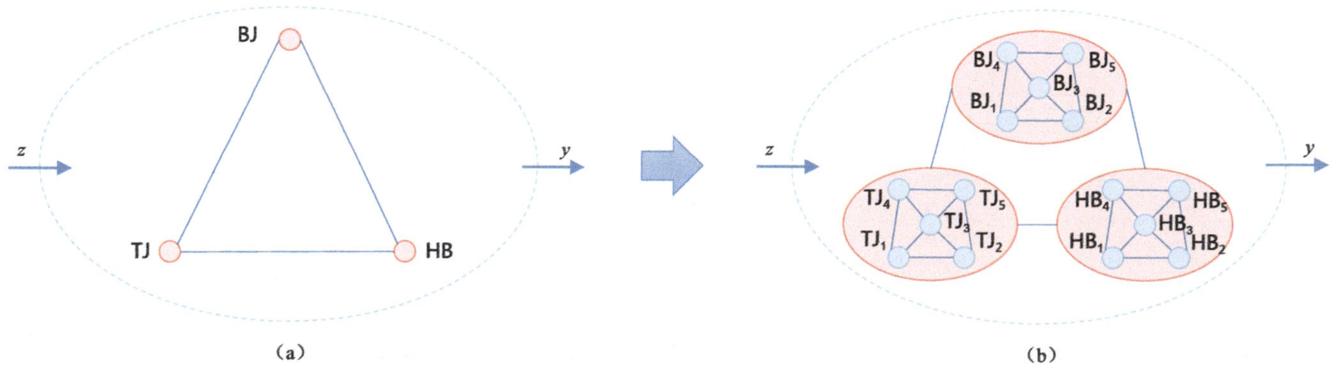

Fig. 5.19 Illustration of the overall network model for the three cities in the Beijing-Tianjin-Hebei (BJ, Beijing; TJ, Tianjin; HB, Hebei) urban agglomeration, and the same model with each city subdivided at the sector level (*Note* z and y represent flows into the system from the external region and from the system into the external region, respectively. Sector descriptions: 1-Agriculture; 2-Industrial; 3-Construction; 4-Transportation; 5-Other Services)

on the flows, thereby providing more comprehensive guidance for plans to reduce energy consumption and waste emission throughout the agglomeration.

5.4 Simulation of Network Characteristics

5.4.1 Network Structure Simulation

Network analysis researchers have developed many tools for quantifying the characteristics of a network, such as the network's central tendency, tightness, ease, or complexity of flows of materials and energy, core-periphery structure, and cohesive subgroups. Each of these characteristics or groups of characteristics provides different insights into the overall network, such as whether one node dominates the flows (and is therefore central to the network) and another node lies at the periphery of the network, where it exerts little influence on the flows.

Network analysis begins by constructing an adjacency matrix (***A***), which defines which nodes are connected by flows (i.e., are adjacent to each other based on the flows, even if they are not physically adjacent in the real world). $\boldsymbol{A} = [a_{ij}]_{n \times n}$, where a_{ij} is the adjacency between nodes *i* and *j*, *n* is the number of nodes, and $a_{ij} = 1$ if $f_{ij} \neq 0$ (i.e., if a flow exists between the two nodes), else $a_{ij} = 0$. ***A*** therefore reflects the pattern and number of direct paths between nodes of the network (Borrett and Patten 2003). Based on the network adjacency matrix, the distribution of the pathways in the urban metabolic system can be analyzed. From this matrix, the network's centrality, density, and average distance indexes can be calculated using the UCINET social network analysis software (Borgatti et al. 2002).

In addition to the centrality characteristics, which reveal the concentration of flows through one node, we can calculate the degree of network connectedness, which is based on both the network's density and the average distance between nodes. Network density represents the number of direct connections between nodes, and the average distance represents the mean number of nodes a flow passes through before it reaches its destination. The network's core-periphery structure reflects the relevance of the network nodes (Snyder and Kick 1979).

5.4.1.1 Centrality Analysis

In this section, my discussion of centrality analysis will focus on the degree centrality and the betweenness centrality of the nodes.

The degree centrality ($C_{D,i}$) for node *i* is an index that measures the node's control range (i.e., the number of other nodes that receive flows only from this node) and ability to communicate with other nodes through the flows it provides or receives. The higher the degree centrality, the more influence the node exerts on the rest of the system and the stronger its ability to obtain or transfer local resources. Because the

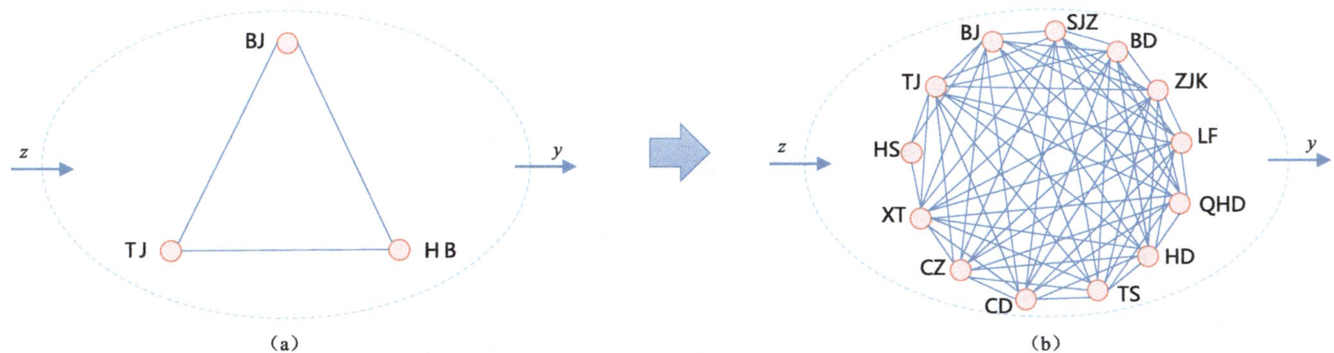

Fig. 5.20 An illustration of examining an energy metabolic network at two scales: three cities or regions and two of the original regions plus the third region divided into 11 key cities (*Note* BD-Baoding; BJ-Beijing; CD-Chengde; CZ-Cangzhou; HD-Handan; HB-Hebei; HS-Hengshui; LF-Langfang; QHD-Qinhuangdao; SJZ-Shijiazhuang; TJ-Tianjin; TS-Tangshan; XT-Xingtai; ZJK-Zhangjiakou)

flows in networks are directional, it is necessary to divide the degree centrality into input degrees ($C_{\mathrm{ID},i}$) and output degrees ($C_{\mathrm{OD},i}$). These represent (respectively) the proportions of the total nodes from which node *i* receives and to which it sends inputs. The input degree and output degree therefore reflect the proportions of the network's nodes that are affected by a given node. The degree centrality is the standardized average of the input and output degrees, which is calculated by summing the two degree centralities, dividing the sum by 1 less than the total number of nodes, and then dividing the result by 2 to obtain a simple arithmetic average to create a value from 0 to 1, and then using the following equation:

$$C_{\mathrm{D},i} = \left(\frac{C_{\mathrm{ID},i} + C_{\mathrm{OD},i}}{n-1}\right) \Big/ 2 \tag{5.9}$$

where $C_{\mathrm{D},i}$ represents the degree centrality of node *i*; $C_{\mathrm{ID},i}$ is the input degree of node *i*; $C_{\mathrm{OD},i}$ represents the output degree of node *i*; and *n* represents the network scale (i.e., the number of nodes). The greater the degree centrality, the more central the node is in the network. The index reflects the node's clustering ability and its influence on the structure and stability of the network (Scott 2000).

A node's degree of betweenness centrality is expressed as the number of shortest paths that pass through the node. It is used to measure the extent to which the node controls the flow of resources between pairs of nodes; that is, it represents how strongly the node mediates the flows among other nodes. If a node lies on the shortest path between a pair of nodes, the node has a high degree of betweenness centrality and plays an important role in the network connectivity.

If the node with high degree of betweenness centrality is removed from the network, this removal is likely to disrupt the flows between pairs of nodes that formerly passed through the removed node. As a result, a node with a high degree of betweenness centrality strongly affects the network's structure even if the node's degree centrality is not high (Freeman 1979). In Freeman's study of network characteristics, star networks (which are dominated by clusters of nodes that connect to a single central node but no other nodes) had the largest degree of betweenness centrality, and the largest possible degree of betweenness centrality for such networks is $(n^2 - 3n + 2)/2$. The betweenness centrality is calculated as follows:

$$C_{\mathrm{B},i} = \frac{2\sum_{j}^{n}\sum_{k}^{n} b_{jk}(i)}{n^2 - 3n + 2} \tag{5.10}$$

where $C_{\mathrm{B},i}$ represents the betweenness centrality of node *i*; *n* represents the network scale (number of nodes), and $b_{jk}(i)$ represents the probability that node *i* lies in the shortest path between nodes *j* and *k*.

5.4.1.2 Network Density and Average Distance

Network density is a parameter that directly describes the degree of connectedness of a network. It equals the ratio of the actual number of relationships (*L*) to the maximum number of relationships (connections between nodes) that could theoretically exist (Scott 2000). The calculation formula is

$$D = L/[n(n-1)] \tag{5.11}$$

If each node is directly connected to all other nodes within the network, the density of the network reaches its maximum. Density is therefore a basic indicator of how complete a network is (i.e., the percentage of theoretical maximum connections achieved). The maximum number of relationships in the network equals the overall number of pairs in the network, which is $n(n - 1)$, where n represents the number of nodes (the "network scale"). The higher the network density, the stronger the robustness of the network's structure; that is, if one node is removed, the flows that relied on this node may be able to transfer through a different node. Density values theoretically range between 0 and 1, but in practice, the maximum network density is generally 0.5 (Liu 2004). The closer the density is to 1, the more relationships exist among members of the network (Scott 2000). Density also reflects how actively a member participates in exchanges with other members of the network.

Network density focuses on paths with a length of 1 (i.e., paths that do not pass through an intermediate node), whereas the average distance accounts for longer paths and therefore includes paths that pass through one or more intermediate nodes and is therefore generally greater than 1 (Watts and Strogatz 1998). Thus, the two variables reflect the ease with which a node can connect to other nodes. The average distance is calculated as:

$$\mathrm{AL} = \frac{\sum_{i=1}^{n}\sum_{j=1}^{n} d_{ij}}{n(n-1)} \tag{5.12}$$

where AL represents the average path length; d_{ij} represents the distance between node i and node j; and n represents the network scale (number of nodes).

However, the average length is usually calculated only for the shortest length from a given node to another node, and therefore represents a "shortcut".

5.4.1.3 Core-Periphery Structure

Using the UCINET social network analysis software (Borgatti et al. 2002), we can identify whether a node belongs to the core or the periphery of the network by analyzing the core-periphery structure of the network. The aim of this analysis is to determine which nodes are located at the center of the network and which nodes are located at the periphery. It also shows whether the members of the core and

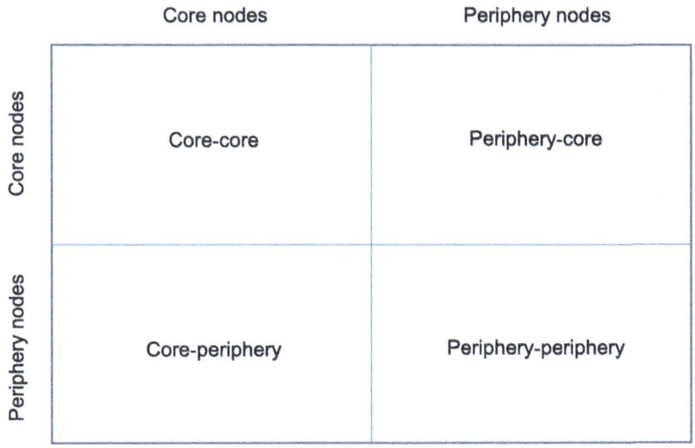

Fig. 5.21 Illustration of four blocks in adjacency matrix ***A*** and four kinds of relationships among the nodes

peripheral positions are interconnected. This structure is an idealized typical model where the rows and columns of the adjacency matrix ***A*** are divided into four blocks (three classes). The first column of the matrix, to the left of the main diagonal, reflects exchanges between core nodes. The second row of the matrix, to the right of the main diagonal, reflects exchanges between peripheral nodes. The remaining cells of the matrix reflect exchanges between the core and periphery nodes (Scott 2000). We can classify the relationships among the nodes into four categories: core-core, core-periphery, periphery-core, and periphery-periphery (Fig. 5.21) by using the blocked adjacency matrix A', which comes from dividing the adjacency matrix A into the abovementioned parts (blocks). Furthermore, we can identify which are the dominant relationships of the system. We also created a "core sub-network" for core-core relations, a "periphery sub-network" for peripheral-peripheral relations, and a "core-periphery sub-network" for core-periphery and periphery-core relations, respectively. After that, we computed and analyzed the degree of connectivity among the sub-networks. The core-periphery structure analysis and the density metrics are useful ways to characterize the different structural properties of social networks. In other words, they can represent the degree of connectivity and integrity of the entire network, as well as the position of its nodes (i.e., which nodes are core members, and which are peripheral members).

5.4.2 Network Function Simulation

Network function simulation mainly analyzes the complex features of the flow distribution and functional relationships among nodes and is based on the environ analysis methods from ecological network analysis (Fath and Borrett 2006). Environ analysis is like structured path analysis or traditional input–output analysis, but it assumes that the network is stable; that is, it assumes that the network has reached an equilibrium in which each node's inputs equal its outputs (Xia et al. 2016). In the case of network nodes, the inputs from other nodes and the external regions are dissipated or stored, and eventually flow into other nodes or into the external region. The input flow from other nodes to node i is represented by f_{i*}, the output flow from node i to other nodes is represented by f_{*i}, the input flow from the external region to node i is represented by z_i, and the output flow from node i to the external region is represented by y_i.

In an urban metabolic network (Fig. 5.22), the inputs and outputs for each node are generally unbalanced (i.e., there is a net flow). If input $(f_{i*} + z_i) >$ output $(f_{*i} + y_i)$, then the balance equation is input $(f_{i*} + z_i) =$ output $(f_{*i} + y_i) + S$, where S represents storage (i.e., a stock), and this situation represents the dissipative properties of this node [i.e., some of the stock must be dissipated (lost) to permit conservation of mass]; conversely, if input $(f_{i*} + z_i) <$ output $(f_{*i} + y_i)$, then the balance equation is input $(f_{i*} + z_i) + S =$ output $(f_{*i} + y_i)$, and represents the accumulation of stock in this node. These calculations include the flows of inputs of materials and energy from the external regions; the production and consumption of the main biomass products and industrial products; and the discharge and treatment of pollutants within the urban administrative boundary.

For urban metabolic systems, node outputs generally include a decrease of the cumulative storage, pollutant emissions, and dissipation losses. Data is generally available for node inputs (raw materials, semi-finished products), but comprehensive data for node outputs are generally more difficult to obtain. Therefore, the total input quantity $T_{i(\text{in})}$ is typically used to represent the node's flux T_i. Based on this analysis, it's possible to determine the utility of the flows between nodes (+ for inputs, − for outputs, or 0 for no flow) for each pair of nodes through a process called network utility analysis.

Network utility analysis focuses on the proportions and distributions of positive, negative, and neutral ecological relationships among nodes, which can be considered based on the metaphor of ecological relationships among the components of an ecosystem and the degree of mutualism (a good result) that results from these

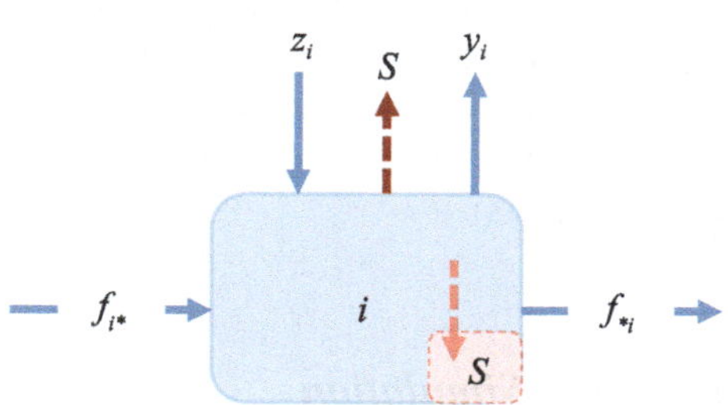

Fig. 5.22 Illustration of the conservation of mass for each sector of the urban system (*Note* i represents any node in the network; f_{i*} represents the total input that sector i obtains from other sectors; f_{*i} represents the total outputs that sector i provides to other sectors; and S represents the dissipation or increase in stock of the sector)

relationships (Zhang et al. 2017a, 2017b). I will discuss this form of analysis in the next section.

5.4.2.1 Ecological Relationship Analysis Based on Utility

The nature of the ecological relationships among nodes can be revealed by means of utility analysis; that is, this can be done by examining the utility (benefit) that a node receives through its relationship with another node. The network structure developed using the network analysis techniques described earlier in this chapter is relatively abstract, because it focuses on the existence of flows and does not consider the quantity of the actual flows. However, if the elements of the adjacency matrix that indicate the existence of a flow ($a_{ij} = 1$) are replaced by the corresponding flow values (f_{ij}), then the direct flow matrix $\boldsymbol{F}$ can be created. Let us define T_i as the sum of the flows from other nodes in the network to node i (f_{ij}) and the boundary inputs into node i from the external regions (z_i). A dimensionless direct utility intensity matrix ($\boldsymbol{D}$) can then be computed, in which matrix element d_{ij} represents the utility of the flow from node j to node i, which can be expressed as:

$$d_{ij} = \left(f_{ij} - f_{ji}\right)/T_i \tag{5.13}$$

$$T_i = \sum_{j=1}^{n} f_{ij} + z_i \tag{5.14}$$

From matrix $\boldsymbol{D}$, a dimensionless integral utility intensity matrix $\boldsymbol{U}$ can be computed from the following power series (Patten 1991, 1992):

$$\boldsymbol{U} = \left(u_{ij}\right) = \boldsymbol{D}^0 + \boldsymbol{D}^1 + \boldsymbol{D}^2 + \boldsymbol{D}^3 + \cdots + \boldsymbol{D}^m + \cdots = (\boldsymbol{I} - \boldsymbol{D})^{-1} \tag{5.15}$$

where "integral" represents the cumulative utility of both direct and indirect flows (see Sect. 5.4.2.3 for details), $\boldsymbol{D}^k$ represents the flow along a path of length k between nodes, and $\boldsymbol{I}$ represents the identity matrix. For this approach to work, the power series must converge. Convergence can be confirmed by determining that all eigenvalues of $\boldsymbol{D}$ have a magnitude less than 1 (Patten 1991). $\boldsymbol{D}^0$ represents the self-feedback of flows within each node, $\boldsymbol{D}^1$ represents the utility of the direct flows between any two nodes in the network along a path of length 1 (i.e., with no intermediate nodes), $\boldsymbol{D}^2$ represents the utility of indirect flows that travel along pathways of length 2 (i.e., with one intermediate node), and $\boldsymbol{D}^m$ (for $m > 1$) reflects the indirect flow utility along paths of m steps. u_{ij} represents the integral dimensionless value of d_{ij}, and $\boldsymbol{U}$ represents the integral utility intensity of the flows between the nodes in the network and reflects the intensity and pattern of the system's use of materials and energy through multiple pathways (Fath and Patten 1998).

Using the integral utility matrix $\boldsymbol{U}$ and the direct utility matrix $\boldsymbol{D}$, the indirect utility can be calculated:

Table 5.9 Relationship types in a natural ecosystem based on network utility analysis

	+	0	−
+	(+, +) Mutualism	(+, 0) Biased mutualism	(+, −) Exploitation
0	(0, +) Unprofitable mutualism	(0, 0) Neutral	(0, −) Amensalism
−	(−, +) Control	(−, 0) Biased damage	(−, −) Competition

Note Values in brackets represent (su_{ij}, su_{ji}), which represents the sign of the utility that results from flows from node *j* to node *i* and from node *i* to node *j*, respectively: + represents a beneficial flow, − represents a harmful flow, and 0 represents a neutral flow

$$\text{Indirect} = \boldsymbol{U} - \boldsymbol{I} - \boldsymbol{D} \tag{5.16}$$

We can designate the sign of the utility of all elements in $\boldsymbol{U}$ as the relationship utility sign matrix, sgn($\boldsymbol{U}$), with su_{ij} as the elements (+, −, or 0) in the sign matrix (Fath 2007). There are nine possible relationships between a pair of nodes based on the utility of their relationship (Table 5.9). In general, the signs in the main diagonal of sgn($\boldsymbol{U}$) are positive, which means that each node is self-mutual and receives a self-promoting positive benefit from being part of the network (Patten 1991).

Although a detailed exploration of all nine relationships is possible, I will focus on the three most important ones: mutualism, exploitation, and competition. As the "control" relationship has the same function as the exploitation relationship but in the opposite direction, therefore, the exploitation relationship here includes the control relationship. If (su_{ij}, su_{ji}) = (+, −), node *i* exploits node *j*. This is to say that node *i* benefits from this relationship (receiving more utility than it transmits to node *j*), while node *j* suffers a loss (receiving less utility than it transmits to node *i*). This is similar to the unequal relationship (i.e. exploitation) in natural ecosystems. On the contrary, if (su_{ij}, su_{ji}) = (−, +), node *i* is exploited or controlled by node *j*. Both exploitation and control relationships represent a reliance between two nodes. Since this relationship accelerates the exchange of products, by-products or waste, it is essential in the early development of urban ecosystems. Although in the short term it may result in benefits for one and losses for the other, in the long term the ecological benefits will increase. If (su_{ij}, su_{ji}) = (−, −), then node *i* competes with node *j*, resulting in lower resource utilization efficiency for both sides. Although this is a negative situation in the short term, competition may be essential to promote the system's long-term development because it encourages both nodes to improve their efficiency and to look for ways in which they can cooperate (Li et al. 2012). If (su_{ij}, su_{ji}) = (+, +), it means that the two nodes have a mutualism relationship, where all of the nodes will benefit from it. Mutualism plays an important role in reaching sustainable development.

5.4.2.2 Mutualism and Synergism Analysis

We can define a mutualism index (M) based on the values in matrix $\boldsymbol{U}$. The function of this index is to describe the relative benefits that nodes receive from the network as a whole. It may also reflect the balance between positive and negative utilities within a network and for a node to a certain extent. But it cannot reflect the real utility. Thus, it is essential to calculate the positive and negative utilities of the network and its nodes by using the elements of the integral utility matrix (U), and to explain the reasons for utility changes based on detailed information about the utility flow between nodes:

$$M = S_+/S_- \tag{5.17}$$

where $S_+ = \sum_{ij} \max\left[\mathrm{sgn}\left(u_{ij}\right), 0\right]$ and $S_- = \sum_{ij} -\min\left[\mathrm{sgn}\left(u_{ij}\right), 0\right]$, , and represent the numbers of positive and negative relationships (respectively) with other nodes in the network. Here, "max" represents the maximum of each pair of values and "min" represents the minimum of each pair of values.

$M > 1$ indicates that there are more relationships with positive utility than negative utility and that the system is mainly mutualistic. The higher M value implies a higher level of mutualism within the system. On contrary, $M < 1$ means that there are more relationships with a negative utility, which reflects a low level of mutualism. Based on the positive and negative values of the elements in matrix $\boldsymbol{U}$, we can calculate the magnitude of the utilities (benefits) for each node and for the system as a whole by using a synergism index (S) (Patten 1991, 1992).

$$S = (B_+) + (B_-) \tag{5.18}$$

where $B_+ = \sum_{ij} \max\left(u_{ij}, 0\right)$ is the sum of the flows with positive utility, and $B_- = \sum_{ij} \min\left(u_{ij}, 0\right)$ is the sum of the flows with negative utility; "max" means the maximum and "min" means the minimum of the two values in brackets.

If $S > 0$, the sum of the positive flows is greater than the sum of the negative flows, resulting in a net positive utility. In contrast, if $S < 0$, which means that the sum of the negative flows is greater than the sum of the positive flows, leading to a net negative utility. If $S = 0$, which represents that there is no net utility.

5.4.2.3 Network Integral-Flow Analysis

By using network integral-flow analysis we can recognize the flows' distribution as well as the dominant nodes of the system. We can also find out how indirect flows (flows which reach a node through one or more intermediate nodes) affect the direct flows. On the basis of the direct flows (f_{ij}) and the flow into node i from the environment (z_i), we can use T_i to represent the aggregation of flows between nodes and of the boundary inputs into node i. Then, input-oriented flows from node j to node i (g'_{ij}) are defined as follows (Zhang et al. 2015b):

$$g'_{ij} = f_{ij}/T_i \tag{5.19}$$

From the direct flow intensity matrix $\mathrm{G}' = (\mathrm{g}'_{ij})\boldsymbol{G}' = \left(g'_{ij}\right)$, we can calculate the dimensionless integral flow intensity matrix $\mathrm{N}' = \left(\mathrm{n}'_{ij}\right)\boldsymbol{N}' = \left(n'_{ij}\right)$ by using the power series as follows:

$$\boldsymbol{N}' = \left(n'_{ij}\right) = \left(\boldsymbol{G}'\right)^0 + \left(\boldsymbol{G}'\right)^1 + \left(\boldsymbol{G}'\right)^2 + \left(\boldsymbol{G}'\right)^3 + \cdots + \left(\boldsymbol{G}'\right)^m + \cdots = \left(\boldsymbol{I} - \boldsymbol{G}'\right)^{-1} \tag{5.20}$$

here, the self-feedback matrix $(\mathrm{G}')^0\left(\boldsymbol{G}'\right)^0$ represents flows that originate from and return to a node, $(\mathrm{G}')^1\left(\boldsymbol{G}'\right)^1$ represents direct flows between two nodes (i.e., flows which reach a node without going through an intermediate node), $(\mathrm{G}')^{\mathrm{m}}\left(\boldsymbol{G}'\right)^m$ represents the indirect flows of length m between nodes, and $\boldsymbol{I}$ is the identity matrix. Note that although Eq. (5.20) resembles Eq. (5.15) in form, $\boldsymbol{G}$ represents the actual flows, whereas $\boldsymbol{D}$ in Eq. (5.15) represents net flows. Convergence can be confirmed by determining that all eigenvalues of G have a magnitude less than 1 (Patten 1991).

Figure 5.23 shows the direct and indirect paths using a four-node network as an example. Metabolic path length k refers to the number of nodes that a flow must past through between the starting node and the final node. If $k = 1$, the path is direct, with no intermediate nodes, whereas if $k > 1$, the path is indirect and passes through at least one intermediate node. If we take node 4 as an example, it has input flow z_4 from and output flow y_4 to the external region, but also receives a direct flow from node 1 (metabolic path length $k = 1$) and an indirect path flow ($k = 2$) from node 1 through node 2, as well as another indirect path flow ($k = 3$) from node 1 that passes through nodes 2 and 3. Summing up the possible paths from node 1 to node 4, we can calculate the integral flow of node 1 to node 4.

Based on the integral flow intensity matrix $\boldsymbol{N}'$ and the direct flow intensity matrix $\boldsymbol{G}'$, the indirect flows between nodes can be calculated (Zhang et al. 2015b) to allow a comparison of the indirect and direct flows along each path:

$$\mathrm{Indirect} = \boldsymbol{N}' - \boldsymbol{I} - \boldsymbol{G}' \tag{5.21}$$

5.4.2.4 Node Contribution Analysis (Push and Pull Indexes)

The integral flow intensity matrix $\boldsymbol{N}'$ contains column vectors $n'_j = \left(n'_{1j}, n'_{2j}, \ldots, n'_{nj}\right)$ and row vectors $n'_i = \left(n'_{i1}, n'_{i2}, \ldots, n'_{in}\right)$. In this notation, $n'_j =$

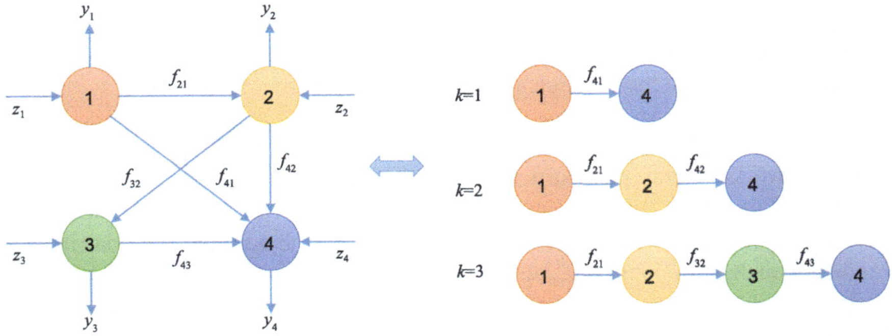

Fig. 5.23 Direct and indirect flows in the urban transfer network (*Note* f_{ij} represents a flow from node j to node i, z_i represents flows into node i from the external region, y_i represents flows from node i to the external region, and k represents the path length)

$\left(n'_{1j}, n'_{2j}, \ldots, n'_{nj}\right)$ reflects the contribution of component j of the network to the other components, whereas $n'_i = \left(n'_{i1}, n'_{i2}, \ldots, n'_{in}\right)$ reflects the contribution from other components of the network to component i. By calculating the sum of each column and row of matrix $\boldsymbol{N}'$, we can obtain the relative contribution weights $W_j = \sum_{i=1}^{n} n'_{ij}$ and $W_i = \sum_{j=1}^{n} n'_{ij}$. $W_j = \sum_{i=1}^{n} n'_{ij}$ reflects the contribution of component j to the whole network, whereas $W_i = \sum_{j=1}^{n} n'_{ij}$ reflects the contribution from the whole network to component i. Thus, two kinds of relative contribution weights (W) can be computed.

W_j reveals the impact of component j on the system, and therefore represents its ability to push the system in a given direction (its "push influence"), whereas W_i represents the ability of component i to respond to the system and can therefore be considered to represent its response to the system (its "pull influence"). Both kinds of weights are determined by the integral flows that occur within the network. These weights can then be applied to the actual measured values of the flows. This lets us determine the contribution from each component to the urban metabolic system and the support provided to every component from material and energy flows in the system. The overall capability can therefore fully reflect the status and functions of each component in the urban metabolic system and can be used to characterize the system's ecological hierarchy.

5.4.2.5 Network Hierarchical Analysis

The emphasis of hierarchical analysis is to identify the upstream and downstream distribution of network nodes, which is generally based on the types of ecological relationships between components of a network and the industrial chains they represent but supplemented by subjective judgments in some cases that I will describe later in this section. The position of each node is then calculated by combining the utility matrix (calculated using the integral flow intensity matrix) with the weight of each node to simulate the ecological hierarchy of the network.

1. Hierarchical analysis based on the types of ecological relationships

The network's hierarchical structure is determined using the results of the utility analysis and the integral-flow analysis. In this way, it's possible to define the ecological relationships among the system's components and determine the contribution weights of each component (i.e., their overall contribution to the system). The utility analysis described in previous sections can be used to define the nature of the relationships among the components, and these relationships will, in turn, serve as the basis for determining the position of each component in the overall hierarchical structure. The resulting structure is named the "hierarchy".

Theoretically, each component's position in the hierarchy can be identified from sgn ($\boldsymbol{U}$). For example, if $(su_{ij}, su_{ji}) = (+, -)$, component i is higher in the hierarchy than component j; if $(su_{ij}, su_{ji}) = (-, +)$, component j is higher than component i in the hierarchy; if $(su_{ij}, su_{ji}) = (-, -)$, components j and i are at the same level of the hierarchy; and if $(su_{ij}, su_{ji}) = (+, +)$, component j is far from component i in the hierarchy, and their relative positions are determined by their relationship with other components in the hierarchy. By describing all pairs of relationships in this manner, the upstream and downstream relationships among all components of the network can be determined, and this will reveal each component's position in the hierarchy.

However, this approach may lead to inconsistent results, because of the presence of recycling paths within the network. We can adopt two principles to resolve this problem. First, since there may be different relationships between two components when seen from different perspectives, we can adopt a "majority first" principle, which means that we choose the relationship that is most common. Second, we can choose exploitation and control relationships before mutualism relationships, and if there are any contradictions, we can then choose a competition relationship. The hierarchical structure of urban metabolic networks generally has four forms (Fig. 5.24). The flow of materials and energy is from the bottom of the hierarchy (the producers) towards the top of the hierarchy (the final consumers). The pyramid form is most balanced, since the producers within the system can supply all higher levels of the hierarchy, whereas the inverted pyramid is least balanced (least self-sufficient), the barbell form has too heavy a top consumer layer, and the spindle shape has too heavy a primary or secondary consumer level. Recycling components (reducers) are problematic because they represent both final consumers (since they reduce materials into raw materials at the end of life of these materials) and primary producers (because they provide raw materials for other primary producers and for consumers). For simplicity, we can consider them to be final consumers, and place them at the top of the hierarchy, but in future research, it will be necessary to find a more satisfactory solution that accounts for their dual role as consumers and producers.

2. Hierarchical analysis based on the pattern of industrial chains

The hierarchy can also be developed based on knowledge of the meaning of the relationships between the industries or sectors in an urban ecological network. For example, if the mining sector produces raw materials such as iron ore that serve

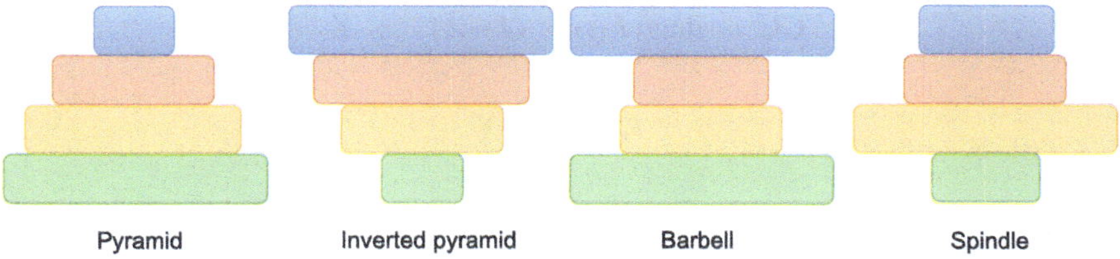

Fig. 5.24 Illustration of the hierarchical structures based on the distribution of ecological relationships in a network

Fig. 5.25 Illustration of hierarchical structures based on knowledge of the industrial chain

as inputs for other sectors, it is a producer; if the steel industry uses that ore as an input, it is then a primary consumer of that ore, and uses it to generate iron and steel as intermediate products; and if the construction or manufacturing sectors use that iron and steel as inputs to create durable goods (stocks), then these sectors are final consumers. The position of each node in the hierarchy can be determined based on knowledge of these industrial chains, and this allows us to assign ecological roles (producer, consumer, and reducer) to each metabolic actor (Fig. 5.25). As in the previous section, the nodes at the bottom of the hierarchy represent producers, followed by primary consumers, and the final consumers and reducers occupy the top of the hierarchy. Again, the reducers can be listed as a single stratum, but whether to place them at the top of the hierarchy as final consumers [Fig. 25a] or the bottom of the hierarchy as producers [Fig. 25b] is still a problem that must be solved.

5.4.3 Network Path Simulation

Using the input–output model, we can consider the carbon transfer network that arises from global trade as an example to illustrate the significance of tracing network paths and identifying the transfer type. Based on the World Input–Output Database (http://www.wiod.org/home), the Eora Global Supply Chain Database (https://www.worldmrio.com/), and other databases, we can use input–output analysis to obtain a country-level consumption-based CO_2 transfer matrix (**CC**) that reveals the flows of carbon from a global perspective:

$$\mathbf{CC} = \operatorname{diag}(\boldsymbol{F}_{\mathrm{C}}) \times (\boldsymbol{I}-\boldsymbol{A})^{-1} \times \boldsymbol{Y}_k \tag{5.22}$$

$$F_{\mathrm{C}} = \mathbf{CP}/\boldsymbol{X} \tag{5.23}$$

$$\boldsymbol{A} = \left(a_{ij}\right) = x_{ij}/X_j \tag{5.24}$$

where $\boldsymbol{F}_{\mathrm{C}}$ represents the global carbon dioxide emission-intensity matrix (i.e., the CO_2 emission per unit of GDP), $\boldsymbol{I}$ represents the identity matrix, $\boldsymbol{A}$ represents the direct consumption coefficient matrix, $\boldsymbol{Y}_k$ represents the final consumption matrix for country k, **CP** represents the carbon dioxide emission matrix for each country's sectors, $\boldsymbol{X}$ represents the total value output for each country's sectors, x_{ij} represents the value flows from sector i to sector j, and X_j represents the total value flow received by sector j. Hereafter, I will refer to the transfers in **CC** as "CO_2 transfers", since we will convert the mass values of product flows into CO_2 flows based on conversion factors.

This matrix comprises the consumption perspectives of a country, including the domestic production and consumption matrix ($\boldsymbol{D}$), the import matrix ($\boldsymbol{M}$) that satisfies consumption, the export matrix ($\boldsymbol{N}$), and the final consumption matrix ($\boldsymbol{C}$). Table 5.10 shows the distributions of the matrices $\boldsymbol{M}$, $\boldsymbol{N}$, and $\boldsymbol{C}$, in which any element in $\boldsymbol{M}$ and $\boldsymbol{N}$ can be represented as the path of the flow from primary production sector S_1 to final production sector S_2. We can subdivide the trade paths among the sectors into five types according to the locations of S_1 and S_2 (both domestic, both foreign, or one domestic and one foreign) and final consumption of their flows (domestic or foreign). On this basis, the industrial paths can be divided into five types: import part M_1 (dark blue cells in the table), from a foreign country's S_{F1} to the domestic S_{D2} for domestic consumption C_{11}; import part M_2 (light blue cells in the table), from foreign S_{F1} to foreign S_{F2} for domestic consumption C_{21}; export part N_1 (dark green cells in the table), from domestic S_{D1} to foreign S_{F2} for foreign consumption C_{12}; export part N_2 (orange cells in the table), from domestic S_{D1} to domestic S_{D2} for foreign consumption C_{22}; and the secondary return (first export and then import) part M_3 (light green cells in the table), from domestic S_{1D} to foreign S_{F2} for domestic consumption C_{31}. M_1 and M_2 make up the import part M, whereas N_1, N_2, and C_{31} make up the export part N (Meng et al. 2014).

Figure 5.26 shows a framework for internal and external industrial linkages that result from import and export activities. Each arrow in the figure represents flows from a primary production sector to a final production sector. The figure shows details of the differences in the sector-associated paths in the import (M) and export (N) components of the framework that support final consumption in a country. In this framework, arrows that end in the domestic country represent imports, whereas arrows that end in a foreign country represent exports. The transfers among sectors can be expressed as the flow from primary production component S_1 to final production component S_2 (which represents transformation of an input into the output for another component), according to whether the component is domestic or foreign.

Table 5.10 Carbon transfer pathways for a country's imports and exports from a global perspective

		Domestic country			Foreign country 1			Foreign country 2		
		S_{D1}	S_{D2}	FD_D	S_{F11}	S_{F12}	FD_{F1}	S_{F11}	S_{F2}	FD_{F2}
Domestic country	S_{D1}		N_2			M_3			N_1	
	S_{D2}			C_{31}			C_{12}			C_{22}
Foreign country 1	S_{F11}		M_1							
	S_{F12}			C_{11}						
Foreign country 2	S_{F2}								M_2	
	S_{F2}			C_{21}						

Note M represents imports, N represents exports, C represents consumption, FD represents domestic final consumption, and S_1 and S_2 represent combinations of domestic (SD) and foreign (SF) parts of the system

Based on this method, it's possible to calculate the total amount of CO_2 transfer in a country from a final consumption perspective, and by repeating this calculation in consecutive years or periods, it's possible to analyze the dynamic trends in these transfers in the context of CO_2 transfers related to global trade. We can then analyze the domestic and foreign contribution structures that underlie CO_2 transfers related to a country's international trade. On this basis, we can identify the key sectors for each country that contribute the most total CO_2 transfers related to trade, and the key sectors that participated in the CO_2 transfers due to local production and international trade. We can use the same criteria to screen out countries and sectors that do not support consumption as strongly, and to determine the spatial distributions of the flows among different countries or continents. We can also analyze the relationships between the key domestic and foreign trade routes and the key sectors in each country.

Invisible flows of carbon ("carbon leakage"), such as those embodied in international trade, have traditionally not been included in greenhouse gas inventories. The failure to account for these flows can have a negative impact on efforts to mitigate the effects of greenhouse gases, so it is therefore urgently necessary to account for these flows within national mitigation systems. Based on the methods described in this section, it becomes possible to locate the key countries involved in this leakage, but also provides a basis to account for their net CO_2 transfers. Nations disagree over their responsibilities for carbon emission targets. The transfer share can therefore be used to assign responsibility in future negotiations, and the emission targets can then be recalculated. This issue will play an important role in future efforts to mitigate global climate change.

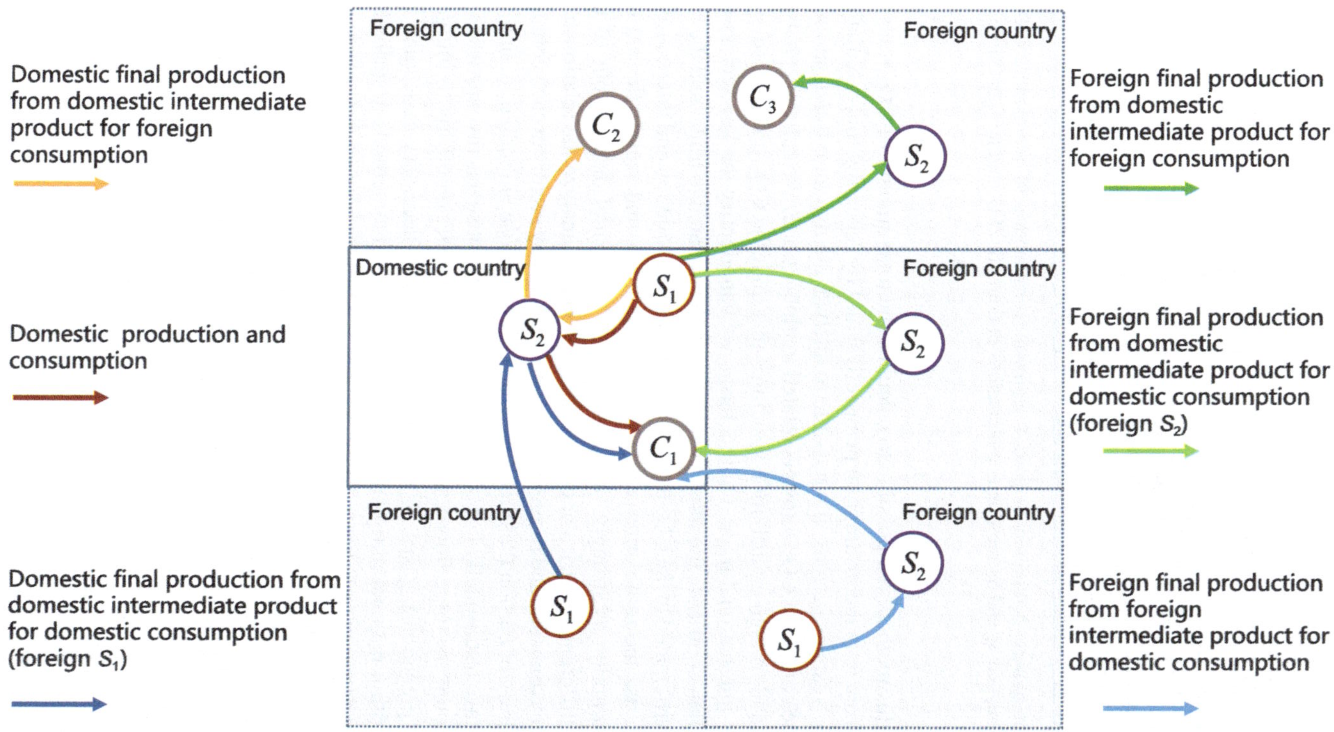

Fig. 5.26 Illustration of the framework for carbon transfer paths that result from international trade (*Note S* nodes represent primary and secondary producers, and *C* nodes represent consumers)

References

Alberti M, Marzluff J (2004) Ecological resilience in urban ecosystems: linking urban patterns to human and ecological functions. Urban Ecosystems 7(3):241–265
Borgatti SP, Everett MG, Freeman LC (2002) UCINET for Windows: software for social network analysis. Analytic Technologies, Harvard, MA
Borrett SR, Patten BC (2003) Structure of pathways in ecological networks: relationships between length and number. Ecol Model 170(2–3):173–184
Chen ZM (2011) Analysis of the flow of ecological elements in the embodiment of the world economy. Peking University, Beijing, China (in Chinese)
Duchin F (2009) Input-output economics and material flows. In: Suh S, Handbook of input-output economics in industrial ecology. Springer, Dordrecht, the Netherlands, pp 23–41
Fath BD (2007) Network mutualism: Positive community-level relations in ecosystems. Ecol Model 208(1):56–67
Fath BD, Borrett SR (2006) A MATLAB function for network environ analysis. Environ Model Softw 21(3):375–405
Fath BD, Patten BC (1998) Network synergism: emergence of positive relations in ecological systems. Ecol Model 107(2–3):127–143
Freeman LC (1979) Centrality in social networks: conceptual clarification. Social Networks 1(3):215–239
Grimm NB, Faeth SH, Golubiewski NE et al (2008) Global change and the ecology of cities. Science 319(5864):756–760
Lenzen M (1998) Primary energy and greenhouse gases embodied in Australian final consumption: an input-output analysis. Energy Policy 26(6):495–506
Li ST (2010) 2002 China regional expansion input-output table: compilation and application. Economic Science Press, Beijing, China (in Chinese)
Li SS, Zhang Y, Yang ZF et al (2012) Ecological relationship analysis of the urban metabolic system of Beijing, China. Environ Pollut 170(11):169–176
Li J, Zhang Y, Liu NY et al (2018) Flow analysis of the carbon metabolic processes in Beijing using carbon imbalance and external dependence indices. J Clean Prod 201(21):295–307
Li YX, Zhang Y, Hao Y et al (2019a) Exploring the processes in an urban material metabolism and interactions among sectors: An experimental study of Beijing, China. Ecol Indic 99(2):214–224
Li YX, Zhang Y, Yu XY (2019b) Urban weight and its driving forces: a case study of Beijing. Sci Total Environ 658(6):590–601
Liu JY, Zhang ZX, Xu XL et al (2009) Analysis on the spatial pattern and driving forces of China's land use change in the early 21st century. Acta Geogr Sin 64(12):1411–1420 (in Chinese)
Liu J (2004) Introduction to social network analysis. Social Sciences Academic Press, Beijing, China (in Chinese)
Meng B, Peters G, Wang Z (2014). Tracing CO_2 emissions in global value chains (2014-12-24) [2022-4-5]. https://papers.ssrn.com/sol3/papers.cfm?abstract_id=2541893
Patten BC (1992) Energy, emergy and environs. Ecol Model 62(1):29–69
Patten BC (1991) Network ecology: indirect determination of the life-environment relationship in ecosystems. In: Higashi M, Bums T, Theoretical studies of ecosystems: the network perspective. Cambridge University Press, New York, pp 288–351
Scott J (2000) Social network analysis: a handbook. Sage, London, UK
Shi MJ, Zhang ZY (2012) China's provincial input-output model and regional economic relations. China Science Publishing & Media Ltd., Beijing, China (in Chinese)
Snyder D, Kick EL (1979) Structural position in the world system and economic growth, 1955–1970: a multiple-network analysis of transnational interactions. Am J Sociol 84(5):1096–1126
Watts DJ, Strogatz SH (1998) Collective dynamics of 'small-world' networks. Nature 393(6684):440–442
Xia LL, Fath BD, Scharler UM et al (2016) Spatial variation in the ecological relationships among the components of Beijing's carbon metabolic system. Sci Total Environ 544(3):103–113

Xia LL, Zhang Y, Wu Q et al (2017) Analysis of the ecological relationships of urban carbon metabolism based on the eight nodes spatial network model. J Clean Prod 140(1):1644–1651
Xia LL, Liu Y, Wang XJ et al (2018) Spatial analysis of the ecological relationships of urban carbon metabolism based on an 18 nodes network model. J Clean Prod 170(1):61–69
Xu XL, Pang ZG, Yu XF (2014) Analysis method and application of temporal and spatial information of land use cover change. Scientific and Technical Documentation Press, Beijing, China (in Chinese)
Zhang Y, Yang ZF, Yu XY (2009) Ecological network and emergy analysis of urban metabolic systems: model development, and a case study of four Chinese cities. Ecol Model 220(11):1431–1442
Zhang Y, Yang ZF, Fath BD (2010a) Ecological network analysis of an urban water metabolic system: model development, and a case study for Beijing. Sci Total Environ 408(20):4702–4711
Zhang Y, Yang ZF, Fath BD et al (2010b) Ecological network analysis of an urban energy metabolic system: Model development, and a case study of four Chinese cities. Ecol Model 221(16):1865–1879
Zhang Y, Liu H, Li YT et al (2012) Ecological network analysis of China's societal metabolism. J Environ Manage 93(1):254–263
Zhang Y, Liu H, Chen B (2013) Comprehensive evaluation of the structural characteristics of an urban metabolic system: model development and a case study of Beijing. Ecol Model 252(5):106–113
Zhang Y, Xia LL, Xiang WN (2014a) Analyzing spatial patterns of urban carbon metabolism: a case study in Beijing, China. Landsc Urban Plan 130(5):184–200
Zhang Y, Liu H, Chen B et al (2014b) Analysis of urban metabolic processes based on input-output method: model development and a case study for Beijing. Frontiers of Earth Science 8(2):190–201
Zhang Y, Zheng HM, Fath BD et al (2014c) Ecological network analysis of an urban metabolic system based on input-output tables: model development and case study for Beijing. Sci Total Environ 468–469(1):642–653
Zhang Y, Zheng HM, Yang ZF et al (2015a) Multi-regional input-output model and ecological network analysis for regional embodied energy accounting in China. Energy Policy 86(11):651–663
Zhang Y, Zheng HM, Yang ZF et al (2015b) Analysis of the industrial metabolic processes for sulfur in the Lubei (Shandong Province, China) eco-industrial park. J Clean Prod 96(11):126–138
Zhang Y, Lu HJ, Fath BD et al (2016a) A network flow analysis of the nitrogen metabolism in Beijing, China. Environ Sci Technol 50(16):8558–8567
Zhang Y, Xia LL, Fath BD et al (2016b) Development of a spatially explicit network model of urban metabolism and analysis of the distribution of ecological relationships: case study of Beijing, China. J Clean Prod 112(2):4304–4317
Zhang Y, Zheng HM, Chen B et al (2016c) Ecological network analysis of embodied energy exchanges among seven regions of China. J Ind Ecol 20(3):472–483
Zhang Y, Zheng HM, Yang ZF et al (2016d) Urban energy flow processes in the Beijing-Tianjin-Hebei (Jing-Jin-Ji) urban agglomeration: combining multi-regional input-output tables with ecological network analysis. J Clean Prod 114(3):243–256
Zhang Y, Wu Q, Wang XJ et al (2017a) Analysis of the ecological relationships within the CO_2 transfer network created by global trade and its changes from 2001 to 2010. J Clean Prod 168(23):1425–1435
Zhang Y, Zheng HM, Lu HJ (2017b) Research progress of urban ecological network analysis. Acta Ecol Sin 37(12):4258–4267 (in Chinese)
Zheng HM, Fath BD, Zhang Y (2017) An urban metabolism and carbon footprint analysis of the Jing-Jin-Ji regional agglomeration. J Ind Ecol 21(1):166–179
Zheng HM, Wang XJ, Li MJ et al (2018) Interregional trade among regions of urban energy metabolism: A case study between Beijing-Tianjin-Hebei and others in China. Resour Conserv Recycl 132(5):339–351

Chapter 6
Regulation and Optimization of an Urban Metabolism

In previous chapters, we have reviewed various ways to define the characteristics of an urban metabolism and quantify the flows. Although this research is interesting for its own sake, the understanding that such analyses can provide could be beneficial for urban planners and managers to regulate and optimize a city's metabolism and consequently to make the metabolism more ecologically sustainable. In this chapter, we will discuss some procedures to achieve that goal. One useful approach is to break complex, compound measurements into their component parts, such as different sectors or industries, focused on production or consumption, and so on. We can then determine the contributions of the driving factors to the metabolism. This process is called "factor decomposition", and relies on developing models that quantify the influences of various socioeconomic factors on urban metabolic processes.

6.1 Factor Decomposition Models

As we have seen in the previous chapters, cities can be expressed using the metaphor of a living organism. Similar to a human body, cities often have areas with deficient or excess "heat". (Here, *heat* refers to a concept from traditional Chinese medicine, rather than the Western thermodynamic concept.) The metabolic rate and throughput of materials or energy are low when there is insufficient heat, whereas cities with excess heat have an excessively vigorous metabolism and excessively high throughput. These differences lead to differences among the city's metabolic actors and among cities. Meanwhile, some cities are production-oriented, because they consume huge materials and energy for transforming material in-use stock, whereas others are consumption-oriented, because they consume some dissipative material and energy for satisfying household consumption. Different management decisions should be considered for these varying city types based on their different urban characteristics.

Y. Zhang, *Urban Metabolism*,
https://doi.org/10.1007/978-981-19-9123-3_6

Urban metabolism research can effectively support urban planning, design, and regulation by analyzing the urban metabolic processes using characteristic indexes that quantify metabolic problems. They can also quantify the extent and direction of changes in various socioeconomic factors and the resulting impacts on metabolic processes, which will help decision-makers propose practical and effective control measures to mitigate problems and improve the efficiency of urban ecological management. Determining the importance of the various factors that influence metabolic processes is mainly based on decomposition analysis (Zhang et al. 2011). Factor decomposition can quantitatively decompose changes in processes such as carbon emission into the contributions of individual factors using various weight-determination methods. Factor decomposition techniques include structural decomposition analysis (Chen et al. 2019) and index decomposition analysis (Zhang et al. 2011). Although both methods are stand on historical data from no less than two time-intervals or periods, and are derived from changes in the key factors (the ones that contribute most to the changes), index decomposition requires less data and is more broadly applicable. Of the indexes that are available, the logarithmic mean Divisia index (LMDI) (Ang 2005) is commonly used because of its robust theoretical foundation, strong adaptability to a range of situations, and ability to provide "perfect" decomposition; in which, no unexplained residual terms appear in the results (Ang 2004, 2005).

6.1.1 Decomposition Model for an Urban Carbon Metabolism

Because of the importance of reducing emission of greenhouse gases, carbon is a common focus of factor decomposition research. The purpose of optimizing urban carbon transformation processes and increasing a city's carbon absorption capacity is to reduce carbon emission and mitigate its impact on global climate change. This is particularly important in China, where the author conducted the original research that is the basis for this chapter (Zhang et al. 2011). Energy-related carbon emissions in China have grown at its fastest pace, driven by the country's unprecedented economic growth.

To mitigate such emissions, it is essential to evaluate the trends (upward or downward) and the magnitudes (tonnes of carbon emitted) of the factors that causes carbon emission through assessing the influential parameters. Based on the outputs of this analysis, the key factors from the perspective of urban production or consumption processes can be identified. Therefore, accurate models (potentially including a regionalization model) for the effects of these factors on carbon emissions can be developed. Such models provide a basis for decision-makers and allow them to reduce carbon emission by improving policies.

Using the LMDI method (Fig. 6.1), we decomposed energy-related carbon emissions into five different driving factors: population, gross domestic product (GDP), economic structure, energy consumption, and the combination of fuels that are consumed. These factors can be grouped into a scale effect caused by GDP and

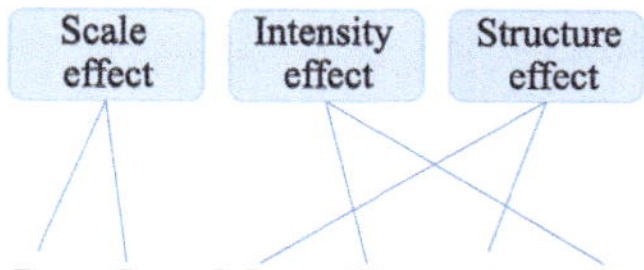

$$C = \sum_{ji} P \times \frac{\text{GDP}}{P} \times \frac{Q_j}{\text{GDP}} \times \frac{E_j}{Q_j} \times \frac{E_{ji}}{E_j} \times \frac{C_{ji}}{E_{ji}} = \sum_{ji} P \times G \times \text{QS}_j \times \text{EI}_j \times \text{ES}_{ji} \times R_{ji}$$

Fig. 6.1 The LMDI decomposition model for energy-related carbon emission (*Note* C represents the total quantity of energy-related carbon emissions; P is the population; GDP is the gross domestic product; Q_j represents the gross value of production of sector j; E_j refers to the energy consumption of sector j; E_{ji} shows the consumption of energy type i by sector j; C_{ji} stands for the carbon emission of energy type i by sector j; G express the per capita GDP; QS_j represents the economic structure (i.e., the proportion of GDP accounted for by the gross domestic product of sector j); EI_j is the energy intensity of sector j (i.e., the amount of energy consumed per unit GDP); ES_{ji} is the fuel mix, and represents the proportion of the total energy consumption by sector j that is accounted for by consumption of energy type i; and R_{ji} represents the carbon-emission coefficient for energy type i in sector j)

population, an intensity effect caused by energy intensity (the energy consumed per unit of output), and a structure effect caused by the economic structure and fuel mix. Referring to these factors, we can construct a perfect-decomposition model (i.e., one that leaves no unexplained residuals) for variation in energy-related carbon emissions that will calculate the directions and the magnitudes of the contributions of the five factors to any changes in carbon emissions (Zhang et al. 2011).

The changes in carbon emissions from the base year of a study (year 0) to the targeted year (year T) can be expressed as follows:

$$\Delta C = C_T - C_0 = \Delta C_{\text{P}} + \Delta C_{\text{G}} + \Delta C_{\text{QS}} + \Delta C_{\text{EI}} + \Delta C_{\text{ES}} + \Delta C_{\text{R}} \qquad (6.1)$$

where ΔC represents the change in carbon emissions from year 0 to year T; C_T represents the carbon emissions in year T; C_0 represents the carbon emissions in year 0; ΔC_{P} represents the impact of population, ΔC_{G} represents the impact of GDP, ΔC_{QS} represents the impact of the economic structure, ΔC_{EI} represents the impact of energy intensity, ΔC_{ES} represents the impact of the fuel mix, and ΔC_{R} represents the impact of the carbon-emission coefficient on changes in carbon emissions.

Although it's best to obtain carbon-emission coefficients for each energy source based on local empirical data for the study area, it's acceptable to use the default values obtained from the Intergovernmental Panel on Climate Change (IPCC).[1] The carbon-emission coefficients for each energy source will remain constant (i.e., the average carbon content of each type of fuel does not change), thus it's not necessary to consider the impact of these coefficients on carbon emissions. (An obvious exception would be if a country or region changed to a fuel type with a different carbon content during the study period, such as replacing high-sulfur coal with low-sulfur coal.) In general, $\Delta C_{\text{R}} = 0$. Additive decomposition of LMDI can then be carried out as

[1] IPCC: https://www.ipcc.ch/data/.

follows:

$$\Delta C_{\mathrm{P}} = \ln\left(\frac{P_T}{P_0}\right)\sum_{ji}\frac{C_{T,ji} - C_{0,ji}}{\ln C_{T,ji} - \ln C_{0,ji}} \tag{6.2}$$

$$\Delta C_{\mathrm{G}} = \ln\left(\frac{G_T}{G_0}\right)\sum_{ji}\frac{C_{T,ji} - C_{0,ji}}{\ln C_{T,ji} - \ln C_{0,ji}} \tag{6.3}$$

$$\Delta C_{\mathrm{QS}} = \ln\left(\frac{\mathrm{QS}_{T,j}}{\mathrm{QS}_{0,j}}\right)\frac{C_{T,ji} - C_{0,ji}}{\ln C_{T,ji} - \ln C_{0,ji}} \tag{6.4}$$

$$\Delta C_{\mathrm{EI}} = \ln\left(\frac{\mathrm{EI}_{T,j}}{\mathrm{EI}_{0,j}}\right)\frac{C_{T,ji} - C_{0,ji}}{\ln C_{T,ji} - \ln C_{0,ji}} \tag{6.5}$$

$$\Delta C_{\mathrm{ES}} = \ln\left(\frac{\mathrm{ES}_{T,j}}{\mathrm{ES}_{0,j}}\right)\frac{C_{T,ji} - C_{0,ji}}{\ln C_{T,ji} - \ln C_{0,ji}} \tag{6.6}$$

where 0 and T denote the values for year 0 and year T, respectively. These factors are subsequently used to clarify the scale, intensity, and structure effects.

6.1.2 Refine the Decomposition Model for the Social and Economic Factors

By accounting for production and consumption activities, we can quantify the carbon emissions associated with the total energy consumption of key sectors in a city. Here, the author will choose six material production sectors and two household sectors for which data was available during the research of Zhang et al. (2013a). To do so, we must decompose the emissions into several components. First, we consider the economic activity of each component of the city's economic structure, since energy consumption is proportional to a component's economic output. It is also crucial to understand the effect of the city's population on carbon emissions; however, because the urban and rural populations have very different situations, we will analyze them separately. The primary, secondary, and tertiary sectors have diverse characteristics, and so do their energy consumption patterns due to the significance of the economic structure. To define each sector's technical level of energy utilization processes sustainment, we used a crucial factor called energy intensity, which stands for the energy consumption per unit product output or population. Energy consumption structure, another similar factor, is also used in our research, which plays a critical role in carbon emissions (Zhang et al. 2013a).

The energy-related carbon emissions, which stem from the urban production sectors and rural and urban household sectors, are generally decomposed into eight influencing factors in our LMDI model, which are as follows (Zhang et al. 2013a): ① economic activity (i.e., the economic output, as GDP), ② economic structure (i.e., the proportion of each sector's contribution to total economic output), ③ energy intensity (i.e., energy consumption per unit of product output or population), ④ the energy mix consumed by the production sectors (i.e., the proportion of energy provided by each of the 22 fuels for which data are available in Chinese statistical databases), ⑤ population size, ⑥ the urban and rural population distribution (i.e., urban and rural households as a proportion of the total population), ⑦ per capita energy consumption, and ⑧ the energy mix consumed by the two (urban and rural) household sectors. Therefore, we decomposed urban energy-related carbon emissions (C) as follows:

$$C = \sum_{ji} \mathrm{GDP} \times \frac{Q_j}{\mathrm{GDP}} \times \frac{E_j}{Q_j} \times \frac{E_{ji}}{E_j} \times \frac{C_{ji}}{E_{ji}} + \sum_{ki} P \times \frac{P_k}{P} \times \frac{E_k}{P_k} \times \frac{E_{ki}}{E_k} \times \frac{C_{ki}}{E_{ki}} \tag{6.7}$$

where C_{ji} refers to the carbon emitted by the usage of energy i in production sector j; C_{ki} stands for the carbon emissions from the consumption of energy i by urban or rural residents ($k = 1$ or 2, respectively); E_j represents production sector j's energy consumption; E_{ji} means the usage of a specific energy type i in production sector j; E_k defines the energy consumed by urban or rural residents (k); E_{ki} means the usage of energy i owing to urban or rural residents (k); GDP is short for gross domestic product and is stated in constant prices (using the official government inflation rate to convert GDP to the standardized values at the beginning of the study period); i stands for various energy types ($i = 1, 2, 3, \ldots, 22$); j shows the production sectors ($j = 1, 2, 3, \ldots, 6$); k represents the household sectors ($k = 1$ for urban households or 2 for rural households); P represents the total population; P_k presents the urban or rural population (k); and Q_j shows the production output value of production sector j. Equation (6.7) can be simplified as follows:

$$C = \sum_{ji} \mathrm{GDP} \times S_j \times \mathrm{EI}_j \times \mathrm{ES}_{ji} \times R_{ji} + \sum_{ki} P \times \mathrm{SS}_k \times \mathrm{EII}_k \times \mathrm{ESS}_{ki} \times R_{ki} \tag{6.8}$$

where EI_j represents the energy intensity of production sector j; EII_k represents the per capita energy consumption of urban or rural households (k); ES_{ji} (energy mix) represents the proportion of total energy consumption by production sector j accounted for by consumption of energy type i; ESS_{ki} (energy mix) represents the proportion of total energy consumption by urban or rural households (k) accounted for by consumption of energy type i; GDP represents the gross domestic product; P represents the population for urban and rural households ($k = 1$ and 2, respectively); R_{ji} represents the carbon-emission coefficient for energy type i in production sector j; R_{ki} represents the carbon-emission coefficient for energy type i of urban or rural households; S_j (economic structure) represents the proportion of GDP accounted for by the output production output for production sector j; and SS_k (the urban and rural population distribution) represents the proportion of the total population accounted for by urban and rural populations (k).

As noted earlier, we can assume that the carbon-emission coefficients for each energy source remain constant over time, and this lets us ignore the impact of these coefficients on carbon emissions. The changes in carbon emissions from the base year (year 0) to the subsequent year (year T) can be expressed as follows:

$$\Delta C = C_T - C_0 = \Delta C_{\mathrm{G}} + \Delta C_{\mathrm{S}} + \Delta C_{\mathrm{EI}} + \Delta C_{\mathrm{ES}} + \Delta C_{\mathrm{P}} + \Delta C_{\mathrm{SS}} + \Delta C_{\mathrm{EII}} + \Delta C_{\mathrm{ESS}} \tag{6.9}$$

where ΔC represents the change in carbon emissions from year 0 to year T; C_0 expresses the carbon emissions in year 0; C_T shows the carbon emissions in year T; ΔC_{EI} illuastrates the impact of changes in energy intensity, ΔC_{EII} refers to the impact of changes in the per capita energy consumption, ΔC_{ES} represents the impact of changes in the energy mix of the production sectors, ΔC_{ESS} shows the impact of changes in the energy mix of the household sectors, ΔC_{G} represents the impact of changes in economic activity, ΔC_{P} represents the impact of changes in population, ΔC_{S} refers to the impact of changes in economic activity, and ΔC_{SS} stand for the impact of changes in the urban and rural population distribution structure.

Additive decomposition of LMDI can be carried out as follows:

$$\Delta C_{\mathrm{G}} = \ln\left(\frac{G_T}{G_0}\right) \sum_{ji} \frac{C_{T,ji} - C_{0,ji}}{\ln C_{T,ji} - \ln C_{0,ji}} \tag{6.10}$$

$$\Delta C_{\mathrm{S}} = \ln\left(\frac{S_{T,j}}{S_{0,j}}\right) \sum_{ji} \frac{C_{T,ji} - C_{0,ji}}{\ln C_{T,ji} - \ln C_{0,ji}} \tag{6.11}$$

$$\Delta C_{\mathrm{EI}} = \ln\left(\frac{\mathrm{EI}_{T,j}}{\mathrm{EI}_{0,j}}\right) \sum_{ji} \frac{C_{T,ji} - C_{0,ji}}{\ln C_{T,ji} - \ln C_{0,ji}} \tag{6.12}$$

$$\Delta C_{\mathrm{ES}} = \ln\left(\frac{\mathrm{ES}_{T,ji}}{\mathrm{ES}_{0,ji}}\right) \sum_{ji} \frac{C_{T,ji} - C_{0,ji}}{\ln C_{T,ji} - \ln C_{0,ji}} \tag{6.13}$$

$$\Delta C_{\mathrm{P}} = \ln\left(\frac{P_T}{P_0}\right) \sum_{ki} \frac{C_{T,ki} - C_{0,ki}}{\ln C_{T,ki} - \ln C_{0,ki}} \tag{6.14}$$

$$\Delta C_{\mathrm{SS}} = \ln\left(\frac{\mathrm{SS}_{T,k}}{\mathrm{SS}_{0,k}}\right) \sum_{ki} \frac{C_{T,ki} - C_{0,ki}}{\ln C_{T,ki} - \ln C_{0,ki}} \tag{6.15}$$

$$\Delta C_{\mathrm{EII}} = \ln\left(\frac{\mathrm{EII}_{T,k}}{\mathrm{EII}_{0,k}}\right) \sum_{ki} \frac{C_{T,ki} - C_{0,ki}}{\ln C_{T,ki} - \ln C_{0,ki}} \tag{6.16}$$

$$\Delta C_{\mathrm{ESS}} = \ln\left(\frac{\mathrm{ESS}_{T,ki}}{\mathrm{ESS}_{0,ki}}\right)\sum_{ki}\frac{C_{T,ki} - C_{0,ki}}{\ln C_{T,ki} - \ln C_{0,ki}} \tag{6.17}$$

6.1.3 Classification Model for Energy-Related Carbon Emission

We can classify the relationship between the driving factors that stimulate carbon emission and the resulting emission in the form of a graph. The economic development and available energy resources vary considerably among the regions in the graph. Given these differences, regions in the figure should not be assessed solely on the basis of their gross carbon emissions; on the contrary, it is essential to clarify the contribution of critical factors driving these emissions. Furthermore, prioritizing the contributions of these factors only determines the factors that are currently dominant or minor with no revealing of whether a region's carbon emissions are increasing or decreasing.

Simple ranking thus is not appropriate for classifying multiple regions of the graph. In order to illustrate the differences among the regions in the graph, it is essential to categorize these regions according to the proportion of stimulating factors (i.e. factors raising emissions) as well as their changes in carbon emissions. A two-dimensional graphical model is built to reach this purpose, in which the proportion of stimulatory factors is used as the vertical coordinate and the rise or fall of carbon emissions serves as the horizontal coordinate (Fig. 6.2).

According to the decomposition model mentioned above, there are four or eight factors that influence carbon emissions associated with energy. The vertical axis in Fig. 6.2 refers to the proportion of the stimulating factors related to carbon emission,

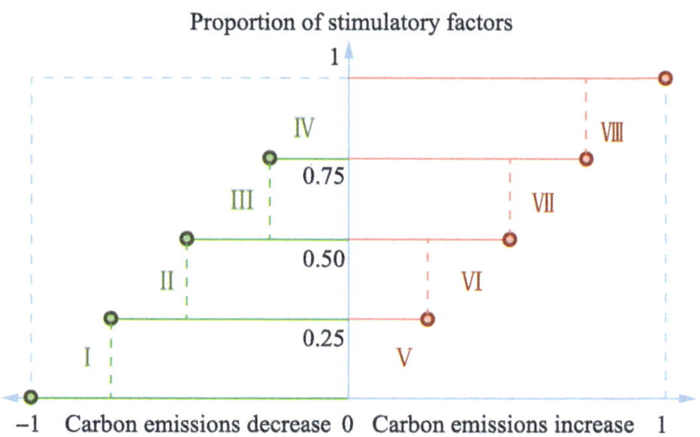

Fig. 6.2 A management model for classification of carbon emission effects

and in this example, can have values as 0, 0.25, 0.50, 0.75, or 1. The horizontal axis in the graph shows whether regional carbon emissions are rising or falling over the research duration. (The carbon emission values were standardized by dividing all values by the corresponding maximum value during the study period, producing a value that ranges between -1 and $+1$). The areas of the graph can therefore be classified into eight categories (from I to VIII) based on the proportions of the factors that increase emissions and whether a region's carbon emissions increased or decreased during the study period. In type I areas of the graph, all factors decrease carbon emissions, leading to the largest decrease in carbon emissions. As the number of inhibitory factors decreases, the number of factors that increase emissions will increase, and the classification number increases, representing progressively greater adverse impacts; thus, type VIII areas of the graph, in which all factors increase carbon emissions, represent the worst situation, with the greatest carbon emissions.

On this basis, we can classify areas of the graph into categories for each region of the study area to reveal differences in the spatial characteristics of the driving factors using the eight-category regionalization model. Based on the results of this analysis, we can then propose potential strategies for reducing each region's carbon emissions, thereby providing a solid basis for coordinating regional development and adapting government policy to mitigate energy-related carbon emissions (Zhang et al. 2011).

6.1.4 Decomposition Model for an Urban Nitrogen Metabolism

Nitrogen is a key element because of its abundance and also presence in biomass, including food, and in products such as fertilizers that are heavily used in human systems. The purpose of optimizing the urban nitrogen metabolic process is to reduce eutrophication due to nitrogen losses into bodies of water during the high nitrogen consumption caused by human activities. We will similarly use the LMDI method to construct a complete decomposition model for the factors responsible for changes of urban anthropogenic nitrogen consumption. In this model, we can identify the effects of six key factors: ① the nitrogen content of each material, ② the material consumption intensity, ③ the material consumption structure, ④ the industrial structure, ⑤ the per capita GDP, and ⑥ the population (Zhang et al. 2020). It is noteworthy rather than using the carbon model in the previous section, the author has chosen a different model to show the flexibility of the LMDI method (i.e., its ability to cope with different modeling situations).

The direction of the effects would be examined as well to determine whether these factors promoted or inhibited nitrogen consumption. The calculation formula for nitrogen consumption (N) is as follows:

$$N = \sum_{ji} N_{ji} = \sum_{ij} P \times \frac{G}{P} \times \frac{G_j}{G} \times \frac{M_j}{G_j} \times \frac{M_{ji}}{M_j} \times \frac{N_{ji}}{M_{ji}} \tag{6.18}$$

where G represents the regional GDP (the real GDP, adjusted for inflation to the value at the start of the study period); G_j stands for the product output by the j-th sector; M_j is the material consumption by the j-th sector; M_{ji} shows the consumption of the i-th material by the j-th sector; N_{ji} presents the amount of nitrogen in the i-th material from the j-th sector; and P expresses the population. This equation can be simplified as follows:

$$N = \sum_{ji} N_i \sum_{ji} P \times R \times \mathrm{IS}_j \times \mathrm{ME}_j \times \mathrm{MS}_{ji} \times F_{ji} \tag{6.19}$$

where F_{ji} is the nitrogen content of material i in sector j, and most of the changes in this parameter are affected by changes in the composition of each material; IS_j is the industrial structure, which represents the ratio of the output value of the j-th sector to the total output value; ME_j is the material intensity, which represents the material consumption per unit of product output value of the j-th sector; MS_{ji} is the material consumption structure, which expresses the proportion of total consumption for the i-th material in the j-th sector; P is the population; and R is the per capita GDP (the real GDP, adjusted for inflation to the value at the start of the study period).

The changes in anthropogenic nitrogen consumption from the base year (time = 0) to the target year (time = T) can be expressed as:

$$\Delta N = N_T - N_0 = \Delta N_{\mathrm{P}} + \Delta N_{\mathrm{R}} + \Delta N_{\mathrm{IS}} + \Delta N_{\mathrm{ME}} + \Delta N_{\mathrm{MS}} + \Delta N_{\mathrm{F}} \tag{6.20}$$

where ΔN presents the change in total anthropogenic nitrogen consumption from year 0 to year T; N_T is the anthropogenic nitrogen consumption in year T; N_0 refers to the anthropogenic nitrogen consumption in year 0; and for the changes in anthropogenic nitrogen consumption, ΔN_{F} represents the changes caused by changes in the nitrogen content of the material, ΔN_{IS} shows the changes in the industrial structure, ΔN_{ME} illustrates changes in the material intensity, ΔN_{MS} stands for changes in the material consumption structure, ΔN_{P} represents changes in the population, and ΔN_{R} is the changes in per capita GDP. The equations obtained by means of the LMDI decomposition method are as follows:

$$\Delta N_{\mathrm{P}} = \ln\left(\frac{P_T}{P_0}\right) \sum_{ji} \frac{N_{T,ji} - N_{0,ji}}{\ln N_{T,ji} - \ln N_{0,ji}} \tag{6.21}$$

$$\Delta N_{\mathrm{R}} = \ln\left(\frac{R_T}{R_0}\right) \sum_{ji} \frac{N_{\mathrm{T},ji} - N_{0,ji}}{\ln N_{T,ji} - \ln N_{0,ji}} \tag{6.22}$$

$$\Delta N_{\mathrm{IS}} = \ln\left(\frac{\mathrm{IS}_{T,j}}{\mathrm{IS}_{0,j}}\right) \sum_{ji} \frac{N_{T,ji} - N_{0,ji}}{\ln N_{T,ji} - \ln N_{0,ji}} \tag{6.23}$$

$$\Delta N_{\mathrm{ME}} = \ln\left(\frac{\mathrm{ME}_{T,j}}{\mathrm{ME}_{0,j}}\right) \sum_{ji} \frac{N_{T,ji} - N_{0,ji}}{\ln N_{T,ji} - \ln N_{0,ji}} \tag{6.24}$$

$$\Delta N_{\mathrm{MS}} = \ln\left(\frac{\mathrm{MS}_{T,ji}}{\mathrm{MS}_{0,ji}}\right)\sum_{ji}\frac{N_{T,ji} - N_{0,ji}}{\ln N_{T,ji} - \ln N_{0,ji}} \tag{6.25}$$

$$\Delta N_{\mathrm{F}} = \ln\left(\frac{F_{T,ji}}{F_{0,ji}}\right)\sum_{ji}\frac{N_{T,ji} - N_{0,ji}}{\ln N_{T,ji} - \ln N_{0,ji}} \tag{6.26}$$

These equations can be employed to analyze the scale effect (population), intensity effect (per capita GDP, material intensity), and structural effect (nitrogen content of the material, material consumption structure, industrial structure) generated by the selected factors that affect urban anthropogenic nitrogen consumption.

In contrast to the analysis described for carbon in the previous section, it's also possible to examine the contributions of each interaction to the total effect. (This shows how a similar analysis for the change in flows can be extended in a different direction by using the results to support an additional calculation.) The contribution of each effect to the total characterizes the magnitude of the effect, and the sign indicates the direction of the effect (positive is promotion of N consumption, negative is inhibition). The contribution of a given factor (calculated from the LMDI results) is divided by the sum of the absolute values of the contributions from all factors (which total is 1), so the contribution is standardized to fall within the range of [−1, 1], which we can instead convert to a percentage value.

6.1.5 Decomposition Model of Material Metabolism

In the first two decomposition analyses described for this chapter, the author focused on individual elements (carbon and nitrogen). Here, aim is to illustrate how the decomposition approach can be used for materials. Identifying the factors that affect the urban material metabolic processes is the key to promoting urban sustainable development. We will again use the LMDI decomposition methodology to analyze the driving forces behind the changes in urban material consumption. The direct material consumption (DMC) provides a measure of the materials consumed by socioeconomic activities within a city's administrative boundaries. DMC has been shown to be strongly related to the city's economic level, population, material consumption structure and intensity, and industrial structure (Li et al. 2019). In this section, we will examine the relationships between urban DMC changes and the following driving factors: the material consumption structure (i.e., the division of materials into subcategories of the total material consumption), material consumption intensity (consumption per unit GDP or per capita), industrial structure (division of consumption among components of the urban system), economic activity, and population (Li et al. 2019). We used these factors to decompose the changes in urban DMC considering the following model:

$$\mathrm{DMC} = \sum_{ji} \mathrm{DMC}_{ji} = \sum_{ji} P \times \frac{G}{P} \times \frac{G_j}{G} \times \frac{M_j}{G_j} \times \frac{M_{ji}}{M_j} \tag{6.27}$$

where M_j refers to material consumption by sector *j*; M_{ji} is the consumption of the *i*-th material ($i = 1, 2, 3, 4$, and 5, representing biomass, mineral ores, non-metallic minerals, industrial products, and fossil fuels, respectively) by sector *j*; *G* represents urban total GDP; G_j denotes the GDP of sector *j*; and *P* is the population. We can simplify Eq. (6.27) as follows:

$$\mathrm{DMC} = \sum_{ji} \mathrm{DMC}_{ji} = \sum_{ji} P \times R \times S_j \times \mathrm{MI}_j \times \mathrm{MS}_{ji} \tag{6.28}$$

where DMC_{ji} represents direct consumption of material *i* by sector *j*; MI_j denotes the material consumption intensity of sector *j* (i.e., the consumption per unit GDP); MS_{ji} is the ratio of consumption of the *i*-th material to the total material consumption for sector *j*; *P* is the population; *R* is the ratio of GDP to the population (i.e., urban per capita GDP, the real GDP, adjusted for inflation to the value at the start of the study period); and S_j stands for the share of total GDP accounted for by sector *j*.

The corresponding change in DMC (here, represented by M = mass) is calculated as follows:

$$\Delta M = M_T - M_0 = \Delta M_\mathrm{P} + \Delta M_\mathrm{R} + \Delta M_\mathrm{S} + \Delta M_\mathrm{MI} + \Delta M_\mathrm{MS} \tag{6.29}$$

where M_T represents the total DMC at time *T* (the end of a period), M_0 represents the total DMC at the start of a period, and Δ refers to the change between times *T* and 0 for MI, MS, *P*, *R*, and *S* for the corresponding variables in Eq. (6.28). The contribution of each factor to the change in urban DMC is calculated using the following equations:

$$\Delta M_\mathrm{P} = \ln\left(\frac{P_T}{P_0}\right) \sum_{ji} \frac{M_{T,ji} - M_{0,ji}}{\ln M_{T,ji} - \ln M_{0,ji}} \tag{6.30}$$

$$\Delta M_\mathrm{R} = \ln\left(\frac{R_T}{R_0}\right) \sum_{ji} \frac{M_{T,ji} - M_{0,ji}}{\ln M_{T,ji} - \ln M_{0,ji}} \tag{6.31}$$

$$\Delta M_\mathrm{S} = \ln\left(\frac{S_{T,j}}{S_{0,j}}\right) \sum_{ji} \frac{M_{T,ji} - M_{0,ji}}{\ln M_{T,ji} - \ln M_{0,ji}} \tag{6.32}$$

$$\Delta M_\mathrm{MI} = \ln\left(\frac{\mathrm{MI}_{T,j}}{\mathrm{MI}_{0,j}}\right) \sum_{ji} \frac{M_{T,ji} - M_{0,ji}}{\ln M_{T,ji} - \ln M_{0,ji}} \tag{6.33}$$

$$\Delta M_\mathrm{MS} = \ln\left(\frac{\mathrm{MS}_{T,ji}}{\mathrm{MS}_{0,ji}}\right) \sum_{ji} \frac{M_{T,ji} - M_{0,ji}}{\ln M_{T,ji} - \ln M_{0,ji}} \tag{6.34}$$

6.2 Decoupling State Criteria

Decoupling was first proposed by Organization for Economic Co-operation and Development (OECD). It is used to measure the relationship between economic development and material consumption, stock accumulation, and pollution emissions; coupling suggests that one or more of the latter three factors increase with increasing development, whereas decoupling suggests that they decrease or increase more slowly relative to economic development. The degree and direction of the decoupling can characterize the inconsistency between driving factors and variables of interest (OECD 2002). Common decoupling models include the OECD decoupling index method (OECD 2002), the Tapio (2005) elasticity analysis method, and econometric analysis (Xiao 2011).

Tapio (2005) proposed an analysis based on elasticity to study the relationships between economic development and the transport capacity and carbon dioxide emissions in Europe, with the goal of advancing the development of decoupling theory. Tapio used the concept of "elasticity" to dynamically characterize the decoupling relationship among variables. The approach was more granular than the OECD approach in describing the relationships between indicators and more convenient for determining the base period for the indicators (Lei 2018). Tapio characterized the decoupling state using ranges of elasticity, and proposed eight decoupling categories, including weak decoupling, strong decoupling, expansive decoupling, recessive decoupling, expansive coupling, and recessive coupling (Table 6.1). The judgment thresholds of 0.8 and 1.2 for the decoupling elasticity index (DEI) were derived empirically; an elasticity between 0 and 0.8 was defined as weak decoupling or weak negative decoupling, whereas a value between 0.8 and 1.2 was defined as expansive or recessive coupling.

Reclassifying the decoupling states as different forms of decoupling and coupling can simplify this system into six categories (Fig. 6.3). During economic growth (i.e., $\Delta SE > 0$), coupling can be divided into coupling, relative decoupling, and absolute decoupling according to the relative changes in resource consumption and pollution emissions. Coupling means that the growth rates of socioeconomic indicators are smaller than the growth rates of resource consumption or pollution emissions, and the elasticity value is greater than or equal to 1. Relative decoupling refers to stable socioeconomic growth, and although resource consumption or pollution emissions increase, their growth rates are less than those of the socioeconomic indicators, and the elasticity value is between 0 and 1. Absolute decoupling defines as the sustainable socioeconomic growth, with resource consumption or pollution emissions decreasing during economic growth, with an elasticity value less than 0. During economic shrinkage (i.e., $\Delta SE < 0$), coupling can be divided into negative decoupling, recessive coupling, and recessive relative decoupling. Negative decoupling identify the negative socioeconomic growth, but resource consumption or pollution emissions still increase and the elasticity value is less than 0. Recessive coupling stands for a situation in which the rate of decrease of socioeconomic indicators is smaller than the rate of decrease of resource consumption or pollution emission, and

Table 6.1 Judgment criteria for the decoupling state of cities

Decoupling state		Decoupling elasticity index (DEI)	Rate of change in resource consumption or pollution emissions (ΔRP/RP)	Rate of change in socioeconomic indicators (ΔSE/SE)
Decoupling	Strong decoupling	DEI < 0	< 0	> 0
	Weak decoupling	0 ≤ DEI < 0.8	> 0	> 0
	Recessive decoupling	DEI > 1.2	< 0	< 0
Coupling	Expansive coupling	0.8 < DEI ≤ 1.2	> 0	> 0
	Recessive coupling	0.8 < DEI ≤ 1.2	< 0	< 0
Negative decoupling	Weak negative decoupling	0 ≤ DEI < 0.8	< 0	< 0
	Expansive negative decoupling	DEI > 1.2	> 0	> 0
	Strong negative decoupling	DEI < 0	> 0	< 0

Note DEI represents the decoupling elasticity index of Tapio (2005). RP shows the value of a resource consumption or pollution emission indicator and SE refers to the value of a coupled or decoupled socioeconomic indicator

the elasticity value is between 0 and 1. Recessive relative decoupling is a situation in which the rate of decrease of the socioeconomic indicator is greater than the rate of decrease of resource consumption or pollution emission, and the elasticity value is greater than or equal to 1. While the socioeconomic growth is negative, resource consumption or pollution emissions are also decreasing. If RP represents the resource consumption or pollutant emission and SE represents a socioeconomic variable (e.g., GDP), the reduction of ΔRP is smaller than that of ΔSE and the elasticity value is between 0 and 1. The reduction of ΔRP is larger than that of ΔSE and the elasticity value is greater than or equal to 1.

There are two major approaches to study a system's decoupling status: ① using a fixed base period as a reference to calculate the decoupling index (e.g., Tapio's DEI) to describe the changes since that base period for each year in the study period; ② using the previous year as a base reference to calculate the decoupling index for the current year. The former method can reflect the relative changes in the long-term relationships between economic growth, resource consumption, and pollution, but when changes in the driving and related variables are relatively synchronized, the

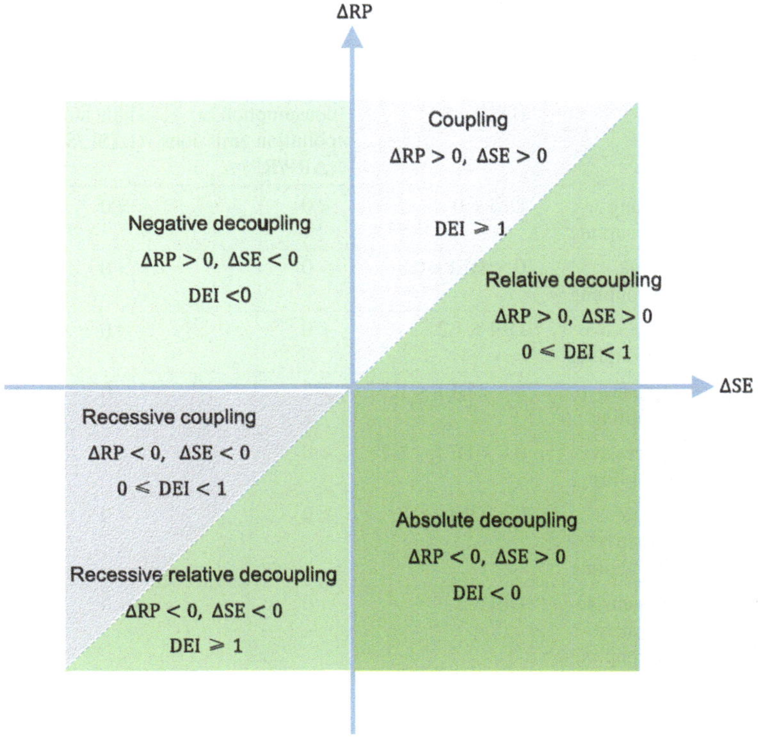

Fig. 6.3 Judgment criteria for the decoupling categories (*Note* DEI is Tapio's (2005) decoupling elasticity indicator, ΔRP is the change in resource consumption or pollution emissions, and ΔSE shows the change in socioeconomic indicators)

method ignores the changes of the structural characteristics that lead to changes in decoupling trends. The latter method measures the annual decoupling index, which aims to reflect the decoupling characteristics of the driving and related variables in each year. This can facilitate the identification and analysis of the related main influencing factors (Liu et al. 2014; Zhang et al. 2013b).

After calculating the urban metabolic flows and stocks using material flow analysis (discussed in Chapter 4), the decoupling model can be used to analyze the relationships between these flows or stocks and the socioeconomic indicators, mainly to assess the convergence effects (synchronous or asynchronous characteristics) for dynamic changes of the key metabolic indicators and socioeconomic indicators to determine whether decoupling or coupling is occurring. The key metabolic indicators are DMC, the in-use material stock, and various indexes that can be derived from this data. The socioeconomic indicators include GDP, population, per capita GDP, and investments in technology.

Using Tapio's (2005) decoupling model, we can analyze the coupling relationship between material metabolic characteristics and socioeconomic development. Based on the urban DMC and per capita GDP, a decoupling model can be established.

DMC represents the environmental pressure created by an urban metabolism, and per capita GDP presents the degree of socioeconomic development. Tapio's DEI is an elasticity index, and equals the ratio of the rate of change in DMC between the base period and the target year to the rate of change in per capita GDP over the same period. DEI is therefore used to measure the degree of separation between material consumption and per capita GDP, and is calculated as follows:

$$\mathrm{DEI} = \left(\frac{\Delta \mathrm{DMC}}{\mathrm{DMC}}\right) \Big/ \left(\frac{\Delta I}{I}\right) \tag{6.35}$$

where ΔDMC represents the difference of DMC between the target year and year 0, and ΔI represents the corresponding difference for the socioeconomic indicators.

The same approach can be used to explore the decoupling between the changes in an urban system's fixed stocks relative to the changes in the socioeconomic parameter. The DMC variable would then be replaced by a material stock (MS) variable, but the rest of the analysis would not change. Tapio's decoupling model can be used to analyze the correlation between urban in-use stocks (other indicators that can be derived from this data) and GDP (other indicators that can be derived from this data). MS represents how resource-intensive a city is, GDP reflects the degree of economic development, and DEI is the elasticity index, that shows the ratio of the rate of change in MS between the base period and the target year to the rate of change in GDP over the same period.

$$\mathrm{DEI} = \left(\frac{\Delta \mathrm{MS}}{\mathrm{MS}}\right) \Big/ \left(\frac{\Delta G}{G}\right) \tag{6.36}$$

where ΔMS represents the change of MS between the target year and year 0, and ΔG represents the change in GDP during the same period.

The correlation between the driving variable (GDP growth) and the related variable (stock growth) can be divided into three states: relative to the growth of the socioeconomic indicator (e.g., GDP), the stock growth in the same time period can be greater than or equal to (coupled), less than (relatively decoupled), or decreasing (absolutely decoupled).

6.3 Center of Gravity Model

In analyzing an urban metabolism, it can be useful to determine where certain activities are concentrated. Hence, one approach is to estimate the center of gravity of a particular process such as energy consumption or energy production; if there is a great distance between these two factors, this suggests the possibility of planning to bring them closer together. This planning would be useful in the context of energy because it is expensive to transport energy from where it is produced to where it is consumed (e.g., to create the required transportation infrastructure), and this transport can create additional problems such as emission of greenhouse gases (e.g., for road or rail transport) or environmental damage (e.g., due to leakage from pipelines). Using a center of gravity model and geographical information system (GIS) software, it's possible to quantify how a region's energy production and consumption centers changed during a given period both in terms of their positions and in terms of the magnitude of their movements, making it possible to calculate both the distances and the rates of these changes. Such changes can reveal regional differences in the relationship between supply and demand. To demonstrate how this works, the author will illustrate an analysis of the spatial distribution and centers of gravity for regional energy production and consumption and how they have changed over time. Since the research on which this section was based (Zhang et al. 2012) occurred in China, same country is considered in this example. The analysis starts with the assumption that each administrative region at a provincial level (including Chinese provinces, autonomous regions, and province-level municipalities) is located on a homogeneous plane (Aboufadel and Austin 2006); that is, the analysis was conducted in two dimensions, without accounting for the third dimension of elevation. In addition, It is assumed that the energy production and consumption by each region is concentrated in the provincial capital. Although this is clearly a simplifying assumption, it ensures consistency among regions that provide different levels of detail in the available data, and can be justified by the fact that most of a province's energy production and consumption are concentrated within or near the capital as a result of the rapid urbanization that has occurred throughout China in the past several decades. Each provincial capital city then acts as a particle on the plane, with weights assigned to the city based on the quantities of energy production and consumption. This simplification is also necessary because finer-grained data (e.g., at a county level) is not yet available throughout China.

The position of the center of gravity can then be calculated using a combination of the geographical coordinates of the capital cities and their corresponding energy production and consumption weights. Any differences in the positions of the centers for production and consumption represent imbalances between the two parameters, and differences in the movement of these centers of gravity over time represent changes in the balance over time. The position of the center of gravity can be expressed in terms of longitude and latitude, and can be calculated as follows:

$$X_t = \frac{\sum M_{ti} x_i}{\sum M_{ti}} \tag{6.37}$$

$$Y_t = \frac{\sum M_{ti} y_i}{\sum M_{ti}} \tag{6.38}$$

where X_t and Y_t represent the longitude and latitude coordinates (respectively) of the center of gravity in year t; M_{ti} shows the amount of energy production or consumption by region i in year t; and x_i and y_i stands for the longitude and latitude coordinates (respectively) of the capital city of region i.

In the original paper, my research group obtained a 1:4 million scale map of provincial administrative boundaries from the Resources and Environment Remote Sensing Database of China, and then used the spatial search and orientation functions of version 9.3 of the ArcGIS GIS software[2] to generate a database of provincial longitude and latitude coordinates (He et al. 2011). We then associated the energy production and consumption data with the map. Subsequently, we used Eqs. (6.37) and (6.38) to calculate the coordinates of the centers of gravity for energy production and consumption. Then the data loading function of the GIS software is used to find the centers of gravity in each year in the study period, and connected each center in turn to define the paths followed by these centers from year to year. This analysis revealed that some major consumptions centers, such as Beijing and Shanghai, were located quite far from their primary sources of energy, creating a large local gap between supply and demand. Although that gap can be reduced slightly by increasing the city's energy-use efficiency, it may be necessary to establish additional power sources closer to the city.

6.4 System Dynamics Simulation Model

The previous section shows that urban metabolism is spatially dynamic at scales larger than those of individual cities. Simulating these and other dynamics can provide additional insights into urban metabolism. After identifying the factors that influence the metabolic system, a system dynamics simulation model can be established. The model reflects the network of nonlinear causal relationships among the metabolic actors. The concept of system dynamics was first proposed by professor W. Forrester from the Massachusetts Institute of Technology based on systems theory (Forrester 1961), but represented an advance because of integrating cybernetics and information theory with systems theory. This discipline studies information feedback mechanisms. Although system dynamics is widely used because of its fast analysis, simple model construction, and ability to handle high-order and nonlinear problems, it also has obvious shortcomings. For example, when establishing the model equations that describe the internal relationships in a system, the equations are often affected by the modeler's understanding of the system's behavior (Forrester 1971).

Based on system dynamics, the author will establish a simulation model to support optimization of a city's industrial structure and its human carrying capacity. To do so, first the relationships between socioeconomic factors and the city's resource environment (specifically, pollution mitigation; Sect. 6.4.1) are comprehensively considered. Next, a similar approach to examine the factors that affect a city's human carrying capacity in terms of the city's ability to feed its residents (Sect. 6.4.2) is utilized. By drawing causal feedback diagrams and system flowcharts to establish the dynamics models, it will be discussed that how this approach can provide decision-makers with simulation programs that can assess different development scenarios (Gao et al. 2003).

[2] http://www.esri.com/.

Table 6.2 Index system for the system dynamics simulation model

Variable	Parameter
State variables	Pollution emission, population, GDP
Rate variables	Pollution rate, pollution purification rate, birth rate, death rate, rate of GDP growth, proportion of GDP invested to generate GDP growth
Auxiliary variables	Deaths caused by pollution, pollution generated during production, environmental carrying capacity
Constants	Proportions of primary, secondary, and tertiary industries; environmental investments; investments in science and technology

6.4.1 Optimization Model for a City's Industrial Structure

The dynamic simulation model for optimization of the urban industrial structure focuses on key actors, such as the city's residents (population), economy (GDP), and environment (pollution emission). Using the model, it is possible to analyze the impact of industrial structure adjustments on the status of these actors under different scenarios to support urban management and planning. The model divides the urban metabolic system into three subsystems: nature, society, and the economy. The natural subsystem will be described using emission indicators for exhaust gases, wastewater, and solid waste. The social subsystem will be explored using population as an important state indicator. The economic subsystem will be discussed by the changes in GDP and its structure (its production by primary, secondary, and tertiary industries). This will let us fully consider the impact of changes in the industrial structure on the development of the three subsystems. The selected social, economic, and environmental indicators are shown in Table 6.2.

The three subsystems are interconnected and influence each other through the coupling among the variables in Table 6.2. Changes in any one subsystem will affect the other subsystems, resulting in cyclical causal feedbacks. The main feedback loops are shown in Fig. 6.4. An increase in population and GDP will increase environmental pollution, which in turn will adversely affect the number of births of human beings, resulting in a decreasing population; however, as the quality of the environment decreases, the amount of available resources will also decrease, which will slow GDP growth. Changes in the industrial structure will also affect the amount of pollutants produced, which will also impact population growth.

The relationships among the variables and constants are shown in Fig. 6.5. Based on the causality diagram in Fig. 6.4, we can see the positive and negative feedback chains and loops, and use that knowledge to establish a dynamic simulation program using a dynamic programming language such as Dynamo[3]. According to the urban resource and environmental conditions, the city's industrial structure should be adjusted to promote the harmonious development of the natural, social, and economic subsystems. At the same time, the model also focuses on some factors, such as the population's educational level, introduction of new technologies, and investments

[3] https://dynamobim.org/.

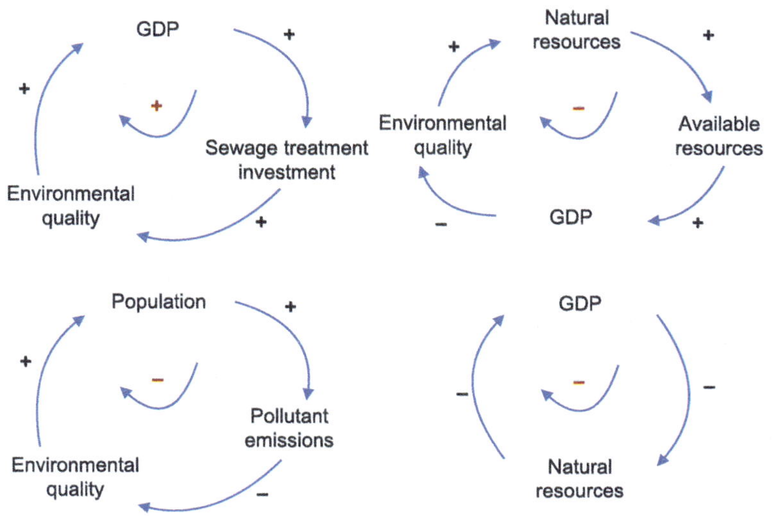

Fig. 6.4 Causality diagram and feedback loops for the dynamic simulation model (*Note* + indicates that increasing a factor will result in an increase of the downstream factor; – means that increasing a factor will result in a decrease of the downstream factor)

in science and technology. The natural subsystem has a close connection with the scientific and technological levels, which are in turn influenced by the education level, technology introduction, and technology investment. The level of science also has an impact on the pollution purification rate, which is related to the time required for purifying pollution and the pollution level, and the pollution generation rate also depends on the production and consumption conditions associated with different technology levels. There are cyclical relationships among these multiple indicators.

The author assumed three factors that could be regulated as auxiliary variables: adjusting the proportion of the tertiary industries to represent industrial restructuring, adjusting the technology level to represent increased environmental protection, and investments in science and technology to represent increased pollutant purification efficiency. The model then integrated these three important variables with the state variables, rate variables, and other auxiliary variables. To develop scenarios, it's possible to choose from one to three of these three control variables. For example, we can compare the status quo (i.e., take no action) with an ideal plan that comprehensively regulates the values of various variables and with a sub-optimal plan that focuses only on socioeconomic development or protection of the ecological environment. In doing so, we can support the development of a more scientific and reasonable action plan to provide an important reference for decision-makers to improve urban resources and environmental conditions and socioeconomic development, and especially optimization of the city's industrial structure (Zhang and Yu 2003).

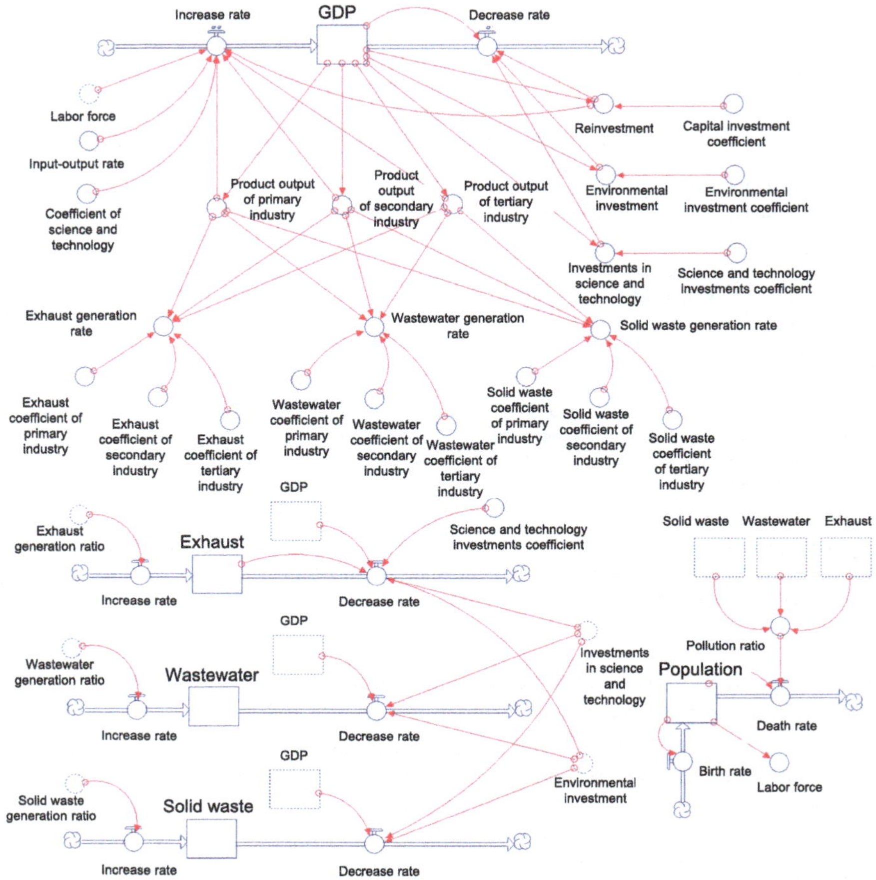

Fig. 6.5 Illustration of the dynamic simulation model for optimization of a city's industrial structure

6.4.2 *Optimization Model for Human Carrying Capacity*

Human carrying capacity represents the number of people that a city or its land can continuously support in a certain area. It is determined by the productive capacity of the land (i.e., its ability to supply the resources and services humans need to survive) and the environment's resilience against the pressure imposed by these people (i.e., its ability to absorb the pressure these humans impose without experiencing progressive degradation). The purpose of research on a city's human carrying capacity is to maintain the ecological balance rather than increasing the population at the expense of damaging the ecological environment. It's not possible to expand the population beyond a certain point, no matter how much technology we apply to the problem. For simplicity, let's assume that this carrying capacity depends only on the area of cultivated land. By simulating the causal feedbacks between the population and the amount of cultivated land, we can establish an optimization model for the human carrying capacity (Wang et al. 2003). The model includes two state variables: the area of cultivated land and the population. In addition, the model includes rate variables and auxiliary variables such as the birth rate, death rate, rate of land reclamation for agriculture, the city's population policy, and food production per unit area of cultivated land. The resulting model can reflect the land's production potential and the human carrying capacity by focusing on the quantity and quality of the cultivated land.

Population growth will inevitably increase the food demand, leading to a larger gap between demand and production, so we must increase the food supply to meet the needs of the city's growing population. We can accomplish this by increasing the area of cultivated land or increasing the yield per unit area of this land. However, there is a tradeoff; increasing the number of people will cause an increase in the area of construction land to build homes and other resources (e.g., hospitals, markets) for these people. Because the area of land is fixed, there will be competition between agriculture and construction: if the area of one land use increases, the area of the other land use must decrease. The basic causal feedback loops are shown in Fig. 6.6.

Based on the causality diagram, we can identify various positive and negative feedback chains and loops, and establish a dynamic simulation program using a programming language such as Dynamo (Fig. 6.7). We can also predict the development trend of the urban population and dynamic changes in the area of cultivated

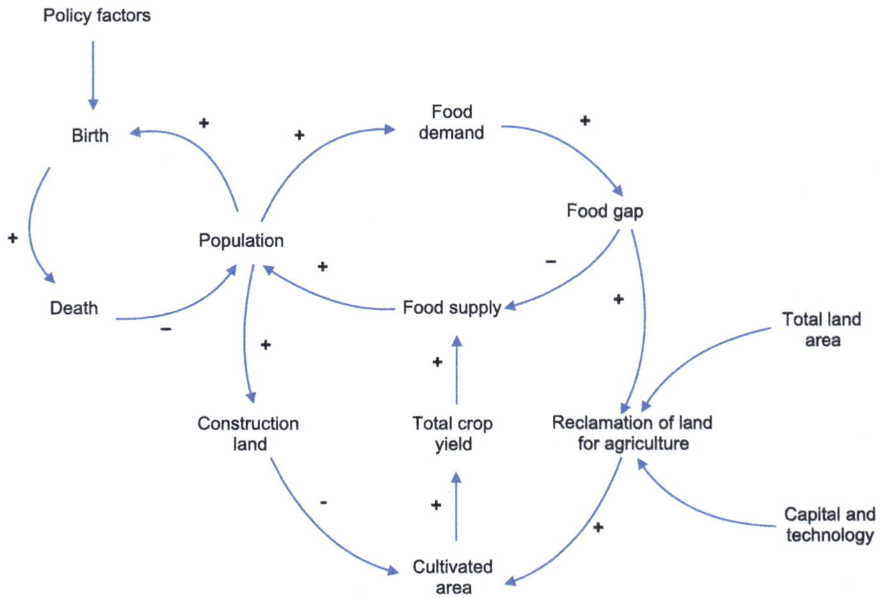

Fig. 6.6 Illustration of the causal feedbacks that determine the human carrying capacity of a city (*Note* + indicates that increasing a factor will result in an increase of the downstream factor; − means that increasing a factor increase will result in a decrease of the downstream factor)

land through the control parameters. It is possible to then define different scenarios for the relationships between humans and the city's land uses. By comparing the scenario results, the natural population growth, and develop plans to moderate, regulate, and optimize the human carrying capacity, thereby providing an objective basis for achieving sustainable urban development can be described.

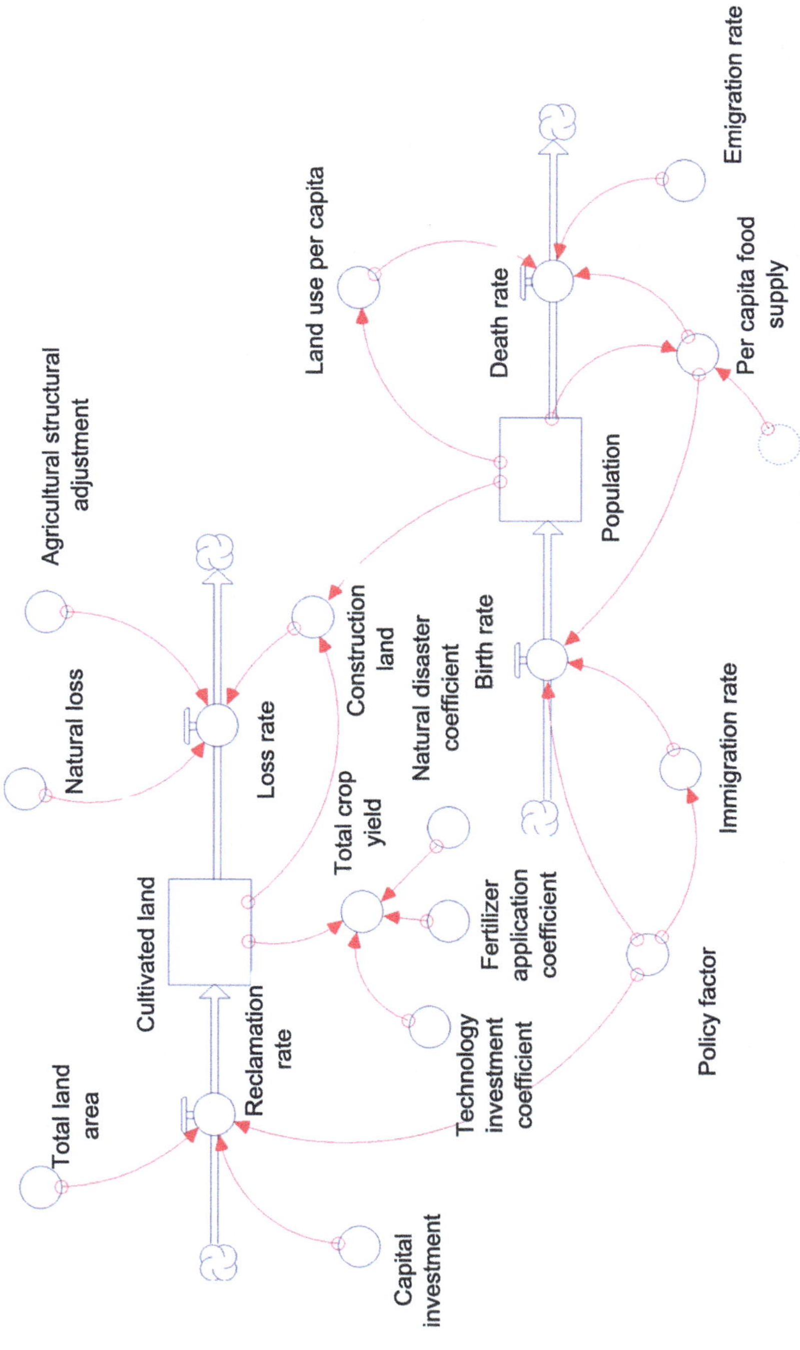

Fig. 6.7 Dynamic simulation model for optimizing the human carrying capacity of a city

References

Aboufadel E, Austin D (2006) A new method for computing the mean center of population of the United States. Prof Geogr 58(1):65–69

Ang BW (2004) Decomposition analysis for policymaking in energy: which is the preferred method? Energy Policy 32(9):1131–1139

Ang BW (2005) The LMDI approach to decomposition analysis: a practical guide. Energy Policy 33(7):867–871

Chen JL, Yuan HM, Tian X et al (2019) What determines the diversity of CO_2 emission patterns in the Beijing-Tianjin-Hebei region of China? an analysis focusing on industrial structure change. J Clean Prod 228(15):1088–1098

Forrester JW (1961) Industrial dynamics. The MIT Press & John Wiley and Sons, New York

Forrester JW (1971) Urban dynamics. The MIT Press, Cambridge, MA

Gao CK, Shang JC, Wang RX et al (2003) Simulation of Changchun city water environment system. Adv Water Sci 14(6):725–730 (in Chinese)

He YB, Chen YQ, Tang HJ et al (2011) Exploring spatial change and gravity center movement for ecosystem services value using a spatially explicit ecosystem services value index and gravity model. Environ Monit Assess 175(1–4):563–571

Lei HX (2018) Study on the relationship between energy consumption, carbon emission and economic growth in Shanxi province. Yunnan: Yunnan University of Finance and Economics, MSc Dissertation (in Chinese)

Li YX, Zhang Y, Yu XY (2019) Urban weight and its driving forces: a case study of Beijing. Sci Total Environ 658(6):590–601

Liu AD, Zeng HX, Liu WJ (2014) Empirical study of decoupling relationship between carbon emissions and export trade in China. China Popul Resour Environ 24(7):73–81 (in Chinese)

OECD (Organization for Economic Co-operation and Development), (2002) Indicators to measure decoupling of environmental pressure from economic growth. OECD, Paris, France. (2002-5-16)[2021-12-1]. http://www.oecd.org/officialdocuments/publicdisplaydocumentpdf/?cote=SG/SD(2002)1/FINAL&docLanguage=En

Tapio P (2005) Towards a theory of decoupling: degrees of decoupling in the EU and case of road traffic in Finland between 1970 and 2001. Transp Policy 12(2):137–151

Wang QF (1995) Advanced system dynamics. Tsinghua University Press, Beijing (in Chinese)

Wang JH, Gu YX, Sun LY (2003) Research on system dynamics model of man-land relationship. Syst Eng Theory Pract 23(1):128–131 (in Chinese)

Xiao X (2011) Research on the spatiotemporal changes of carbon emissions in Jiangsu cities over the past 15 years. Nanjing University, Nanjing (in Chinese)

Zhang Y, Yu XY (2003) System dynamics optimization simulation study on the environmental impact of industrial structure in Changchun City. Econ Geogr 23(5):681–685 (in Chinese)

Zhang Y, Zhang JY, Yang ZF et al (2011) Regional differences in the factors that influence China's energy-related carbon emissions, and potential mitigation strategies. Energy Policy 39(12):7712–7718

Zhang Y, Zhang JY, Yang ZF et al (2012) Analysis of the distribution and evolution of energy supply and demand centers of gravity in China. Energy Policy 49(10):695–706

Zhang C, Cai WH, Yu TS (2013a) Regional economic growth and carbon productivity. China Industrial Economics 5:18–30 (in Chinese)

Zhang JY, Zhang Y, Yang ZF et al (2013b) Estimation of energy-related carbon emissions in Beijing and factor decomposition analysis. Ecol Model 252(S1):258–265

Zhang XL, Zhang Y, Fath BD (2020) Analysis of anthropogenic nitrogen and its influencing factors in Beijing. J Clean Prod 244(2):118780

Part III
Applications

Based on the theoretical, technical, and methodological framework for urban metabolism described in previous chapters, this section describes the analysis of urban metabolic processes for materials, energy, and specific elements such as carbon, nitrogen, and sulfur, as well as the application of these techniques at the scales of cities, urban agglomerations, regions, and countries. In doing so, I demonstrate the validity of the urban metabolism model and how it has improved the theoretical foundations of urban metabolism research.

Based on the stock-flow approach, which combines analyses of flows of the abovementioned substances and the resulting changes in their urban stocks, my research group analyzed material metabolic processes, calculated urban weights (i.e., the actual physical weight of a city's materials), analyzed the characteristics of urban material metabolic networks, identified the factors that most strongly influence material consumption, and provided empirical support for efforts to optimize urban weights and adjust the city's metabolic structure. We also extended the analysis of energy metabolism processes from the tracking of energy flows to the analysis of emergy flows to permit unified measurements of materials, energy, and information flows (i.e., flows with different units of measurement). From that starting point, we moved on to flows of embodied energy (i.e., the equivalent energy carried by flows of economic value). I describe each of these studies to demonstrate how the techniques from previous chapters can be applied to generate real-world insights.

Analyzing network models with different precision makes it possible to identify the features of energy flows and transfers. In this context, I describe the efforts of my research team to enrich the literature on case studies of urban, inter-city, and regional energy metabolism research. By applying network analysis and input-output analysis to analyze a city's carbon metabolism processes, it becomes possible to comprehensively consider the flows of a substance such as carbon in physical form and the carbon transfers embodied in exchanges of goods and services. It enables us construct spatially explicit models of urban carbon metabolism networks, including carbon transfers, by quantifying the carbon emission or absorption that results from land use and land cover change during urbanization. The resulting model provides an important reference for adjusting the carbon reduction targets for cities, urban agglomerations, countries, and the world.

Based on the theory of complex urban ecosystems, we developed a nitrogen metabolic network model that focused on food and energy consumption, which enable us analyze the characteristics of their consumption structures and functional relationships. By introducing the theory of food chains within networks, which define the trophic levels of the network's metabolic actors, we can simulate their ecological hierarchy, and identify the roles of the metabolic components by analyzing the structure and external drivers of nitrogen consumption, thereby providing empirical support for comprehensive management decisions on how to adjust nitrogen consumption and reduce nitrogen emission.

I conclude the book by analyzing the sulfur metabolism within an eco-industrial park designed to maximize the reuse of waste and byproducts to minimize their release into the city's environment. By tracing the relationships between industries in the park, it was possible to clarify the industrial symbioses within the park's network and simulate their ecological characteristics (such as the distribution of flows and functional relationships). The results provide an important basis for the ecological transformation of industrial parks and promote a circular economy that emphasizes recycling and flows that resemble natural ecosystems more closely.

Chapter 7
Analysis of Material Metabolic Process: Urban Weight

One of the consequences of material flows is that they create a physical weight, both from the flows and the stocks, and that weight exerts direct pressure on the Earth beneath the city. That pressure serves as an interesting proxy for the stress a city places on its environment. In this chapter, I will build on previous chapters to describe how these weights are determined and their consequences for urban metabolism. Here and in subsequent chapters, I will also move from a purely theoretical discussion of these topics to providing actual data on flows and weights to illustrate the importance of metabolic research and how it can be used to support real-world policy development and management of an urban ecosystem.

7.1 Urban Weight Analysis from a Flow Perspective

Cities create large benefits, as they are the main places for human social innovation and wealth creation, but the accelerating process of global urbanization has led to huge problems related to excessive resource consumption and the resulting environmental pollution. The global urbanization rate (i.e., the proportion of the world's population that lives within a city) is expected to proliferate, increasing from 52% in 2010 to around 65% in 2050. By 2050, the consequent urban material consumption is predicted to increase to 90 Gt, 2.25 times the value of 40 Gt in 2010 (Swilling et al. 2018). China's current urbanization rate is 56.1%, and the urbanization rate of Beijing, China's capital, maybe as high as 87%. As a result, Beijing's resource consumption is increased. Despite having only 1.6% of the national population, Beijing consumes 4.1% of the country's gasoline, 7.6% of its natural gas, and 5.6% of its liquefied petroleum gas; as a result, it produces 4.1% of China's domestic waste and 2.1% of its wastewater (BBS 2016). To promote the sustainable development of a city's urban resources, it is necessary to measure the city's weight. This is the only way to obtain a reliable understanding of resources being consumed (i.e., for

Y. Zhang, *Urban Metabolism*,
https://doi.org/10.1007/978-981-19-9123-3_7

all substances and materials combined), at what rate, and by whom. For this knowledge, it's helpful to use the techniques I described in previous chapters to treat cities as living organisms. This lets us explore the urban material flows and the underlying processes by taking advantage of the analogy with the metabolism of a living organism.

The material flow analysis is an important tool for measuring the material exchanges between a city's socioeconomic system and its natural environment. It provides details of the nature and the magnitude of the flows within an urban metabolism; it has become a standard method of urban metabolism study. Furthermore, this approach can trace the sources, flows, and destinations of materials within an urban system (Brunner 2007). In the research on which this section is based (Li et al. 2019a), my research group first constructed a sector-scale accounting framework for material flows to measure the urban weight by analyzing the amount and structure of urban resource consumption (Fig. 7.1). Next, by defining the details of the material flows during a city's metabolic processes, we determined the contributions of the key metabolic components to the urban weight. Using the logarithmic mean Divisia index (LMDI) framework described in detail in Chapter 6, we established a quantitative relationship between urban material consumption changes and the driving factors, clarified the contributions of the different factors to urban weight, and identified where it would be necessary to improve urban resource-utilization efficiency. To make this approach concrete, we used Beijing's data for a case study. In this chapter, I will review the methods and describe the results they produced.

We obtained data from 2000 to 2015 at an annual resolution from the Beijing Statistical Yearbook, China Energy Statistical Yearbook, China Agricultural Yearbook, China Industrial Yearbook, and other relevant statistical datasets. Most countries will provide similar statistical resources that can be used to support a material flow analysis. However, details of their contents and resolution will differ from those in China. The units of measurement were all unified in mass terms by using appropriate conversion factors. For example, where the statistics recorded the number of units, we used available data on the weight per unit; where the statistics recorded the economic value of a flow, we used the market price of the material to estimate the number of units represented by that flow and then converted that number into a total weight using the weight per unit. After the unification, we obtained the final material accounting data for Beijing's socioeconomic system components. To observe changes over time, we divided the study period into three parts (2000–2005, 2005–2010, and 2010–2015) and then examined the relationships between Beijing's urban weight changes and the factors that influenced these changes during the three periods.

7.1.1 Analysis of Urban Flows' Weight and Its Structure

A wide extensive range from 206.6 Mt (equivalent to 15.1 t/person and 6.5 t/10,000 RMB of GDP) to 359.6 Mt (16.5 t/person and 1.6 t/10,000 RMB of GDP) belonged to Beijing's direct material consumption (DMC) during 15 years we focused on

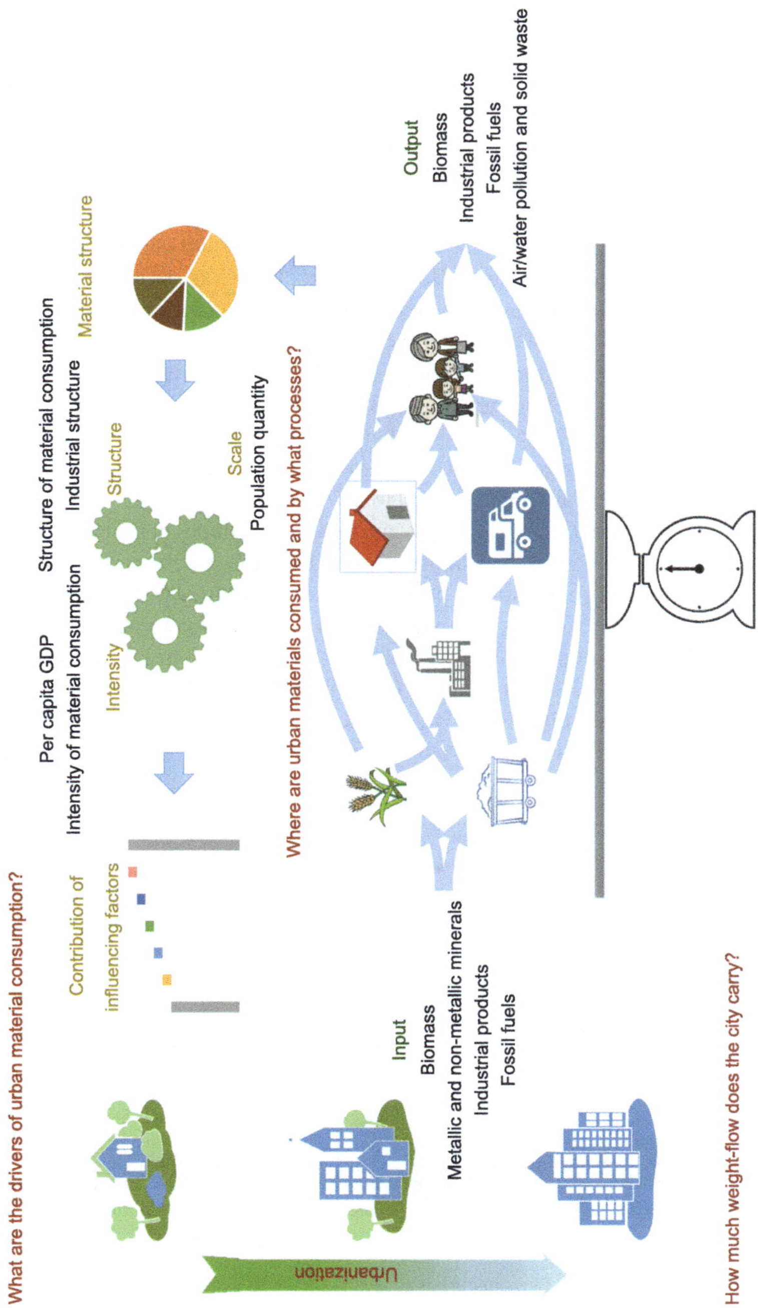

Fig. 7.1 An analytical framework for urban weight analysis from the perspective of metabolic flows

(Fig. 7.2). The reason that both the GDP and the population were changed during the study period, the amounts per RMB and per person did not increase proportionally. Non-metallic minerals (overwhelmingly consisting of construction materials), which accounted for about 45% of DMC, drove changes in the structure and magnitude of DMC. This significant increase exists for these two reasons: rapid urban development due to urbanization and the construction of venues for the 2008 Olympics and associated infrastructure. There was a slight decrease in non-metallic minerals between 2006 and 2009. And this trend is consistent with the effects of the ended housing boom due to the global financial crisis and national and global recessions, which represents the decrease in the use of materials such as sand and gravel associated with urban construction activities. However, this effect was short-lived and the mass of non-metallic minerals was recovered after 2009. The consumption of fossil fuels, accounting for about 20% of DMC, increased slowly during 15 years period, meanwhile, its contribution to the variation of the DMC was much smaller than that of the non-metallic minerals. For the industrial products, its weight increased obviously (to about 50 Mt) from 2006 to 2009, whereas the weight for metallic minerals decreased significantly (to about 16 Mt). The consumption of renewable materials (i.e., biomass) decreased from 7.5% in 2000 to 3.7% in 2015.

As Beijing's local extraction of resources and production fluctuated over time, the total material consumption increased significantly. There was a noticeable reduction in the proportion of total consumption provided by extraction or production inside Beijing, especially for the biomass component. This is a common trend that accompanies urbanization and urban economic growth in both developing and developed countries. Beijing depended increasingly on extracting biomass and fossil fuels from the environment outside the city (Fig. 7.3). As discussed in previous chapters, I have defined Beijing's internal environment as the city's administrative boundary. More

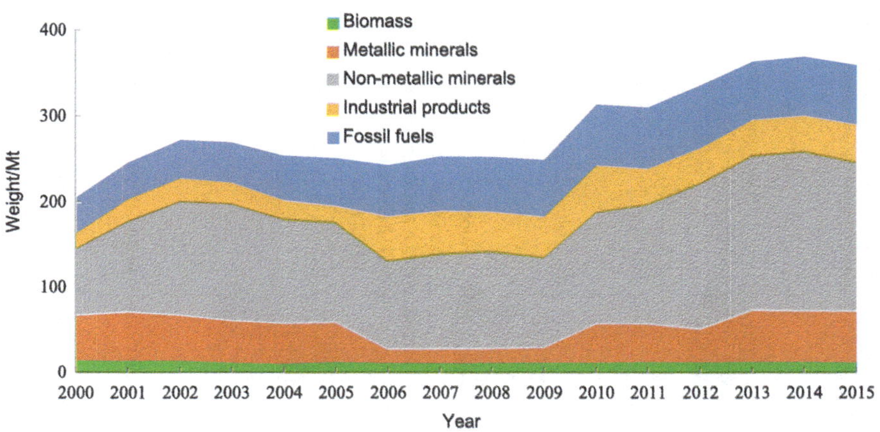

Fig. 7.2 Changes in the weights of key materials in Beijing and the associated structural composition from 2000 to 2015 (*Note* Weights in a given material category equal the width of the colored area; the top of each colored area represents the cumulative total)

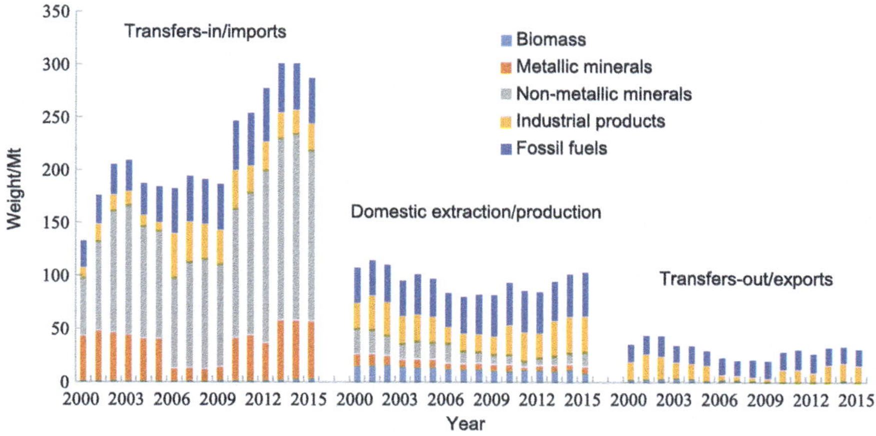

Fig. 7.3 Production, import, and export structures of materials in Beijing from 2000 to 2015

than 80% of the biomass and 50% of Beijing's fossil fuels were obtained outside its administrative boundary. The dramatic decrease in biomass extraction (about 70%) resulted from decreased planting of cereal and vegetable crops. The extraction of metallic and non-metallic minerals, and the production of industrial products, were significantly reduced by 2008, primarily due to Beijing's restrictions on mining, processing, and manufacturing activities before and during the Olympics. These activities consume large amounts of energy and require the combustion of fossil fuels, leading to significant air pollutant emissions.

Beijing imported large amounts of metallic, non-metallic minerals, industrial products, and fossil fuels outside the city's boundary (Fig. 7.3). These imports accounted for about 65% of total consumption during the study period. The most significant imports were for non-metallic minerals, which accounted for more than 40% of Beijing's total imports. Imports of industrial products increased from 10.3 Mt in 2000 to 26.5 Mt in 2015, which indicates that the demand for processed products increased because of the increasing population and accelerating economic development. The 11th Five-Year Plan (2006–2010) promoted that high-consumption industries should be transferred from Beijing to neighboring regions (e.g. Tianjin and Hebei) to reduce Beijing's environmental pressure. It is just the decreased scale of Beijing's internal industrial production that has promoted imports of these materials from outside. However, there is a sharp decrease in imports of metallic minerals between 2005 and the period from 2006 to 2009, from about 40 Mt to around 11 Mt. The reason is related to changes before the 2008 Olympics. One reason is that many industries that consume metallic minerals for industrial production, mainly used as raw materials, were moved outside of Beijing. The other is that Beijing Municipal Government strictly controlled the production scale of Beijing's industry sectors, especially of highly-polluting industries such as metallurgy, in order to ensure excellent environmental conditions during the Olympic Games.

Massive materials inputs met the city's needs, meanwhile increasing its weight. Some of the urban weight were stored in stocks, such as primary building materials and industrial products. The remainings were discharged from the city as exports and waste, which are named by direct material outputs. In fact, building materials for an overwhelming proportion (e.g. concrete, cement, steel, and other minerals), were essential constituents of Beijing's building stock, and they had a limited effect on variations in exports and waste emissions.

Beijing's material consumption was equivalent to 5.8 times that of transfers-out and exports (Fig. 7.3), and Beijing was always a net importer of material consumption. For fossil fuels, the output remained relatively stable (at about 15 Mt), but net input increased to 3.2 times compared with its initial level (from 8.6 Mt in 2000 to 27.5 Mt in 2015). In 2009, Beijing's industrial exports obtained their minimum value due to restrictions on raw materials for industrial production. Although exports have risen slightly since 2010, Beijing was still changing from a net exporter of industrial products (which can be seen between 2000 and 2005) to a net importer (which can be seen between 2006 and 2016). Biomass exports decreased due to a reduction in biomass extraction. Beijing, therefore, changed from a net biomass exporter (which can be seen between 2000 and 2007) into a net importer (which can be seen between 2008 and 2015).

Air pollutants were the dominant flow in terms of waste and pollution, accounting for more than 70% of direct material outputs (Fig. 7.4). The primary source of air pollutants was fossil fuel combustion, clearly illustrating the significance of reducing these emissions by all sectors. Manufacturing accounted for the most enormous amount of total air pollutant emission (at about 66%), followed by Transportation (at about 23%). Beijing is important as its status as air transportation and national railway hub, with railway passenger volume throughput of 3.7 billion/year and airport passenger volume throughput of 1 billion/year. In addition, a large number of personal vehicles, more than 5.6 million in 2015, belonged to Beijing.

Figure 7.4 also shows an interesting result that the total waste emission was droved by the increase of solid waste. In general, large cities' high-level economic activities need significant inputs of goods and materials. Except for the immobilized materials, the remainder is transformed chiefly into waste in the city's stock (Kennedy et al., 2007). Solid waste emission grew by 3.7 times compared with the level in 2000, and this rate was faster than the population growth rate (an increase of 0.6 times) compared with the 2000 level during the study period. Due to the increasing incomes of urban residents, their higher standard of living resulted in a larger demand for various goods, such as food, clothing, transportation, and household and telecommunication. Therefore, the increased environmental pressure from household waste emissions was the reason for changes in consumer behavior.

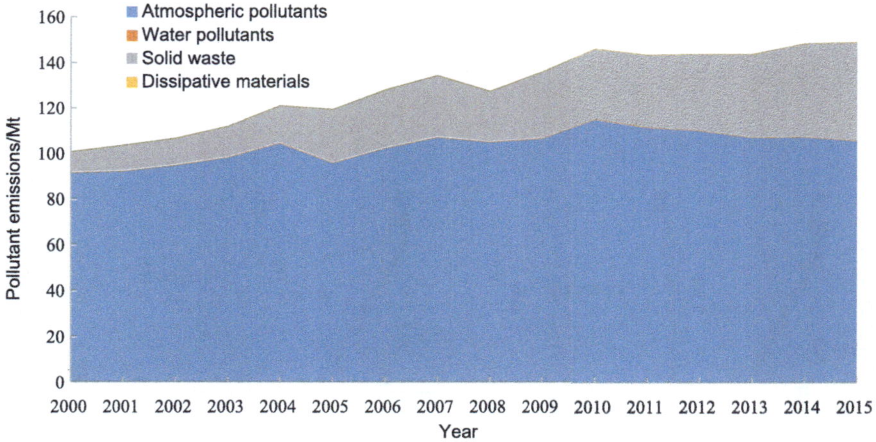

Fig. 7.4 Quantities and structure of Beijing's waste emissions from 2000 to 2015 (*Note* Emissions in a given pollutant category equal the width of the colored area; the top of each colored area represents the cumulative total. Dissipative materials represent materials that are deliberately dissipated into the environment because of their inherent characteristics. Examples include inorganic and organic fertilizers and pesticides. At the scale used in this graph, the quantities of water pollutants and dissipative materials (<0.5 Mt) are too small to be visible)

7.1.2 Contributions of the Metabolic Components

Illustrating the flows among the city's metabolic components helps to reveal the vital metabolic processes (Fig. 7.5). The main inputs for Beijing's urban system were through the Construction, Manufacturing, and Mining sectors, with negligible flows through the remaining sectors. through the metabolic processes. Among the above sectors, the Construction sector contributed the maximum flows through the metabolic processes, and therefore the Construction sector was the dominant importer, receiving about 47% of Beijing's total material imports. Imports by Construction, specifically, increased from 51.9 Mt to 153.2 Mt between 2000 and 2015 (nearly 3 times the 2000 value), in which non-metallic minerals accounted for the overwhelming majority of 90% and biomass and industrial products accounted for 10%. In a word, there was a massive consumption of materials in the Construction sector. However, most material inputs of the Construction sector were immobilized in stock, and only a tiny part was discharged as waste.

Manufacturing was closely connected with both the external region and other sectors by large flows, which utilized around 55% of resources extracted within Beijing and 32% of imported resources. The primary imported materials of the manufacturing sector, which reached minimum levels around 2008, included metallic minerals (about 63% of the total) and non-metallic minerals (about 26% of the total). Manufacturing was also Beijing's largest source of exports, accounting for 54% of the total exports. Although Beijing's industrial production was mainly export-oriented, Manufacturing remained an important supplier for the Construction and

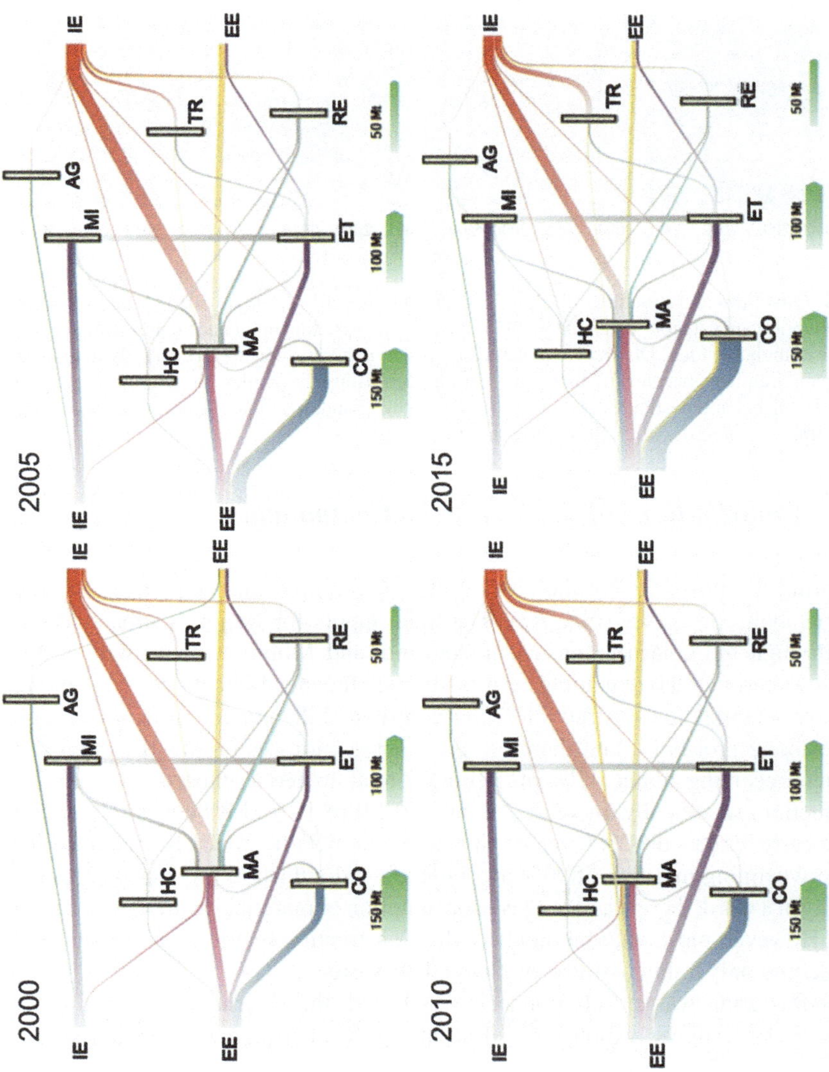

Fig. 7.5 Flows of materials among Beijing's main socioeconomic sectors at four points during the study period (Nodes: AG-Agriculture; CO-Construction; EE-External Environment (the area beyond Beijing's administrative boundaries); ET-Energy Transformation; IE-Internal Environment (the environment within Beijing's administrative boundaries); HC-Household Consumption; MA-Manufacturing; MI-Mining; RE-Recycling; TR-Transportation)

Household Consumption sectors (Li et al. 2019a), contributing 35% of Beijing's demand for industrial products. Manufacturing, in addition, was the largest waste producer (about 67% of total waste emission) and created massive air pollution (at 60 Mt in 2015), causing tremendous pressure on the city's environment.

However, there was a continuing downward trend in Manufacturing waste during the study period. Mining played a leading role in the maximum resource exploitation Specifically speaking, Almost 80% of the extraction and production inside Beijing flowed into Mining; the remainder was achieved by Manufacturing and Agriculture. For Mining's material exploitation, it was mainly composed of coal-based fossil fuels (about 71% of the total exploitation by Mining), and non-metallic minerals (about 29%). Biomass was supplied to other sectors for consumption and export by Agriculture, whereas metallic minerals extracted inside Beijing were used as raw materials for Manufacturing. Although Beijing's economic level improved continuously with the increasing population, the extraction of all materials has decreased (Li et al. 2019a).

Energy Conversion and Recycling sectors produced a marked effect in transferring materials. Fossil fuels from domestic reserves in Beijing or imports mainly flowed through the Energy Conversion sector, which supported the city's consumption activities and exports. After accounting for the waste emissions directly into the air, the solid waste and wastewater, which were generated by urban production (about 60% of the total waste emission) and living (about 40%), were collected by the Recycling sector. After treatment, there was a relatively high proportion of the waste (39.2%) reused by Manufacturing. And the remaining waste was released into the environment.

7.1.3 Identification of the Driving Forces Behind the Urban Weight

Using the LMDI method, it could be discovered that the material consumption intensity (i.e., the amount of the material consumed per 10,000 RMB of output) was the only inhibitory effect compensating for the increase in consumption of materials(Table 7.1). Changes in these contributions are illustrated in Fig. 7.6. Beijing's material consumption intensity decreased throughout the study period, from 6.5 t/10,000 RMB in 2000 to 1.6 t/10,000 RMB in 2015, and this trend can be explained by the government's policies to improve resource utilization efficiency. For example, Beijing's government invested heavily in workers, physical resources, and financial support, and meanwhile cooperated with research organizations and enterprises, to enhance material-saving technology, then it could spread throughout the whole production chain. Moreover, Beijing's government promoted the transformation of traditionally coal-reliant production and living patterns towards production and living patterns relying on cleaner energy(e.g. natural gas and electricity) by providing financial support. As a result, the energy-use efficiency of Beijing was improved, but even with the background of rising consumption, the inhibitory effect of the efficiency improvements was weakening due to the fact of slow-growing improvements.

The economic activity effect has been the most significant driver among the factors which contributed to material consumption. The per capita GDP in Beijing increased from 23,000 RMB in 2000 to 106,000 RMB in 2015. On the one hand, the increasing urban economic level has promoted the development of the Construction and Manufacturing sectors, because there was a higher material demand in Beijing to support the economic growth. On the other hand, the lifestyle of urban residents changed with their income growth, which means more significant resource inputs and more waste emissions. However, it is worth saying that the growth rate of per capita GDP has gradually decreased, and this means that the corresponding effect on material consumption has weakened. As a microcosm for China's cities, Beijing also reflects an ongoing profound transformation of the Chinese economy. The economic growth rate has slowed, but it has been accompanied by demands to improve the quality of the economic development (Fig. 7.7).

Another critical driver of Beijing's increasing material consumption was the population effect. From 2000 to 2015, Beijing's population grew by 59.2%, reaching 21.7 million people in 2015, and therefore it is predictable that the consumption of urban resources has increased with the growing demand of the population. In the 15 years we focused on, the population growth was the fastest, with a 27.6% increase from 2005 to 2010, about twice as the other two periods. Beijing's population growth has slowed since 2010 due to the government successfully enforcing a population control policy. However, it is projected that Beijing's population will increase because of the attractiveness and inertia that makes prominent demographic trends change slowly. Thus, more resources should be allocated in Beijing to reduce the pressure on the environment created by a large and growing population.

Although the effects of the industrial structure and material consumption structure were finite on Beijing's material consumption, a noticeable shift in structure began in the industrial during the study period. It is evident that rapid economic growth and urbanization have dramatically improved urban residents' incomes in China, undoubtedly causing more expenditure on service industries and businesses. And the demand for energy and industrial products related to these sectors therefore increased. Meanwhile, the government tried to promote an industrial transition to reduce pressure on the environment, in which industrial enterprises with high consumption of resources and high emissions of waste are transferred to both neighboring and distant regions outside Beijing's administrative boundaries.

This has dramatically decreased the number of enterprises and their production scale. As has occurred during the development of other megacities, two primary sectors, Agriculture and Mining, have been replaced by other sectors due to their importance decreasing. Despite these apparent changes, the structural change only slightly influenced Beijing's increasing material consumption as the overall effects of a minor decrease.

Table 7.1 Contribution of influencing factors (mass [ΔM] and proportion of the total) to the change of direct material consumption (DMC) in Beijing

	ΔM_{MS}		ΔM_{MI}		ΔM_{S}		ΔM_{EA}		ΔM_{P}		ΔM /Mt
	Contribution /Mt	C/%	Contribution /Mt	C/%	Contribution /Mt	C/%	Contribution /Mt	C/%	Contribution /Mt	C/%	
2000–2005	−0.2	−0.07	−85.1	−40.10	2.3	1.10	99.2	46.75	25.4	11.98	41.7
2005–2010	1.1	0.46	−96.4	−38.92	14.8	5.97	73.0	29.46	62.5	25.20	54.9
2010–2015	−0.6	−0.68	−8.3	−10.29	20.0	24.63	20.0	24.63	32.3	39.76	63.3
2000–2015	0.4	0.08	−189.9	−35.18	37.1	6.88	192.2	35.61	120.1	22.26	160.0

(Note C is the contribution (proportion of the total); ΔM_{MS}, ΔM_{MI}, ΔM_{S}, ΔM_{EA}, and ΔM_{P} are the changes in the material consumption structure, material consumption intensity, industrial structure, economic activity (per capita GDP), and population size; ΔM is the change in total consumption)

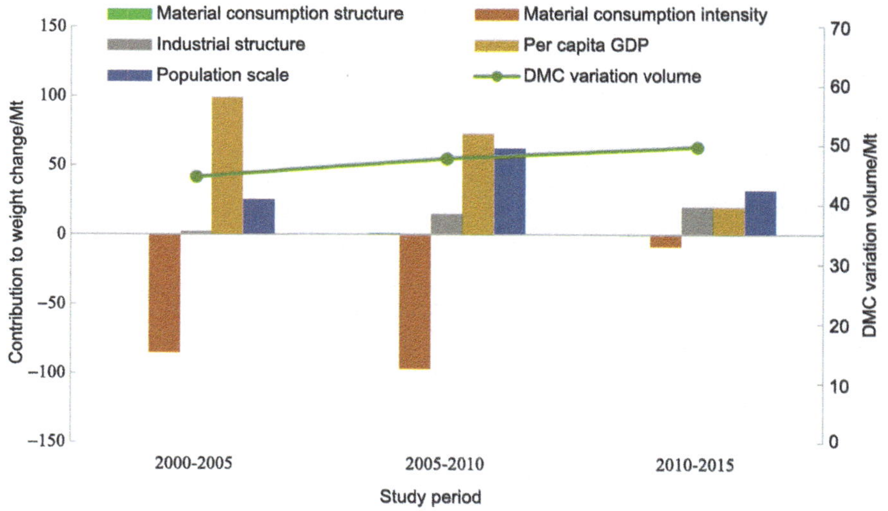

Fig. 7.6 Decomposition of the factors responsible for the total change in direct material consumption (DMC) in Beijing from 2000 to 2015 (*Note* Positive values represent factors whose contribution increased consumption; negative values represent factors whose contribution decreased consumption)

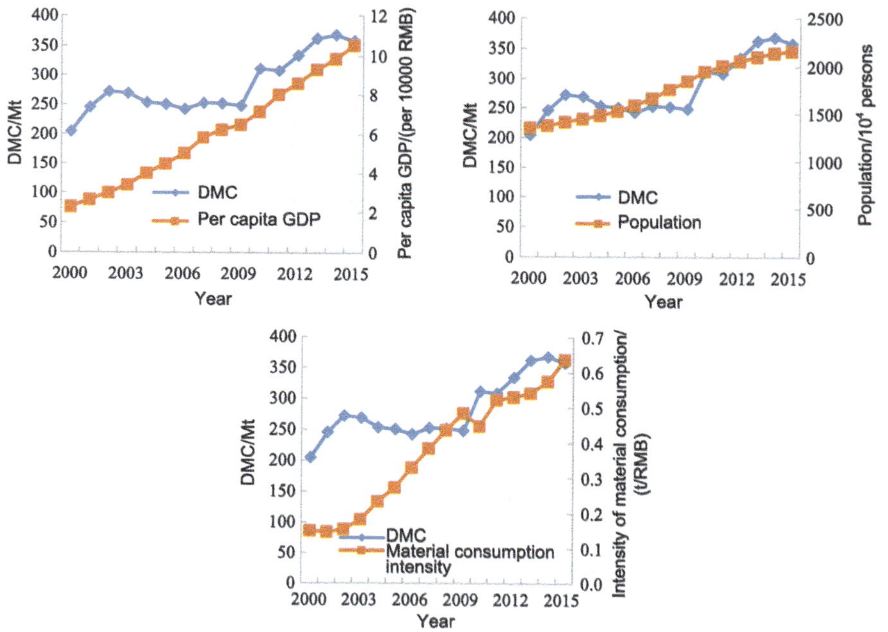

Fig. 7.7 Changes in the direct material consumption (DMC), population, per capita GDP, and material consumption intensity (DMC/10,000 RMB) during the study period

7.1.4 The Significance of Measuring Urban Weight from the Flow Perspective

Measuring the weight of cities helps us to reveal the nature and magnitude of their resource demand and the resulting environmental pressure. This knowledge tells planners and policy developers that they can promote urban development more sustainably by improving higher resource-utilization efficiency. Material flow analysis has been developed as a relatively mature methodology to support this research in both theoretical and practical applications (Fischer-Kowalski et al. 2011). It has been used to measure urban weights (Swilling et al. 2018) and the related environmental impacts (Sastre et al. 2015; Kennedy et al. 2007; Li et al. 2019a). In addition, decomposition analysis, such as the LMDI method, can reveal relationships between urban weight changes and socioeconomic factors, thereby revealing the most important driving forces.

Of course, some shortcomings also exist in the above sections. First, electricity and water consumption are not included in the flow analysis framework, leading to a lower resource consumption than the actual value. Second, it is necessary to define the administrative boundary of Beijing as the research boundary. With higher data quality and spatial resolution, we can overcome the data limitation, and it will be possible to determine the system boundary more precisely and increase analysis accuracy.

Third, we should consider as many factors as possible under the LMDI framework to improve the accuracy of identification of driving forces responsible, thereby providing more accurate and reliable information for urban material consumption and better support for policy development.

7.1.5 Comparisons with Other Cities

It compares the urban weights of different cities that can help reveal how they demonstrate in providing social and economic services. To classify the benefits of such comparisons, we compared Beijing's material consumption with those of 18 other typical cities (Fig. 7.8). Beijing's per capita material consumption was at 16 t/person as the highest value, and only 4 of the 18 cities (Lisbon, Limerick, Leipzig, and Dalian) had higher values. The lowest per capita material consumption was 3.5 t/person for Cape Town, which is even less than 25% of Beijing's value, followed by several European cities, whose value was less than about 10 t/person. The cities with the highest per capita consumption levels were mainly dominated by industrial enterprises, such as Dalian, which had the highest value at 25.8 t/person. Comparatively, some European cities depending heavily on tourism, services, and modern industry, such as Madrid, Paris, Hamburg, and Vienna, have high resource-use efficiency and low per capita material consumption. Other European cities depending on traditional industries, such as Lisbon and Leipzig, have a significant demand created by their industrial production and therefore have high consumption value. Compared to European and American countries, the per capita material consumption of three Chinese cities was relatively high because China's fast-growing economic development has led to rapid growth in industries and resource consumption. Handan consumes large amounts of fossil fuels and metallic minerals due to its prime industries, such as steel, coal, and electricity, causing an extremely high material consumption intensity at 11.05 t/10,000 RMB. However, Dalian relies on equipment manufacturing and the petrochemical industry and has a relatively low material consumption at 4.9 t/10,000 RMB. Beijing's material consumption intensity was only 2.9 t/10,000 RMB because urban construction and material consumption related to the lives of its residents were the main components of its material consumption. Meanwhile, Beijing's material consumption efficiency has gradually increased with decreasing inputs of materials per unit output value in recent years. Our comparison shows a fundamental result that industrial structure differs among cities, and this can be demonstrated by different changes in urban resource consumption and environmental pressure.

There are many vital factors affecting the change in material consumption, one of which is the intensity of material consumption. The increasing rate of material consumption has been reduced by reducing this intensity (i.e., increasing efficiency). It happens that there is a similar case about the factors that influenced Chinese material consumption found by Wang et al. (2017). In addition, some scholars have analyzed the roles of similar factors described in this chapter in energy consumption and found that the material consumption intensity also influenced energy consumption and was one of the critical factors responsible for changes in NO*x* emission (Wang et al. 2017,

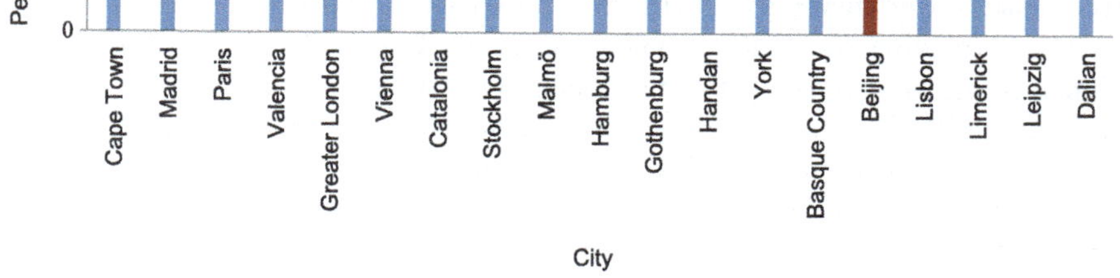

Fig. 7.8 Comparison of the weight of cities in the material flow analysis framework, expressed as per capita direct material consumption (DMC) (Data sources: Beijing, present study; Cape Town (Hoekman and von Blottnitz 2017); Dalian (Bao 2010); Greater London (Chambers et al. 2002); Handan (Lou and Shi 2008); Limerick (Browne et al. 2011); Lisbon (Niza et al. 2009); Paris (Barles 2009); Madrid, Valencia, Catalonia and Basque Country (Sastre et al. 2015); Malmö, Stockholm, and Gothenburg (Rosado et al. 2016); Vienna, Hamburg, and Leipzig (Hammer and Giljum 2006); and York (Barrett et al. 2002))

2014; Li 2016; Jia 2014; Li et al. 2019a). In a word, improving the consumption intensity can significantly decrease both material and energy consumption.

7.1.6 *Diagnosis of and Solutions to Material Metabolism Problems in Beijing*

In the context of socioeconomic development, the material consumption of a city must be placed, and this conclusion can be proved by the evidence we found that strongly influences Beijing's key sector (Construction). Before 2008, Beijing's increasing resource consumption of the Construction sector was mainly driven by infrastructure, which supported the population growth and the construction of venues growing for the Beijing Olympic Games. However, this sector was particularly strongly affected after the 2008 global financial crisis shocked Beijing's economy, demonstrating apparent decreases in consumption of building materials from 2008 to 2010.

Manufacturing was another vital sector, as its main impact on material consumption resulted from energy production by the combustion of fossil fuels, so Beijing is still facing air pollution-the dominant environmental challenge. Beijing's industrial energy supply structure has main components of fossil fuels such as coal and oil, and they produce more atmospheric pollution than cleaner fuels. However, CO_2 emissions have not continued to increase in Beijing with the rapid growth of consumption of fossil fuel, which is different from the results that have been observed in both developed and developing countries. There are two reasons for this improvement. Firstly, Beijing has made significant progress in raising energy efficiency and using cleaner energy. Secondly, Beijing Municipal Government takes action to restrict some industries with the worst pollutant production, which have been moved outside of Beijing's administrative boundaries.

In addition, remarkable success has been achieved via the "2 + 26 cities" air pollution control program in northern China has achieved, with Beijing as the center (Lang 2020). A delightful phenomenon should be noted that most enterprises in Beijing, and surrounding cities have begun to adopt cleaner natural gas and electricity instead of traditional fossil fuels, and this trend should continue. It is the above actions taken by governments that resulted in greatly reduced air pollution emissions and significantly improved air quality in Beijing. Future research in Beijing and its surrounding regions must test whether the government's policy will continue to provide long-term control of the severe air pollution problem (Li et al. 2019a).

7.2 Urban Weight Analysis for Beijing from the Perspective of Stocks

Changes in in-use stocks, including cars, buildings, machinery, and factories, can drive changes in urban weight through the process of urbanization. There is a large part of the stock of buildings, which provides services for humans, formed by mineral processing and manufacturing from the lithosphere (Chen and Graedel 2015; Pauliuk and Müller 2014). Global in-use stocks increased by about 23 times in the last century, which contributed to a four-time increase in the global urbanization rate (Krausmann et al. 2017). It is undoubtful that stocks increased far faster than the urban population. The accumulation of stocks has resulted in consuming enormous quantities of materials and energy, and actual data as global annual material extraction growing from 7 Pg in 1900 to 78 Pg in 2010 can prove the above conclusion (Krausmann et al. 2017).

Beijing is a global megacity and capital of China, and during the past 40 years since China's implementation of the Reform and Opening Up in 1978, its material stocks have rapidly increased. Meanwhile, the urbanization rate increased by 1.55 times, the urban residential area increased by 15 times, and, even more strikingly, the number of vehicles increased by nearly 90 times. Along with the socioeconomic development, a large stock of buildings, infrastructure, and manufactured products accumulated during the Reform and Opening Up, causing severe resource shortages and environmental pollution. Therefore, in order to solve these problems and provide scientific support for setting management targets and developing effective resource recycling strategies, it is necessary to perform the in-use stock account and analyze its composition and dynamic changes during the past years.

Using material flow analysis, we calculated changes in Beijing's in-use stocks from 1978 to 2015. According to the research on which this section was based conducted by my research team (Fu et al., 2019), an accounting system was developed with 3 major categories (buildings, infrastructure, and manufactured products), 11 subcategories, and 54 accounting items (Fig. 7.9), to analyze Beijing's urban weight changes. The overall in-use stock and key sub-categories based on Beijing's characteristics were selected to analyze the coupling or decoupling between in-use stock and the underlying factors (i.e., whether the stocks showed the same trend as the driving factors or not). Based on the above characteristics, we analyzed the relationships between the stocks and the driving factors. Coupling and decoupling mechanisms can be analyzed, which leads to the changing rates of these factors converging or not. The resulting knowledge can support decision-making for stock management.

The raw data were collected mainly from national and regional statistical yearbooks, including the China Construction Industry Statistical Yearbook, China Urban Construction Statistical Yearbook, China Traffic Yearbook, and Beijing Statistical Yearbook. It is very significant to have a weight analysis, and most countries will have similar databases available to support this work. The material intensity parameters (the constants which were used to change the raw data into weight values)

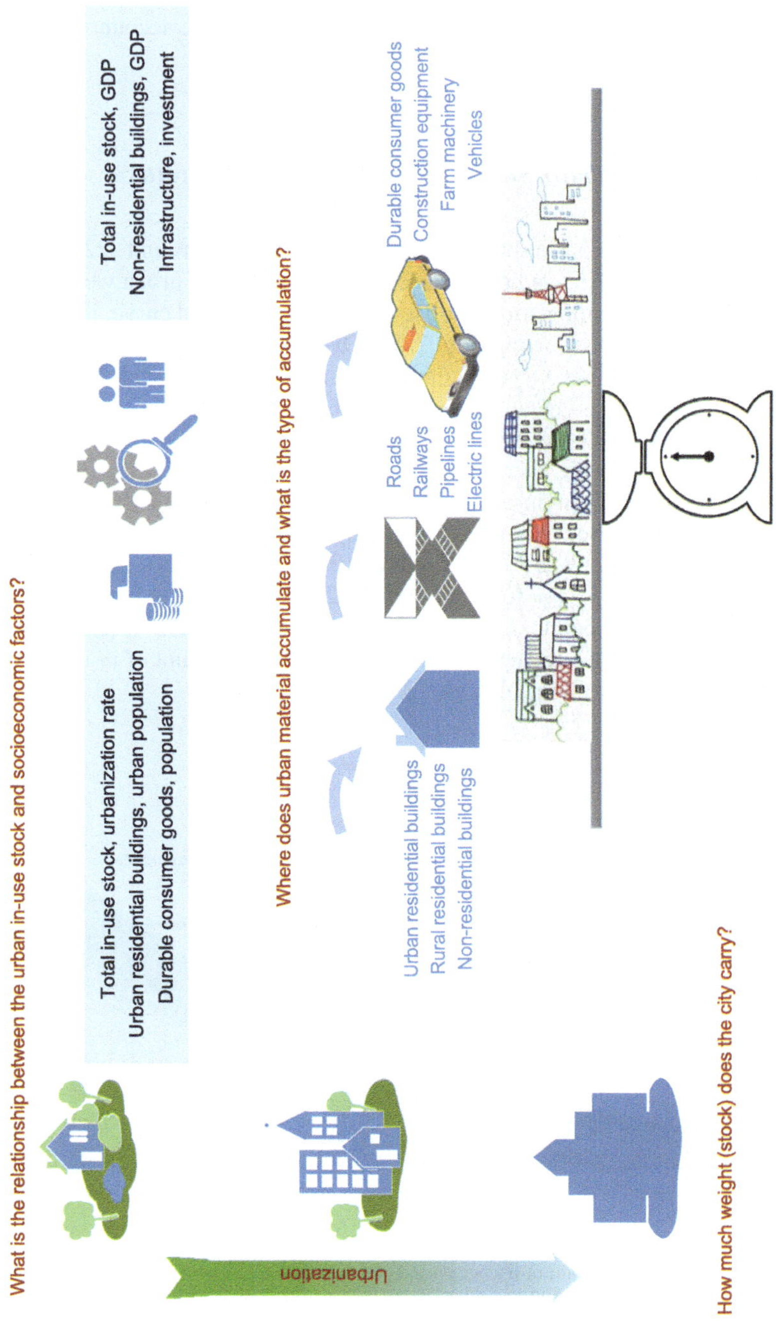

Fig. 7.9 Analytical framework for analyzing the urban weight from a stock perspective

mostly came from previous studies whose research regions were in Beijing, China, or Europe. In the analysis conducted by our research team (Fu et al. 2019), at the same time, the basis for this section, we obtained data on specific accounting items from 1978 to 2015.

7.2.1 Analysis of Urban Stocks' Weight and Its Structure

Beijing's total in-use stock had a sharping increase ranging from around 222.7 Mt (in 1978) to 1926.4 Mt (in 2015), indicating a nearly ninefold rise of urban weight, and we can see the trend of Beijing's total in-use stock as an S-shaped curve, indicating that the increasing rate of is slowing (Fig. 7.10). Overall, the average annual growth rate was 6%. During the first 11 years, the in-use stock increased twofold, and the average annual growth rate decreased by half from 9.5% in 1978 to 4.7% in 1990 (Fu et al. 2019). From 1990 to 2005, the in-use stock increased twofold, but at a relatively stable rate, and the average annual growth rate of the in-use stock was 5.5%. From 2005 to 2007, the in-use stock jumped from approximately 1001.7 Mt to 1295.6 Mt. Therefore, there was an exceptionally high average annual growth rate of 14.7%. After 2007, there was a slower increase in total in-use stock in 2015, with the rate decreasing from 15.1% to 0.7%. And in 2008 occurred, the most a rapid decrease by 7 percentage points in the increasing rate, then in 2012, it had a slight increase to 2.6%, then decreased again in 2015. Overall, the amount of in-use stock only decreased in 1990 by about 1.5% (6.9 Mt) (Fu et al. 2019).

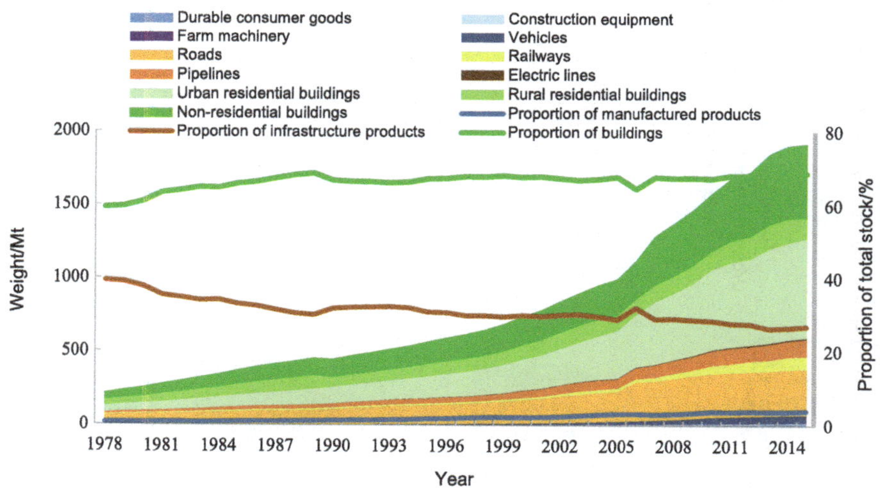

Fig. 7.10 Trends for Beijing's total in-use stock and changes in the proportions of the total in the three major stock categories from 1978 to 2015 (*Note* Weights in a given material category equal the width of the colored area; the top of each colored area represents the cumulative total)

There was a similar trend between the three stocks and the total in-use stock (Pearson's $r = 0.99$; Fig. 7.11). Beijing's weight increased by ninefold, which was mainly because of the tenfold increase in the building stock, accounting for 59.6 to 68.9% of the total in-use stock. Compared to the building stock, the infrastructure stock had a slower increase by sixfold, accounting for 26.4% to 39.6% of the total. Although the most remarkable growth of the three stocks was the stock of the manufactured product, whose value increased to 41.4 times the value in 1978, it only accounted for 2.2% of the total in-use stock, so the stock of the manufactured product had a weak overall effect on the total in-use stock. On the other hand, the building stock was the main driver of the overall in-use stock, and it had an average annual rate of 6.4%; in contrast, the growth rate of infrastructure, which had a specific time lag after changes in the total in-use stock, was at an average annual rate of 4.9%, and obviously, this number is lower than that of the building stock. The main exception was the dramatic increase in infrastructure after 2005, which resulted from preparations to host the 2008 Olympics (Fu et al. 2019).

At the beginning of the study period, the infrastructure stock grew slower than the building stock. In 2006, there was a peak at 24.8% and a fluctuating trend of the infrastructure stock during the next 5 years, averaging around 5.4% per year. The growth of building stock, coincidentally, followed a similar situation, also showing a lag. Specifically, the building stock peaked at 20.6% in 2007 and then fluctuated with an average annual growth rate of 5.3% until 2015. For manufactured products, the accumulation increased most from 15.6 Mt in 1999 to 62.7 Mt in 2010, nearly by 4 times, and it had an annual growth rate of 13.5%. Thereafter, its growth rate decreased significantly, along with a decrease in the two construction categories, at an annual growth rate of 4.6% (Fu et al. 2019).

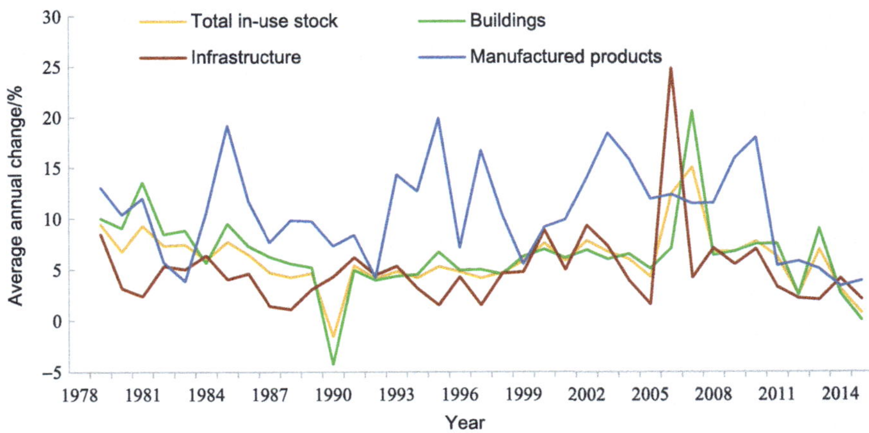

Fig. 7.11 Average annual rates of change in Beijing's total in-use stock and in the three major stock categories from 1978 to 2015

7.2.2 Structural Analysis of the Stock Subtypes

7.2.2.1 Building Subtypes

Urban residential stock (40.9% of the total building stock, on average) and non-residential stock (38.7%, on average) mainly influenced the trend of the building stock (Fig. 7.12) because they have been the most prominent contributors as an average ratio of 30.4 and 25.8% respectively, to the entire in-use stock. The proportion of urban residential buildings increased. However, the proportion of rural residential buildings had a converse changing trend from 29.5% in 1978 to 10.3% in 2015. And there is an exciting phenomenon that compared to urban and rural residential buildings, the proportion of non-residential buildings remained relatively stable.

The composition of the building stock showed three distinct phases. During the first 11 years, the ratio of urban residential, rural residential, and non-residential building stock to total in-use stock, which was at an average value of 35.7, 24.8, and 39.5%, respectively, and their growth rates of them, were an average value of 8.1, 7.8, and 8.5%, respectively, were showing relatively similar characteristics (Fig. 7.12). In fact, the amounts of urban buildings, both residential and non-residential buildings, had a similar trend to that of the total building stock since 1989, whereas three significant decrements happened in the rural building stock, which caused an undoubtful fact that the overall increase and its impact on the entire stock were much lower than those of the other subtypes.

Beijing's urbanization process could explain this phenomenon. It is because of the urbanization process in Beijing that the rural area was converted into urban areas, with the change of stock from rural stock to urban stock. From 1989 to 2006, the growth of non-residential buildings, which at an annual average growth rate of 4.8%, was lower than that of urban residential buildings, which at an annual average growth rate of 7.4%. After 1999, the gap between the growth of residential and non-residential buildings enlarged, with the real estate industry changed by the Chinese government from a traditional system in which the government purchased housing and allocated it to citizens to the gradual implementation of a private real estate market. Thereafter, China's real estate industry entered a market-driven phase in which urban residential buildings dominated by affordable housing increased sharply (Fu et al. 2019). In 2007, a peak occurred in the average annual growth rates of both urban residential and non-residential buildings, 20.0% and 27.8%, respectively. Later, there was a decrease in both the growth for urban residential and non-residential buildings, of 5.9% and 5.3%, respectively, in the last 8 years of this research. For the rural building stock, the first decrease of it occurred at around 25 Mt (28.1%) in 1990, and hence the total in-use stock declined. Again, the stock dropped in 2005 (around 12 Mt, or 11.0%) and 2015 (by nearly 30 Mt, or 17.1%). However, the dropping stock only decelerated the overall increasing building stock rather than decreasing this stock due to their small proportions of the total (Fu et al. 2019).

Fig. 7.12 Amount, share, and average annual growth rate of Beijing's building stock from 1978 to 2015

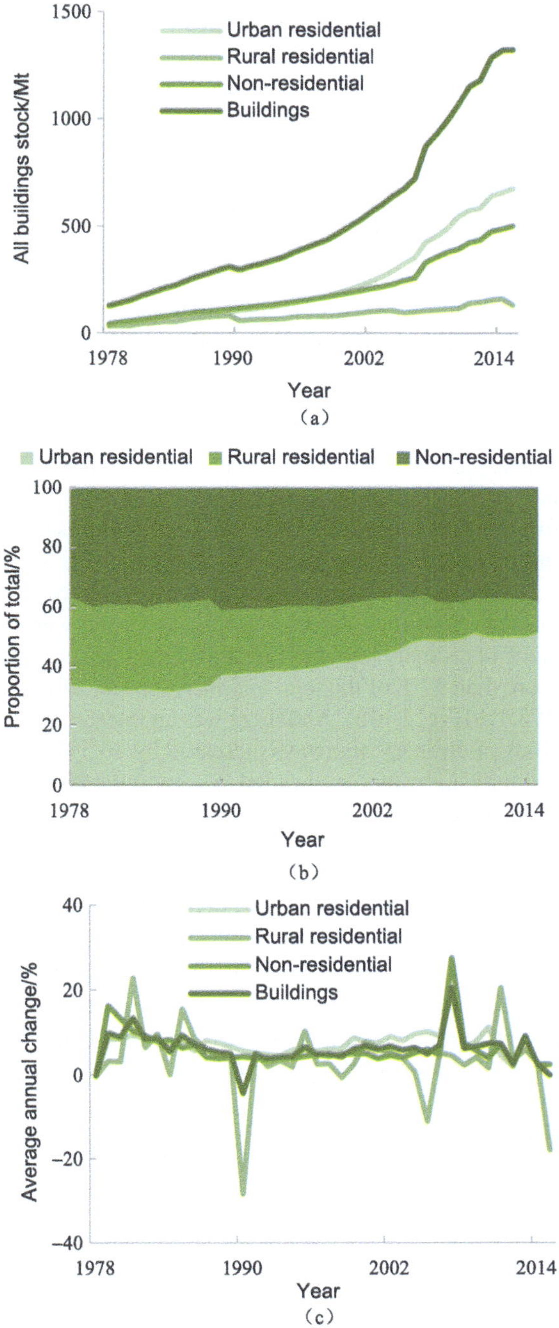

7.2.2.2 Infrastructure Subtypes

Roads ranked third in the proportion of the total in-use stock, 20% on average (Fig. 7.13), whose changes were consistent with the total infrastructure stock changes during the study period. From 1978 to 2005, the stock of roads was highest at 5.2 times that of pipelines at the initial time and increased slowly, which was at an average annual rate of 4.1%. And in 2006, there was a dramatic increase of 60 Mt (a proportion of 29.9%) of road stock, and this caused an evident increase in the stock of total infrastructure. Compared to the above phenomenon, there was a decrease in growth, with an annual average increase of 1.9% from 2007 to 2015. For the railway stock, it had an outstanding contribution at 17% of the total infrastructure stock, nearly 89.5 Mt (which was equivalent to the value of the pipeline stock in 2015) because it increased significantly at an annual average rate of 15.9%, particularly since 2007. The pipeline stock had a much faster increasing rate of 9.6% on an annual average during the first 8 years, and then it had the same trend as the total infrastructure stock but varied within more comprehensive limits during the following years. The electric line stock accounted for only 1.2% of the infrastructure and, therefore, only weakly influenced the total in-use stock (Fu et al. 2019).

Considering the amount of the in-use stock for 15 infrastructure accounting items, ordinary highways, expressways, and subways were the top three [Fig. 7.14a], and the stock of ordinary highways occupied the most enormous amount of in-use stock with more than 37% of the total and increased steadily, at an average annual growth rate of 3.4% [Fig. 7.14b]. And there was an interesting result is that the amount of in-use stock of ordinary highways increased by 40.5% in 2006 because the construction of ordinary highways was needed to support the 2008 Olympics. In total, we should owe the change of the in-use stock of road infrastructure to urban road development as the expressways and highways, which are parts of a city, are responsible to necessary to meet the need for rapid transportation of goods and materials between cities. Beijing had been building expressways in the following 19 years since 1990, resulting in the stock increasing rapidly by more than 24 times with an average annual increase of 18.5% from around 2.8 Mt in 1990 to 71.1 Mt in 2009. Subsequently, the growth of expressways decreased to an average annual rate of 1.8%. Subways are a practical choice to relieve traffic pressure (Fu et al. 2019). Although Beijing was the first city to build a subway system in China, at the initial stage, there was a slow increase in Beijing's subway stock. After 2006, it rose dramatically and up to nearly 3 times compared to the value of 2006, with an average annual growth rate of 12.8%, and then in 2015, Beijing's subway stock exceeded the expressway stock, so that it became the 2nd most significant transportation component with the proportion of 15.7% of the total in-use stock. During the middle and later years of this study, the proportion of ordinary highways decreased from nearly 70% in 1978 to 37% in 2015 due to the fact of the rapid accumulation of expressways and subways (Fig. 7.15).

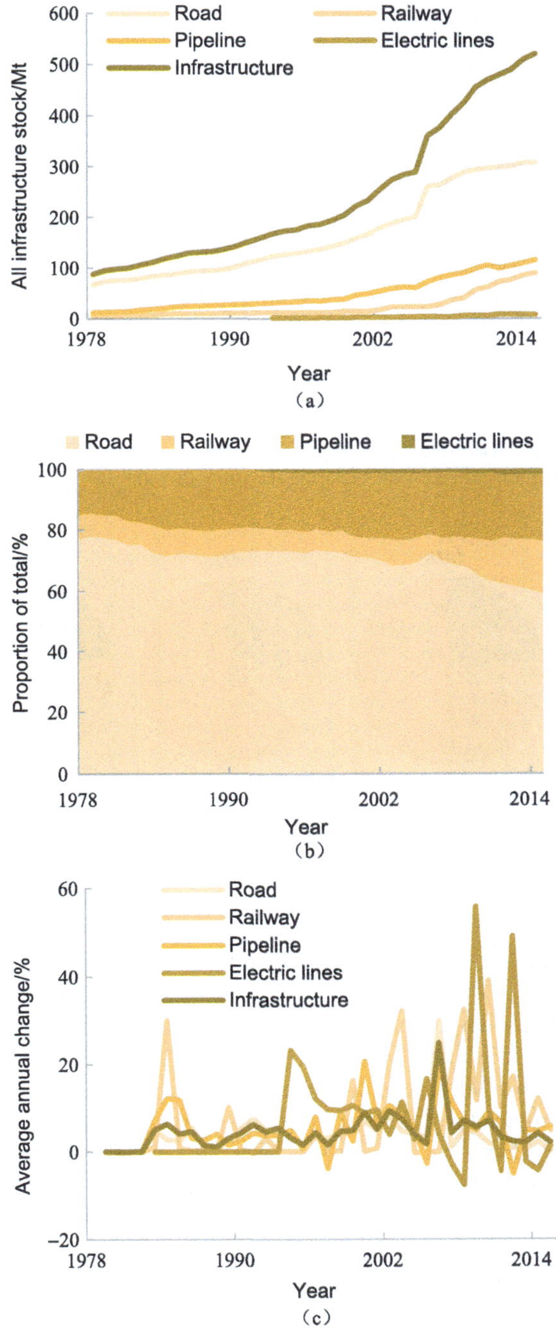

Fig. 7.13 Amounts, proportions of the total, and average annual growth rates of Beijing's infrastructure stock from 1978 to 2015

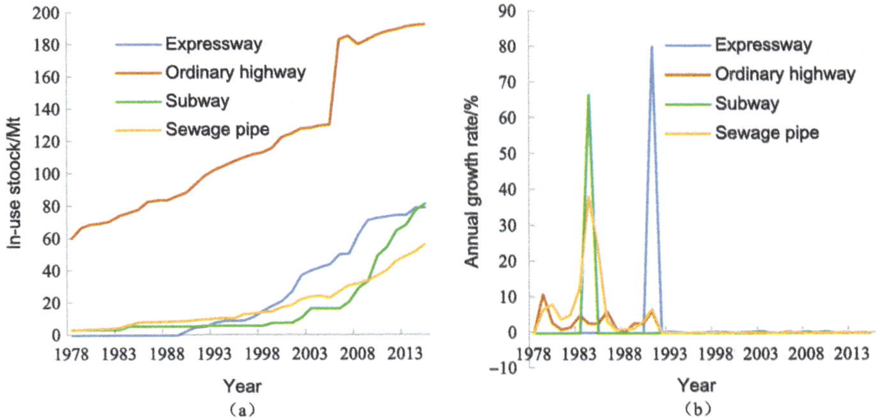

Fig. 7.14 Changes in Beijing's in-use stocks and the annual growth rate of the total stock of manufactured products during the study period

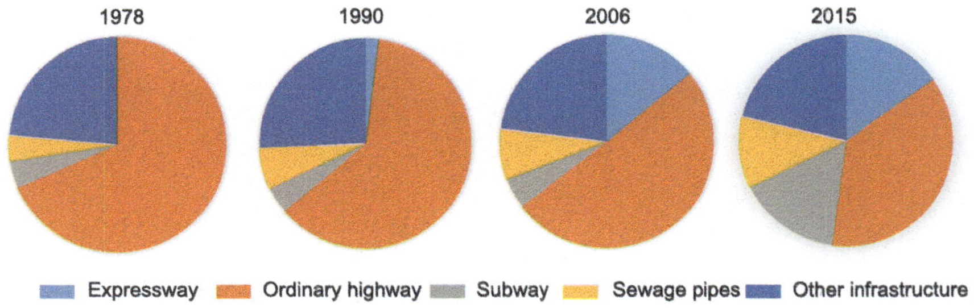

Fig. 7.15 Changes in the proportions of each component of Beijing's manufactured products (*Note* "Other infrastructure" includes urban roads, general railways, and water pipelines)

7.2.2.3 Manufactured Products

There is a similar trend between the total manufactured products stock and the vehicle stock, as Fig. 7.16 shows. Before 1987, the stock of the manufactured product was less than 5 Mt, in which construction equipment was the dominant stock as the average proportion of 57% of the total, and the growth rate of the stock of the manufactured product was at an annual average rate of 5.9%. However, the growth rate of the vehicle stock, which was at annually 13.2% on average, was greater than that of the manufactured products, and this brought the opportunity for the vehicle stock to gradually become the dominant component as the proportion of 60% since 1988. The growth rate of construction equipment was at an average rate of 3.7% annually, while vehicles were at an average of 14.8% annually. However, the latter has a trend that it raised from about 2.2 Mt in 1988 to 45.4 Mt in 2010, nearly an increase of about 20 times, and then its growth rate decreased to 3.8% annually from 2011 to 2015. In addition, durable consumer goods had a significant proportion of 37% of the total manufactured product stock. However, it had a similar trend as the vehicles that it rose from hardly any weight in 1978 to approximately 11 Mt in 2009, and then its growth rate decreased to an average of 1.9%.

Commercial passenger vehicles and household cars were the main items in the stock of the manufactured product (Fig. 7.17). From 1978 to 2010, the commercial passenger vehicle stock increased rapidly, reaching 38.3 Mt, about 81 times, and an annual average growth rate of 14.7% (Fu et al. 2019). During the next 5 years, it grew slowly, showing an average yearly rate of 3.6%, and this caused a slight increase of around 7.4 Mt from the proportion of 24.5% in 1978 to 57.9% in 2015 (Fig. 7.18).

Household cars were one of the leading manufactured products, but its stock emerged gradually only in the late 1990s with the actual low utilization of advanced vehicles and inadequate infrastructure (e.g., roads and parking lots) to support car use, and it increased as different growth rates for two stages: from 1997 to 2004 it increased from almost 0 to 3.1 Mt; from 2005 to 2015 it increased steadily with an average annual rate of 15.2%. As a result, household cars in 2015 became the second-largest of the in-use manufactured products stock, with a proportion of 18.6%.

In addition to vehicles, the stock of bulldozers and color TVs also changed greatly during the study period (Fig. 7.19). At the beginning of the study, the bulldozer stock was second only to commercial passenger vehicles (at 20.2% of the total in 1978). Still, its growth rate (an average of 4.3% annually from 1980 to 2015) was much lower than commercial passenger vehicles. In 1992 and 1999, the growth rate decreased greatly (by 10.2% and 17.3%, respectively). As a result, the proportion of bulldozer stock shrank to only 2.7% of the total manufactured products by 2015. In 1978, China approved the first TV production line. Since then, China's TV industry has proliferated. However, due to high prices that led to low sales, the limited consumer demand for TVs, and other conditions, its stock was deficient at the beginning of the study, accounting for no more than 1% of the total before 1985. From 1985 to 1992, Chinese citizens replaced black and white TVs with color TVs, and the stock of TVs increased from 0.1 Mt to 0.4 Mt (an increase of 4 times). The share of color TVs in the stock of manufactured goods peaked in 1992, at 6.1% of the total. Subsequently, with the rise of computers, mobile phones, and other electronic products, combined with the rapid development of the Internet, the stock of color TVs grew more slowly (at an average rate of 4.3% annually), and the overall proportion was decreased (to 1.4% in 2015).

7.2.3 Changes in the Relationship Between the Weights and Socioeconomic Factors

It is necessary to explore the reasons for stock accumulation and changes in this accumulation (Fishman et al. 2016). After selecting the in-use stock and the other accounting indicators, we can analyze their relationships with key driving forces such as urbanization (movement of rural residents into cities), population, GDP, road

Fig. 7.16 Amounts, proportions of the total, and average annual growth rates of Beijing's manufactured stock from 1978 to 2015

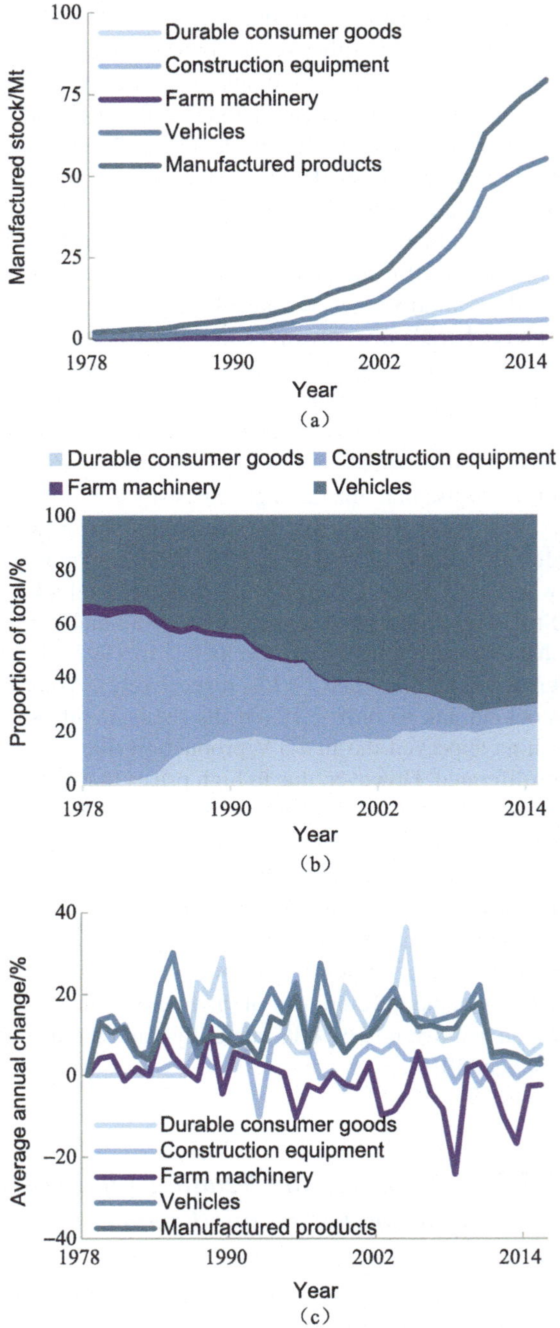

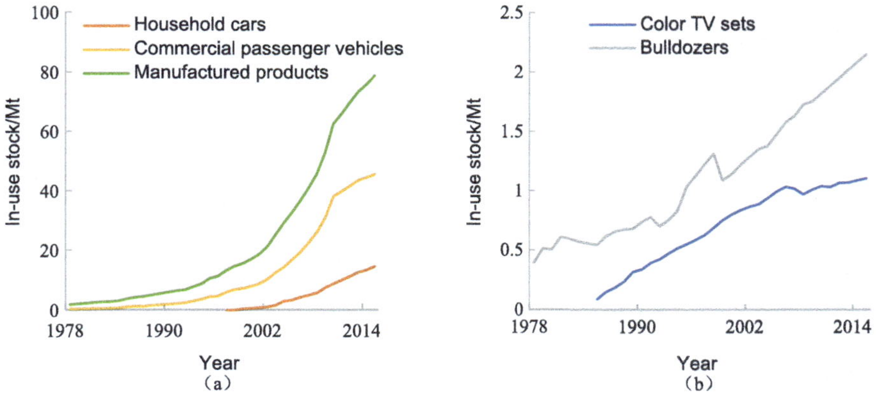

Fig. 7.17 In-use stocks of Beijing's main goods in the manufactured products category

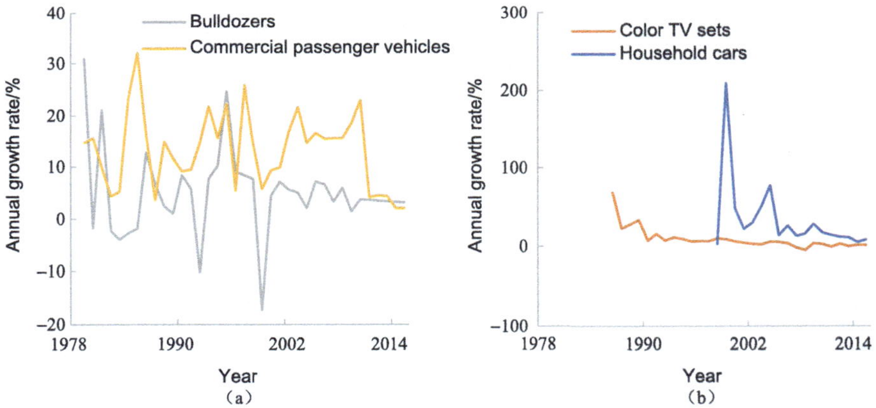

Fig. 7.18 Changes in the annual growth rate of the total stock of each of the major components of Beijing's manufactured products category

investments, and other socioeconomic factors. The urbanization rate and population directly affect the demand for the stock since demand is proportional to the number of people. GDP is derived from the economic benefits of the stock based on its production function; on the other hand, it represents funds that can be reinvested in the maintenance of existing stock and the construction of new stock. Therefore, we can relate the urbanization rate and GDP to the overall in-use stock. The population is closely related to the stock of residential buildings, as this building stock is created to house these people. Residents require basic shelter, so increasing the population increases the demand for housing stock. However, when the density of residential buildings is too high, the functions of big cities become more challenging, and the population growth will be increasingly restrained. Non-residential buildings include offices, workshops and factories, retail space, and warehouses. These buildings are

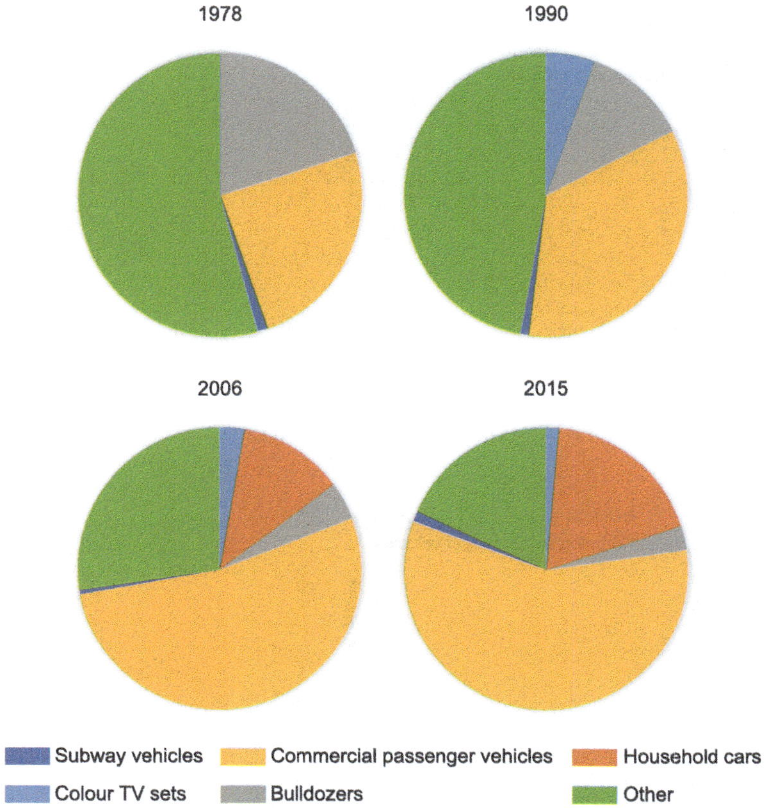

Fig. 7.19 Changes in the proportions of each major component of Beijing's manufactured products category Note: "Other" includes goods-carrying vehicles, public trams, and refrigerators

the primary sources of economic output, so their stock size is closely related to GDP. Road construction funds come from the local government's investment in roads, so the road investment directly affects the change of road stock. The consumer base determines the demand for manufactured products, so we can select population factors to study the relationship between the population and the stock of manufactured products.

Based on these categories, we can analyze changes in the growth elasticity (i.e., the difference in stock per unit change in a driving force) of the driving forces (such as GDP growth) and the weight (stock growth) during the same period. In this analysis, a stock can grow at least as rapidly as the analyzed socioeconomic factor (coupling), or less rapidly (relative decoupling), or it can even decrease (absolute decoupling) as the socioeconomic level grows. Coupling indicates that in-use stock increases together with the socioeconomic indicators, whereas decoupling can make development more sustainable by reducing the pressure on the environment.

In the research that was the basis for this section (Fu et al. 2019), my research group's analysis revealed that the urbanization rate and the overall weight (the total in-use stock) exhibited coupling, whereas GDP was relatively decoupled from weight (Fig. 7.20). During the study period, Beijing's urbanization level experienced two rapid increases in 1990 and 2005, leading to three distinct phases. Interestingly, during these phases, the growth rate of Beijing's urbanization gradually slowed, whereas the total in-use stock increased more rapidly. As the urbanization rate increased by 1%, the total in-use stock per unit increased over time by 35.0 Mt before 1990, by 77.4 Mt between 1990 and 2005, and by 321.0 Mt after 2005. So it could be seen that the rise of the total in-use stock gradually became less sensitive than the urbanization rate, that is, decoupling. In addition, the in-use stock of the GDP was continuously increasing, but the former rose slower than the latter, and by 2015 the former had increased by 35.2 times compared to the 1978 level, while the latter increased by only 7.7 times compared to the 1978 level. And compared to the relationship between the in-use stock and the urbanization rate, there is a similar result for the relationship between the in-use stock and the GDP, that is, decoupling. Specifically, the data could prove that the in-use stock per 10^8 RMB of GDP growth increased gradually slowly, from 2.0 Mt before 1990 to 0.06 Mt after 2005.

The relationships between the stock subtypes and the population and economic factors were also analyzed. The urban residential building stock and the stock of the manufactured product were coupled with the urban and total populations, respectively (Fig. 7.21) (Fu et al. 2019). Compared to the above two subtypes of stock, non-residential building stock was decoupled with GDP, and road infrastructure stock was also decoupled with the investment in roads (Fig. 7.22); that is, there existed the difference between the relative change rates for each pair of variables. With the urban population increasing by 2.9 times compared with the initial value from 4.8×10^6 to 18.8×10^6, the stock of urban residential buildings increased by about 14.1 times compared with the initial value from 45.1 Mt to 682.7 Mt, and the phenomenon

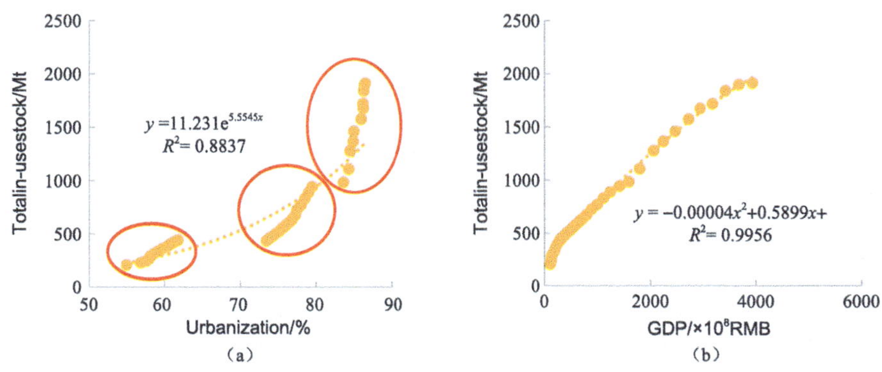

Fig. 7.20 Relationships between Beijing's total in-use stock and the urbanization rate and the GDP (*Note* All GDP values were adjusted to 1978 values using the government's official published inflation rates. Red ellipses represent periods during Beijing's urbanization process with different growth trends)

existed with the background of population growth because the government wanted to meet the increasing demand for places to let people live.

The stock of Beijing's manufactured products increased by 39.8 times compared to the initial value in the study period presenting a continuous upward trend with the total population, including both urban and rural populations, increased to 2.5 times the initial value, and it is essential for supporting the increasing need of Bejing's household consumption, industrial production, and construction. In fact, there was an increasing decoupling between the population and the stocks; that is, the growth of the former caused a more pronounced increase of the latter (around 45.5 Mt per 10^6 people), and this amount is nearly 7 times greater than the increase in the stock of manufactured goods (around 5.9 Mt). This indicates that the increase in urban residential buildings was more sensitive than the increase in manufactured goods to population growth (Fig. 7.21) (Fu et al., 2019).

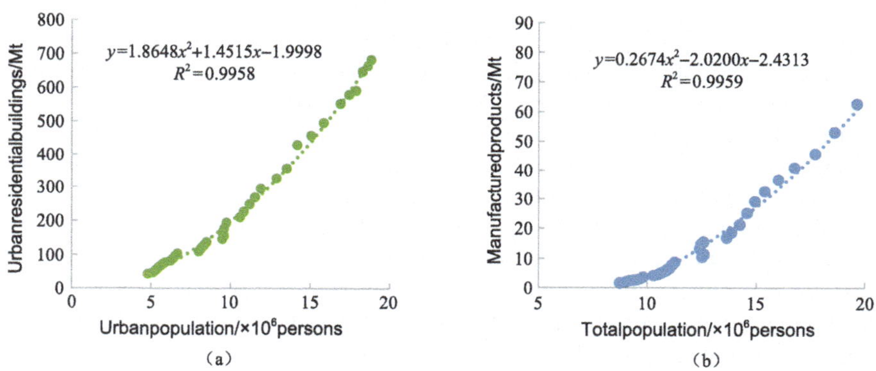

Fig. 7.21 Relationships between Beijing's urban residential building stock and the urban population and between the manufactured products stock and the total population

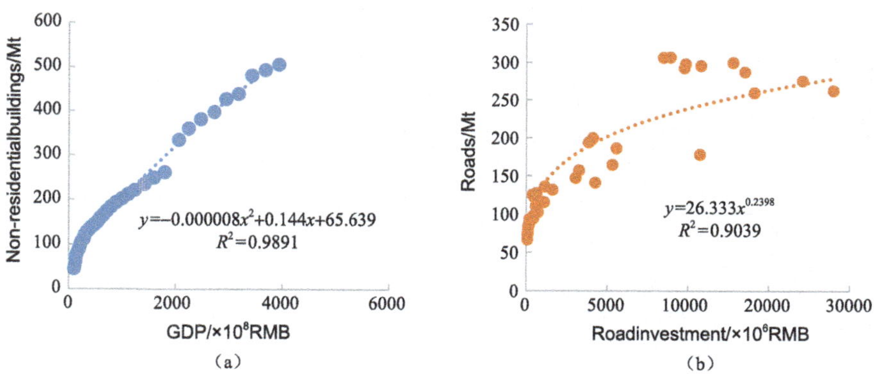

Fig. 7.22 Relationships between Beijing's non-residential building stock and GDP and between the road stock and road investment (*Note* All GDP values were adjusted to 1978 values using the government's official published inflation rates)

There was a relatively gradual decoupling between non-residential buildings and GDP (Fig. 7.22). However, there was a gradual decrease in its change per unit of GDP growth. When GDP dropped in the range from 3.8×10^{10} RMB to 20.0×10^{10} RMB, decoupling was most pronounced, and non-residential building stock grew by only 0.09 Mt, with the GDP growing by 1×10^8 RMB. When GDP decreased below that range, the stock of non-residential buildings grew by 0.32 Mt, and when increasing above that range, it rose again (0.11 Mt).

There was a non-linear relationship between road infrastructure investment and years (Fig. 7.22). Due to the long cycle of road construction, the investment cannot be quickly converted into stock, and the data could prove that mainly before 2003, the increase per unit investment (per 10^6 RMB) resulted in the road stock increasing by 0.02 Mt, and in this period, the investment was below 5.5×10^9 RMB. Afterward, due to the quality of road engineering design and the equipment used to build the roads improving during the study period, and the punctual time to complete to built the road infrastructure before the end of the study period, the increase of the stock slowed, and the data could prove that since 2009 the stock remained relatively stable at around 300 Mt, and in this period the investment was between 8.5×10^9 and 13.5×10^9 RMB. Comparatively, when the investment was $> 14.0 \times 10^9$ RMB, the investment in this period was mainly used for maintenance and upgrading of existing roads, and the data could prove that the curve increased slowly to about 260 Mt in this period. That is, the road stock created per unit investment was one order of magnitude lower than that at the start of the study period, and the growth was slow (Fu et al. 2019).

7.2.4 *The Significance of Measuring Urban Weight from the Stock Perspective*

Urban weight represents the pressure on the Earth's surface exerted by the mass of materials that accumulate within a city (i.e., its stocks of these materials), traditionally expressed in mass units. The in-use stock mediates the interactions between natural and human capital, so its amount can also represent the degree of sustainability (Fishman et al. 2016). This weight means an important impact on the lithosphere since the lithosphere must physically support the city's weight and provide the materials required for urban construction. The stocks and flows that are used to characterize urban weight are closely related. The amount of accumulated stock determines the scale of the consumed flows, which in turn affects the throughput of materials and energy, and exerts pressures and impacts on the natural environment (Krausmann et al. 2017). It is necessary to study the saturation value of the stock (i.e., the maximum amount that can accumulate in a given area) to explore the flows and find potential ways to improve efficiency and facilitate circulation of the materials that create and maintain the stock after consumption of the flows. Understanding these flows and their effects on the stock will enable management decisions (Liu et al. 2012).

Therefore, to define the weight of the city in ways that support improved urban management, it is essential to research stock accounting and the degree of stock saturation (Liu et al. 2012). Research results in this section should be extended to consider the material stocks in specific sectors or industries, or the stocks of specific elements, thereby deepening our understanding of in-use stocks in both the built environment and manufactured products. This will provide a more comprehensive picture of the urban weight and the composition of the stocks (Fischer-Kowalski 2011; Fischer-Kowalski et al. 2011).

In the research that was the basis for this section (Fu et al. 2019), my research group obtained the stocks of various materials in the built environment using calculations based on the material intensity. In the future, it would be desirable to combine material flow analysis with the life cycle assessment to understand the inputs, outputs, and cycling of different elements (Feng et al. 2020). It will improve our understanding of the weight of Beijing's in-use stocks and the relationships between urban flows and stocks. Another goal for future research will be to combine the analysis results in the present study with GIS technology to spatialize the distribution of stocks in the built environment and the stocks of manufactured products. However, this may require field research to provide the necessary data. Analyzing the distribution of these stocks among Beijing's communities could guide the transformation of the current consumption model to account for the distribution of the city's population and income, thereby providing additional support for improving the spatial distribution of urban resources.

7.2.5 Comparison with Other Research

Beijing's material stock is enormous, but the problem for it is very important from a global perspective since it has not yet come to its head, and its density is relatively high. The research conduct by Krausmann et al. (2017) proved that the in-use stock per unit land area was about 6.2 kt/km^2 in 2015 for the whole world, while the value was 117.4 kt/km^2 for Beijing, which was about 18 times larger than the global average. However, we could find an exciting result that Beijing's stock per unit area was less than or close to that of other mature cities. For example, its building stock per unit area was 55% of that of Melbourne, Australia, and what made this difference was that the proportion of rural (green) land is exceptionally high outside Melbourne, while most office, education, retail, and other buildings clustered in the downtown area covering only 36.2 km^2. Compared to Melbourne, some of Beijing's factories and university campuses have located outside the main urban area to avoid congestion. Another example is Rio de Janeiro, Argentina, whose residential building stock per unit area was 152% of that of Beijing (Condeixa et al. 2017). Although Rio de Janeiro and Beijing are both megacities in newly industrialized countries with high construction demand to provide adequate living space, the contradiction between population and lands was more severe in Beijing. The population of Rio de Janeiro was equivalent to 82% of that in Beijing, while the lands was less than 8%

of that in Beijing. In addition, for the metropolitan area of Taibei, the concrete and cement stock per unit area was equivalent to that in Beijing (Wang et al. 2018). This is because the population of the metropolitan area of Taibei has an order of magnitude smaller than Beijing, while the area, which is only 14% of Beijing, narrowed the gap between the stock per unit area of the two cities.

We can also compare the present study results with those of other studies of buildings and infrastructure to confirm the validity of the current research. Guo et al. (2014) calculated that Beijing's road stock was 9.2 t/person in 2001, slightly lower than the present results for the same period (11.9 t/person). This is mainly because the road categories in their study were defined according to the administrative and technical levels; that is, the roads were not covered by a single standard, which may lead to some stock being excluded from the underlying data. In the study that was the basis for this section (Fu et al. 2019), my research group calculated the expressway, ordinary highway (including all standard highways), and urban road stocks, which permitted a more comprehensive accounting.

The trends for the building and infrastructure stocks in Beijing, Tianjin, and Shanghai reported by Huang et al. (2017) were similar to the present results. From the late 1970s to the mid-1980s, and since the beginning of the twenty-first century, the cumulative rate of increase in these stocks was relatively fast. Although Beijing's stock value in the Huang et al. study (approximately 47.3 t/person) was lower than that in the present study (around 85.1 t/person), this was mainly because the previous study did not account for the stock of manufactured products and some of the infrastructure (i.e., urban roads, subways) that we measured in the present study.

When the urbanization rate came up to about 74%, the farm machinery stock in the United States declined, and the growth rate of construction machinery in the United States decreased (Chen and Graedel 2015). This situation was approximately the same as Beijing, and this situation happened because the United States shifted some of the industrial production capacity to other regions (Chen and Graedel 2015) and introduced more efficient machines, which could decrease the demand for additional machines. Nevertheless, when the urbanization rate of the United States came up to about 75%, there was a gradual decrease in commercial passenger vehicles per capita, while there was still an increase of that in Beijing is still increasing. The reason for the above phenomenon is that the demand for commercial passenger vehicles in the United States and Beijing had a diverse presentation caused by differences in population growth and demand price elasticity (Cameron and Trivedi 2005). When the urbanization rate exceeded 75%, population growth showed a slow pace, with the demand for cars decreasing in the United States, while this trend was the opposite in Beijing. This is because the automobile industry in the United States was a significant industry with a smaller price elasticity of demand. At the beginning of the late period of urbanization, cars were a necessity in the United States because many rural residents did not have access to public transportation, whereas, in Beijing, cars were a luxury (Fu et al. 2019). Subsequently, there was a gradually stable trend of per capita car ownership in the United States during a continuous rapid upward trend of the demand for commercial passenger vehicles in Beijing. This is because the status of cars as luxury products has maintained stable, caused by the gradual completion of the urban infrastructure construction and the increasing purchasing power with economic development. In addition, after the urbanization rate came up to 75%, these trends may be affected by the following two factors in Beijing (the per capita disposable income growth rates and per capita GDP) because they changed much more significantly than those in the United States.

In addition, the per capita sewage pipeline stock increased about 8 times during the study period because Beijing's urbanization rate increased by 55% (Fig. 7.23). However, from 1984 to 2012 for Japan, during which time its urbanization rate increased by about 15%, this value was less than 78%, and every 1% increase in this value increased Japan's per capita sewage pipeline stock by 0.52 Mt for the population with access to sewage service (Lwin et al. 2017). Comparatively, when the urbanization rate exceeded 78%, the per capita sewage pipeline stock decreased by 0.05 Mt per 1% increase in urbanization, then increased slightly (Fu et al. 2019). This depends on Japan's sewage system with careful and logical development during the early years of this urbanization period, leading to a significant reduction of the need for additional infrastructure during later years. At the same time, there was a decrease in the per capita stock due to the rapid improvement of the material-use efficiency and people who owned the existing sewer service. Compared to Japan, Beijing's sewage pipeline system just began construction shortly after 1978 when the Reform and Opening Up was implemented. Indeed, with accelerating urbanization, more residents owned this municipal infrastructure, but the construction was still not enough and rapid due to the slower pace than the rapid demand.

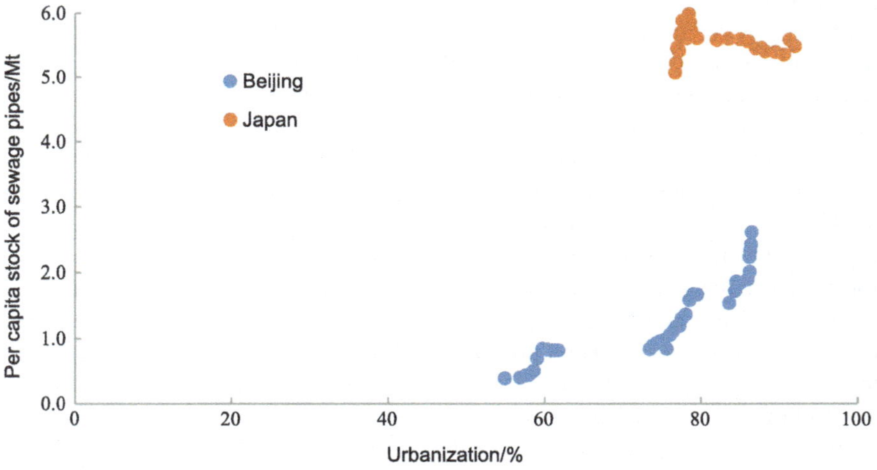

Fig. 7.23 Changes in the per capita sewage pipe stock in Beijing and Japan as a function of the urbanization rate

7.2.6 Diagnosis of Metabolic Disorders in Beijing from a Stock Perspective and Recommended Solutions

Beijing'S overall weight in 2015 was as much as 9 times the level in 1978. The population factors were strongly coupled with the stock, while the economic factors, such as GDP and investment, were gradually decoupled. Beijing's three most extensive stocks, whose growth had slowed by the end of the study period, were urban residential buildings, non-residential buildings, and roads. In contrast, the stocks of subways and primary vehicles (commercial passenger vehicles and household cars) experienced rapid growth (Fu et al. 2019). In fact, there was a close relationship between the change in Beijing's in-use stocks and some socioeconomic factors, such as the economic development level, major international events, and government policies and regulations, and it could be seen from the increasing building stocks between 2003 and 2006, which was caused by the construction of the venues and support facilities for the 2008 Beijing Olympics.

However, the increasing rate of urban residential building stocks fell sharply in 2008 after 2007 when it came up to the peak, and it returned to normal levels. In fact, there were three reasons explaining this phenomenon. The first reason was that under the background of housing prices rapidly rising and real estate investment quickly enlarging, a targeted regulation ("National Six-points Package") was adopted to mitigate speculation of real estate by strictly limiting additional development and tightening the supply of credit. Due to a time lag in the implementation of the policy, there was an actual peak of urban residential building stocks. The second reason was that the Olympics ended in 2008. The third reason was the 2008 global financial crisis since investors became more cautious. In addition, there were fluctuations in the growth rate of the building stock towards the end of the study period, and the 2008 global financial crisis could also explain this phenomenon. As monetary policy began easing again at the end of 2008, investment in real estate development increased again (Fu et al. 2019). Nevertheless, there were both trends of restraint and recovery for the real estate market, respectively. Specifically, the restraint trend was that less speculation led to slower price increases, while the recovery trend was that the more affordable real estate caused by its low increase in cost led to increased sales.

The road stock accumulated rapidly during the 9th Five-Year Plan (1996–2000) and the 10th Five-Year Plan (2001–2005) due to the financial increase of investment in Beijing's road infrastructure in order to create a modern metropolis. At the same time, the focus shifted to the outer suburbs, and there was a peak in the growth rate of the infrastructure stock in 2006 when the transportation infrastructure construction prepared for the Olympics was just at a critical stage. Then, in the 11th Five-Year Plan (2006–2010), there was a priority for public transportation development with large subway and rail transit projects finished, leading the infrastructure stock increasing rapidly. Subsequently, the subway stock increased in the 12th Five-Year Plan (2011–2015) with subway transport routes improving, and we should owe this progress to the increasing public transit investment with the proportion of 75.7% of the municipal total transport infrastructure investment. The vehicle technology and the economic level both considerably improved after entering the twenty-first century, and there was a continuing increasing trend of the stocks of commercial passenger vehicles and household cars because Beijing's disposable personal income was rising in the 10th and 11th Five-Year Plan periods, and people had more money to purchase of new vehicles. At the same time, the stock of road infrastructure presented a parallel growth to the commercial passenger vehicles and household cars that both the stock of ordinary highways and expressways, which had a proportion of more than 20% of the total infrastructure stock, grew quickly.

In the future, we put forward three suggestions to reduce the growth of fast-growing stocks and Beijing's environmental load. The first suggestion is that Beijing's government should create and enforce rigorous recycling standards related to the existing large stocks recycled, which should depend on life cycle analysis and provide policy support. Actually, there are many material stocks created by construction with high recycling potential. There was a long history in developed countries of recycling construction waste, with a series of laws and regulations promoting this trend, resulting in the recycling rate of construction waste being more than 90%, while this value in China was 5% (Du 2012). Therefore, the enterprises should strictly obey the regulations and adopt advanced waste treatment techniques, and all processes, including demolition, classification, transportation, processing, and production, should be limited within the standards due to more money and more land for recycled construction materials than natural materials. Enterprises that process these construction waste resources should be given reasonable financial and land support (Fu et al. 2019). The second suggestion is that manufacturing enterprises should concentrate on the development of their technical skills and manufacturing processes for vehicles and other products, and these actions are related to the decreasing material intensity. In fact, there is an enormous potential for fast-growing stocks to decrease, and existing data can prove this conclusion that China's manufacturing labor productivity is 85% lower than that of the United States and Japan (ILO 2018). The last suggestion is that we can optimize the change in stock caused by the import of products from outside Beijing through strict control, and governments should also continue discouraging new car purchases and promote the city's extensive and efficient subway system (Fu et al. 2019).

7.3 Identification of Key Entities in Beijing's Material Metabolism

In the view of urban metabolism, different metabolic actors form a network of relationships, so the tools of ecological network analysis can be used to analyze the interactions among metabolic actors (Fath and Borrett 2006), by which we can make an analogy of the exchanging processes of materials and energy between an urban network and an ecological network. Different sectors in the urban socioeconomic system are seen without exception as producer, predator, and prey species in an ecological network, and the relationships between different sectors are regarded as flows among them, and therefore ecological network analysis can reveal details of urban metabolic processes by analyzing the flow distribution and complicated relationships among components of the network.

In the study that was the basis for this section (Li et al. 2019b), my research group developed a network model of 8 components with the help of ecological network analysis (Fig. 7.24). We then used the model to reveal the characteristics of flows.within the network and the distribution of ecological relationships among the components (Fath and Patten 1999). We also identified key sectors and pathways within the network by analyzing the associations among sectors (Fath and Borrett 2006) and the contributions of the metabolic actors, represented by nodes in the network, to the overall metabolism (Browne et al. 2011; Eurostat 2016). The results guide future regulation of urban metabolism (Zhang et al. 2014).

Research data was obtained from the Beijing Statistical Yearbook, China Mining Yearbook, China Energy Statistical Yearbook, and other relevant datasets. Most countries will have similar databases that can be used to support such research. After the original data collection was completed, we performed the unification of the final accounting items in mass units using conversion factors obtained from the research literature. We analyzed the metabolic actors and network characteristics of Beijing's urban material metabolism to identify the key actors in three years (2001, 2008, and 2015) to cover the periods before and after the 2008 Olympics, with 2015 representing the most recently available data at the time of our original study.

7.3.1 Relevance Analysis

The integrated flow intensity (in Sect. 5.4.2.3) in an ecological network represents the total direct plus indirect (embodied) flows of material between metabolic actors, so this variable can accurately show the strength of the relationships between different nodes. We could find that there was a decrease in integrated flow intensity by 37.4% and a relatively balanced range of flow distribution between 10 and 20% (Fig. 7.25), except for Transportation (a maximum proportion less than 10%) and Construction (a maximum of 5%) (Li et al. 2019b). However, between 2001 and 2015, there was a significant increase in flow intensity in the sectors related to the Transportation sector by 52.8%.

It is possible that for the integrated flow intensity, the Internal Environment and the Recycling, which have a proportion of about 18% and about 19% of the total, respectively, are the main reasons for the rapid decrease of integrated flow intensity of the whole network, because their integrated flow intensity decreased by 48.0% and 36.0%, respectively, more than that of the whole. Household Consumption and Agriculture, which accounted for about 10% and 14% of an overall reduction in flow intensity, decreased by 55.9% and 41.5%, respectively (Table 7.2).

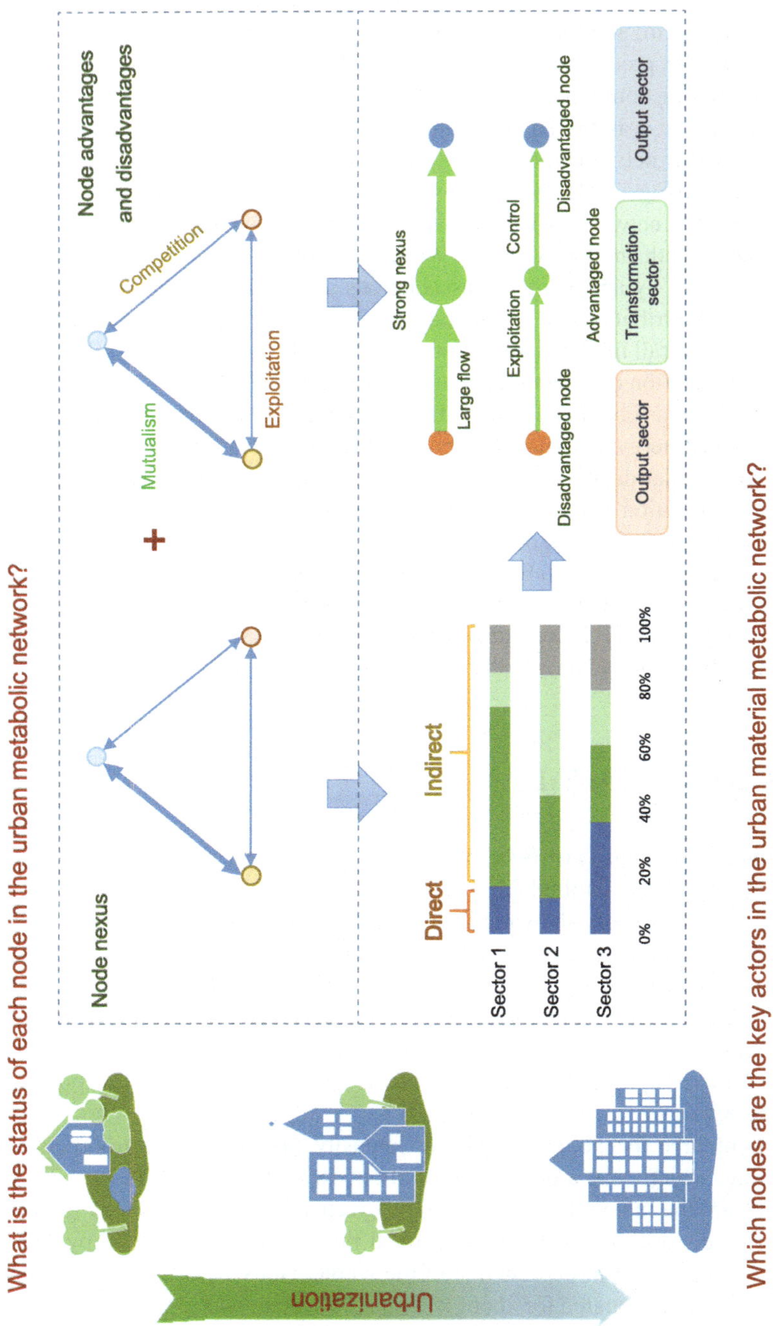

Fig. 7.24 Analytical framework for the identification of key metabolic agents in an urban metabolic network

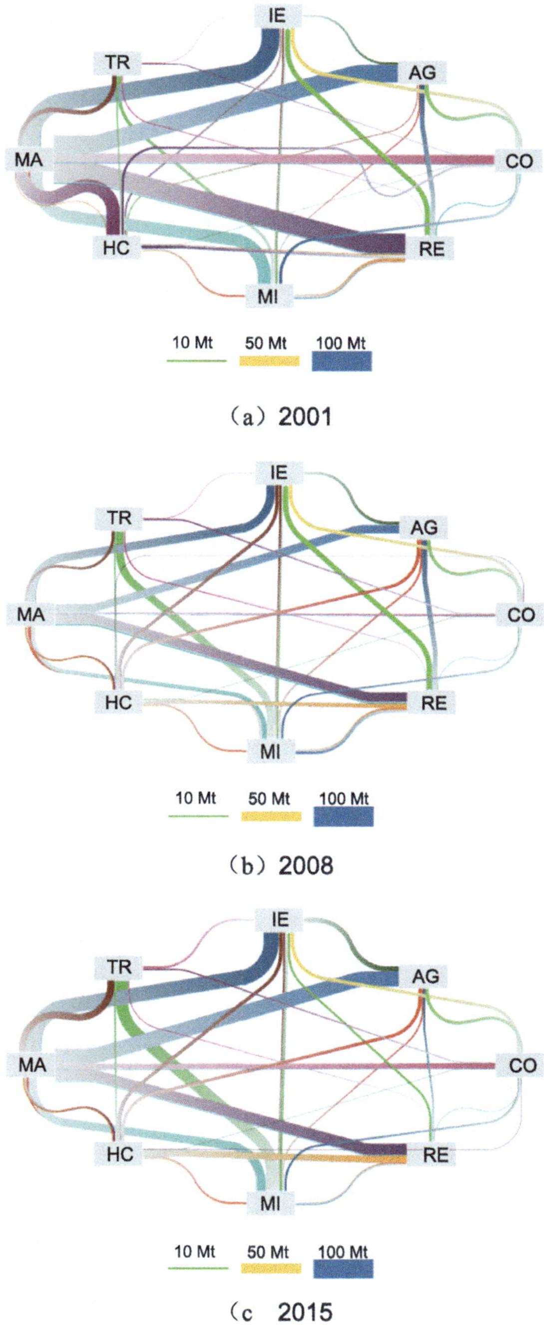

Fig. 7.25 Integrated (direct plus indirect) flows through Beijing's metabolic system network in 2001, 2008, and 2015 (*Note* AG-Agriculture; CO-Construction; IE-Internal Environment; HC-Household Consumption; MA-Manufacturing; MI-Mining; RE-Recycling; TR-Transportation)

Table 7.2 Contributions of the key metabolic actors to the total flow *Note* Transportation and Construction are not included because their total contribution was less than 15% of the total throughout the study period

	Average proportion of total/%	Change since 2001/%	Rank of change	Rank of proportion	Score (rank based on the percentage change since 2001)	Score (rank based on the average proportion of the total)	Total score (sum of ranks for change and proportion)	Final rank based on the total score
Internal Environment	19.0	– 36.0	4	1	3	6	9	2
Agriculture	14.4	– 41.5	3	4	4	3	7	3
Mining	12.4	– 29.0	6	5	1	2	3	5
Manufacturing	15.5	– 32.1	5	3	2	4	6	4
Recycling	17.6	– 48.0	2	2	5	5	10	1
Household Consumption	10.4	– 55.9	1	6	6	1	7	3

(*Note* Transportation and Construction are not included because their total contribution was less than 15% of the total throughout the study period)

The Manufacturing and Mining sector had a relatively small proportion of 32.1 and 29.0%, respectively, and their contributions to the overall decrease of the flow were also small. However, their flow accounted for a relatively large proportion of about 16 and 12%, respectively. Outputs dominated recycling with a ratio of 83, and 72% of the total outputs were exported to the Agriculture, Mining, and internal Environment. Compared to the Recycling sector, Manufacturing was dominated by outputs with a proportion of 65, and 64% of the total outputs were exported to the Agriculture, Recycling, and Internal Environment.

Agriculture and Mining were the two main sectors of inputs, with a proportion of about 84 and 60% of the total, respectively, and their outputs, which accounted for 81 and 82% of their total inputs, respectively, were from Manufacturing, Recycling, and Internal Environment. The inputs and outputs of the Internal Environment and Household Consumption sectors were roughly equal, with outputs accounting for about 55 and 53% of their total input plus output, respectively (Li et al., 2019b). For the Internal Environment sentor, the input was mainly from Recycling and Manufacturing, with a proportion of around 68% of its input, and the output was mainly to Mining and Agriculture, with a proportion of around 61% of its output. For the Household Consumption sector, the input came mainly from the Internal Environment, Agriculture, Manufacturing, and Recycling with a proportion of around 90% of its output, and the output was mainly from Internal Environment, Agriculture, and Mining with a proportion of around 72% of the output. Overall, Recycling, the Internal Environment, Agriculture, and Household Consumption influenced the network most strongly, so they can be regarded as the key nodes in the network (Li et al. 2019b).

Indirect flows accounted for more than 70% of the total flows, and it had a significant effect on the integrated flow intensity of all of the metabolic components (Fig. 7.26). There was an evident influence on Recycling, the Internal Environment, Agriculture, and Transportation for the direct flows, which had a proportion of more than 20% of the integrated flow, while there was an apparent influence on Mining, Manufacturing, Construction, and Household Consumption for the indirect flows, which had a proportion of more than 80% of the integrated flow. Among the main metabolic components with high direct intensity, there was only the Transportation sector, whose direct and indirect flow intensities had fluctuations, while the direct flow of the other sectors (Internal Environment, Agriculture, and Recycling sectors) were almost stable at around 1% of the total intensity, affected little on decreasing in their integrated intensity by 29.6, 24.1, and 31.6%, respectively. Some sectors were dominated by indirect inputs, leading to similar trends for both the integrated intensity and the indirect intensity of them, and this result can be proved by Mining and Household Consumption sectors, with their integral intensity demonstrating a slow downward trend of 33.3 and 65.2%, respectively. At the same time, the integral intensity of the Manufacturing sector first increased and then decreased; while the value of the Construction sector was relatively stable at about 1.6%.

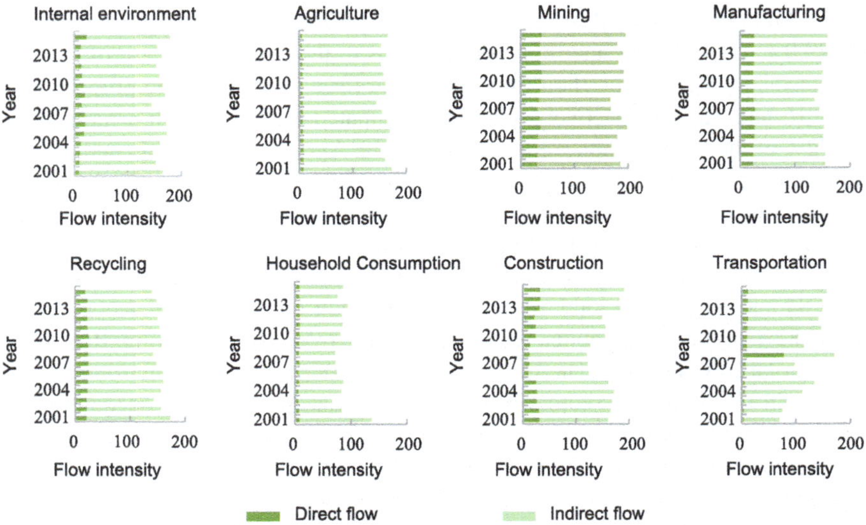

Fig. 7.26 Changes in the direct and indirect flow intensities for the eight components of Beijing's metabolic flow network from 2001 to 2015

7.3.2 Analysis of the Ecological Relationships

It is also significant for identifying the critical nodes to analyze the distribution of ecological relationships among different sectors, not just to analyze the distribution of the flows. For the ecological relationships among different sectors, we have defined and described them in Chapter 5, and furthermore, we can define 28 pairs of ecological relationships among eight sectors in Beijing's ecological network. Figure 7.27 summarizes the distribution of the relationships that remained stable from 2001 to 2015 (Li et al. 2019b). Among four types of ecological relationships in Beijing's ecological network, the number of exploitation and control relationships was 16, with the most significant proportion of 57.1%, and therefore it was dominant in Beijing's metabolic system; the number of mutualism relationships was 8, with the second largest proportion of 28.6%; and the number of competition relationship was 4, with the most minor proportion of 14.3%. Specifically, the relationships that were related to the Manufacturing, Household Consumption, and Internal Environment sectors had a proportion of 47% of the total, and in these relationships, exploitation relationships held two-thirds, including agriculture exploiting all three sectors, Construction exploiting Internal Environment and Manufacturing, Mining exploiting Internal Environment, and Recycling exploiting Manufacturing. In addition, there were other exploitation relationship such as the Internal Environment exploiting Transportation and Recycling, Manufacturing exploiting Transportation and Mining, and Household Consumption exploiting Transportation and Construction. From the perspective of the exploitation and control relationships, Agriculture and Construction sectors had

	IE	AG	MI	MA	RE	HC	CO	TR	Exploitation	Mutualism	Competition
IE	+	-	-	+	+	+	-	+	5	0	0
AG	+	+	-	+	+	+	-	+	4	4	1
MI	+	-	+	-	+	+	+	+	3	3	1
MA	+	-	+	+	-	-	-	+	5	1	1
RE	-	+	+	+	+	+	-	-	3	3	2
HC	+	-	-	-	-	+	+	+	5	1	1
CO	+	+	+	+	-	-	+	+	4	2	1
TR	-	+	+	-	-	-	+	+	3	3	1

Exploitation Control Competition Mutualism

Fig. 7.27 Distribution of the ecological relationships among the eight sectors in Beijing's ecological network for the stable relationships from 2001 to 2015 (Components: AG-Agriculture; CO-Construction; HC-Household Consumption; IE-Internal Environment; MA-Manufacturing; MI-Mining; RE-Recycling; TR-Transportation)

a proportion of 25%, in which 75% were exploitation relationships and 25% were control relationships, and the Construction exploited Agriculture.

For the mutualism relationships, Mining and Transportation, which had the most significant proportion of them, both were mutually beneficial to Construction. At the same time, there was also a mutualism relationship between Mining and Recycling, and Transportation had a mutualism relationship with Agriculture. Recycling, Agriculture, and the Internal Environment also accounted for about 43% of the mutualism relationships (Li et al. 2019b). For the above three sectors, there were mutualism relationships between Agriculture and Recycling, between the Internal Environment and Manufacturing, and between the Internal Environment and Household Consumption, respectively. For the competition relationships, Recycling, Construction, and Transportation had a proportion of 50% of them, while Mining, Manufacturing, and Household Consumption were the components of the remaining 50%. Overall, the sectors which played exploitation and control roles obviously were Internal Environment, Manufacturing, and Household Consumption, which were treated as crucial sectors due to the status of critical links in the flows of resources and pollutants in Beijing's urban ecological network.

7.3.3 Identifying the Key Actors

The key nodes in Beijing's ecological network can be identified by both network-flow analysis and relationship analysis, and by the former method, we can draw a conclusion that Recycling, Internal Environment, Agriculture, and Household Consumption components were the key nodes, while the latter method, the Internal Environment, Household Consumption, and Manufacturing sectors were the critical nodes (Table 7.3). So combining the results of the two analyses, the Internal Environment and Household Consumption were regarded as critical metabolic components in Beijing's urban metabolism due to their enormous contribution of material inputs and outputs. In addition, they had a prominent role in connecting the other sectors, as they played vital roles in influencing the resource and waste flows through the metabolic network (Li et al. 2019b). For example, the Internal Environment sector had enormous pressure from both control and exploitation relationships due to the fact that it can both absorb pollutants and waste from outside itself, and provide resources outside of itself, such as biomass and minerals. Another sector that had a significant effect on the material flows of outside itself was Household Consumption, which receives the resources provided by outside itself at the end of the industrial chain and at the same time produces Household pollution and waste emissions. Therefore there should be some regulations of the flows through this sector to reduce material consumption and waste emissions (i.e., promote urban dematerialization). Overall, future decision-makers should focus on the Internal Environment and Household Consumption components.

Recycling and Agriculture, which lied between other sectors in a long flow pathway, functioned as "transit points" in Beijing's urban metabolic network, and therefore it had a strong effect on the network flows. However, there were less exploitation or control relaitonships related to them. In fact they are very significant in the urban biomass supply and material recycling. For the former, the Recycling sector, waste produced in urban areas is received and disposed in this sector and then some changed into renewable resources (e.g., secondary metals) that was entered into urban socioeconomic sectors, and the others was changed into non-recyclable

Table 7.3 Identification of key nodes in Beijing's metabolic system

Sector	Flow analysis	Relationship analysis (no. of relationships)	
		Exploitation	Control
Internal Environment	Equal importance of inputs and outputs	2	3
Household Consumption	Equal importance of inputs and outputs	2	3
Recycling	Output-dominated	2	1
Agriculture	Input-dominated	4	0
Manufacturing	Output-dominated	2	3

material that was released into the environment in this sector, while for the latter, the Agriculture sector, it supports urban life by biomass resources with the material consumption such as fertilizers, pesticides, and plastic films, at the same time it needs the natural resources from the Internal Environment. Therefore, policymakers should weaken the effect of future possible losses in the processes associated with the Recycling and Agriculture sector. In addition, though Manufacturing's impact on network flows was not large, its exploitation and control effects on other sectors remain significant (Li et al. 2019b), due to its role of both utilizing natural resources (such as minerals, fossil fuels, and biomass), and providing products to outside itself. Indeed, Manufacturing is it is a major sector, and there should be some decisions aiming at increasing the efficiency of material use, such as taking technologically innovative actions in industrial production.

7.3.4 Conclusions and Prospects

The indirect flows had a large proportion of the network's overall integrated flow intensity, and the Recycling, Internal Environment, Agriculture, and Household Consumption sectors played a dominant role in overall integrated flow intensity. Exploitation and control relationships, which mainly related to the Manufacturing, Household Consumption, and Internal Environment sectors, had a large proportion of the whole ecological relationships. Policymakers should take decisive actions, such as improving the efficiency of industrial material utilization and decreasing pressure on the environment, demand, and emissions caused by household consumption, material recycling losses, and agricultural production, to control the negative effect of these flows and, furthermore, achieve sustainable urban development.

Inevitably, there are many uncertainties in the processes of distributing flows to different components of Beijing's metabolic network according to their characteristics in material utilization. For example, we used data related to the physical quantity of each flow based on conversion coefficients, which contain certain inherent inaccuracies, and perhaps data based on empirical determination of the actual flows could improve the reliability. In addition, it would be helpful to use finer divisions of the physical categories and components of the network to allow more detailed monitoring of the flows. Given the current lack of a suitable accounting framework for urban material flows (Li et al. 2019b), experts and scholars in related fields can apply this accounting framework, and then its advantages and disadvantages are identified to future improve this framework, which will also help us to strengthen the depth of analysis and the ability to compare metabolic types of different cities. This framework is now in its early state, but it seems to show great potential to describe an urban metabolism, including the system's structural features and the balance between input and output flows with further development, which is helpful for urban planners to develop urban ecological management strategies.

References

Bao ZM (2010) Material flow analysis of the environmental-economic system of Dalian. Dalian University of Technology, Dalian, China (in Chinese)

Barles S (2009) Urban metabolism of Paris and its region. J Indus Ecol 13(6):898–913

Barrett J, Vallack H, Jones A, et al. (2002) A material flow analysis and ecological footprint of York (2002–1–1) [2022–1–23]. https://www.researchgate.net/publication/257494298_A_Material_Flow_Analysis_and_Ecological_Footprint_of_York.

BBS (Beiijng, Bureau of Statistics) (2016) Beijing statistical yearbook. China Statistics Press, Beijing, China (in Chinese)

Browne D, O'Regan B, Moles R (2011) Material flow accounting in an Irish city-region 1992–2002. J Clean Prod 19(9–10):967–976

Brunner PH (2007) Reshaping Urban Metabolism. J Indus Ecol 11(2):11–13

Cameron AC, Trivedi PK (2005) Microeconometrics: Methods and applications. Cambridge University Press, Cambridge, UK

Chambers N, Heap R, Jenkin N, et al. (2002) City limits: a resource flow and ecological footprint analysis of Greater London (2002–1–1) [2022–7–28]. https://library.uniteddiversity.coop/Measuring_Progress_and_Eco_Footprinting/Ecological_Footprinting/City%20Limits%20London.pdf

Chen WQ, Graedel TE (2015) In-use product stocks link manufactured capital to natural capital. Proceed Nat Acad Sci USA 112(20):6265–6270

Condeixa K, Haddad A, Boer D (2017) Material flow analysis of the residential building stock at the city of Rio de Janeiro. J Clean Product 149(7):1249–1267

Du B (2012) Construction waste recycling network system and model construction. Nanjing University of Technology, Nanjing, China (in Chinese)

Eurostat (2016) Economy-wide material flow accounts (EW-MFA). (2015–3–1) [2021–12–1]. https://www.mendeley.com/catalogue/economywide-material-flow-accounts/

Fath BD, Borrett SR (2006) A MATLAB function for network environ analysis. Environ Model Software 21(3):375–405

Fath BD, Patten BC (1999) Review of the foundations of network environ analysis. Ecosystems 2(2):167–179

Feng HB, Liyanage DR, Karunathilake H et al (2020) BIM-based life cycle environmental performance assessment of single-family houses: Renovation and reconstruction strategies for aging building stock in British Columbia. J Clean Product 250(6):119543

Fischer-Kowalski M (2011) Analyzing sustainability transitions as a shift between sociometabolic regimes. Environ Innov Societal Trans 1(1):152–159

Fischer-Kowalski M, Krausmann F, Giljum S et al (2011) Methodology and indicators of economy-wide material flow accounting: state of the art and reliability across sources. J Indus Ecol 15(6):855–876

Fishman T, Schandl H, Tanikawa H (2016) Stochastic analysis and forecasts of the patterns of speed, acceleration, and levels of material stock accumulation in society. Environ Sci Technol 50(7):3729–3737

Fu CL, Zhang Y, Yu XY (2019) How has Beijing's urban weight and composition changed with socioeconomic development. Sci Total Environ 675(3):98–109

Guo Z, Hu D, Zhang F et al (2014) An integrated material metabolism model for stocks of urban road system in Beijing. China. Sci Total Environ 470–471(3):883–894

Hammer M, Giljum S (2006) Materialflussanalysen der Regionen Hamburg. Wien Und Leipzig. Natur Und Kultur 7(2):62–78

Hoekman P, von Blottnitz H (2017) Cape Town's metabolism: Insights from a material flow analysis. J Indus Ecol 21(5):1237–1249

Huang C, Han J, Chen WQ (2017) Changing patterns and determinants of infrastructures' material stocks in Chinese cities. Resourc Conser Recyc 123(8):47–53

ILO (2018) Labour productivity—ILO modelled estimates. Geneva, Switzerland: International Labour Organization. (2018–11–15) [2021–12–30]. https://www.ilo.org

Jia JJ (2014) Energy consumption and carbon emission influencing factors based on LMDI. Tianjin University, Tianjin, China (in Chinese)

Kennedy C, Cuddihy J, Engel-Yan J (2007) The changing metabolism of cities. J Indus Ecol 11(2):43–59

Krausmann F, Wiedenhofer D, Lauk C et al (2017) Global socioeconomic material stocks rise 23-fold over the 20th century and require half of annual resource use. Proceed Nat Acad Sci USA 114(8):1880–1885

Lang J (2020) Research on air pollution control status and air governance of "2 + 26" cities. IOP Conf Series: Earth Environ Sci 450(1):012050

Li RC (2016) Decomposition analysis and comparison of carbon emission influencing factors in Beijing, Tianjin and Hebei. Hebei University of Geosciences, Shijiazhuang, China (in Chinese)

Li YX, Zhang Y, Hao Y et al (2019a) Exploring the processes in an urban material metabolism and interactions among sectors: An experimental study of Beijing, China. Ecol Indicators 99:214–224

Li YX, Zhang Y, Yu XY (2019b) Urban weight and its driving forces: a case study of Beijing. Sci Total Environ 658(6):590–601

Liu G, Bangs CE, Müller DB (2012) Stock dynamics and emission pathways of the global aluminium cycle. Nat Clim Chang 3(4):338–342

Lou Y, Shi L (2008) Material flow analysis of Handan city. Environ Res 21(4):201–204 (in Chinese)

Lwin CM, Dente SMR, Wang T et al (2017) Material stock disparity and factors affecting stocked material use efficiency of sewer pipelines in Japan. Resources, Conser Recycl 123(8):135–142

Niza S, Rosado L, Ferrão P (2009) Urban Metabolism. J Indus Ecol 13(3):384–405

Pauliuk S, Müller DB (2014) The role of in-use stocks in the social metabolism and in climate change mitigation. Glob Environ Chang 24(1):132–142

Rosado L, Kalmykova Y, Patrício J (2016) Urban metabolism profiles: An empirical analysis of the material flow characteristics of three metropolitan areas in Sweden. J. Clean Product 126(13):206–217

Sastre S, Carpintero O, Lomas PL (2015) Regional material flow accounting and environmental pressures: The Spanish case. Environ Sci Technol 49(4):2262–2269

Swilling M, Hajer M, Suh S, et al. (2018) The weight of cities: resource requirements of future urbanization. Nairobi, Kenya: United Nations Environment Programme, International Resource Panel. (2018–1–31) [2022–4–12]. https://www.researchgate.net/publication/327035481_The_Weight_of_Cities_Resource_Requirements_of_Future_Urbanization

Wang W, Liu X, Zhang M et al (2014) Using a new generalized LMDI (logarithmic mean Divisia index) method to analyze China's energy consumption. Energy 67(4):617–622

Wang Y, Chen PC, Ma HW et al (2018) Socio-economic metabolism of urban construction materials: a case study of the Taibei metropolitan area. Resources, Conser Recyc 128(1):563–571

Wang Z, Feng C, Chen J et al (2017) The driving forces of material use in China: an index decomposition analysis. Resour Policy 52(2):336–348

Zhang Y, Zheng HM, Fath BD et al (2014) Ecological network analysis of an urban metabolic system based on input-output tables: Model development and case study for Beijing. Sci Total Environ 468–469(1):642–653

Chapter 8
Analysis of a City's Energy Metabolism

In Chapter 7, I discussed the importance of material flows and the resulting changes in a city's stock and weight. However, all metabolic processes, including material flows, depend on energy. Because energy is still provided mostly by burning fossil fuels, leading to serious atmospheric pollution and promoting climate change, achieving urban sustainability requires planning and management to reduce energy consumption. In this chapter, I will build on my previous descriptions of urban metabolic processes to consider the flows of energy through an urban metabolism. I will also explore the effect of scale (i.e., changes in the number of metabolic actors or cities included in the analysis) on the results of such analyses. Because most of the research on which this chapter is based was conducted for China, I will use Chinese examples to make the abstract theory more concrete.

8.1 Analysis of Energy Metabolic Processes

As a result of China's continuous and rapid socioeconomic development since China opened to the west in 1978, energy shortages have become increasingly prominent. Although China is a major energy producer and consumer, there is a big gap between its energy supply and demand. Energy conservation and consumption reduction have always been important considerations in China's Five-Year Plan cycle, and analyzing urban energy consumption from a metabolic perspective is an effective approach to find weak links in a city's energy utilization, thereby supporting urban planning to strengthen these links. Such research has great theoretical and practical significance to support more sustainable utilization of urban energy by analyzing a city's energy metabolic processes and identifying the key links in its energy metabolism. China is self-sufficient in terms of its raw coal consumption, but this has led to severe atmospheric pollution, and regarding other energy types, there is a large gap between supply and demand. Securing China's energy supply while also moving towards cleaner energy sources is the key to sustainable development, and has been the focus of both international and Chinese attention. In addition to national energy security,

Y. Zhang, *Urban Metabolism*,
https://doi.org/10.1007/978-981-19-9123-3_8

regional energy security cannot be ignored. Some Chinese provinces have a severe imbalance between supply and demand because of excessive consumption or surplus production. These imbalances in energy supply and demand within China make it important to understand the problems that arise from the spatial patterns of energy supply and demand, the relationships between supply and demand, and how the patterns have evolved over time.

The patterns of regional energy production and consumption profoundly affect the processes in energy metabolism, leading to different metabolic characteristics. In the research that was the basis for this section (Zhang et al., 2012), my research group used geographical information system (GIS) technology to analyze the evolution of the patterns of raw coal, crude oil, natural gas, and electricity production and consumption in China based on their centers of gravity from 1997 to 2009. As I noted in Chapter 6, the center of gravity approach can reveal gaps between supply and demand that should be reduced or eliminated to improve the efficiency of a metabolic system. The results of our analysis revealed movements of the centers of gravity for energy supply and demand during China's socioeconomic development and spatial variations in the gap between energy supply and demand. By understanding the reasons behind these patterns, it will become possible to provide guidance for the development of policies intended to improve China's national and regional energy security (Zhang et al., 2012).

In addition, we developed an ecological network model of an urban energy metabolic system by collecting, sorting, and analyzing the available energy consumption data in 2006. The model can reflect the whole energy consumption process, from resource development, exploitation, transformation, and consumption to the recovery of waste energy. Using the model, we developed a new method for studying the internal characteristics of an urban energy metabolic system, to provide both theoretical and practical methods for optimizing and managing energy flows in Chinese cities (Zhang et al., 2010a). In this section, I will describe how the results change when the scale of the analysis changes (Fig. 8.1). To do so, I will describe network analysis based on a small number of nodes (metabolic actors) and then repeat the analysis with a larger number of nodes to illustrate the different insights that can be obtained.

To illustrate how the status of an urban energy metabolic system can be analyzed by means of ecological network analysis, I chose four major Chinese cities (Beijing, Tianjin, Shanghai, and Chongqing) to serve as examples and developed a model with 5 or 17 nodes (metabolic actors) to support analysis of these cities. The data used in this analysis mainly came from the statistical yearbooks of the studied cities, the *China Energy Statistical Yearbook*, and the literatures published at home and abroad. Most countries can provide comparable data to support such analyses. I also used GIS technology to analyze the spatial distribution of raw coal, crude oil, natural gas, and electricity production and consumption in China's 30 provinces. Furthermore, I analyzed how this distribution evolved based on the movement of their centers of gravity from 1997 to 2009. At the time of the original research, data were available from 1997 to 2009, so I will focus on that period. The production of raw coal, crude oil, and natural gas was used to represent indigenous primary energy

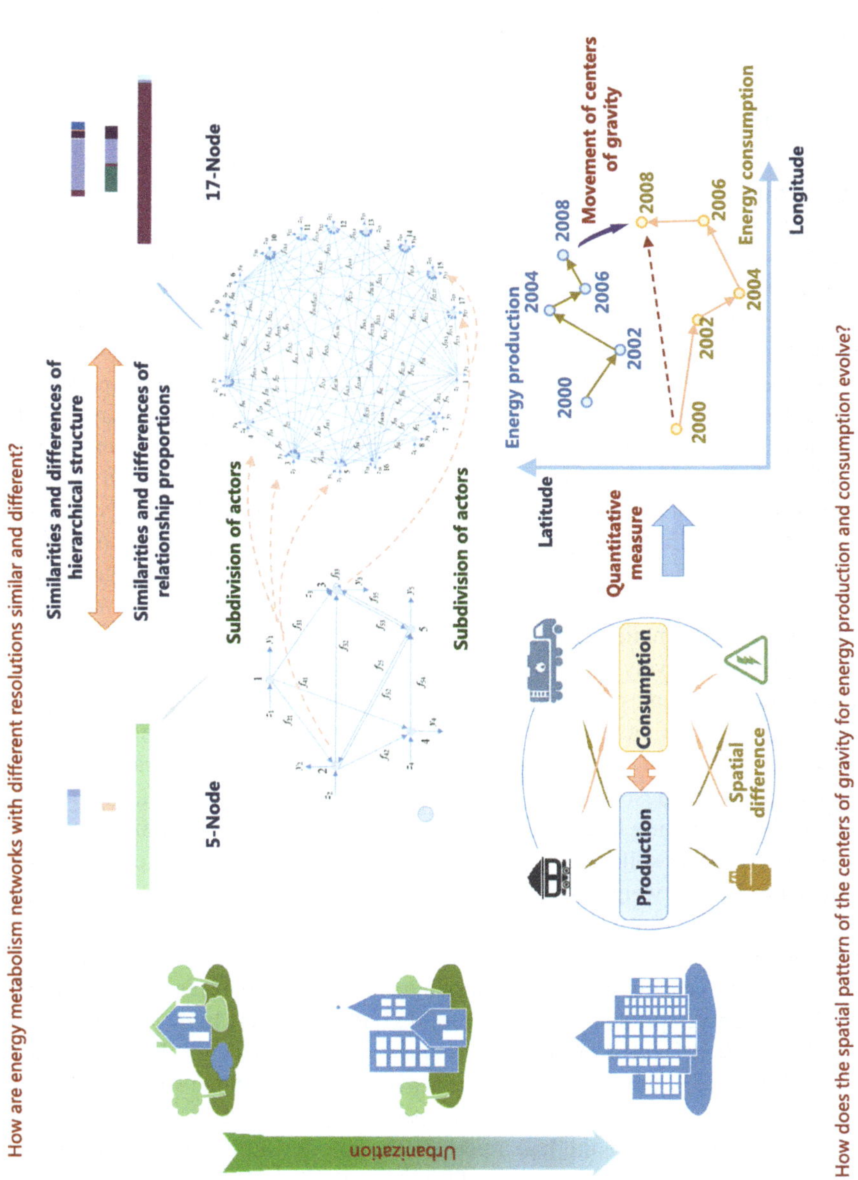

Fig. 8.1 An analytical framework for describing a metabolic network's characteristics and changes in energy supply and demand spatial patterns

production, and the consumption of these energy types was used to represent inputs for the transformation and final consumption sectors. The production of electricity increasingly includes hydroelectric, nuclear, wind, and thermal power production. The consumption of electricity was used to represent the final consumption.

8.1.1 Analysis of a Metabolic Network

8.1.1.1 Analysis of a Metabolic Hierarchy

In the 5-node network, we can use the energy metabolic flows among the nodes to construct integral flow matrices for Beijing, Shanghai, Tianjin, and Chongqing. (Here, "integral" means that we will include both direct flows and indirect flows embodied in the transfers of materials between nodes.) From these matrices, we can calculate the weight matrices for Beijing, Shanghai, Tianjin, and Chongqing. We can then standardize the values to fall within a range from 0 to 1 so that the values for all nodes in a city total 1.0 and can be directly compared with the corresponding values for the other cities. Based on these results, we can define the role, status, and function of each component of the network. Chongqing had the most incredible energy production weighting (the Energy Exploitation and Energy Recovery sectors), with Tianjin, Beijing, and Shanghai following, which suggests that the city has a high energy availability capacity (Fig. 8.2). Beijing had the greatest primary consumer weighting (the Energy Transformation sector), followed by Tianjin, Shanghai, and Chongqing, which implies a relatively high investment in the field of energy transition in Beijing. Shanghai had the greatest secondary consumers weighting (Other Production and Household Consumption sectors), with Beijing, Tianjin, and Chongqing following, which suggests that Shanghai's energy requirements are comparatively high. Shanghai has the greatest recycling weighting, with Chongqing, Tianjin, and Beijing (which had no recycling of energy) following, which reflects Shanghai's comparatively self-regulatory solid ability. Shanghai had the most significant consumption by Other Production weightings, with Beijing, Tianjin, and Chongqing following. Beijing had the greatest Household Consumption weighting, with Shanghai, Chongqing, and Tianjin following.

A robust urban energy metabolism structure should be in the shape of a pyramid, with adequate energy producers at the pyramid's base supporting higher-level components. The second level is primary consumers (here, the Energy Transformation sector), while the tip of the pyramid comprises secondary consumers of lesser weight (Burns, 1989). However, Shanghai had an upside-down pyramidal ecological structure, while Beijing and Tianjin had a periodic upside-down pyramidal structure. In comparison, the environmental system of Chongqing was in the shape of an irregular pyramid, but with approximately equal shares of producers and secondary consumers and a smaller weight of primary consumers. The urban energy metabolic system differs from the natural system because of its lower self-regulatory capacity. All four cities in the 5-node analysis weighted $\leq 5\%$ for the Recycling component.

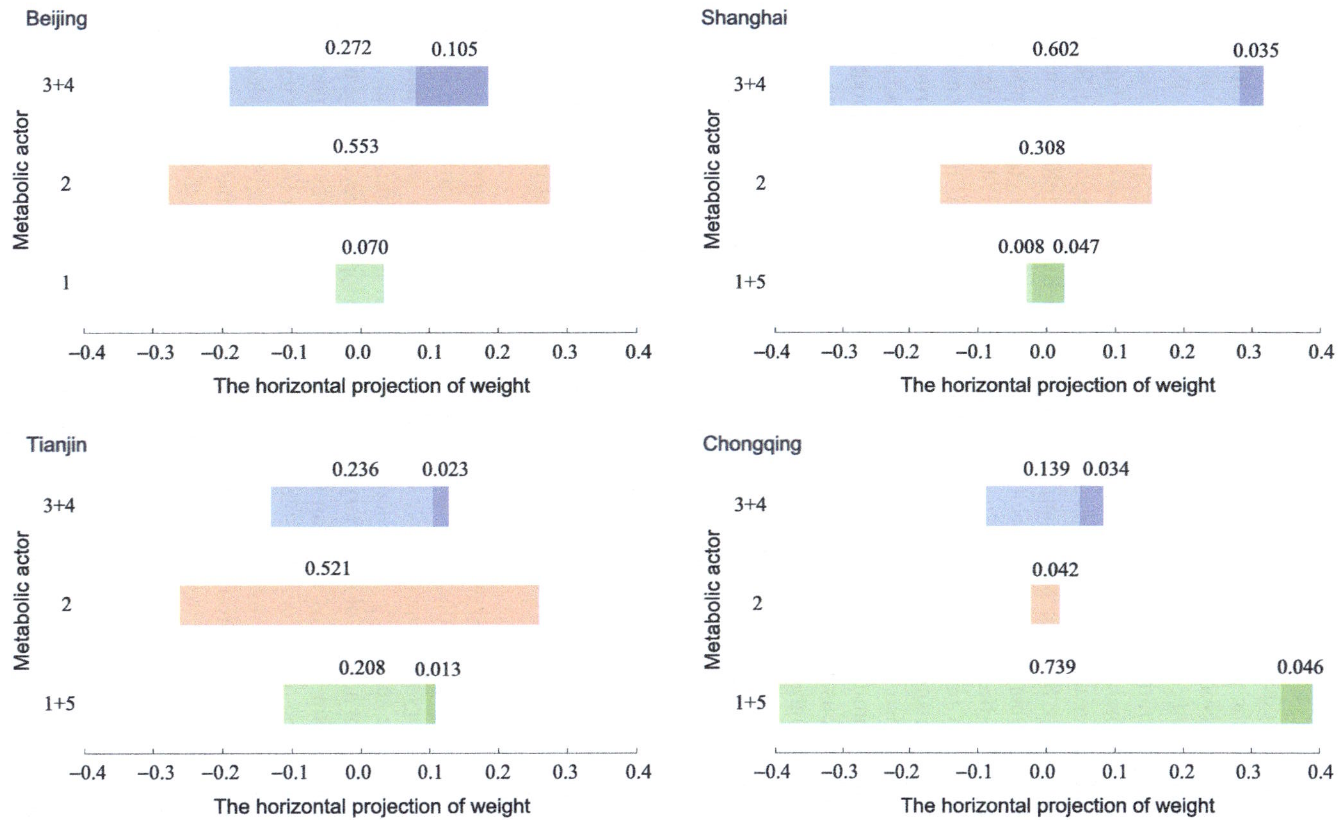

Fig. 8.2 Illustration of the structure (weights) of the 5-node energy metabolic model for four Chinese cities (*Note* Beijing has no Recycling sector. The lengths of the bars have been standardized so that each bar or segment of a bar represents the proportion of the total flow (which equals 1.0). Components: 1-Energy Exploitation sector; 2-Energy Transformation sector; 3-Other Production sector; 4-Household Consumption sector; 5-Recycling sector)

We can repeat the same analytical process for the same cities but with larger and more complex networks. For example, if we subdivide the nodes from the 5-node network, we can create a 17-node network, as shown in Fig. 8.3. Note that in the more complex model, we split the energy stocks apart from the consumers and saw them as part of the producer component, which would result in a different hierarchical structure. Note that Beijing lacked a Recycling component, only Chongqing had a Washed Coal component, Shanghai lacked a Coal Products component, Tianjin lacked Gas Generation and Coal Products components, and Chongqing lacked Oil Refining, Gas Generation, and Coal Products components. Chongqing had the highest weight for the three nodes involved in energy production (Energy Exploitation, Recycling, and Energy Stocks), with Tianjin, Shanghai, and Beijing following, which reflects the high energy delivery capacity of the city. Beijing had the greatest primary consumers (Coal-fired Power, Heat Supply, Washed Coal, Coking, Oil Refining, Gas Generation, and Coal Products) weighting, with Shanghai, Tianjin, and Chongqing following, which suggests that Beijing invests more in the Energy Transformation sector. Beijing also had the greatest secondary consumers (Agriculture; Industry; Construction; Communication; Transportation; Wholesale, Retail, Accommodation and Catering; Household Consumption; and Other Consumption) weighting, with Chongqing, Shanghai, and Tianjin following, which indicates that Beijing has a relatively high need for energy. Tianjin had the greatest Energy Stocks weighting, with Shanghai, Chongqing, and Beijing following, which shows that the city has a high level of energy self-sufficiency. Shanghai had the greatest Recycling weighting, with Tianjin, Chongqing, and Beijing (which had no Recycling component) following, which reflects that Shanghai has strong self-regulatory capability; however, all four cities weighted $\leq 10\%$ for the Recycling component. Each city's Recycling component is lightly weighted, which implies that quite a lot of space is available within the system to increase energy recovery (e.g., to recover waste heat from industrial processes).

The natural metabolic system should have a structure whose shape is a pyramid, with adequate energy producers at the base of the pyramid supporting the minor activity of primary consumers (here, the Energy Transformation sector), and the movement of the secondary consumers at the tip is even negligible. with an even more insignificant activity of secondary consumers at the top. However, since urban energy metabolism systems are man-made structures, their energy flow patterns differ from natural ones. Therefore, their designs are also different from the natural ecosystem. Based on the nutritional levels of producers, consumers, and decomposers (I include them as producers because they recycle energy in the system) in the urban energy metabolic system, as well as the relative weights of the components, the eco-structure of the urban metabolism system can be expressed in a way that how it is similar to the pyramidal structure of the natural ecosystem. As is shown in Fig. 8.3, the energy generators at the base of the pyramid (Energy Exploitation, Recycling, and Energy Stocks) provide support to the primary energy consumers (Coal-fired Power, Heat Supply, Washed Coal, Coking, Oil Refining, Gas Generation, and Coal Products) at the second tier, who in turn support secondary consumers

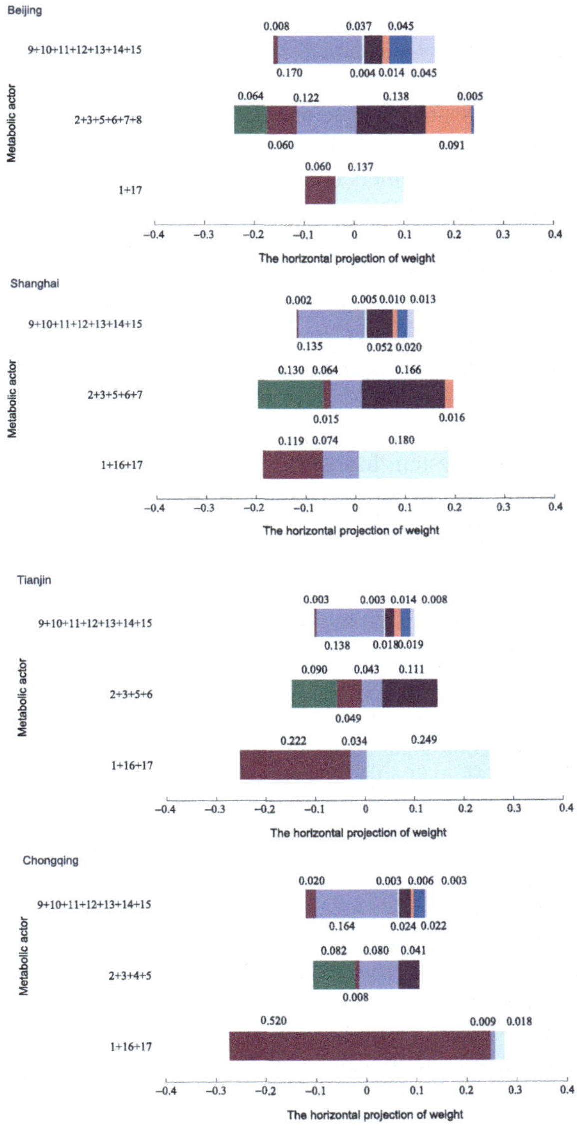

Fig. 8.3 Ecological structure (weights) of the 17-node model of the energy metabolic systems of the four Chinese cities (*Note* The lengths of the bars have been standardized so that each bar or segment of a bar represents the proportion of the total flow (which equals 1.0). Components: 1-Energy Exploitation; 2-Coal-fired Power; 3-Heat Supply; 4-Washed Coal; 5-Coking; 6-Oil Refining; 7-Gas Generation; 8-Coal Products; 9-Agriculture; 10-Industry; 11-Construction; 12-Transportation; 13-Wholesale, Retail, Accommodation, and Catering; 14-Household Consumption; 15-Other Consumption; 16-Recycling; 17-Energy Stocks)

(Agriculture; Industry; Construction; Transportation; Wholesale, Retail, Accommodation and Catering; Household Consumption; and Other Consumption) at the summit of the pyramid. According to the distribution of net transactions in pairs in the system, Beijing's ecological structure is similar to an irregular but mainly upside-down pyramid, in which the share of primary consumers is the largest, with secondary consumers following, while the last is the weight of producers. Chongqing and Shanghai present irregularly shaped structures, but with Chongqing having a roughly pyramidal structure, Shanghai has a higher weight of primary consumers than secondary consumers and producers. In contrast, Chongqing has a higher share of secondary consumers than primary consumers but a lower share than producers. In comparison, the structure of Tianjin resembles a natural pyramid, with producers having the most enormous weights, primary consumers having smaller weights, and secondary consumers having even smaller weights.

The results for the 5-node and 17-node models show that the urban metabolic hierarchies of the four cities differed greatly from that of a natural ecosystem. In the 5-node model, Beijing, Shanghai, and Tianjin presented an irregular inverted pyramid pattern with excessively high weights for the consumers, whereas Chongqing showed an irregular pyramidal structure, with a better balance between energy production and consumption. This can be attributed to the fact that Chongqing has abundant energy resources, and there is rich natural gas storage in this region (one-quarter of China's natural gas is produced here). Because of these energy resources, Chongqing's energy metabolic system has many producers. In the 17-node model, the subdivision of the original five nodes did not produce major differences from the results for the 5-node model; the total contribution of each level of the hierarchy was approximately equal to the contribution before this subdivision. Therefore, the weights of the producers, primary consumers, and secondary consumers did not change as a result of subdivision into a more complex system model.

However, cities had different structures because the 17-node model split the Energy Stocks apart from the consumers and saw them as part of the producer component. This more sophisticated depiction of the form of the system suggested that the ecological structure of Beijing was still an irregular upside-down pyramid with a high dependence on outside energy sources. In contrast, the ecological structure of Chongqing is still a distinctive pyramid with less reliance on external energy sources, but the similarity to the pyramidal structure of natural ecosystems is not perfect. However, the structures of Tianjin and Shanghai were quite different from those in the 5-node model: Shanghai was much closer to an irregular pyramidal structure, while Tianjin had a natural pyramid-like structure. These changes were due to the different roles the Energy Stock components played in the two models. This result demonstrates how the subdivision of a system into more parts with more complicated flow networks and thus, a more accurate description of the metabolic structure of the system can alter the system's structural characterization. Nevertheless, in this example, the general conclusion remained the same.

The overall descriptions of each system did not change significantly between the two models. For instance, Chongqing has the most tremendous energy production weight in both models, with Tianjin following; in these two models, Beijing had the most significant weight of primary consumers, while Chongqing had the slightest weight; Shanghai had the most excellent weight of Recycling component, indicating a comparatively strong self-regulatory capability, yet Beijing did not have Recycling component in either of these models. But these two models also had something different. Taking the weights of energy production, primary consumer, and recycling components as illustrations, the importance of energy production is higher in Beijing than that in Shanghai in the 5-node model. At the same time, contrasting results were obtained in the 17-node model. The primary consumer weight is higher in Tianjin than in Shanghai in the 5-node model, contrary to the results obtained in the 17-node model. The importance of Recycling in Chongqing was higher than that in Tianjin when using the 5-node model, while the contrary result was obtained in the 17-node model. Although the relative rankings of some components changed, the overall difference between the two models is not significant. This suggests that when a system is becoming more and more segmented, the weight of a given sector would not significantly affect the overall results, though it may change the weights for individual components in some cities.

Such differences in the weighting of the ecological components suggest that the 5-node model does not require as much complex work as the 17-node model (e.g., high data collection requirements, high cost, and time demands) to present the overall characteristics of the system well. In most cases, minor differences between the results of the two models did not have a significant impact on the main features of the system. However, more details in the 17-node model unveiled the differences in the secondary consumers' weights which could not be seen in the 5-node model. This variation in secondary consumer weighting was primarily due to a change in the role (and weighting) of the Energy Stocks component. The 17-node model also offered the possibility for more detailed comparisons. For instance, we identified that the weight of Energy Stocks was highest in Tianjin, which indicates that energy is stored in other components, with Shanghai, Beijing, and Chongqing following, which represents the greater energy self-sufficiency of Tianjin. In planning the city's management, these details can offer significant guidance in establishing policies and specific operational programs. While the 5-node model provides enough high-level guidance to recognize the overall problem (i.e., the unbalanced energy structure), just the 17-node model offers sufficient detail to identify the exact cause of the situation revealed in the 5-node model. Therefore, the more accessible model provides a quicker and cheaper approach to identifying areas that need to be studied more thoroughly by using the more complicated model.

8.1.1.2 Analysis of Metabolic Relationships

In this section, I will continue my discussion of the network models described in previous sections. In the analysis of the metabolic relationships, we again start with the integral utility intensity matrix ($\boldsymbol{U}$) values and the sign utility matrix [sgn($\boldsymbol{U}$)], which I described in detail in Chapter 5. Figure 8.4 shows that Beijing's mutualism index ($M = 2.20$) was higher than that of Shanghai (2.13), and much higher than those of Tianjin (1.50) and Chongqing (0.92). Chongqing's mutualism index was less than 1 because of the greater frequency of exploitation relationships between sectors. Thus, the urban energy metabolic processes of three of the four cities reveal an overall degree of mutualism, with mutualism indexes greater than 1 (Fath, 2007). Still, differences in the ecological relationships among each city's metabolic components lead to large differences in these indexes.

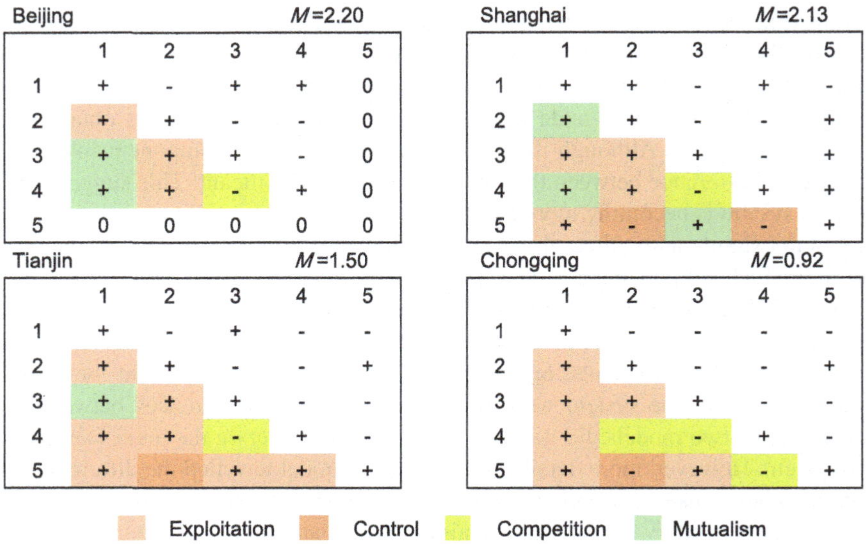

Fig. 8.4 Ecological relationship analysis using the 5-node network models of four Chinese cities (*Note* Symbols (+ or –) are based on the sign utility matrix [sgn(***U***)]. *M* represents the mutualism index for the networks (see Chapter 5 for details). Components: 1-Energy Exploitation sector; 2-Energy Transformation sector; 3-Other Production sector; 4-Household Consumption sector; 5-Recycling sector)

A total of 10 pairs of relationships formed among the five nodes of the cities. Except for Beijing, the combination of exploitation and control relationships in the other three cities accounted for more than 50% of the total, with the highest proportion in Tianjin (80%), versus 60% and 70% for Shanghai and Chongqing, respectively. As Beijing's Recycling component did not exist, and therefore had no associations with other nodes in the 5-node network, neutral relationships accounted for 40% of the network, whereas exploitation or control relationships only accounted for 30%. The highest proportion of mutualism relationships was in Shanghai (30%), followed by Beijing (20%), Tianjin (10%), and Chongqing (0). In Chongqing, all relationships that were neither exploitation nor control were competition relationships (30%), whereas competition relationships in the other three cities accounted for only 10% of the total.

In the 17-node network, the sgn(***U***) matrices for the four cities present 136 pairs of ecological relationships (Fig. 8.5). Beijing, Shanghai, and Tianjin were dominated by exploitation and control relationships, accounting for around 35% of the total (37%, 35%, and 35%, respectively), but Chongqing had the highest proportion of competition relationships, at 47%. The competition relationships in Beijing, Shanghai, and Tianjin accounted for < 35% (29%, 34%, and 25%, respectively). The mutualism relationship was most common in Beijing (11%), versus 8% in Shanghai and Tianjin and only 1% in Chongqing.

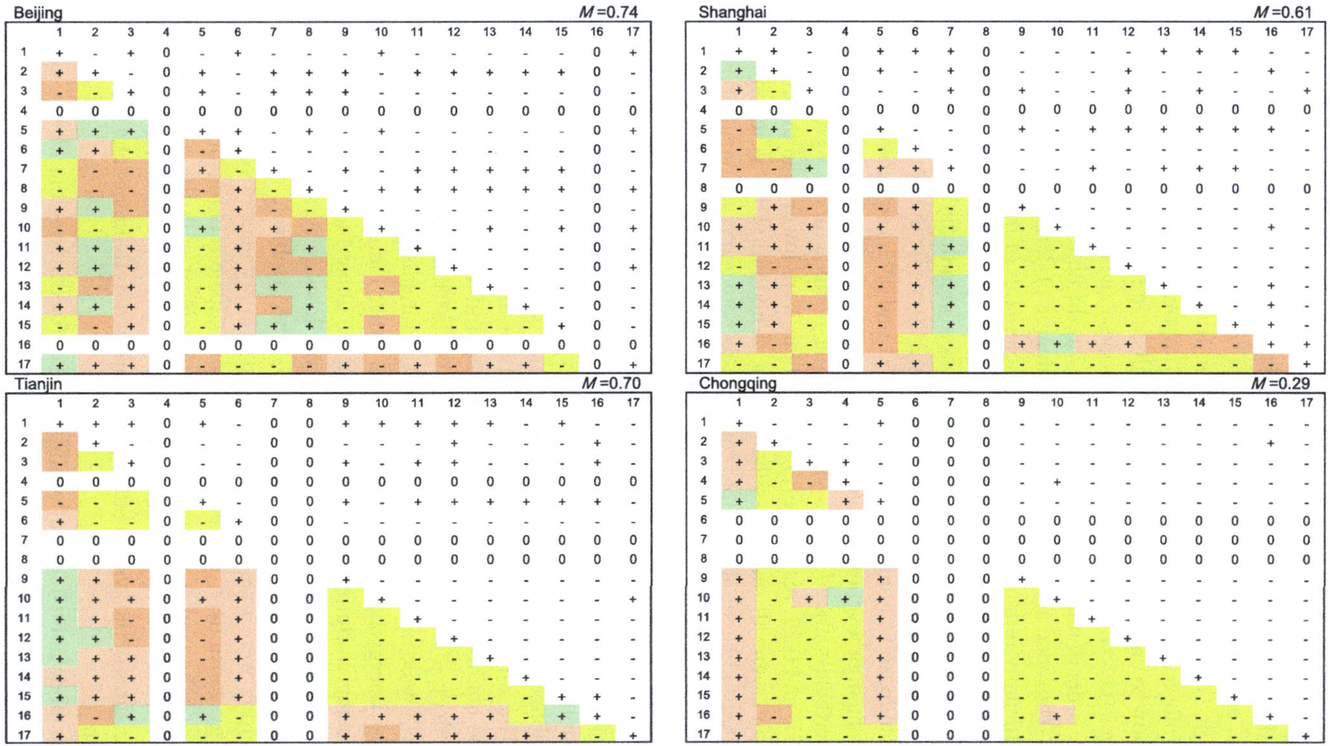

Fig. 8.5 Ecological relationships among components of the 17-node network model for the four cities (*Note* Symbols (+ or –) are based on the sign utility matrix [sgn(***U***)]. *M* represents the mutualism index for the networks (see Chapter 5 for details). Components: 1-Energy Exploitation; 2-Coal-fired Power; 3-Heat Supply; 4-Washed Coal; 5-Coking; 6-Oil Refining; 7-Gas Generation; 8-Coal Products; 9-Agriculture; 10-Industry; 11-Construction; 12-Transportation; 13-Wholesale, Retail, Accommodation, and Catering; 14-Household Consumption; 15-Other Consumption; 16-Recycling; 17-Energy Stocks)

According to the integral utility intensity matrix (***U***) values, Beijing had a higher mutualism index (0.74) than Tianjin (0.70) and Shanghai (0.61) and much higher than Chongqing (0.29). Since the values of all four indicators were less than 1, the components of the urban energy metabolic system show mainly negative correlations with each other, and the degree of network mutualism was low. Though the mutualism indexes give a general picture of the system, it is still essential to check the variations in ecological relationships among the components if city managers want to understand how to enhance the degree of mutualism in their cities.

In both studies (i.e., with different numbers of nodes), the four cities had approximately the same order of mutualism index: Beijing was the highest, while Chongqing was the lowest. This once again indicates that the degree of detail in the subdivision would not significantly influence the general characteristics of the system, and both degrees of detail can be applied to make comparisons between cities. However, the values of the mutualism index of these two studies were significantly different. In the 5-node model, the mutualism index of Beijing was 2.20, while the mutualism index of Shanghai, Tianjin, and Chongqing were 2.13, 1.50, and 0.92, respectively. Thus all four systems had a high level of symbiosis, even the lowest, Chongqing, had a value close to 1. Nevertheless, the M values yielded by the 17-node model were different (0.74, 0.70, 0.61, and 0.29, respectively). All of them were lower than those in the 5-node model and were all much less than 1. The 5-node model was much more densely connected (>32% of all possible connections between components existed) than the 17-node model (<16% of all possible links), so the more complex model had a higher order and lower connectivity (Zhang et al., 2011b).

This dramatic difference shows that the 17-node model provides a comprehensive view of the system, especially in the energy-consuming sectors. For instance, the ecological relationship between secondary energy consumers in Beijing was mainly competitive (Table 8.1). In Beijing, competition relationships comprised 56% of all relationships. The simplified 5-node model could not disclose these relationships, therefore, would not help planners to decrease competition between these sectors.

However, both these two models showed a similar distribution of mutualism relationships. In the 5-node model (Fig. 8.4), there was a mutualistic ecological relationship between the Energy Extraction sector and the Other Industrial and Household Consumption sectors in Beijing, which reflects that the Other Industrial and Household Consumption sectors were largely dependent on flows derived from the energy transition sector. In the 17-node model (Fig. 8.5), the symbiotic ecological relationships in Beijing were mainly located between primary energy consumers and secondary energy consumers. In Beijing's case, mutualistic relationships made up more than 90% of these interactions. This is in agreement with the results of the 5-node model. Therefore, the more detailed segmentation of the model did not hide the symbiotic features of the system but offered more details to understand them with more resolution.

In the 5-node model, only ten relationships were defined among the sectors, so the model revealed less detail about the problem. Therefore, the 5-node model could not reflect internal conflicts within the system about resource usage. Nevertheless, the model gave a straightforward calculation method to obtain simple and clear results.

Table 8.1 Comparison of the values in the sign utility matrices sgn(***U***) for the integral utility for secondary consumers (components 3 and 4 in the 5-node model; components 9 to 15 in the 17-node model), using Beijing as an example

5-node model	Secondary consumers [a]		17-node model	Secondary consumers [a]						
	3	4		9	10	11	12	13	14	15
3	+	–	9	+	–	–	–	–	–	–
			10	–	+	–	–	–	–	+
			11	–	–	+	–	–	–	–
			12	–	–	–	+	–	–	–
			13	–	–	–	–	+	–	–
4	–	+	14	–	–	–	–	–	+	–
			15	–	–	–	–	–	–	+

[a], Components in the 5-node model: 3-Other Production; 4-Household Consumption
Components in the 17-node model: 9-Agriculture; 10-Industry; 11-Construction; 12-Transportation; 13-Wholesale, Retail, Accommodation, and Catering; 14-Household Consumption; 15-Other Consumption

It helped to understand the holistic picture of the system, thus guiding critical aspects of the general problem faced by the city. In comparison, although some cities were missing components resulting in fewer relationships being included,17 the node model still inspected 136 relationships among members. The extra relationships offered specific additional insight into the underlying causes of the holistic problem determined by the simpler model.

8.1.2 *Shifts of the Centers of Gravity for Energy Production and Consumption*

This section builds on the discussion of centers of gravity in Chapter 6. In the study on which this section is based (Zhang et al., 2012), my research group analyzed energy flows in China. Our analysis uncovered both similarities and apparent differences among the tracks, directions, and distances of China's energy supply and demand centers of gravity from 1997 to 2009. Among the eight paths which described the motion of different forms of gravity for energy in this period, five of them (raw coal consumption, crude oil production and consumption, and electricity production and consumption) generally move southwest; two of them (raw coal and natural gas production) move typically northwest, and one (natural gas consumption) generally move southeast (Fig. 8.6). The supply and demand centers of gravity for natural gas were located in the Northwestern part of China, those for electricity were found in the Central, for raw coal production were located in the Northern, those for raw coal consumption were located in the Central, that for crude oil production was found in the Northern. In contrast, crude oil consumption was situated in the Eastern part. The most dramatic shift in the center of gravity is in the production and consumption of natural gas, with crude oil following. In comparison, the raw coal and electricity production and consumption centers changed only moderately compared with natural gas and crude oil. This suggests a relatively stable allocation of demand and supply of raw coal and electricity. The center of gravity of production had moved further than the center of consumption. The resulting shortfall in supply and demand for raw coal, crude oil, and natural gas was significant and grew as time went on. In 2009, the producing centers of raw coal, crude oil, natural gas, and electricity were located west to northwest of the consuming centers, which implies that China's energy production is still concentrated in the west and northwest, while consumption is still concentrated in the east and southeast. All of these patterns relate directly to changes in the distribution of energy resources. They are a consequence of China's energy development strategies (transferring energy from north to south and west to east).

China has abundant coal resources but scarce petroleum and gas resources, which makes China's economic growth largely dependent on coal. Nevertheless, the relative concentration of raw coal production areas and the significant inter-provincial differences have led to prominent contradictions between supply and demand. Among the 30 provinces of China, there are only seven provinces (Shanxi, Inner Mongolia, Guizhou, Yunnan, Shaanxi, Qinghai, and Xinjiang) that produced more coal than they consumed in 2009, while the rest of the provinces imported coal. From 1997 to 2009, the center of gravity for raw coal production was between 111.70°E and 113.11°E and between 35.91°N and 36.41°N in southern Shanxi Province in northern China. In contrast, the consumption center of gravity was between 113.92°E and 114.52°E and between 34.71°N and 34.94°N in northeastern Henan Province in central China (Zhang et al., 2012). Therefore, the center of Raw coal production was located in the northwest of the consumption center, which led to the transportation of coal from north to south and from west to east.

These patterns illustrate the general movement of coal production to the northwest; the general trend followed an azimuth of 286.4°, even though it moved from Changzhi City (1997) to Jincheng City (1998), back to Changzhi City (1999 to 2005), and then to Linfen City (2006 to 2009), the center consistently stayed in Shanxi Province (Fig. 8.6). The movement of the gravity center of raw coal consumption generally followed an azimuth of 242.8° to the southwest. From 1997 to 2002, it was situated between Xinxiang and Kaifeng cities, then it moved from Kaifeng City (2003 to 2007) to Zhengzhou City (2008 to 2009), while the center consistently stayed in Henan Province. In conclusion, the center of gravity of production and consumption both moved westward. However, the northward shift of production and southward change of consumption reflected the expanding gap between supply and demand.

The gravity center of raw coal production moved from 113.03°E and 36.01°N in 1997 to 111.70°E and 36.41°N in 2009, a total distance of 125.9 km to the northwest, at an average rate of 10.5 km/year (Zhang et al., 2012). The consumption center of gravity moved from 114.36°E and 34.94°N in 1997 to 113.92°E and 34.79°N in 2009, a total distance of 42.6 km to the southwest, at an average rate of 3.6 km/year,

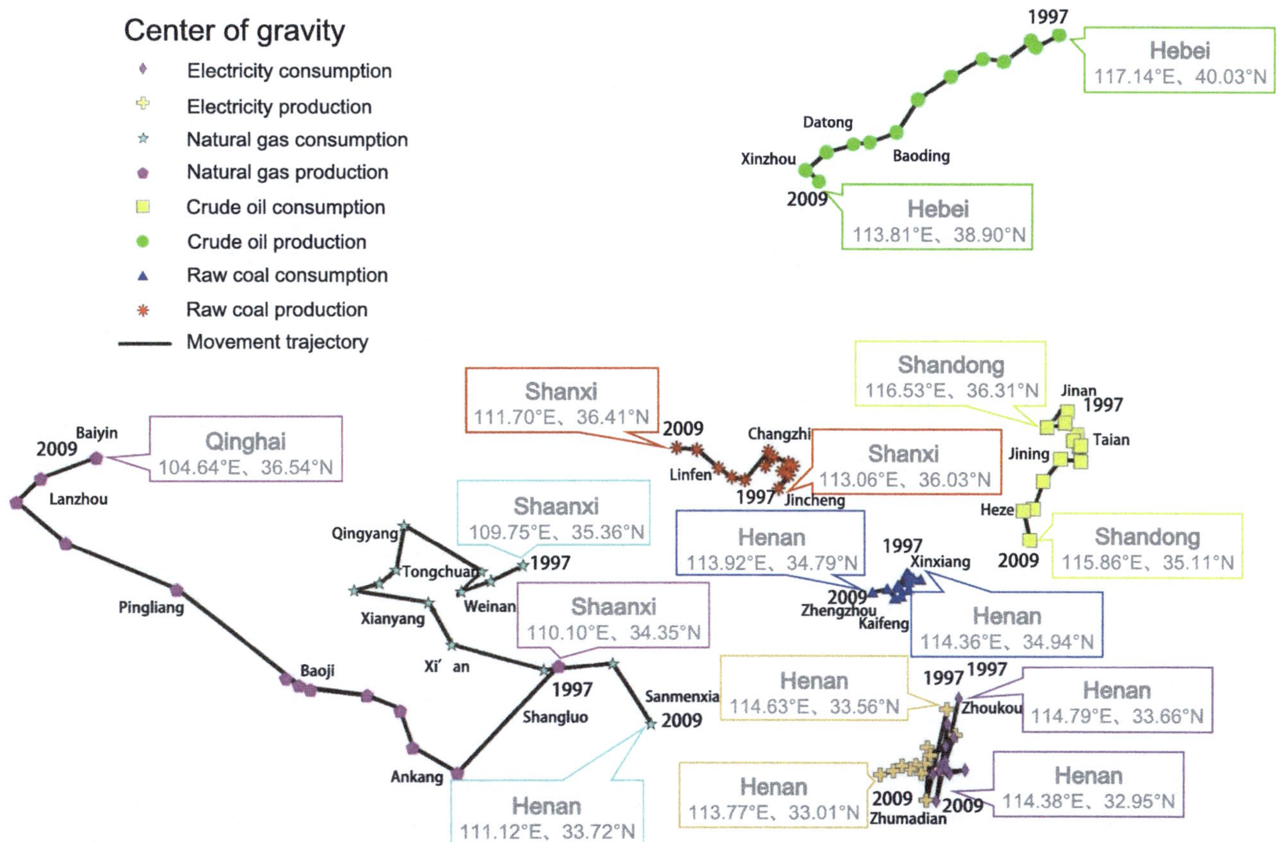

Fig. 8.6 Locations and movement of the centers of gravity for energy production and consumption in China from 1997 to 2009

which equals approximately one-third of the production center's movement (Zhang et al., 2012).

Crude oil consumption is increasing rapidly compared with production. China has become reliant on crude oil imports from 1993 onwards. Presently, China's crude oil-producing areas are dispersed, and inter-provincial differences are remarkable, with a significant imbalance between supply and demand. In 2009, out of 30 provinces, only 5 (Tianjin, Heilongjiang, Shaanxi, Qinghai, and Xinjiang) had oil surpluses; all the rest were forced to import crude oil. From 1997 to 2009, the center of gravity for crude oil production was between 113.66°E and 117.14°E and between 38.90°N and 40.02°N, which lies in central Hebei Province in northern China. In contrast, the center of gravity for consumption was between 115.83°E and 116.63°E and between 35.11°N and 36.35°N, which lies in western Shandong Province in eastern China (Zhang et al., 2012). As a result, the gravity center of crude oil production is located in the north of the consumption center, forming a basic pattern dominated by the north–south transport of oil.

For production, the general movement followed an azimuth of 240.2° toward the southwest (Fig. 8.6). This represents a gradual movement from Langfang City in Hebei Province (1997 to 1998) to Beijing (1999 to 2002), to Baoding City in Hebei Province (2003 to 2005), and then to Datong City (2006 to 2007) and Xinzhou City (2008) in Shanxi Province, then finally back to Baoding City (2009) (Zhang et al., 2012). For consumption, the gravity center of gravity moved southwest following an azimuth of 197.0°. This represents a gradual movement, always within Shandong Province, from Jinan City (1997) to Liaocheng City (1998) and Taian City (1999 to 2003), and then to Jining City (2004), back to Taian City (2005), and finally to Heze City (2006 to 2009) (Zhang et al., 2012).

The crude oil production center of gravity moved from 117.14°E and 40.00°N in 1997 to 113.81°E and 38.90°N in 2009, a distance of 307.4 km to the southwest, at an average rate of 25.6 km/year (Zhang et al., 2012). The consumption center of gravity moved from 116.53°E and 36.31°N in 1997 to 115.86°E and 35.11°N in 2009, a distance of 148.8 km to the southwest, at an average rate of 12.4 km/year, which is about half that of the production center of gravity (Zhang et al., 2012). Therefore, the gap is becoming more and more prominent between the two centers.

China's natural gas production regions are relatively concentrated. Among the 30 provinces, only 9 (Shanxi, Inner Mongolia, Jilin, Heilongjiang, Chongqing, Sichuan, Shaanxi, Qinghai, and Xinjiang) produced more natural gas than they consumed; while other provinces all needed to import gas in 2009. During the study period, the natural gas production center of gravity was located between 103.66°E and 110.10°E and between 33.36°N and 36.54°N (Zhang et al., 2012). These regions are located in the south of Shaanxi Province and the east of Gansu Province in northwestern China. In comparison, the consumption center of gravity was located between 107.72°E and 111.12°E and between 33.72°N and 35.81°N (Zhang et al., 2012). This is located in the center of Shaanxi Province in northwestern China. The gravity center of production was usually located in the west of the consumption center. Therefore, most of the natural gas is delivered from west to east.

The net result was a gradual northwestward shift of the gravity center of natural gas production along the azimuth of 295.6° (Fig. 8.6), firstly within Shaanxi Province, moving from Shangluo City (1997 to 1998) to Ankang City (1999), Xi'an City (2000), and Baoji City (2001 to 2004), and then into Gansu Province, to Pingliang City (2005), Lanzhou City (2006 to 2008), and finally Baiyin City (2009) (Zhang et al., 2012). In comparison, the gravity center of consumption showed a gradual southeastward movement followed by an azimuth of 143.0° within Shaanxi Province, moving from Weinan City (1997 to 1998) to Tongchuan City (1999 to 2000) and then to Qingyang City in Gansu Province (2001 to 2002), and back into Shaanxi Province, to Xianyang City (2003 to 2004), Tongchuan City (2005), Xianyang City (2006), and Weinan City (2007), and finally arrived at Sanmenxia City of Henan Province (2008 to 2009) (Zhang et al., 2012). This pattern shows that the gravity center of natural gas production and consumption moved in opposite directions, which widens the gap between supply and demand.

In 1997, the natural gas production center was at 110.10°E and 34.35°N but moved to 104.64°E and 36.54°N in 2009 (Zhang et al., 2012). This represented a 545.8 km northwestward movement at a rate of 45.5 km/year. The consumption center was at 109.75°E and 35.36°N in 1997 and moved to 111.12°E and 33.72°N in 2009 (Zhang et al., 2012). This represented a 222.6 km southeastward movement at an average rate of 18.5 km/year, which equals 40.7% of the production center's rate. The eastward pull toward consumption centers was much weaker than the westward pull toward production centers (Zhang et al., 2012).

Every province in China produces electricity, but almost half of them have a demand that exceeds their self-sufficient capacity: in 2009, 13 of the 30 provinces (Beijing, Tianjin, Hebei, Liaoning, Shanghai, Jiangsu, Zhejiang, Jiangxi, Shandong, Henan, Hunan, Guangdong, and Chongqing) needed to import electricity from other areas. During the study period, the electricity production center of gravity was between 113.77°E and 114.63°E and between 32.93°N and 33.55°N, which lies in southeastern Henan Province in central China, as was the consumption center of gravity, which was between 114.38°E and 114.79°E and between 32.94°N and 33.66°N (Zhang et al., 2012). Thus, the center of gravity of production is close to the center of consumption (actual to the west), leading to the transmission of electricity from west to east.

The net impact of these shifts was the southwestward movement of the electricity production center, which followed an azimuth of 226.9° (Fig. 8.6). The movement happened only within Henan Province from Zhoukou City (1997) to Zhumadian City (1998 to 2009). The movement of the consumption center was also southwestward, following an azimuth of 199.3°. The movement happened only within Henan Province from Zhoukou City (1997 to 1999) to Zhumadian City (2000 to 2009).

The gravity center of electricity production represented a 99.3 km southwestward movement at a rate of 8.3 km/year, from 114.63°E and 33.55°N in 1997 to 113.77°E and 33.01°N in 2009. The consumption center represented an 87.6 km southwestward movement at a rate of 7.3 km/year, from 114.79°E and 33.66°N in 1997 to 114.38°E and 32.95°N in 2009; this was 88.2% of the production center's movement.

8.1.3 Advantages of Models with Different Precision

Through ecological network analysis, my group built the basic urban energy metabolic network model that we described earlier in this chapter and applied it in analyzing the urban energy metabolic system in four Chinese cities. The approach lets us analyze the system's inherent processes from the perspective of energy flows, thereby providing insights into the ecological relationships between model components (Marzluff et al., 2008). This method innovatively extends the traditional black-box model of urban metabolism, which only accounted for total flows and will be increasingly used in future studies of other urban systems (Fath and Borrett, 2006).

The different levels of segmentation of the system (into 5 and 17 nodes) provided a different understanding of the system's details; both of them yielded a similar view of the system's general characteristics. Even though the 17-node model was more complicated than the 5-node model, it offered more detailed insight into the problems. The 5-node model helped uncover the holistic problem and facilitate comparisons among different cities. However, the 17-node provided insights into critical details of those general issues and their differences among the four cities, thereby providing better guidance for particular endeavors to resolve those issues. In particular, the more detailed the system segmentation by dividing special functions into different sectors, the more practical advice can be provided for departmental classification. The refined division amplified the weight of exploitation and competition relationships in the system, thus emphasizing the conflicts between the constituent elements of urban resource use hidden under the simpler model. This means that, in a simple system model, a sector that appears to be dominated by mutualism may involve a competitive relationship. However, this relationship will not be identified until the subdivision of the industry. From this perspective, the more detailed segmentation can help to clarify the relationship among and inside sectors.

Based on this comparison, it's evident that, compared with the 17-node model, the 5-node model is more concise and can offer a similarly practical overview of the general situation of the system but requires less time and resources to perform the analysis. Thus, it can be used for judging a system in general, exposing the areas where the most severe problems are present but not revealing the details. In comparison, creating a 17-node model is costlier and more time-consuming but can uncover sufficient information about the characteristics of the system to make particular issues visible so that managers can concentrate on the operational actions to address them. Thus, the degree to which system components are detailed in systems research should be determined by the demands driving the study: the simpler model is appropriate for gaining an initial understanding of the general situation and can guide the development of more detailed models. This will provide better support for regulatory decisions.

8.1.4 Diagnosis of Urban Energy Metabolism Problems and Potential Solutions

The hierarchical structures of energy metabolism in the four cities mostly resembled an inverted pyramid, with the weight of the producers lower than that of the consumers, resulting in a severe energy imbalance. The overweighting of primary or secondary consumers (i.e., much greater than that of producers) in Beijing, Shanghai, and Chongqing indicate that addressing this imbalance would require either a reduction in energy consumption or an increase in energy production. Unlike those three cities, Tianjin had a more excellent balance for its producers, overtaking its primary and secondary consumers. By dividing each of these higher-level components into smaller components, my group also gained enough detailed information about the relationships between the components to offer practical management suggestions for solving these problems.

According to the conclusions drawn from our results, we can make some suggestions for improving energy efficiency in these four cities. Chongqing needs to enhance the efficiency of energy use by secondary consumers. Tianjin and Shanghai need to limit the current overdevelopment of the energy conversion sectors. In contrast, Beijing needs to reshape its energy structure and establish a Recycling component that currently does not exist to promote the use of recycled energy. All cities should reduce the over-exploitation of local energy by primary and secondary consumers and decrease energy competition between Other Industries and Household Consumption sectors. Over-exploitation of external environmental resources by Other Industries and Energy Transformation sectors should also be reduced, together with the promotion of local energy recovery and renewable energy development.

It is necessary to work out measures to control the harmful effects of the competitive relationship between Heat Supply and Coal-fired Power sectors, Heat Supply and Oil Refining, and other Energy Transformation sectors. For instance, the Heat Supply sector should improve the use of refinery gas, a by-product of oil refining, and should reduce the proportion of input materials like fuel oil and diesel oil. Besides, the construction of cogeneration projects should also be accelerated. Third, it's necessary to increase coal imports from outside the city. At the same time, the energy consumption patterns should be changed from coal to natural gas to alleviate the pressure of the demand for coal products, gas production, and coking with raw coal as the primary input raw material. Meanwhile, energy efficiency (providing the same level of energy services with less energy) should also be improved.

8.1.5 Spatial Patterns of Supply and Demand for the Energy Types

The shifts in the routes of centers of gravity from 1997 to 2009 show that the transportation of coal from west to east and north to south was the fundamental pattern of raw coal production and consumption. This pattern has a very close relationship with energy resource endowments and regional economic development strategies. Raw coal resources have a broad distribution in China. Still, they are mainly concentrated in North and Northwest China. The main raw coal consumption areas are in economically developed zones such as North China, East China, and Central China. There were significant differences between raw coal resource stocks and consumption, resulting in notable differences between raw coal origins and consumption locations.

To summarize, Shanxi Province was the leading raw coal-producing and consuming province in 1997, while the significant raw coal-producing regions in 2009 changed to Inner Mongolia, Shanxi, and Shaanxi, with consumption concentrated in the east of the producing regions. Raw coal originating from the northern and western regions was mainly shipped to the highly developed areas of the southeast coast of China, for instance, Jiangsu, Zhejiang, and Guangdong. This long-term transportation pattern of coal from west to east and north to south is of great importance to the economic growth of China.

The shift in the center of gravity of crude oil production and consumption from 1997–2009 indicated a fundamental pattern of crude oil transportation from north to south. Crude oil resources are broadly dispersed in China instead of being evenly distributed. These resources are largely allocated in Northeast, North, and Northwest China, while the main consumption regions were economically developed regions such as North, East, Central, and South China. There were clear differences in the spatial distribution of crude oil resources and consumption (Zhang et al., 2012).

To summarize, the biggest crude oil-producing areas in 1997 were Heilongjiang and Shandong provinces, while the biggest consumers were Liaoning; in 2009, due to the development of new oil fields, crude oil production was more broadly dispersed, and the consumption of crude oil was mainly in the southeast. The result of these trends was that oil was mainly transported from the north to the south. Up to 2009, crude oil pipes had stretched to 17,000 km, which is more than three times the distance between China's north and south (5,500 km).

The shift in the center of gravity of natural gas production and consumption from 1997–2009 indicated that natural gas delivery was mainly west to east. China's natural gas resources are relatively concentrated but unbalanced with respect to consumption centers (Zhang et al., 2012). The main reserves of natural gas are in the northwest of China, while the main consuming regions are located in the economically developed regions of North, East, South, and Central China.

The center of gravity for both natural gas supply and demand had moved a long way but in reverse directions. The significant increase in natural gas production in the West led to a major westward movement of production centers. At the same time, the center of gravity of consumption shifted to the southeast. However, it was not as dramatic as the center of gravity of production. Two pipe projects are bringing huge volumes of natural gas from the west to the east, enhancing China's ability to meet its gas demand in the east.

From 1997 to 2009, the shift in the center of gravity of electricity production and consumption eventually led to electricity transmission from west to east. The dominant forms of electricity production are coal, hydroelectric and nuclear power. Production of coal-generated power mainly occurs in coal-producing areas such as northern and northwestern China, and the production of hydroelectric power mainly occurs in southwestern China. In contrast, nuclear power primarily exists in southern and eastern China (Zhang et al., 2012). The main electricity-using regions fall in economically developed areas such as North, East, Central, and South China. As a result, there were significant differences between the production and consumption of electricity resources.

To summarize, in 1997, both production and consumption of electricity were centralized in North China and Guangdong Province, while in 2009, they were concentrated in North China, East China, and South China. There was a widespread overlap in the distribution of electricity production and consumption in 1997 and 2009. The electricity generated in the west was delivered to the developed areas on the southeast coast, with the Beijing-Tianjin-Tangshan area, eastern China (Shanghai, Jiangsu, and Zhejiang provinces), and Guangdong Province as the major destinations. This is a long-term pattern that is of great importance to China's recent economic growth.

8.1.6 Conclusions Related to the Centers of Gravity for Energy Supply and Demand

According to the results of the center of gravity, we can provide three policy proposals to promote coordinated regional development and optimize the development of regional energy. The Northwest is rich in energy resources and generates an increasingly large amount of energy. Most of the production is used to support the development of the southeastern region. Thus, the Northwest should pay attention to the sustainable development of energy resources to protect the native environment.

The national government has historically placed too much emphasis on the development of the eastern region, and now to remedy this imbalance, the government is encouraging the southeast and coastal regions to come together to support the priority development of the western and central regions. Such assistance is not necessarily based on industrial transfer, which would shift energy-intensive and backward industries from the southeast and coastal regions to the west and central regions. Rather, development should emphasize advanced technology, attract more investment, and improve the literacy and technical skills of workers to enable them to work in more advanced industries. Nevertheless, there is also the possibility of substituting old and low-efficiency production facilities in the Southeast and coastal regions with new and higher-efficiency equivalents in the West, which are usually located closer to energy sources. This industrial transfer will narrow the gap between energy supply and demand and reduce the requirement to transport energy for long distances. Furthermore, the new facility will dramatically reduce overall energy consumption and pollution because it will be more efficient and cleaner in its energy use. The western and central regions should inspire their economic potential through scientific and technological innovation. They should also improve their industrial structure, accelerate their development as well as alleviate their environmental problems by reducing the intensity of energy consumption.

China's development relies heavily on traditional energy sources, most of which must be imported. This raises the issue of energy supply security, which could impact the stability of China's development soon. Besides, conventional energy sources like coal more easily cause severe contamination. Thus, it is an unsustainable way to produce and consume energy. Luckily, China has huge renewable energy potential. Therefore, each province should be more active in developing renewable forms of energy, making the best choices based on natural conditions (e.g., wind versus solar versus water availability). The purpose is to improve the energy structure, balance the supply and demand of energy, and reduce carbon emissions and environmental pollution.

8.2 Analysis of the Characteristics of Urban Emergy Metabolic Networks

There are numerous and complicated material, energy, and monetary flows in an urban system. This makes it challenging to integrate diverse flows with different magnitudes and different units of measurement to calculate the inputs, transformations, and outputs of urban metabolic processes. The solution to this problem is to develop a unified system of measurements that can account for the diversity of these flows. Because the present chapter focuses on energy, I will focus on the closely related emergy analysis method to show how this can be done (Huang et al., 2006).

In the research on which this section is based (Zhang et al., 2009a), my research group developed a set of emergy-based indicators for evaluating the dependence, environmental loading, efficiency, pressure, welfare, and intensity components of urban metabolism, thereby comprehensively accounting for the key processes within the urban metabolism. We also developed a method that integrated emergy analysis with ecological network analysis to study urban metabolism from a combined perspective (Fig. 8.7). This innovation accounts for the complex flows within an urban metabolic system in an integrated manner and supports an analysis of the basic structures of the ecological network, the functional relationships among metabolic actors. Thus, it provides a new approach for describing, optimizing, and regulating urban metabolic systems.

In the research that was the basis for this section (Zhang et al., 2011a, 2009b), my research group used data on flows of materials, energy, and money through Beijing's metabolic system from 1990 to 2004, at 2-year intervals, to evaluate the changes in the system's development status. To illustrate how the status of an urban metabolic system can be analyzed using ecological network analysis, we expanded our analysis from Beijing to include three additional major Chinese cities (Shanghai, Tianjin, and

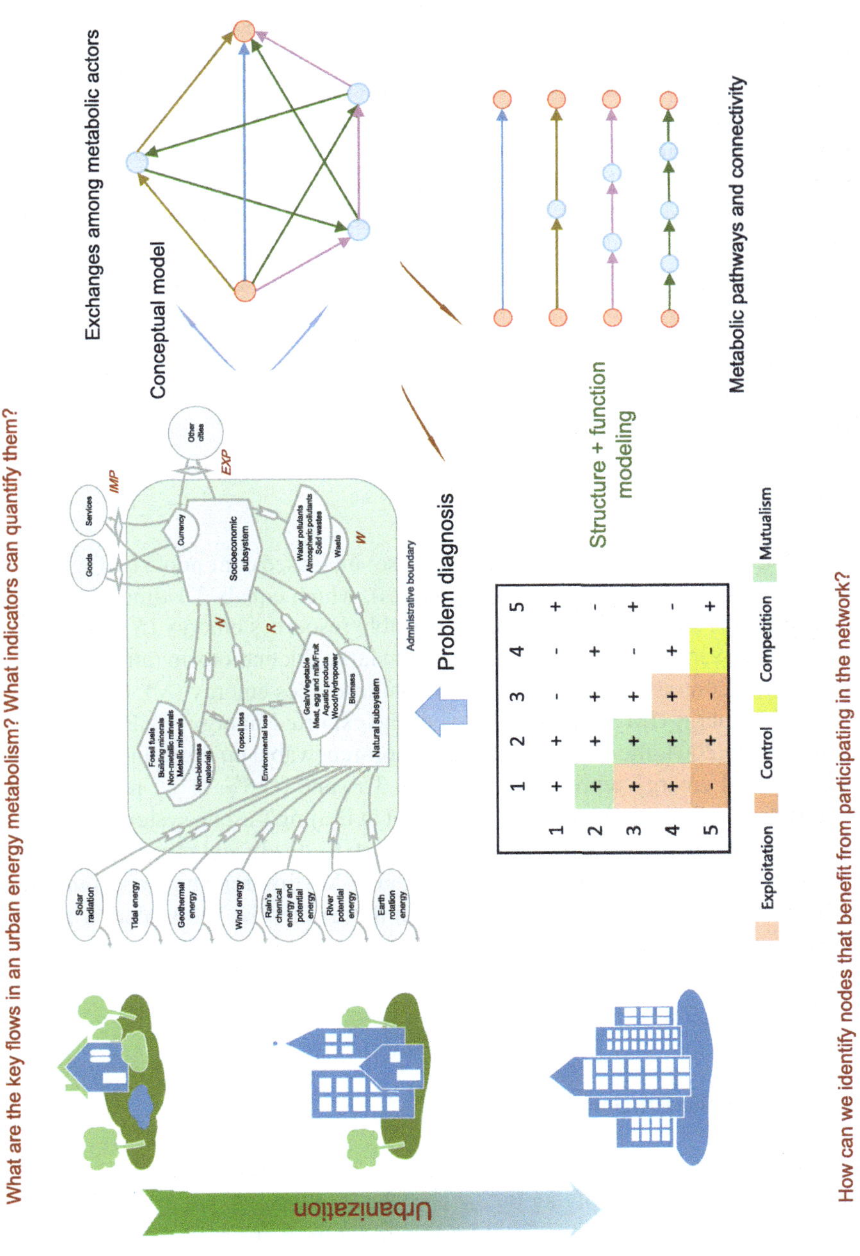

	1	2	3	4	5
1	+	+	-	-	+
2	+	+	+	+	
3	+	+	+	-	+
4	+	+	+	+	-
5	-	+	-	-	+

Fig. 8.7 An accounting and evaluation framework for an urban energy metabolism that combines energy analysis with ecological network analysis

Chongqing) to serve as examples. We developed a measurement model to support the analysis of these cities and then the country, China. The primary data used in our emergy calculations were obtained from the Beijing Statistical Office, the China Energy Statistical Yearbook, the Editorial Committee for the China Environment Yearbook, and the literatures published at home and abroad. Similar data should be available for most cities in other countries, though with different resolution and possibly different categories of flow. Three emergy baselines have been used in emergy studies (Lu and Campbell, 2009): the first one is 9.44×10^{24} seJ/a (Odum, 1996), the second one is 9.26×10^{24} seJ/a (Campbell, 2000) through using the method of Odum (2000), and third one is 15.83×10^{24} seJ/a based on a re-evaluation and subsequent recalculation of energy contributions conducted in the year 2000 (Odum, 2000). Now, The global empower base 15.83×10^{24} seJ/a was used as the basis for calculating the emergy per unit for all kinds of emergy flows (Zhang et al., 2010b).

8.2.1 Metabolic Characteristics

8.2.1.1 Metabolic Scale

From 1990 to 2004, Beijing's U (total emergy use) increased continuously (Fig. 8.8). In 2004, U was 6.51×10^{23} seJ (solar emjoules), which equals 3.1 times the 1990 value. During the same period, flows of renewable resources emergy (R) fluctuated but showed an overall decrease (primarily due to large fluctuations in rainfall during this period). In contrast, flows of non-renewable resources emergy (N), transfers-in and imported emergy (IMP), and transfers-out and exported emergy (EXP) increased continuously. In 2004, the renewable emergy, non-renewable emergy, and transfers-in and imported emergy represented 0.2%, 34.3%, and 65.5% of U, respectively. These results mean that the metabolic throughput in Beijing's urban system increased continuously during this period. This is due, in part, to steady and rapid increases in IMP and EXP. Although IMP remained higher than EXP, this trend may not continue; the IMP value in 2004 had increased to just over three times the 1990 value, whereas the increase in EXP during this period has been nearly 13 times. The operation of Beijing's urban system still clearly depends on non-renewable resources, especially those imported from outside the system.

8.2.1.2 Metabolic Efficiency

Results from existing studies of emergy intensity were converted to 15.83×10^{24} seJ/a baseline through multiplying by the appropriate factors (1.68, 15.83/9.44) for the 9.44×10^{24} seJ/a baseline or 1.71 (15.83/9.26) for the 9.26×10^{24} seJ/a baseline) (Table 8.2). From 1994 to 2004, Beijing's emergy self-support ratio (ESR), which represents the city's ability to support its emergy demand, declined steadily (to 3.4% below its 1990 level; Fig. 8.9). In 2004, Beijing's ESR reached 34.4%, which was

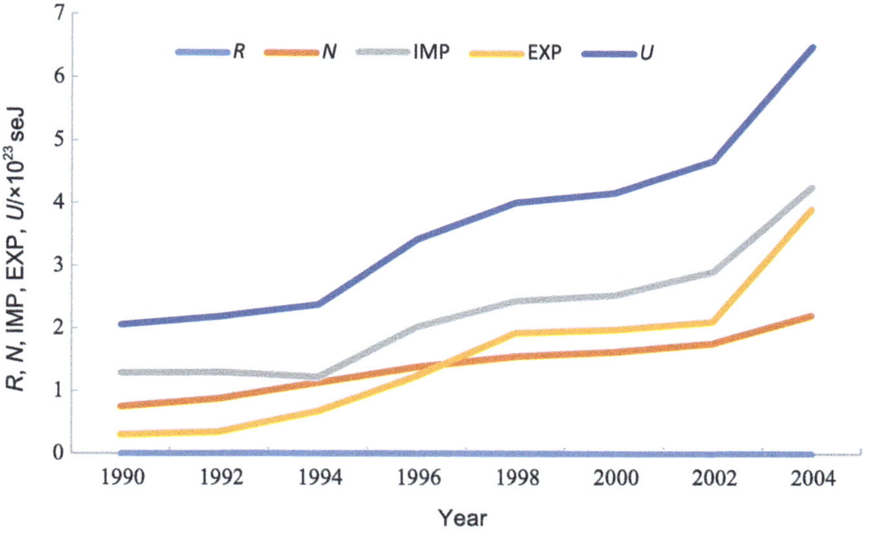

Fig. 8.8 Values of key emergy flows for Beijing's urban ecosystem Components: *R*-renewable resources emergy; *N*-non-renewable resources emergy; IMP-transfers-in and imported emergy; EXP-transfers-out and exported emergy; *U*-total emergy use

lower than Guangzhou (67.0%; Sui and Lan, 2006) and Shanghai (61.0%; Sui and Lan, 2006) in 2002, as well as Ningbo in 2000 (68.1%; Li and Zhang, 2003). Only Hong Kong in 1988 (2.0%; Lan and Odum, 1994) and Macao in 2004 (1.6%; Lei et al., 2008) had a lower ESR value (Table 8.2). Beijing's ESR in 2004 was far lower than the average for China as a whole (81.0% in 2004; Jiang et al., 2008). Compared with other cities, Beijing's high metabolic reliance on external resources implies significantly limited self-sufficiency in the urban system. Beijing has to face the problem of insufficient internal urban resources, and its economic growth is almost completely dependent on external support.

From 1990 to 2004, Beijing's environmental load ratio (ELR, which represents the metabolic load on the environment) experienced significant growth (to 4.6 times its 1990 value; Fig. 8.9). In 2004, Beijing's ELR was 531.23 (Table 8.2), which was remarkably higher than Guangzhou (4.74; Sui and Lan, 2006), Shanghai (17.34; Sui and Lan, 2006), and Ningbo (17.14; Li and Zhang, 2003). Only Hong Kong (2587.60; Lan and Odum, 1994) and Macao (743.00; Lei et al., 2008) had higher ELR values. In 2004, China's general ELR was 9.29, compared to the world average of 1.15 (Jiang et al., 2008). Many authors suggest that when ELR is greater than 10, the environmental load is unsustainably high (Ulgiati and Brown, 1998; Ulgiati et al., 1994). The burden on the local natural environment has remained high. It continues to increase due to the massive amount of external energy input into the city's metabolic system and due to the excessive use of the city's environmental resources without any other factors (such as increasing the use of renewable resources) to moderate it.

Table 8.2 Comparison of emergy evaluation indexes between Beijing (present study), for several major Chinese cities [a], and for China as a whole

	ESR (dimensionless)	ELR (dimensionless)	ED/($\times 10^{13}$ seJ/m^2)		EDR/($\times 10^{13}$ seJ/US$)		EPC/ ($\times 10^{16}$ seJ/person)		Emergy baseline/ ($\times 10^{24}$ seJ/a)
			Reference value	Correction	Reference value	Correction	Reference value	Correction	
Beijing (2004)	34.45	531.23	3.88	3.88	1.22	1.22	4.36	4.36	15.83
Guangzhou (2002)	67.00	4.74	0.19	0.32	0.38	0.65	1.97	3.37	9.26
Shanghai (2002)	61.00	17.34	0.40	0.68	0.46	0.79	2.32	3.97	9.26
Ningbo (2000)	68.12	17.14	0.31	0.52	0.21	0.35	0.54	0.91	9.44
Hong Kong (1988)	2.00	2587.60	5.20	8.89	0.10	0.17	0.98	1.68	9.26
Macao (2004)	1.57	743.00	80.50	137.62	0.24	0.41	4.90	8.38	9.26
China (2004)	81.00	9.29	0.21	0.21	1.21	1.21	1.53	1.53	15.83

[a], *Sources* Guangzhou (Sui and Lan, 2006), Shanghai (Sui and Lan, 2006), Ningbo (Li and Zhang, 2003), Hong Kong (Lan and Odum, 1994), Macao (Lei et al., 2008), and China as a whole (Jiang et al., 2008). Variables: ESR-emergy self-sufficiency ratio; ELR-environmental load ratio; ED-emergy density; EDR-emergy-dollar ratio; EPC-emergy per capita. Boldfaced numbers represent the corrected values based on the emergy baseline

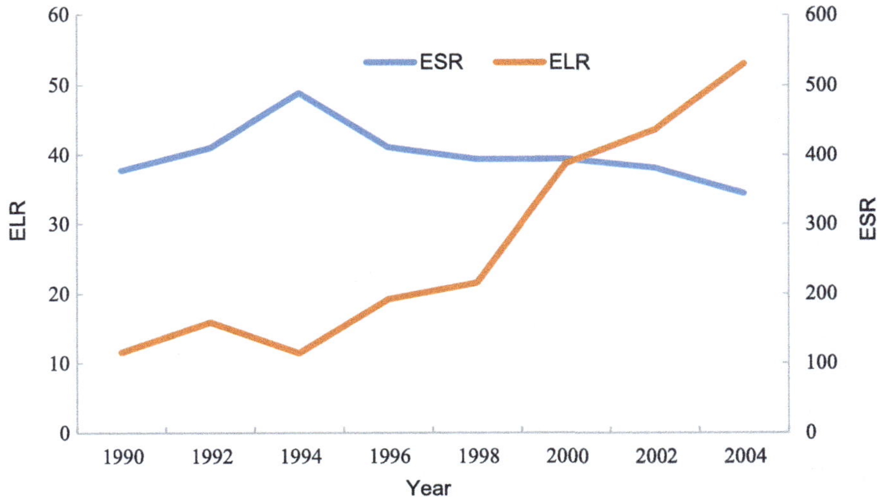

Fig. 8.9 Changes in Beijing's metabolic efficiency from 1990 to 2004 Variables: ESR-energy self-sufficiency ratio; ELR-environmental load ratio

If the ELR remains high for a long time, the city's ecosystem may be irreversibly destroyed.

8.2.1.3 Metabolic Intensity

Beijing'S emergy density (ED), which represents the emergy consumption per unit area, increased continuously throughout the study period, at a compound average annual rate of 8.5%, from 1.23×10^{13} seJ/m^2 in 1990 to 3.88×10^{13} seJ/m^2 in 2004, which represents 3.1 times the 1990 value (Fig. 8.10). Based on ED, Beijing created a higher metabolic pressure on its environment than China as a whole in 2004 (0.21×10^{13} seJ/m^2; Jiang et al., 2008). In 2004, Beijing's ED was also far higher than those of Guangzhou (0.32×10^{13} seJ/m^2; Sui and Lan, 2006), Shanghai (0.68×10^{13} seJ/m^2; Sui and Lan, 2006), and Ningbo (0.52×10^{13} seJ/m^2; Li and Zhang, 2003). Only Hong Kong (8.89×10^{13} seJ/m^2; Lan and Odum, 1994) and Macao (137.62×10^{13} seJ/m^2; Lei et al., 2008) had higher ED values. These results mean that Beijing's urban development is imposing high pressure on the city's ecological environment.

During the 15-year study period, Beijing's emergy per capita (EPC) increased from 1.91×10^{16} seJ/person to 4.36×10^{16} seJ/person, which represents 2.3 times the 1990 value (Table 8.2). Beijing's EPC in 2004 was lower than that of Macao (8.38×10^{16} seJ/person; Lei et al., 2008), but higher than that of Guangzhou (3.37×10^{16} seJ/person; Sui and Lan, 2006), Shanghai (3.97×10^{16} seJ/person; Sui and Lan, 2006), Ningbo (0.91×10^{16} seJ/person; Li and Zhang, 2003), and Hong Kong (1.68×10^{16} seJ/person; Lan and Odum, 1994), and also higher than the average for China as a whole (1.53×10^{16} seJ/person; Jiang et al., 2008). These results mean that

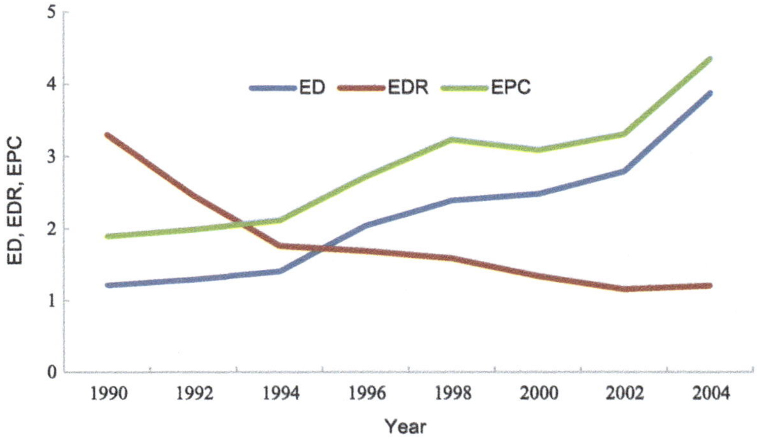

Fig. 8.10 Changes in Beijing's metabolic intensity from 1990 to 2005 Variables: ED-emergy density; EDR-emergy-dollar ratio; EPC-emergy per capita

the emergy consumption by Beijing's residents is relatively high, at approximately 2.8 times the national average level, and awareness of environmental protection by reducing consumption should be improved.

Beijing's emergy-dollar ratio (EDR), which represents the emergy consumed per unit of economic activity, decreased from 3.31×10^{13} seJ/US$ in 1990 to 1.22×10^{13} seJ/US$ in 2004 (Table 8.2), which represents 36.8% of the 1990 value. EDR thus decreased at a compound average annual rate of 7.4% throughout the study period. Beijing's EDR in 2004 was higher than those of the other five cities and close to the national level. Beijing's high EDR means that more of the economy involves direct environmental resources (internal and external) that are not paid for (or paid for at a lower rate than in the other cities) and that the rate of economic growth was lower. Generally speaking, Beijing's metabolic intensity is relatively high, which means that a high environmental cost has accompanied the city's rapid socioeconomic development.

8.2.2 Metabolic Paths and Relationships

8.2.2.1 Metabolic Paths

Network structures can be represented by directed graphs ("digraphs") and adjacency matrices (Zhang et al., 2009a). In digraphs, each node is a point that represents a component of the system, and directional lines between nodes represent a transaction path (i.e., an energy or matter flow in a specific direction) or a relationship. A pathway represents the sequence of nodes and lines over which energy or a substance flows (in transactions) or in which relationships emerge. The metabolic length (k) represents the number of path lines from an initial to a terminal node (Borrett and Patten, 2003). A direct pathway between two nodes has only one line, thus $k = 1$; an indirect pathway that passes through one or more intermediate nodes has $k > 1$ lines. Cycles are an important type of pathway in network structures. A cycle is a pathway that starts and ends at the same node, usually after passing through one or more intermediate nodes (Borrett and Patten, 2003). The number of lines (L) in a metabolic pathway reflects the number of pathways from component i to component j for a given metabolic length(k).

We can establish a simple flow structure for an urban ecological network. In the adjacency matrix $\boldsymbol{A} = [a_{ij}]_{5\times5}$, where 5×5 represents the dimensions of the matrix for a matrix with five nodes, we can construct the adjacency matrices $\boldsymbol{A}_1$, $\boldsymbol{A}_2$, $\boldsymbol{A}_3$, and $\boldsymbol{A}_4$ for $k = 1, 2, 3$, and 4, respectively, and we can define the total adjacency matrices $\boldsymbol{A}_{1\text{–}2}$, $\boldsymbol{A}_{1\text{–}3}$, and $\boldsymbol{A}_{1\text{–}4}$ (Fig. 8.11). Here, $\boldsymbol{A}_{1\text{–}2}$ is the sum of pathway numbers of metabolic length = 1 and 2, $\boldsymbol{A}_{1\text{–}3}$ is the sum of pathway numbers of metabolic length = 1, 2, and 3, and $\boldsymbol{A}_{1\text{–}4}$ is the sum of pathway numbers of metabolic length = 1, 2, 3, and 4. Among the elements in these matrices, a_{ii} represents the number of cyclic metabolic pathways (i.e., pathways that begin and end in node i). Urban metabolic pathway analysis focuses on identifying which metabolic pathways will show a significant increase with increasing metabolic length and the changes in the reachability of paths (i.e., the ability of flows to reach a given node) and in the numerical distribution of metabolic pathways with different lengths.

In theory, the materials that flow through an urban metabolic system can be infinitely cycled, but in practice, constraints imposed by the current level of science and technology and the potentially high economic costs of this cycling mean that some losses are inevitable and that the metabolic path length should not be too long. Thus, in practice, this length is usually limited to 3 or 4. I have, therefore, analyzed situations in which the metabolic path length was 1, 2, 3, or 4. The results show that with increasing metabolic length, the number of metabolic pathways grew exponentially, with $C = 0.2005e^{1.3326\,k}$ (Fig. 8.12), and the reachability of the metabolic pathways grew linearly, with $\ln L = 1.3326\text{ k} + 1.6117$.

This analysis shows that when $k = 1$, the degree of cycling is low, and the metabolic reachability (0.76) is relatively poor. When k increases to 2, 3, and 4, the degree of cycling among the External Environment, the Agriculture sector, and the Industry sector increases continuously; exchanges between the Industry and Agriculture sectors and between the Industry sector and the External Environment become more frequent. The capacity of the Internal Environment to absorb flows from the Agriculture, Industry, and Household Consumption sectors increased considerably, as did the drain on the Internal Environment by these sectors. The capacity of the Household Consumption sector to contribute flows to the Internal Environment, the External Environment, and the Agriculture and Industry sectors increased significantly, and the overall metabolic reachability of the network increased continuously.

L=19 , C=0.76

A_1	1	2	3	4	5
1	0	1	1	1	1
2	1	0	1	1	1
3	1	1	0	1	1
4	1	1	1	0	1
5	0	1	1	1	0

L=72 , C=2.88

A_2	1	2	3	4	5
1	3	3	3	3	3
2	2	4	3	3	3
3	2	3	4	3	3
4	2	3	3	4	3
5	3	2	2	2	3

L=273 , C=10.92

A_3	1	2	3	4	5
1	9	12	12	12	12
2	10	11	12	12	12
3	10	12	11	12	12
4	10	12	12	11	12
5	6	10	10	10	9

L=1035 , C=41.40

A_4	1	2	3	4	5
1	36	45	45	45	45
2	35	46	45	45	45
3	35	45	46	45	45
4	35	45	45	46	45
5	30	35	35	35	36

A_{1-2}	1	2	3	4	5
1	3	4	4	4	4
2	3	4	4	4	4
3	3	4	4	4	4
4	3	4	4	4	4
5	0	1	1	1	0

A_{1-3}	1	2	3	4	5
1	12	16	16	16	16
2	13	15	16	16	16
3	13	16	15	16	16
4	13	16	16	15	16
5	3	2	2	2	3

A_{1-4}	1	2	3	4	5
1	48	61	61	61	61
2	48	61	61	61	61
3	48	61	61	61	61
4	48	61	61	61	61
5	39	48	48	48	48

Fig. 8.11 Relationships between the length of paths (L) in an urban emergy metabolism, the number of paths (values in the matrix), and the reachability of paths (C) (*Note* A_1 to A_4 represent the numbers of paths of length 1 to 4 (respectively) between nodes; $A_{1\text{-}2}$, $A_{1\text{-}3}$, and $A_{1\text{-}4}$ represent the total number of paths of length 1 to 2, 1 to 3, and 1 to 4, respectively. Components: 1-Internal Environment, 2-External Environment, 3-Agriculture, 4-Industry, 5-Household Consumption)

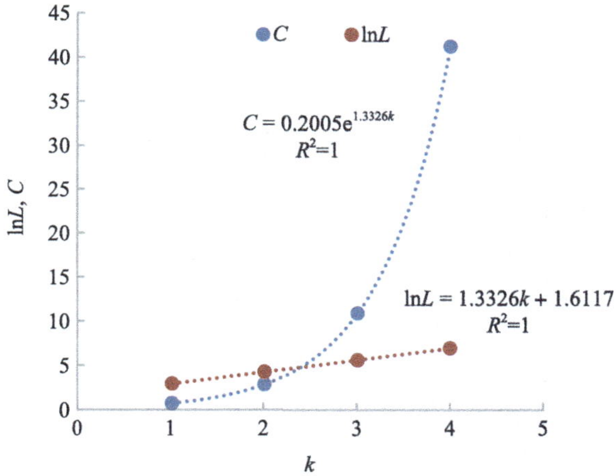

Fig. 8.12 Relationships between the path length (k) and the logarithm of the total number of paths (L) and the reachability of the paths (C)

8.2.2.2 Metabolic Relationships

In the research that was the basis for this section (Zhang et al., 2009a), my research group used a five-component model such as the one described earlier in this chapter. We used the mutual relationships between components to compute the direct utility intensity matrix ($\boldsymbol{D}$) and the integral utility intensity matrix ($\boldsymbol{U}$) for each of the four cities I described earlier in this chapter (i.e., Beijing, Tianjin, Shanghai, and Chongqing). Based on the signs of the elements in matrices $\boldsymbol{D}$ and $\boldsymbol{U}$ for the four cities, it is possible to calculate the sign matrices sgn($\boldsymbol{D}$) and sgn($\boldsymbol{U}$) for each city, which define the relationships between the pairs of sectors in each city (Fig. 8.13).

In the sgn($\boldsymbol{D}$) and sgn($\boldsymbol{U}$) matrices, there are ten pairs of ecological relationships. Based on $\boldsymbol{D}$, the mutualism index of all four cities (M, which equals the ratio of the numbers of positive signs to negative signs) equals 1.0. However, based on the integral utility intensity matrix ($\boldsymbol{U}$), which also accounts for indirect flows, Beijing's mutualism index (2.12) was higher than that of Shanghai (1.50) and much higher than those of Tianjin and Chongqing (1.27). In total, the urban metabolic processes of the four cities reveal an overall degree of mutualism, with mutualism indexes all greater than 1. Still, due to differences in the ecological relationships among the components, these indexes differed among the cities.

The cities had both similar and different inter-component ecological relationships. The relationships between the Agriculture sector and the External Environment and between the Industry sector and the External Environment were similar for the four cities. In contrast, the relationships between the Internal Environment and External Environment and between the Agriculture sector and the Internal Environment differed among the cities; Beijing and Shanghai showed one pattern, whereas Tianjin and Chongqing showed another pattern. For Beijing, the relationships between the Agriculture and Industry sectors, between the Household Consumption sector and the Internal Environment, and between the Household Consumption and Industry

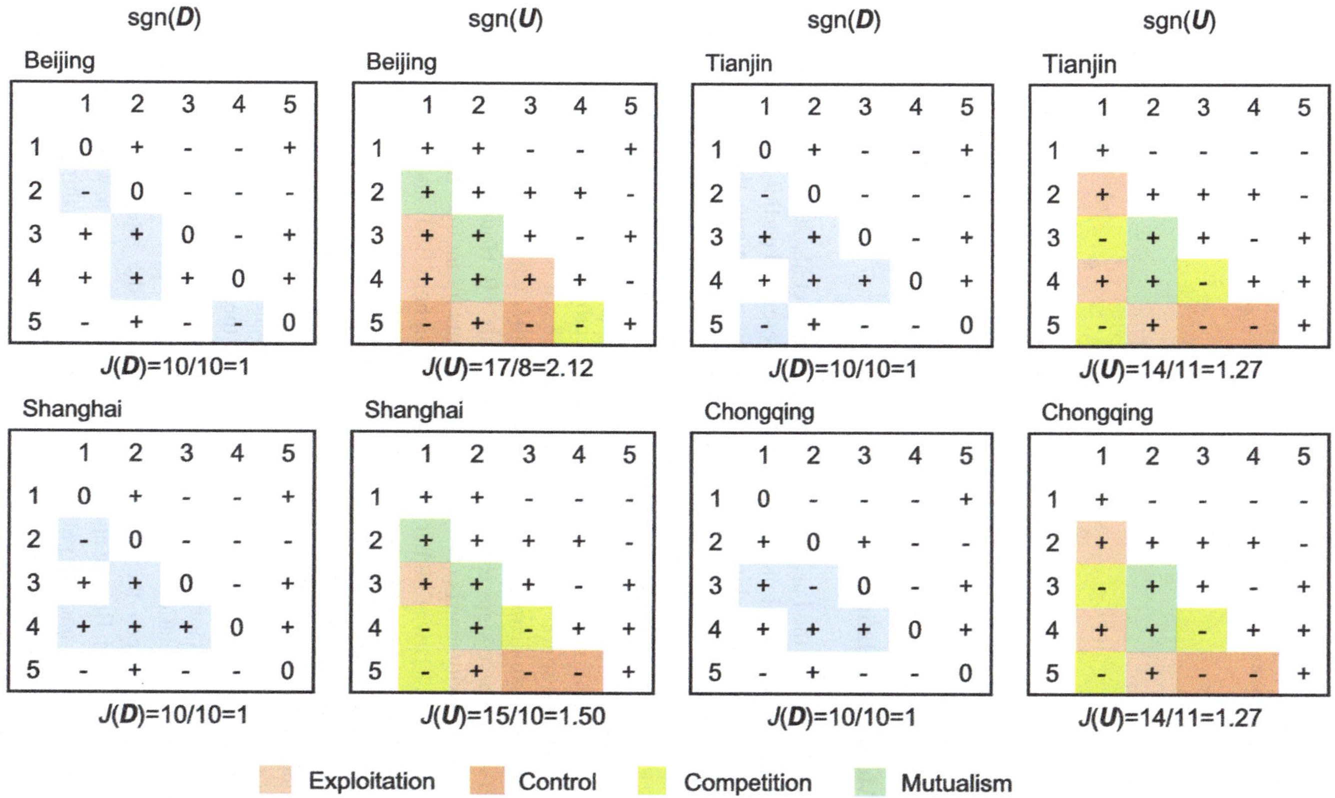

Fig. 8.13 The direct utility sign matrices, sgn(***D***), and the integral (direct plus indirect) utility sign matrices, sgn(***U***), for the four cities (*Note* The ***U*** matrix includes indirect flows that are not present in the ***D*** matrix. Utilities are for the flows from the sector in the top row of the table to the sector in the first column of the table. Components: 1-Internal Environment, 2-External Environment, 3-Agriculture, 4-Industry, 5-Household Consumption. $M(\boldsymbol{D})$ and $M(\boldsymbol{U})$ represent the ratios of positive signs to negative signs in the two matrices)

sectors were clearly different from those of the other cities. For Shanghai, the relationship between the Industry sector and the Internal Environment was different from that in the other cities.

8.2.3 Management Suggestions Based on Beijing's Emergy Accounting Evaluation

Integrating emergy analysis with material metabolism analysis may offer an important contribution to understanding the state of urban metabolic systems. This method allows for a comprehensive assessment of the state of urban development and is important for determining the causes of the deterioration of urban systems and potential ways to protect that ecosystem.

From 1990 to 2004, the metabolic scale, efficiency, and intensity in Beijing remarkably grew. The metabolic processes of the city were overly reliant on non-renewable resources, but there was a continuous decline in resource pressure from outside the city. Beijing had higher metabolic fluxes and a higher metabolic density than the other three cities; it had a lower metabolic efficiency and a higher metabolic intensity. The assessment of these metabolic indicators unveiled the vulnerabilities of the urban metabolic system, thus assisting planners in identifying measures to maintain these urban metabolic processes.

Based on the emergy synthesis described earlier in this section, Beijing's socioeconomic growth relies heavily on outside environmental resources, and the city cannot be self-sufficient. Consequently, its GDP is supported by the emergy bought from the outside. During the solid economic expansion period from 1990 to 2004, Beijing's U, ELR, ED, and EPC continued to rise, while ESR and EDR continued to fall. In comparison with the other five cities (Table 8.2), Beijing's ELR, ED, and EPC were higher than Guangzhou, Shanghai, and Ningbo but lower than Hong Kong and Macao; the results of ESR were opposite, and Beijing had a higher EDR than all other cities. As a result, it is evident that Beijing's urban system is still resource-consuming, and its economic development is increasingly reliant on external resources. Beijing has also exhibited growing pressure on the ecological environment as well as rising consumption of energy, which is closely related to human well-being in complicated ways. In the future, it will be more and more essential to improve the efficiency of Beijing's metabolism.

Beijing needs to bring in creative methods and technologies to increase the capacity of sustainable development based on the current level of development and present study results. For instance, Beijing's ecosystem can achieve sustainable development by taking measures to alleviate the environmental pressures which accompany the city's rapid economic growth, for example, enhancing the conservation and development of renewable resources, reducing the city's reliance on non-renewable resources, promoting the efficient use of non-renewable and imported resources, and improving the reuse and recycling of waste.

8.2.4 Suggestions for Improving the Urban Energy Metabolic Network

By basing this research on ecological network analysis, it was possible to construct a simple network model for the urban metabolism and to use that model to compare the urban metabolic systems of four Chinese cities to demonstrate the approach. The results provided important insights into the relationships between the structure and function of the urban metabolic system and revealed exploitation, control, competition, and mutualism relationships among each city's metabolic components. By combining ecological network analysis with emergy accounting, it was possible to integrate flows with different magnitudes and units of measurement and to analyze the mechanisms inherent in the urban metabolic system from the perspectives of its nutrient structure (i.e., the flows of energy, materials, and money that sustain the system) and the ecological relationships among its components. The result is a significant improvement on the black-box model of urban metabolic research. This method can therefore be adapted for application in future studies of urban metabolic systems.

Although most urban metabolic systems are more complex than the models I have reviewed in this chapter, the simple ecological network models nonetheless provide a useful means of analyzing the main components of an urban system and their functional relationships. Through the ecological network model and pathway analysis, we can discern the types of ecological relationships between the components of the system and can use the results to interpret the problems facing a given urban system. Future ecological network models of urban metabolic systems should be further optimized by subdividing the components and the flows among them to provide a more precise simulation of the complexity of the real urban system.

8.3 Analysis of the Embodied Energy Metabolism Network of the Beijing-Tianjin-Hebei Region

In 2014, integrated development of the Beijing-Tianjin-Hebei region became part of China's national development strategy, and this proved to be a historic opportunity to improve the development of this urban agglomeration to achieve synergies. However, as the industrial structures of Beijing, Tianjin, and Hebei were similar, the potential for synergies was low, and the region's ecological capacity was limited, leading to high ecological and environmental costs and excessively high consumption of energy. The main reason was that the region is a gathering place for many energy-intensive industries, such as cement and steel production, oil refining, and petrochemical manufacturing. The energy consumption of this region accounts for more than 10% of China's total, despite covering only 2.3% of China's total area, and its coal consumption per unit area is 30 times the global average level. Analysis of the energy flow processes in the region will provide insights that support planning to adjust the region's industrial structure and support the integrated development of the region. Furthermore, the capital transferred among the two large cities (Beijing and Tianjin) and 11 key cities in Hebei Province through trade increased from 2002 to 2010, at an average growth rate of 15% each year. However, the capital output from Beijing to Tianjin and Hebei decreased during this period, by 4% and 6% per year, respectively. These trends will cause the quantity of energy consumed in trade to change. Thus, it's important to understand how trade within the Beijing-Tianjin-Hebei region will influence the corresponding energy flows.

In the research on which this section was based (Hao et al., 2018; Zhang et al., 2015, 2016), my research group chose the Beijing-Tianjin-Hebei agglomeration as a case study, and from an urban metabolism perspective, defined the key natural and socioeconomic sectors in each city as nodes in the network so we could analyze the flows among them and build a network model. First, based on multi-regional monetary input–output tables for China, we divided the study region into three overall regions (Beijing and Tianjin cities and Hebei Province), then studied the natural and socioeconomic sectors in the three regions and in the 13 cities in these regions (Beijing, Tianjin, and 11 cities in Hebei Province) to establish an energy flow network. Next, we used ecological network analysis to analyze the energy consumption embodied in the flows among regions and sectors, to analyze the flows among the components of the system, and to define their functional and ecological relationships (Zhang et al., 2016). The results provide insights at different network scales, ranging from 3 to 13 nodes, to identify the roles (i.e., producer or consumer) of the 3 regions, the 13 cities, and five sectors in each region from the perspective of energy exchanges. We obtained different perspectives on the flows by examining the system across multiple scales, thereby providing more comprehensive guidance for plans to reduce energy consumption and waste emission throughout the region (Fig. 8.14).

The energy consumption data for 2002, 2007, and 2010 were obtained from the China Energy Statistics Yearbook and from the China Emission Accounts & Datasets, and the capital data were obtained from multi-regional input–output tables for 2002, 2007, and 2010. Most countries and regions can provide similar data to support this kind of analysis.

8.3.1 Analysis of the Embodied Energy Metabolism of the Nodes

Figure 8.15 compares the embodied (integral) energy consumption of Beijing, Tianjin, and Hebei, which equals the direct plus indirect consumption. For each of these parts of the overall study region, the embodied energy consumption first increased and then decreased. The magnitude and rate of the change were greater for Hebei than for Beijing and Tianjin. In 2007, Hebei's embodied consumption was 5.20 times that in 2002, versus increases of only 1.60 times for Beijing and 2.18 times for Tianjin. From 2007 to 2010, the embodied consumption decreased to

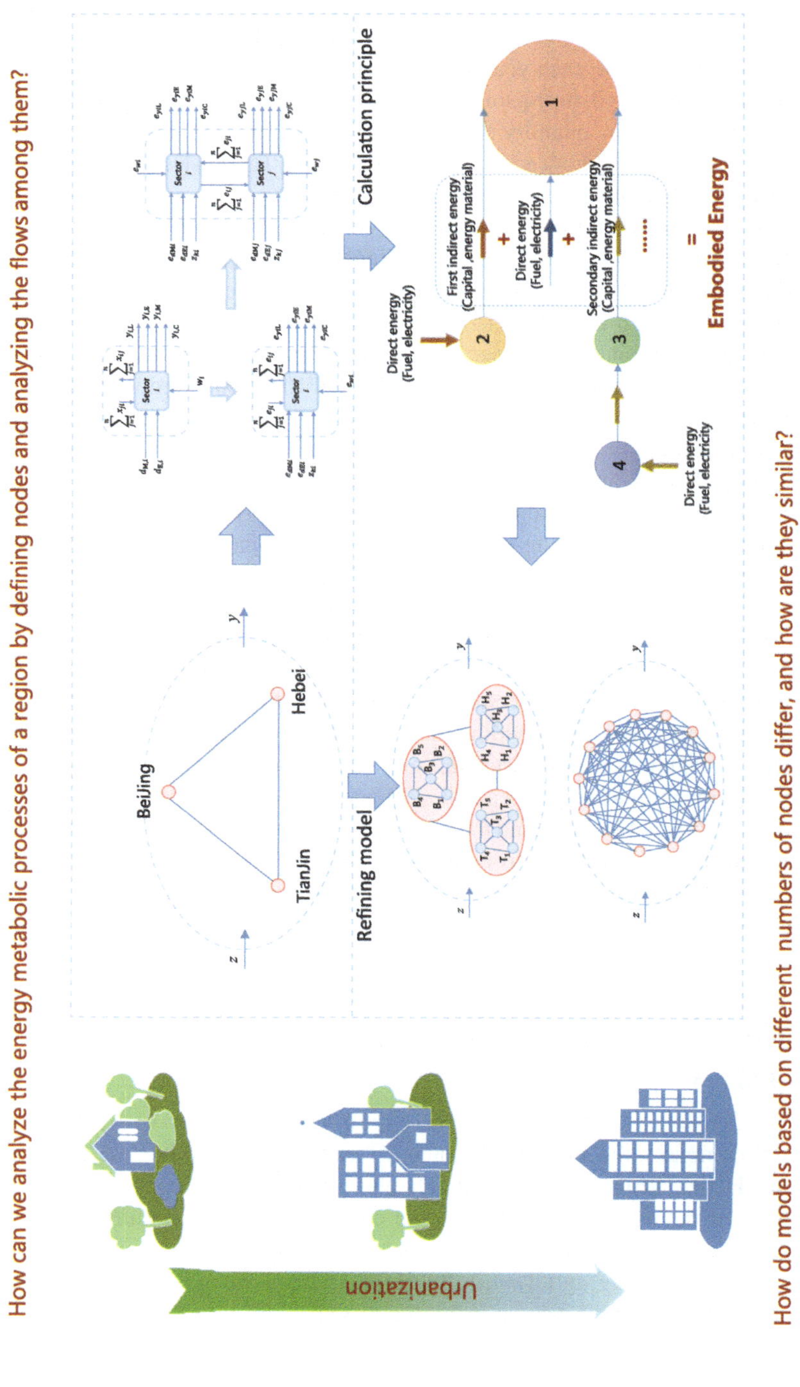

Fig. 8.14 Framework for analysis of the embodied energy metabolism in the Beijing-Tianjin-Hebei agglomeration

0.97 times the 2007 value for Beijing, versus 0.77 for Tianjin and 0.36 for Hebei. Hebei's embodied consumption remained much higher than those of the other cities throughout the study period.

Figure 8.15 also shows the embodied energy consumption of the 13 cities from 2002 to 2010. Beijing's embodied consumption first decreased and then increased, that of Tianjin increased throughout the study period, and the other cities all showed an initial increase, then a decrease. The difference in the trends for Beijing and Tianjin between the 3-node and 13-node networks is because the embodied energy consumption was calculated based on a monetary input–output matrix with 13 nodes derived from a matrix with 3 nodes. By accounting for flows between Beijing and Tianjin and the other 11 cities that were invisible in the analysis based on three nodes, important differences in the metabolic flows were revealed. Division of the data before performing ecological network analysis to account for the embodied energy flows resulted in different trends at these two scales. In 2002, embodied consumption was highest for Beijing, at 2.47 times that of Tianjin, which ranked second. In 2007, Tangshan's embodied consumption ranked first; Zhangjiakou, which consumed the least energy, had embodied consumption equal to only 17.4% of that of Tangshan. In 2010, Tianjin's embodied consumption was the highest, at 6.11 times that of Hengshui, which ranked lowest.

Figure 8.16 shows that direct energy consumption was less than indirect consumption for all sectors in the 3-node model. The indirect energy consumption by Hebei's Industry sector (HB2) was highest in both years. In contrast, indirect consumption by Tianjin's Agriculture sector (TJ1) was lowest. The proportions of indirect consumption by Beijing's and Tianjin's Construction sectors (BJ3 and TJ3) and by Hebei's Agriculture (HB1), Construction (HB3), and Other Services (HB5) sectors were all greater than 90% of the total in both years.

Concerning direct energy consumption, the Industry sector in Hebei Industry had the largest consumption in both years (6.5×10^7 tce and 14.1×10^7 tce in 2002 and 2007, respectively), while the Agriculture sector (TJ1) in Tianjin had the smallest (0.06×10^7 tce). For Beijing, the transportation sector accounted for the largest share of direct consumption in 2007 (42% of the total). Tianjin's Agriculture, Transportation, and Other Services sectors, Beijing's Agriculture sector in 2007, and Tianjin's Transportation sector in 2002 also made up large proportions of the total (28% to 38%).

From 2002 to 2007, direct energy consumption increased in all sectors, while indirect energy consumption declined. Three Agriculture sectors, for example, have experienced the fastest decline in indirect consumption, especially in Beijing; in 2007, Beijing's consumption reached only 20% of the level in 2002. The Agriculture sector in Hebei experienced the slowest decline; consumption in 2007 reached 75% of the 2002 level. The fastest-growing sector of direct consumption was Transportation in Hebei; in 2007, its consumption reached 3.47 times that in 2002. Consumption by Beijing's Transportation sector, Tianjin's Construction sector, and Hebei's Industry sector also increased rapidly, with 2007 values 2 to 3 times those in 2002 (Zhang et al., 2016). These sectors also use a substantial volume of energy generated internally,

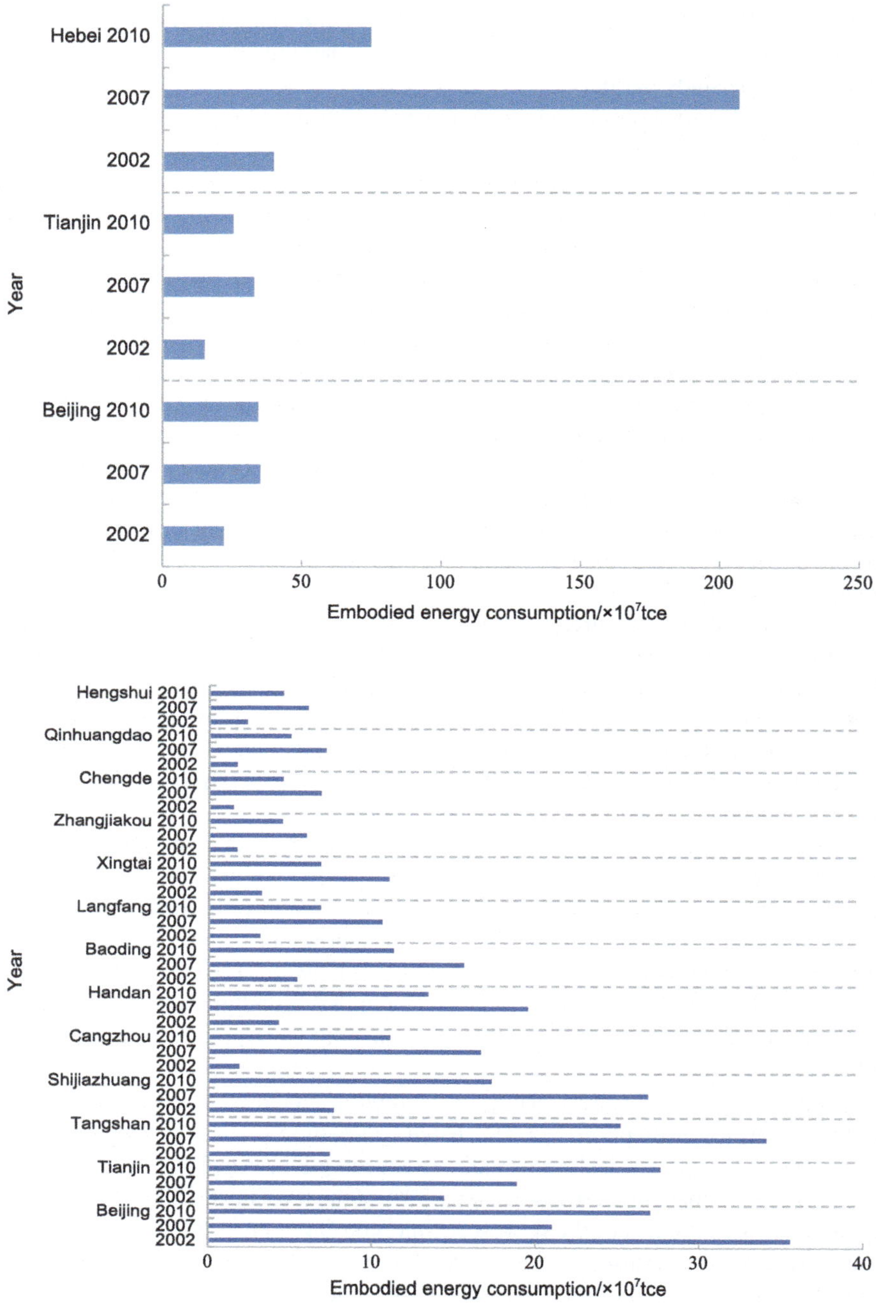

Fig. 8.15 Embodied metabolic flows (tce, tonnes of coal equivalent) for (top) Beijing, Tianjin, and Hebei and for (bottom) the 13 cities in the agglomeration

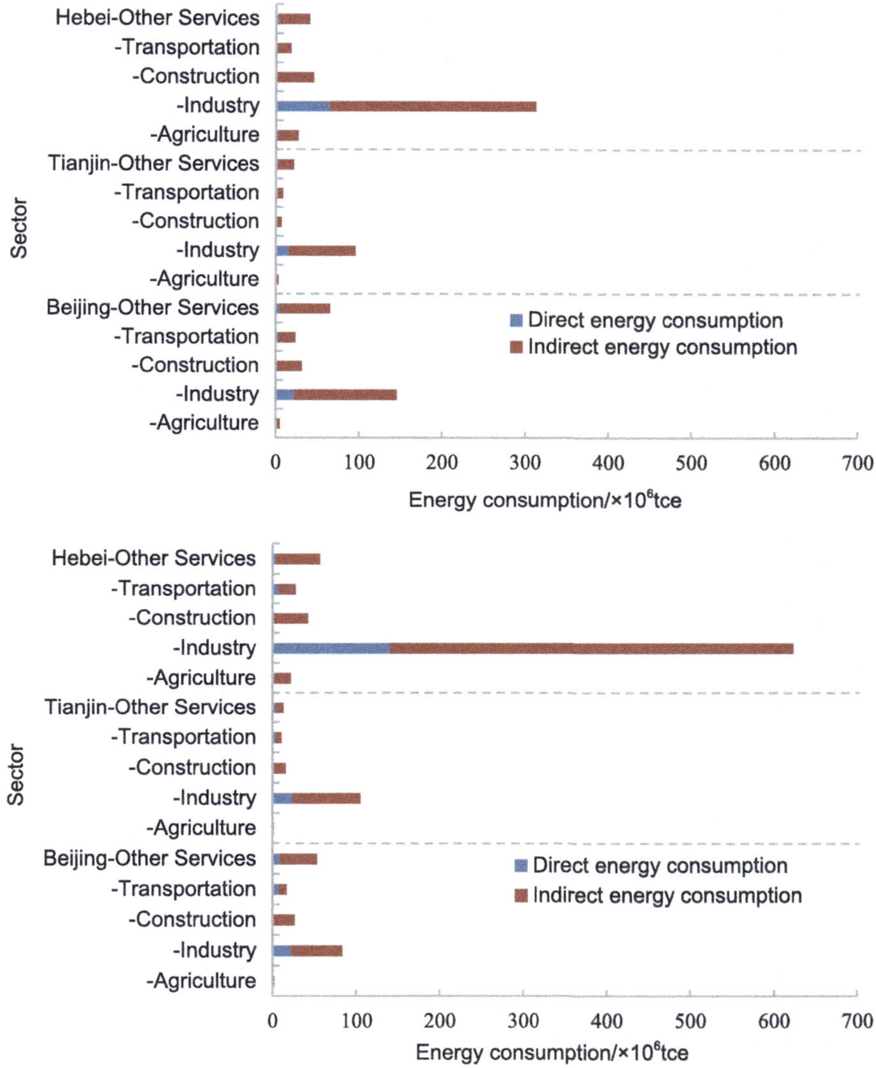

Fig. 8.16 Embodied (direct plus indirect) energy consumption by the sectors in the three regions (tce, tonnes coal equivalent) in 2002 (top) and 2007 (bottom)

ranging from roughly 26% to 84% of total consumption in 2002 and 29% to 93% in 2007.

On a sectoral level, the Industry sector consumed the most energy in the three cities. The reason for this is that the Industry sectors comprised many energy-intensive heavy-industry subsectors. Even though they are energy intensive, they are essential for the development of the region and the technological improvement of the city or province. Energy consumption in the Other Services sector (except for

Hebei Province in 2002 and Tianjin in 2007) and the Construction sector followed closely behind. There has been an increase in the share of the tertiary sector (i.e., the Other Services sector) in the economies of these three regions. In 2007, Beijing's consumption in two sectors (Agriculture and Industry) accounted for less than half of its 2002 level. The reason for this may be that the government relocated many industries, particularly the most polluting ones, to Tianjin and Hebei Province in order to reduce Beijing's air pollution in preparation for the 2008 Olympic Games. For instance, some well-established heavy industries like iron and steel production were shifted from Beijing to Tianjin or Hebei.

8.3.2 Analysis of the Embodied Energy Metabolism of Paths

Figure 8.17 illustrates the embodied energy flows in the 3-node network. The energy flow from Hebei to Beijing was the highest one, with the flows from Hebei to Tianjin following. In 2002, the flow from Hebei to Beijing was highest, reaching 0.91×10^8 tce, while the flows from Beijing to Hebei and from Hebei to Tianjin accounted for only 82.8% and 67.4% of the flow from Hebei to Beijing, respectively. In 2007, the flows from Hebei to Beijing and Tianjin remained the highest and also increased significantly compared to 2002; they were 1.46 and 1.99 times higher than the corresponding flows, respectively. This was followed by the flows from Tianjin to Hebei and from Beijing to Hebei, accounting for 32.5% and 24.1% (respectively) of the flows from Hebei to Beijing. The flows from Hebei to Beijing remained the highest in 2010, with a 33.7% increase over 2007. The second highest was the flow from Hebei to Tianjin, which declined between 2007 and 2010 (reaching only 90.5% of that in 2007), yet it remains higher than the flow from Tianjin to Beijing.

Figure 8.18 illustrates the embodied energy flows among the 13 cities in 2002, 2007, and 2010. The highest flows were from Shijiazhuang to Tangshan in all years, except for 2002, when the highest flow was from Beijing to Tianjin. The next-highest was the flow from Tangshan to Shijiazhuang. In 2002, the highest flow was from Beijing to Tianjin, achieving 0.24×10^8 tce, with the flow from Tangshan to Shijiazhuang following, which accounted for 88.5% of the flow from Beijing to Tianjin. In 2007, all cities experienced an increase in embodied energy flows, with the highest

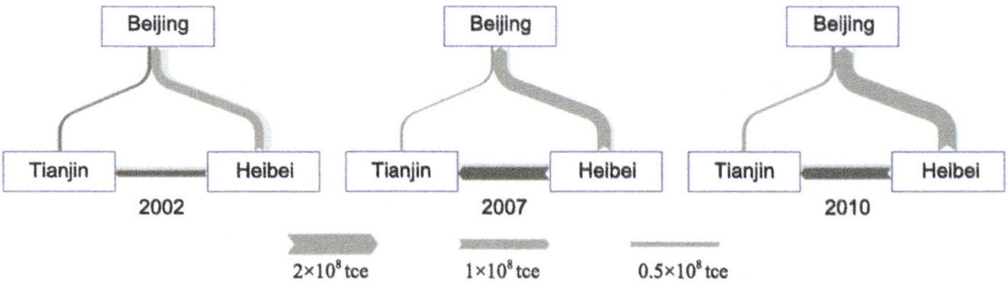

Fig. 8.17 Embodied energy metabolic flows of paths among the three regions

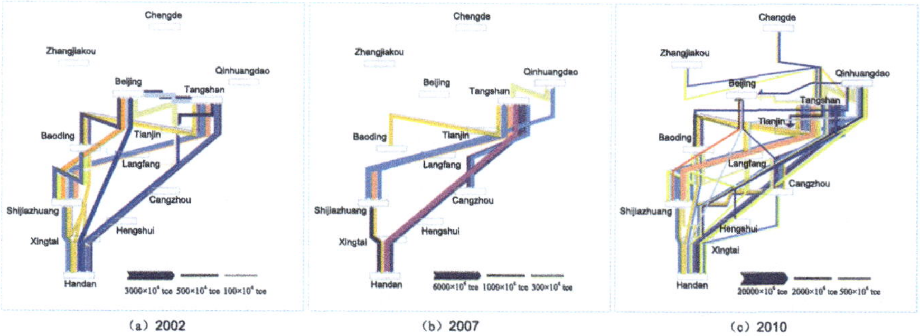

Fig. 8.18 Embodied energy flows in the inter-city paths among the 13 cities in the Beijing-Tianjin-Hebei agglomeration in 2002, 2007, and 2010 (*Note* All flows in 2002 were > 1.0×10^6 tce, versus > 3.0×10^8 tce in 2007 and > 5.0×10^8 tce in 2010)

flow from Shijiazhuang to Tangshan, which reached 52.63×10^8 tce. The second highest flow was from Tangshan to Shijiazhuang, accounting for 98.0% of the highest flow. All flows rose again by 2010, with the highest flow from Shijiazhuang to Tangshan, which reached 128.42×10^8 tce (2.44 times the flow in 2007).

Figure 8.19 shows the embodied energy flows among each region's five sectors in 2002 and 2007. In 2002, the majority of flows greater than 1.0×10^7 tce were located in Beijing and Hebei; among them, the largest flows were between the Industry sector of Beijing and Hebei. The situation changed in 2007, as flows were concentrated in Hebei's Industry sector; the flows between the Industry sectors of Heibei and Tianjin increased, while the flows between Beijing's Industry sector and other sectors reduced. This revealed that the core sector of Heibei's development was the Industry sector. It supplied and accepted massive amounts of energy flows from other sectors both within Hebei and between Hebei and Tianjin. Additionally, it provided more energy to other sectors in 2007 than it did in 2002. This made Beijing and Tianjin more reliant on Hebei. In 2007, more energy was exchanged between the Transportation and Other Services sectors and the Industry sector in Heibei than in 2002. This implied that energy flows in Hebei got larger. Flows greater than 1.0×10^6 tce changed considerably. In 2002, most of the energy flowed into the Industry sector in Beijing, supplying energy to other sectors in Beijing and the Industry sector in Hebei. In 2007, Hebei's flows exceeded that within Beijing and Tianjin, and the interactions between Hebei and other regions' sectors also increased. In comparison, there was a significant decrease in flows along the paths between the Industry sector and other sectors in Beijing.

In 2002, the industrial sectors were also the supplier of energy. The Industry sector in Beijing supplied energy to the downstream sector in Beijing as well as the Industry sectors and other downstream sectors in Tianjin and Hebei. For instance, the Industry sector in Beijing provided energy for the Other Services sector in Tianjin and the Transportation and Other Services sectors in Hebei. In 2007, the Industry sector in Beijing provided less energy than in 2002, and the major recipients were sectors within the region instead of sectors in other regions. The energy flow provided

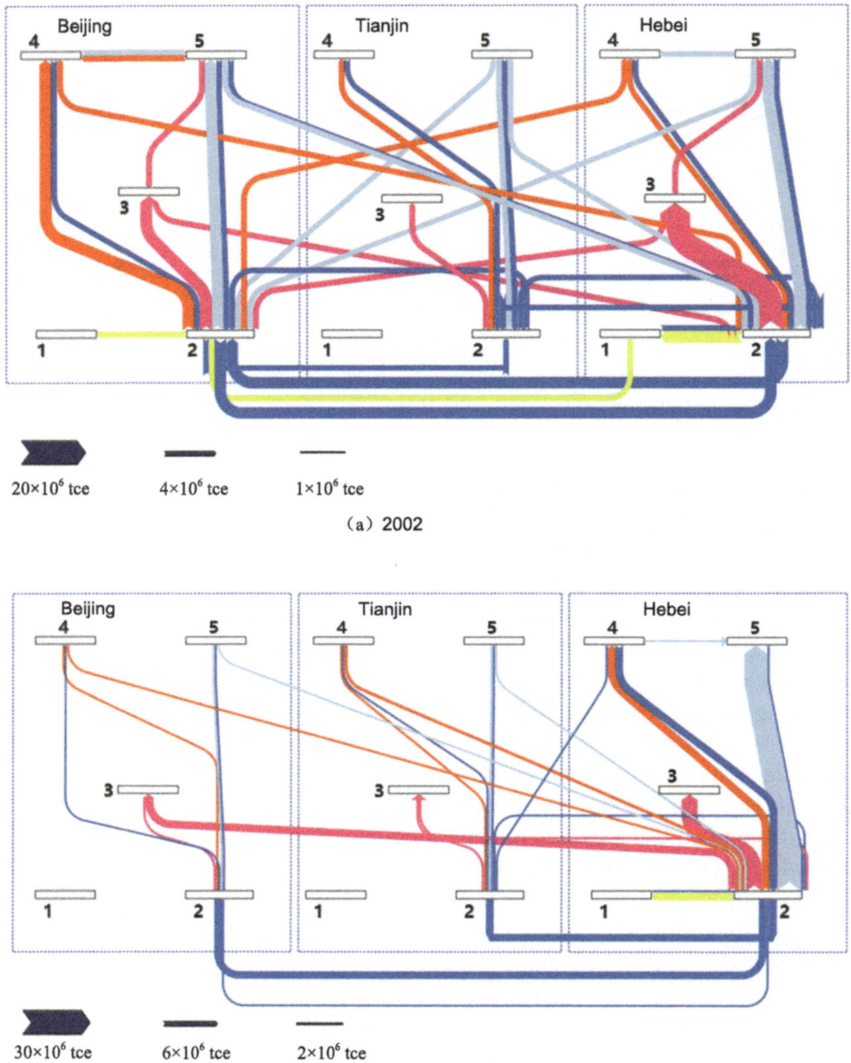

Fig. 8.19 Embodied flows of energy along transmission paths between sectors in the Beijing-Tianjin-Hebei agglomeration in 2002 and 2007 (*Note* The flows along all paths in the figure are greater than 1.0×10^6 tce. Components: 1-Agriculture, 2-Industry, 3-Construction, 4-Transportation, and 5-Other Services)

by the Industry sector in Hebei both ensured the normal operation of the sector in the region and provided energy for the Transportation and Other services sectors in Beijing. The flows from the Industry sector in Hebei remained high in 2007, which also supplied energy to Other Services sectors in Beijing and Tianjin. In 2002, the Construction sectors in Hebei and Beijing both obtained energy from their respective regional sectors and the same sectors in other areas, after which they transported energy downstream to the Other Services sector. Beijing and Hebei obtained energy from each other's Industry sectors. In 2007, the Industry sector in Hebei provided as well as received more energy than in 2002. In 2002, both sectors obtained energy from many upstream sectors in other regions. Beijing relied heavily on energy from other sectors within the region and imports from Hebei to run its Transportation and Other service sectors. The Other Services sector in Tianjin imported energy from Industry sectors in Beijing and Hebei. The Transportation sector in Heibei relied on energy from all sectors within the region, and the energy imported from the Industry sector in Beijing was also lower than in 2002. In 2007, Beijing and Tianjin relied on energy from Hebei's Industry sector to operate their Transportation and Other service sectors. The Transportation and Other Services sectors in Heibei relied heavily on energy from this region, and the energy imported from the Industry sector in Tianjin also decreased compared to 2002.

8.3.3 Relationships Analysis

8.3.3.1 Analysis of Relationships Among the Three Regions

Figure 8.20 illustrates the ecological relationships among regions in the 3-node Beijing-Tianjin-Hebei model. This network contained only three pairs of relationships. There was only a competitive relationship between Beijing and Tianjin in 2007, and there was no mutualism at any point. In all three years, relationships of control and exploitation predominated in the network. The control relationships always existed between Beijing and Hebei and between Tianjin and Hebei and did not change over the study period. This suggests that Hebei supplies Beijing and Tianjin with resources through the embodied energy flows. In 2002, Tianjin exploited energy from Beijing, while in 2010, Tianjin supplied energy to Beijing. This suggested that Beijing was more reliant on the resources of Tianjin and Hebei in 2010 than at the beginning of the study. In 2007, the only competitive relationship between Beijing and Tianjin implied that there was competition between Beijing and Tianjin in terms of exploiting the energy of Hebei.

8.3.3.2 Analysis of the Relationships Among the 13 Cities

Figure 8.21 shows the metabolic relationships among the cities in the 13-node network in 2002, 2007, and 2010. The 13 cities had a total of 78 relationships with

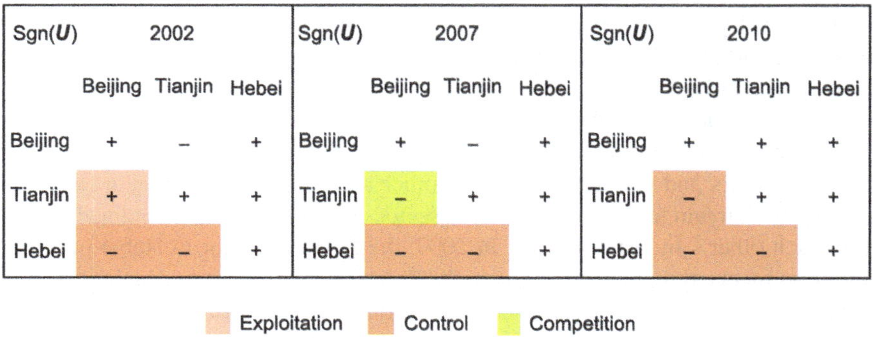

Fig. 8.20 Distribution of ecological relationships based on the sign utility matrix, sgn($\boldsymbol{U}$), for the integrated flows of energy among Beijing, Tianjin, and Hebei (*Note* Utilities are for flows from the region in the top row of the table to the region in the first column of the table)

each other. Like the 3-node network, the mutualism relationships did not exist at any time, and from 2002 to 2010, the relationship changed significantly. In 2002, control and exploitation relationships, which accounted for 87.2% of all relationships, were the dominant players in the network. Beijing and Tianjin exploited energy from 9 and 10 other cities, respectively, while Hengshui was exploited by 9 cities. In 2007, there was only one exploitation relationship (between Qinhuangdao and Hengshui), and there were 46 competition relationships (59.0% of the total); the rest 39.7% were control relationships. The control relationship was mainly associated with Beijing and Tianjin, both of which exploited energy from 11 cities in Hebei; Qinhuangdao was exploited by 12 other cities. In 2010, there was still only one exploitation relationship (i.e., between Qinhuangdao and Hengshui). However, the number of control relationships grew to 47, accounting for 60.3% of the total relationships. Beijing and Tianjin continued to exploit energy from 11 other cities in Hebei, while Qinhuangdao and Hengshui were also exploited by other cities.

From 2002 to 2007, the number of competition relationships first increased and then decreased. In 2002, there were only 10 pairs of competition relationships, they were among Hengshui and Beijing, Tianjin, and Langfang. By 2007, the number of competition relationships had increased to 46, and the proportion of the total number of relationships also reached 59.0%. Most of them were among the 11 cities in Hebei. In 2010, the total number of competition relationships dropped to 30 (38.5% of the total), most of them in Tangshan, Shijiazhuang, Cangzhou, Handan, and Baoding.

8.3.3.3 Relationships Among Sectors

I will focus on the relationships in 2002 and 2007 and their changes between the two years in this section. In 2002, exploitation and control relationships dominated the relationships (Fig. 8.22) between the sectors in Beijing and Tianjin (52% of the total), with competition relationships following (44%), while the mutualism relationship

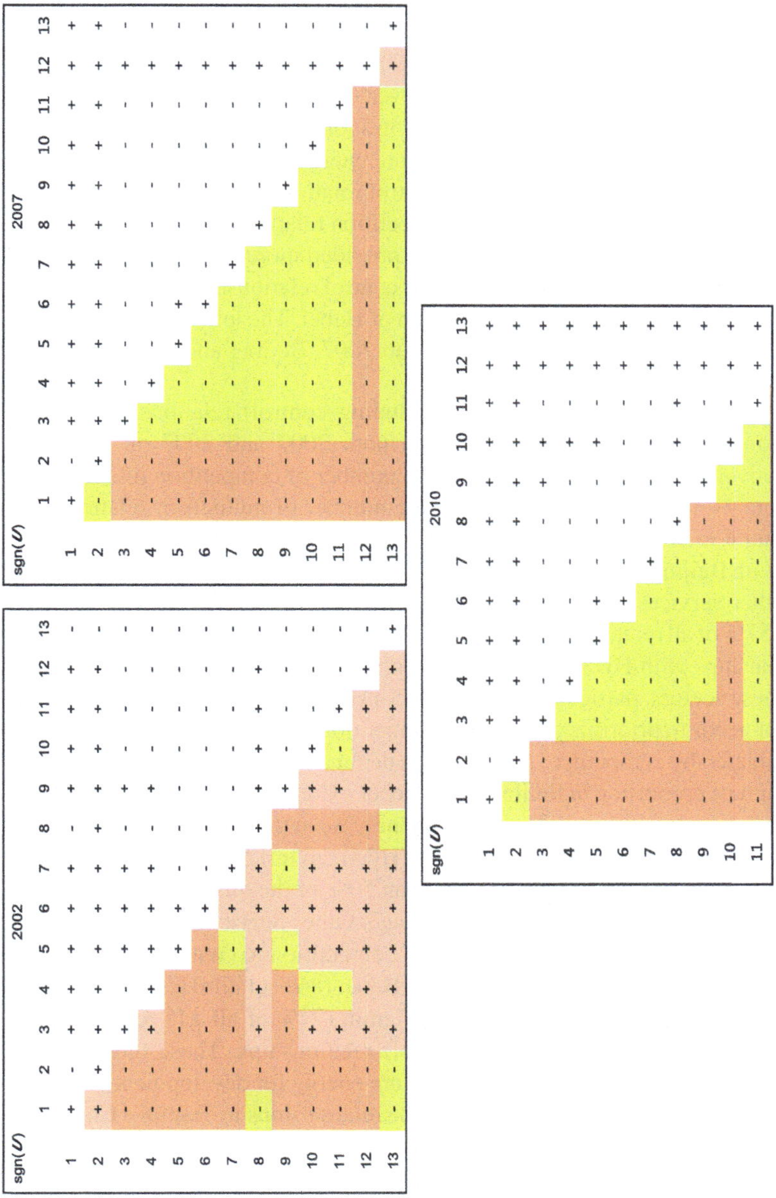

Fig. 8.21 Distribution of ecological relationships among the 13 cities in the Beijing-Tianjin-Hebei urban agglomeration based on the sign utility matrix, sgn(***U***), for the integrated flows of energy (*Note* Utilities are for flows from the region in the top row of the table to the region in the first column of the table. Cities: 1-Beijing, 2-Tianjin, 3-Tangshan, 4-Shijianzhuang, 5-Cangzhou, 6-Handan, 7-Baoding, 8-Langfang, 9-Xingtai, 10-Zhangjiakou, 11-Chengde, 12-Qinhuangdao, 13-Hengshui)

accounted for only a small portion (4%). The growth of exploitation (control) relationships in 2007 (to 68%, approximately evenly distributed between exploitation and control categories) indicates that these regions are increasingly reliant on each other. In 2002, competition relationships (60%) between Beijing and Hebei dominated the relationships, with the exploitation or control relationships (36%) following, and once again, there was only a single mutualism relationship. This means that these two regions were competing for energy. However, in 2007, the dominant place was taken by exploitation or control relationships (56% of the total), among which 14% were the exploitation of Hebei. Heibei sent more energy to Beijing than it received from Beijing. In 2002, the exploitation or control relationship between Tianjin and Hebei made up 68% of the total, with the competition relationship following at 24% and with two mutualism relationships. Tianjin provided more energy to Heibei than it received from Hebei. In 2007, exploitation or control relationships dropped to 60% of the total, 87% of which were the exploitation of Hebei. The left 40% were competition relationships. In conclusion, from 2002 to 2007, Beijing and Tianjin became more dependent on Hebei (Fig. 8.22).

Figure 8.22 reveals that exploitation (including control) relationships among sectors become the dominant form again in both 2002 and 2007, at 54% and 67%, respectively. During these five years, the number of competition relationships declined from 44 to 32 (i.e., by 27%), and the number of mutualism relationships declined from 4 to 3. In 2002, the dominant competition relationships were mainly associated with Beijing's Construction sector (79% of all the relationships with BJ3), Beijing's Other Services sector (64% of all relationships with BJ5), Hebei's Agriculture sector (57% of all relationships with HB1), and Hebei's Construction sector (57% of all relationships with HB3); the rest of sectors had $\leq 50\%$ competition relationships. For these sectors, particularly Beijing's Construction sector (BJ3), their future development needs to prioritize mitigating competition relationships. The number of sectors dominated by competitive relationships declined in 2007, but they still maintained the highest proportion in Beijing's and Hebei's Construction sectors (both with 57% of their total relationships). There was an increase in the number of competition relationships in Hebei's Other Services sector (HB5), up to 57% of the total in 2007. The competition relationships made up less than 50% of the other sectors' relationships; in 2007, the relationship between Beijing's Other Services sector (BJ5) and Hebei's Agriculture sector (HB1) shifted from competition to control.

In 2002, the Industry sectors in Beijing and Hebei (86% of all BJ2 and HB2 relationships) and the Other Services sectors in Tianjin (71% of all TJ5 relationships) concentrated most of the exploitation (control) relationships. These three sectors exploited others more because they required more energy for development and were heavily reliant on other sectors. Over 50% of the relationships in Tianjin's Construction and Transportation sectors (TJ3 and TJ4) were exploitation (control) relationships, even with a higher percentage of TJ4 (64% of its relationships). Meanwhile, Tianjin's Other Services sector (TJ5) was mainly in an exploitation relationship with Hebei's sectors (80% of the total). Therefore, the sectors in Hebei relied heavily on TJ5. More than half of the relationships with Tianjin's Construction sector (TJ3)

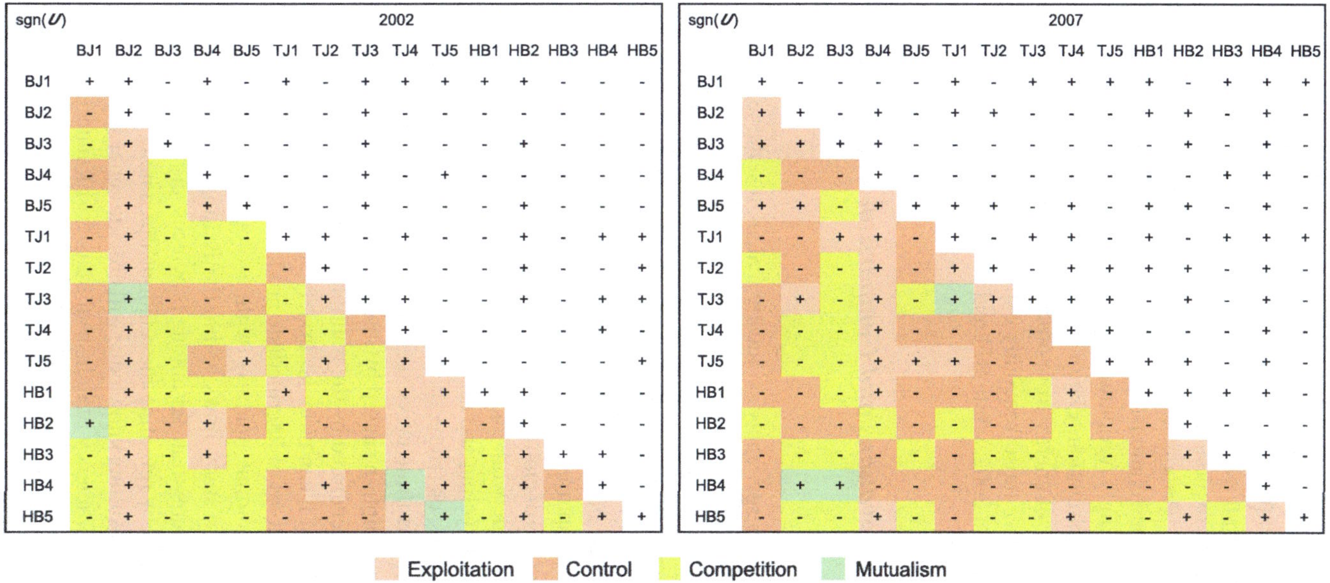

Fig. 8.22 Distribution of ecological relationships among the five sectors in Beijing, Tianjin, and Hebei based on the sign utility matrix, sgn(***U***), for the integrated flows of energy (*Note* Utilities are for flows from the region in the top row of the table to the region in the first column of the table. Cities: BJ-Beijing, TJ-Tianjin, HB-Hebei. Sectors: 1-Agriculture, 2-Industry, 3-Construction, 4-Transportation, 5-Other Services)

were exploitation (control) relationships, which were mainly controlled by Beijing's sectors (80%).

In 2007, most sectors had increases in the number of exploitation relationships. From 2002 to 2007, the exploitation relationships related to Beijing's Transportation sector (BJ4) and Tianjin's Agriculture sector (TJ1) both rose from around 50% to 86%. BJ4 was mainly exploited by other sectors, while TJ1 mainly exploited energy from other sectors. As a result of the changes in competition relationships, the number of exploitation relationships associated with Beijing's Industry sector (BJ2) declined. Both Beijing's Agriculture sector and Hebei's Transportation sector had a high number of exploitation relationships (BJ1 and HB4, 79%). BJ1 mostly played the role of receiver of energy from other sectors, while HB4 always acted as the provider. The respective most frequently exploited sectors differed in 2002 and 2007 (BJ2, TJ2, and TJ5 in 2002, while BJ4 and HB4 in 2007).

8.3.4 Research Innovations and Comparison with Previous Research

Traditional embodied energy analysis mostly calculates the embodied energy consumption based on the expenditure on final consumption, total capital formation, exports, and other data obtained from input–output analysis and is based on the sector's energy consumption coefficient (Sect. 5.3.4). However, this research has rarely considered the indirect energy flows that result from the energy embodied in intermediate products. For example, Li et al. (2014) calculated the per capita embodied energy consumption of 30 provinces and regions in China but did not quantify the flow processes of embodied energy. Although Chen S and Chen B (2015) introduced the ecological network analysis method and studied network characteristics by using an index of cycling and ascendency analysis, they still paid little attention to details of the processes. The ecological network analysis method described in this section can go beyond these previous studies by comprehensively considering the indirect energy consumption caused by intermediate processes and can calculate the energy consumption generated by the multiple transfer paths between industries (Zhang et al., 2014). Thus, it can more accurately simulate the relationships among sectors generated by the flows of intermediate products between sectors.

By integrating input–output analysis with ecological network analysis, my research group was able to study flows within the Beijing-Tianjin-Hebei region to quantify the amount of energy transfer among provinces, cities, and sectors, as well as the flow direction, thereby revealing the relationships among sectors in the urban energy metabolism. The resulting knowledge will provide better support for decision-making and improve the efficiency of industrial transfers of materials and energy to promote energy conservation (Zhang et al., 2015).

8.3.5 Policy Recommendations

The central government of China has implemented plans aimed at promoting more efficient and integrated development in the Beijing-Tianjin-Hebei region. Because of these plans, energy-intensive industries have been gradually shifting out of Beijing since 2007. Some companies kept their headquarters in Beijing, but their production facilities were relocated to Hebei. This reduced the overall energy consumption in Beijing and facilitated the substantial indirect energy flows from Beijing to Tianjin and Hebei, which contributed to balancing the implied energy flows among these regions. For instance, one of China's largest steel conglomerates, Shougang Group, was originally situated in Beijing but subsequently moved all of its production facilities to Tangshan in Hebei while retaining only its research and development department in Beijing. This made a significant contribution to the reduction of Beijing's energy consumption, which declined by 5.4% between 2006 and 2007. Beijing is the largest trading and distribution center for technology and commodity in China. In the future, for the purpose of more balanced energy metabolism, Beijing may provide more technology and information to Tianjin and Hebei through indirect flows.

But there are still some problems. Beijing, as the capital of China, will be more easily tilted in national policy, which may lead to an imbalance of opportunities for Tianjin and Hebei. Relocating industries or enterprises to Hebei will also transfer related environmental issues along with them. For this reason, it is important for Hebei to pay attention to the features of the companies it receives and take action to make sure that they would not exceed their own environmental carrying capacity. Since the Industry sector in Hebei consumes a tremendous amount of energy, it is necessary to take measures to reduce energy consumption in this sector in the future; such as restructuring the energy utilization of industries in this sector and using cleaner forms of energy to replace fossil fuels. For the consumer sector, production technologies should be upgraded to reduce production and energy losses as well as the related waste generation during production activities, thus increasing the efficiency of resources and energy use. Addressing these and other regional issues demands more research at a higher resolution (e.g., at an industry or enterprise level within each sector) to determine the most problematic energy flows and then find ways to mitigate them.

8.3.6 Importance of Multi-Scale Comparative Analysis

Similarities and differences can be found by comparing analyses performed at multiple scales. For example, in the 3-node network, the direct energy consumption of Hebei was 2 to 3 times that of Beijing and Tianjin. In the 13-node network, Beijing had the highest direct energy consumption in 2002, which was 11.6 times higher than Hengshui, whose direct energy consumption was the lowest. In 2007 and 2010, Handan's consumption of direct energy was the highest. From 2002 to 2007, the embodied energy consumption for each node first increased and then decreased in the 3-node network, but in the 13-node network, the trend was reversed. The different direct energy flow matrices that we used caused this difference. The direct and embodied energy flows revealed that in the 3-node network, most of the energy that Beijing and Tianjin received came from Hebei; in the 13-node network, Tangshan was the most significant energy provider for the other 12 cities. Regarding the ecological relationships, the percentage of competition relationships increased in the 13-node model, which was mainly among 11 cities in Hebei. In the 3-node model, exploitation and control dominated the relationships among Beijing, Tianjin, and Hebei. Thus, the results reflected both similarities and differences between the two scales.

The result of the 13-node model is a refinement of the 3-node model. The embodied energy flow from Hebei to Beijing was the highest in the 3-node model; it contributed about 30% of the total energy flow in 2002 and 2007 and was close to 50% in 2010. In the 13-node model, the embodied energy flow from Hebei to Beijing remained the largest in the network, although its share declined from about 13% in 2002 to about 5% in 2007 and 2010. The reason for this decline was that the increase in network size reduced the advantages offered by some paths in the network. In addition, through the 13-node model, we identified that Tangshan, Shijiazhuang, and Handan were the significant contributors to the embodied energy flow from Hebei to Beijing, they together accounted for 47% of the total.

Exploitation and control relationships were dominant in the 3-node model, making up more than 67% of the total relationships. Also, in the 13-node model, they remained dominant in 2002 and 2010, making up more than 61% of the total relationships. In the 13-node model, the proportion of exploitation and control relationships decreased because of the existence of multiple pairs of competition among the 11 cities in Hebei. In the 3-node model, Beijing and Tianjin exploited energy from Hebei, while in the 13-node model, Beijing and Tianjin exploited energy from most cities in Hebei (Hao et al., 2018). In 2002, Beijing had a competitive relationship with Langfang and Hengshui, and Tianjin showed a competitive relationship with Hengshui (Zhang et al., 2015).

There were some differences between the results of the 3-node and 13-node models. The lowest energy flow in the 3-node model was the embodied energy flow from Beijing to Tianjin, which was only about 19% of the total. In the 13-node model, the flow from Beijing to Tianjin contributed more to the overall energy flow. The embodied energy flow from Beijing to Tianjin became the highest, especially in 2002. In 2007 and 2010, the embodied energy flow from Beijing to Tianjin was about 2.5 times higher than the minimum flow. Furthermore, there were different ecological relationships between Beijing and Tianjin in the models of different nodes. In 2010, the result of the 3-node network showed that Beijing exploited energy from Tianjin. However, in the 13-node network, the relationship between Beijing and Tianjin shifted to competition, mainly because both Beijing and Tianjin exploited energy from 11 cities in Hebei.

References

Borrett SR, Patten BC (2003) Structure of pathways in ecological networks: Relationships between length and number. Ecol Model 170(2–3):173–184

Burns TP (1989) Lindeman's contradiction and the trophic structure of ecosystems. Ecology 70(5):1355–1362

Campbell D E, (2000) A revised solar transformity for tidal energy received by the earth and dissipated globally: implications for emergy analysis//Brown M T. Emergy Synthesis: Theory and Application of the Emergy Methodology. Gainesville: The Center for Environmental Policy: 255–263.

Chen S, Chen B (2015) Urban energy consumption: Different insights from energy flow analysis, input-output analysis and ecological network analysis. Appl Energy 138(1):99–107

Fath BD (2007) Network mutualism: Positive community-level relations in ecosystems. Ecol Model 208(1):56–67

Fath BD, Borrett SR (2006) A MATLAB function for network environ analysis. Environ Model Softw 21(3):375–405

Hao Y, Zhang MH, Zhang Y et al (2018) Multi-scale analysis of the energy metabolic processes in the Beijing-Tianjin-Hebei (Jing-Jin-Ji) urban agglomeration. Ecol Model 369(3):66–76

Huang SL, Lee CL, Chen CW (2006) Socioeconomic metabolism in Taiwan: Emergy synthesis versus material analysis. Resour Conserv Recycl 48(2):166–196

Jiang MM, Zhou JB, Chen B et al (2008) Emergy-based ecological account for the Chinese economy in 2004. Commun Nonlinear Sci Numer Simul 13(10):2337–2356

Lan SF, Odum HT (1994) Emergy evaluation of the environment and economy of Hong Kong. Journal of Environmental Science 6(4):432–439

Lei KP, Wang ZS, Ton SS (2008) Holistic emergy analysis of Macao. Ecol Eng 32(1):30–43

Li JL, Zhang RS (2003) Energy analysis for evaluating sustainability of Ningbo ecology-economic system. Geography and Geo-Information Science 19(2):73–76 (in Chinese)

Li Z, Pan L, Fu F et al (2014) China's regional disparities in energy consumption: An input-output analysis. Energy 78(12):426–438

Lu HF, Campbell DE (2009) Ecological and economic dynamics of the Shunde agricultural system under China's small city development strategy. J Environ Manage 90(8):2589–2600

Marzluff JM, Shulenberger E, Endlicher W et al (2008) Urban Ecology: An International Perspective on the Interaction between Humans and Nature. Springer, Boston, MA

Odum HT (1996) Environmental Accounting: Emergy and Environmental Decision Making. John Wiley and Sons, New York

Odum H T, 2000. Handbook of emergy evaluation: A compendium of data for emergy computation issued in a series of folios. Folio No. 2: Emergy of Global Processes. Gainesville (FL): Center for Environmental Policy, Environmental Engineering Sciences, University of Florida.

Sui CH, Lan SF (2006) Emergy analysis of Guangzhou and Shanghai urban ecosystem. Urban Environment & Urban Ecology 19(4):1–3 (in Chinese)

Ulgiati S, Brown MT (1998) Monitoring patterns of sustainability in natural and man-made ecosystems. Ecol Model 108(1–3):23–26

Ulgiati S, Odum HT, Bastianoni S (1994) Emergy use, environmental loading and sustainability: an emergy analysis of Italy. Ecol Model 73(3–4):215–268

Zhang Y, Yang ZF, Yu XY (2009a) Ecological network and emergy analysis of urban metabolic systems: Model development, and a case study of four Chinese cities. Ecol Model 220(11):1431–1442

Zhang Y, Yang ZF, Yu XY (2009b) Evaluation of urban metabolism based on emergy synthesis: A case study for Beijing (China). Ecol Model 220(13–14):1690–1696

Zhang XH, Deng SH, Jiang WJ et al (2010a) Emergy evaluation of the sustainability of two industrial systems based on wastes exchanges. Resour Conserv Recycl 55(2):182–195

Zhang Y, Yang ZF, Fath BD et al (2010b) Ecological network analysis of an urban energy metabolic system: Model development, and a case study of four Chinese cities. Ecol Model 221(16):1865–1879

Zhang Y, Li SS, Fath BD et al (2011a) Analysis of an urban energy metabolic system: Comparison of simple and complex model results. Ecol Model 22(1):14–19

Zhang Y, Yang ZF, Liu GY et al (2011b) Emergy analysis of the urban metabolism of Beijing. Ecol Model 222(14):2377–2384

Zhang Y, Zhang JY, Yang ZF et al (2012) Analysis of the distribution and evolution of energy supply and demand centers of gravity in China. Energy Policy 49(10):695–706

Zhang Y, Zheng HM, Fath BD et al (2014) Ecological network analysis of an urban metabolic system based on input-output tables: Model development and case study for Beijing. Sci Total Environ 468(1):642–653

Zhang Y, Zheng HM, Yang ZF et al (2015) Multi-regional input-output model and ecological network analysis for regional embodied energy accounting in China. Energy Policy 86(11):651–663

Zhang Y, Zheng HM, Yang ZF et al (2016) Urban energy flow processes in the Beijing-Tianjin-Hebei (Jing-Jin-Ji) urban agglomeration: combining multi-regional input-output tables with ecological network analysis. J Clean Prod 114(3):243–256

Chapter 9
Analysis of Carbon Metabolic Processes

In the previous chapters, we have mentioned the importance of carbon as a major component of most energy sources used by cities (i.e., fossil fuels) and a primary constituent of many resources consumed by urban metabolism (e.g., food, paper, or plastics). In this chapter, I will show how the techniques described in the previous chapters can be used to quantitatively describe a city's carbon metabolism, identify carbon metabolic disorders that must be reversed, and help improving the ecological impact of urban metabolism (e.g., to reduce carbon emission).

9.1 Identification of the Key Metabolic Actors in the Urban Carbon System

A report released by the C40 Urban Climate Leadership alliance and the Carbon Disclosure Project organization revealed that cities account for 70% of global carbon emissions (Satterthwaite, 2008). This is a result of high consumption of energy, food, and minerals by cities, where the vegetation can only offset 8% of the carbon emission resulting from this consumption (Escobedo et al., 2010). This imbalance produces serious resource and environmental proble ms, resulting from a disorder of urban carbon flows, including carbon transfers among the production and living sectors, and carbon emission and absorption that results from flows between urban sectors and the atmosphere. To support efforts to mitigate carbon emissions, it is necessary to consider these carbon flows from a metabolic perspective and analyze the degree of disorder in carbon metabolism caused by each sector and by its interactions with other sectors.

To provide a high-resolution and more holistic picture of a city's urban carbon metabolism, my research group chose Beijing as an example, and refined the natural and socioeconomic metabolic actors into various categories with the goal of simultaneously quantifying the carbon flows among the metabolic actors and between these actors and the atmosphere (Li et al., 2018; Xia et al., 2018, 2017a, 2017b; Zhang

Y. Zhang, *Urban Metabolism*,
https://doi.org/10.1007/978-981-19-9123-3_9

et al., 2016, 2015). As described in the previous chapters, Beijing's administrative boundary represented the dividing point between the city's internal and external environments. The internal environment excludes the natural components of the city that lie inside that boundary. We divided the metabolic actors into natural and socioeconomic categories based on the different types of land use and the different socioeconomic actors. We identified a total of 18 metabolic actors, for which there was sufficient data available to be included in the analysis.

Based on the available data, we accounted for carbon metabolic flows (including nine material categories) and evaluated their changes from 1995 to 2015 to analyze the total carbon metabolism, the evolution of these flows, and the resulting structure of the metabolism (Fig. 9.1). We also established an 18-node network model of the city's carbon metabolism, and developed two new indicators: the carbon external dependence index (CEDI), and the carbon imbalance index (CII). CEDI represents the ratio of external resources consumed to internal resources consumed and therefore determines how heavily the city depends on its external environment. CII describes the ratio of carbon emission to carbon absorption and thus represents the magnitude of the imbalance between absorption and emission. Our goal was to detect and comprehensively evaluate any carbon metabolic disorders in Beijing's urban metabolism. We used this approach to identify the main actors that were responsible for the carbon metabolic disorders and support efforts to optimize utilization of carbon-containing materials in Beijing and the development of plans to reduce carbon emissions (Li et al., 2018; Zhang et al., 2015).

The data we used were mainly obtained from the Beijing Statistical Yearbook, China Energy Statistical Yearbook, China Rural Statistical Yearbook, China Environmental Statistical Yearbook, China Plastics Industry Statistical Yearbook, and other data sources, including research literature. Similar data is available for most cities to support such an analysis. Most of the collected data do not exactly align with the metabolic actors and the paths between them. Therefore, it is necessary to determine how to distribute the data among the actors and paths based on the characteristics of the flows, supplemented by practical investigations, data collection, and other means, before it is possible to reproduce the material flow processes with acceptable accuracy (see Zhang et al. [2015] for more details). The carbon accounting included data on the following metabolic substances: energy, minerals, plastics, chemical fertilizer, food, feed, wood and straw, paper products, carbon in water, and waste. Since our goal was to quantify urban carbon transfers, we focused on the materials that were consumed over short periods (within a year), and excluded material consumption to create the urban carbon stock (e.g., durable goods such as buildings, plastic and steel products, and construction materials used in the urban infrastructure). The accumulation of these carbon stocks will have an impact on carbon flow processes, but currently there is insufficient data to determine the magnitude of the stocks and the flows between different actors that create them. In the future, when relevant data becomes available, the interactions between carbon flows and carbon stocks should be analyzed more in-depth to provide an improved basis for optimizing the selection of measures to regulate carbon flows.

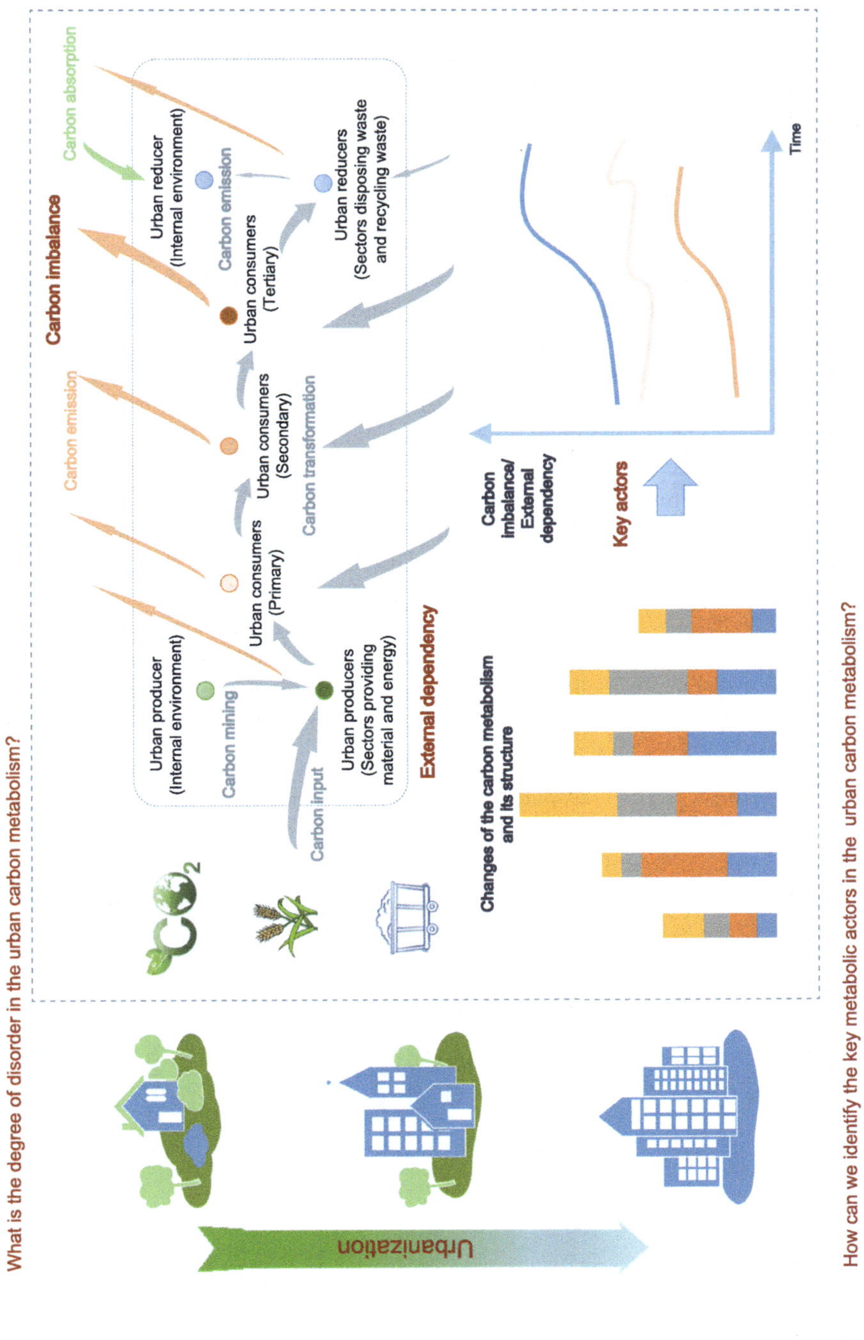

Fig. 9.1 Analytical framework for identifying key actors in the urban carbon metabolism

9.1.1 Changes of the Carbon Metabolism and Its Structure

The inputs and outputs of carbon for Beijing showed an overall increase trend from 1995 to 2015, but with some variation (Fig. 9.2). The carbon inputs increased by 62.4% during the 21-year study period, but the carbon outputs grew faster (by more than 2 times). Both inputs and outputs showed periods of growth and decrease. From 1995 to 2006, the carbon inputs and outputs increased by 49.0% and 73.8%, respectively. From 2006 to 2014, they decreased by 24.6 and 45.2%, respectively (relative to the value in 2006), and then increased by 44.5% and 240%, respectively, from 2014 to 2015.

These changes were mainly driven by energy flows, which accounted for more than 70% of the total throughout the study period, with inputs and outputs of energy increasing by 95.4% and 300%, respectively, compared with the values from 1995. The increase of total inputs also came from food imports. Although the initial proportion of the total flow was small (0.5%), it increased rapidly (to 18.8 times the value from 1995), and hence accounted for 6% of the total flows by 2015. The inputs of plastic, paper products, water, wood and straw, and fertilizer also increased rapidly (to as much as 7.5 times the corresponding values from 1995), and together, they accounted for 4% of the total flows. At the same time, the inputs of minerals and feed, which accounted for 20% and 5% of the total, respectively, at the beginning of the study period showed a decreasing trend; however, by 2015, they had decreased by 90.6 and 56.3%, respectively, which reduced the growth of carbon inputs. Outputs of plastic and paper products, and carbon in water increased to more than 10 times and 2 times, respectively, to the values from 1995. Even though outputs of food fluctuated, by 2015, they increased to 9.9 times the value from 1995. However, because

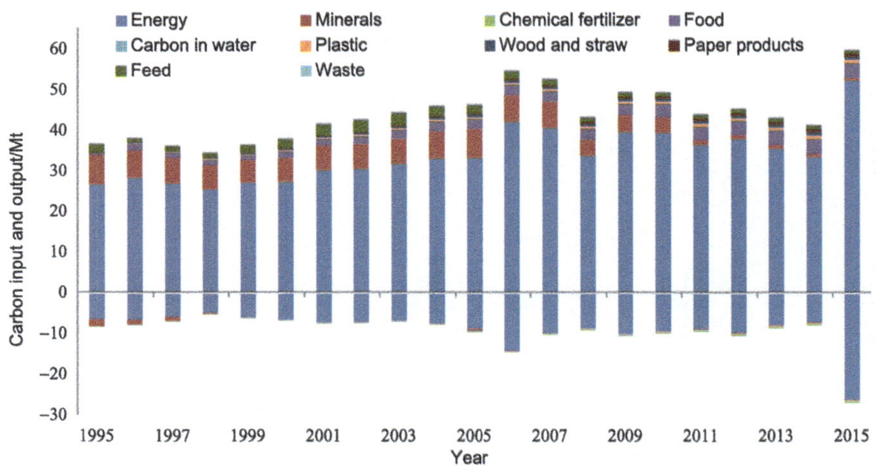

Fig. 9.2 Total carbon inputs and outputs of Beijing's carbon metabolic system from 1995 to 2015 and the associated changes in the material structure (*Note* Positive values represent carbon in input materials; negative values represent carbon in output materials)

the cumulative proportion of the total outputs for plastics, paper, water, and food was < 0.1%, they had little effect on the total carbon output. The amount of waste (about 1% of the total output) was also small, and since its increase was only moderate (by 28.7%), it had little impact on the change in the total carbon output. The decrease of mineral inputs partly offset the increase of carbon inputs. This was because initially, during a period of large-scale construction in Beijing, it accounted for a relatively large proportion of the total input (20.5%) and subsequently considerably decreased (by 96.7%). In 2015, mineral inputs accounted for only 0.2% of total inputs. The decrease (by 23.3%) in the output of wood and straw had little influence on the overall change, due to its small proportion of total outputs (an average value of < 0.1%).

9.1.2 Identification of the Key Actors Based on the Carbon Imbalance Index

CII closely followed the trend for carbon emission, since carbon absorption showed little change (a slight decrease in mass terms) during the study period (Fig. 9.3). Before 2001, CII was not synchronized with carbon emission, but thereafter, it closely followed carbon emissions. The increase of CII resulted from the reduction of carbon absorption being greater than the decrease of carbon emissions. Overall, carbon emissions increased by 7.7%, whereas carbon absorption decreased by 38.1% during the study period. Beijing's CII increased by nearly 74%, indicating that the magnitude of the imbalance greatly increased. This trend was caused by changes in the flows related to the key actors.

Figure 9.4 shows that all carbon metabolic actors had output paths to the atmosphere, whereas < 25% of the actors had inputs from the atmosphere; the latter ones were Crop Cultivation, Forest, Grassland, and Surface Water. Based on the key actors identified in Fig. 9.3, we can see that the carbon emission and absorption were relatively large for Manufacturing, Production and Supply of Electric Power and Heat, Transportation, Other Services, Urban Households, Rural Households, and Crop Cultivation, and that these actors had a large contribution to the changes of CII. The share of carbon emission accounted for Mining, Retail and Catering, and Waste Disposal was small (only 2% of the total values in 1995), but Retail and Catering and Waste Disposal increased to 10 and 4 times of their levels in 1995, respectively, reaching 9% of the total in 2015. Mining was relatively stable before 2007, but then increased to 7.5 times of this level by 2011, and then decreased by 95% until 2013. The contributions of these three actors to the changes of CII were relatively large.

The carbon emission by Manufacturing, Rural Households, and Crop Cultivation generally decreased during the study period. Manufacturing, which accounted for 42% of the total in 1995, remained relatively stable, with only slight fluctuations before 2006. However, in 2011, it decreased, and was overtaken by Production and Supply of Electric Power and Heat. In 2015, carbon emission by Manufacturing was only 20% of that in 2006 (and its proportion of the total decreased to 9% by 2015).

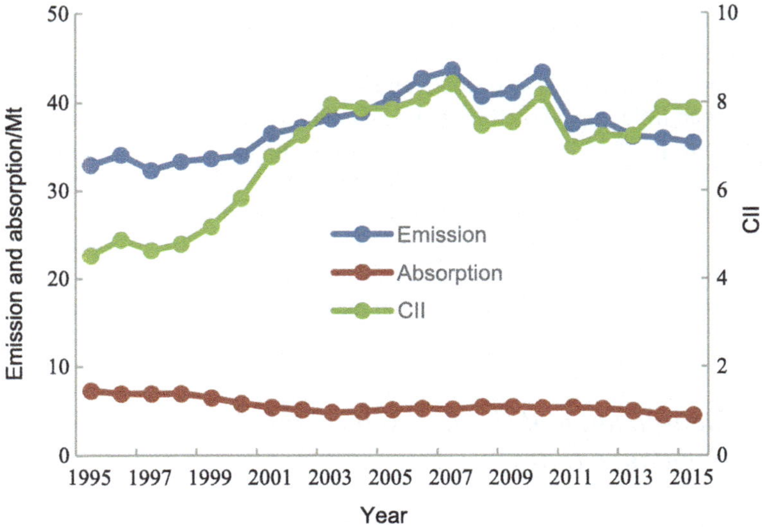

Fig. 9.3 Changes in the value of the carbon imbalance index (CII, which equals to carbon emission divided by its absorption) for Beijing from 1995 to 2015

Before 2005, carbon emission by Rural Households decreased steadily, meanwhile in 2006 it increased significantly, reaching 1.6 times the value in 2005. Further, it decreased to 67% of its level in 2006. Nevertheless, its share of the total during the study period was relatively stable, fluctuating between 4 and 6%. Crop Cultivation (7% of the total in 1995) had a share close to that of Rural Households, but its share has been decreasing, with a reduction to 30% of its level in 1995 by 2015 (when it accounted for 2% of the total). Production and Supply of Electric Power and Heat, Transportation, Other Services, and Urban Households all showed continuous emission growth. The proportion for Production and Supply of Electric Power and Heat (19%) was smaller than that of Manufacturing in 1995. After 2011, it rose to first place (25%), and by 2015, it had increased to 1.4 times its value in 1995, and still accounted for about 25% of the total. Although in 1995 the proportions of the total for Transportation and Urban Households were both less than 10%, by 2015, they increased to 6.7 and 1.4 times their values of 1995, respectively, when they accounted for 18% and 16% of the total. Other Services increased twice as fast, such as Production and Supply of Electric Power and Heat, but because its initial proportion was only 4%, its proportion of the total remained less than 10% in 2015.

Crop Cultivation had the largest carbon absorption (accounting for 49% of the total), however due to shrinking of the Farmland area, its carbon uptake decreased year by year. In 1999, it was overtaken by Forest, and in 2003, it dropped to 35% of its value from 1995 (accounting for only 26% of the total). From 2003 to 2008, it increased, reaching 1.5 times its value in 2003 (accounting for 35% of the total), and further decreased again, reaching only half of its value in 2008 (accounting for 22% of the total). Forest had relatively large carbon absorption, but its total absorption

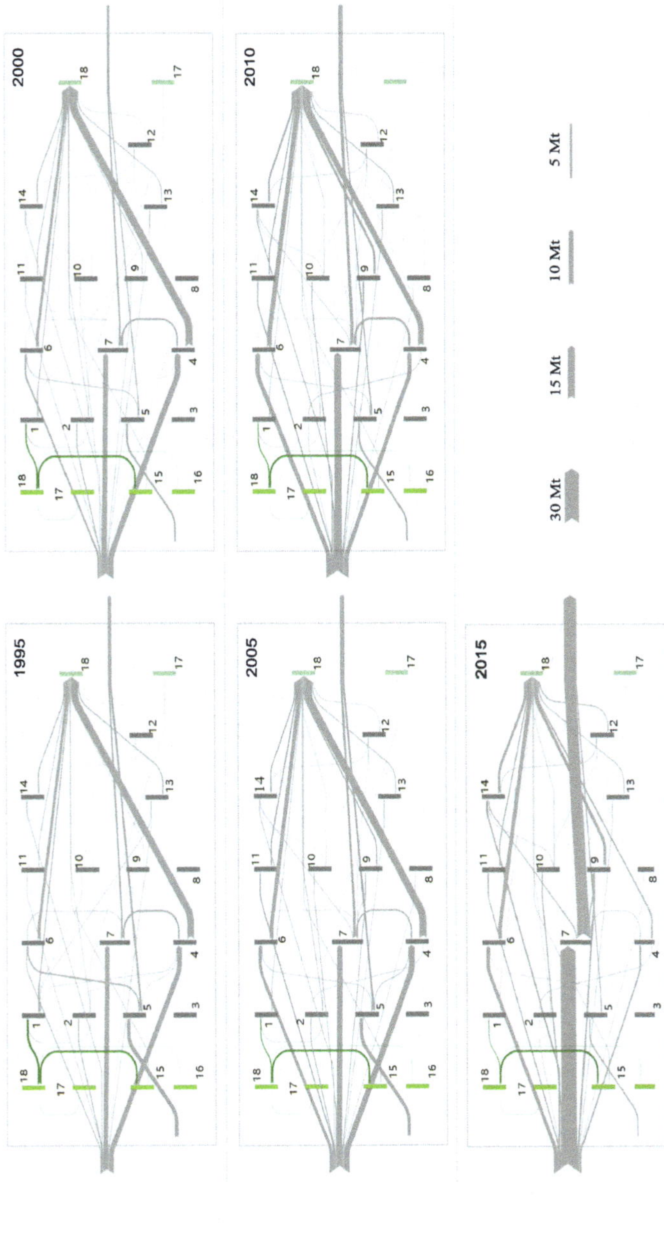

Fig. 9.4 Carbon flows among the metabolic actors in Beijing's carbon metabolic system from 1995 to 2015 (*Note* The path width represents the size of the flow, and the paths connected to the upper and lower halves of an actor represent the input and output paths, respectively. Green and grey rectangles represent the natural and socioeconomic metabolic actors, respectively. Green lines represent carbon sequestration or emission of CO_2, and grey lines represent carbon transmission between the actors. Actors: 1-Crop Cultivation; 2-Animal Husbandry; 3-Fisheries; 4-Manufacturing; 5-Mining; 6-Production and Supply of Electric Power and Heat; 7-Energy Transformation; 8-Construction; 9-Transportation; 10-Retail and Catering; 11-Other Services; 12-Waste Disposal; 13-Rural Households; 14-Urban Households; 15-Forest; 16-Grassland; 17-Surface Water; 18-Atmosphere)

did not change much during the study period. i.e., it decreased by less than 4%. However, with decreasing carbon absorption by Crop Cultivation, the proportion of carbon absorption by Forest increased from 42 to 65%.

9.1.3 Identification of Key Actors Based on the Carbon External Dependence Index

CEDI generally increased, reaching 4 times its value from 1995 by the end of the study period, despite some fluctuation caused by the influence of inputs from both the external and the internal environments (Fig. 9.5). From 1995 to 2004, CEDI showed the opposite trend from that for internal inputs, however it fluctuated more significantly than these inputs. By 2000, the internal inputs decreased to half their values from 1995, whereas CEDI increased to 1.6 times its value in 1995, mainly due to a large increase (by 48.1%) of net external inputs. From 2004 to 2010, CEDI showed fluctuations similar to those of net external inputs, increasing to 1.9 times of the value in 2004 and decreasing to 22.9%, respectively. Yet, CEDI fluctuated more than the net external inputs. This was mainly due to a 56.8% decrease in the internal inputs during this period. From 2010 to 2015, CEDI showed a synchronous downward trend similar to that for the net external inputs (decreasing by 14.1% and 20.7%, respectively), mainly owing to the relatively stable internal inputs.

Except for Waste Disposal, Grassland, and Atmosphere, the other carbon metabolic actors (which accounted for 83% of the 18 actors) all had carbon transfers with the external environment, but only Mining received carbon inputs from the internal environment. Among the key actors identified in Fig. 9.5, the net external inputs for Manufacturing, Production and Supply of Electric Power and Heat, Energy Transformation, Transportation, and Urban Households were large, hence their contributions to CEDI changes were also substantial. Although Retail and Catering and Animal Husbandry had a relatively small share of total inputs in

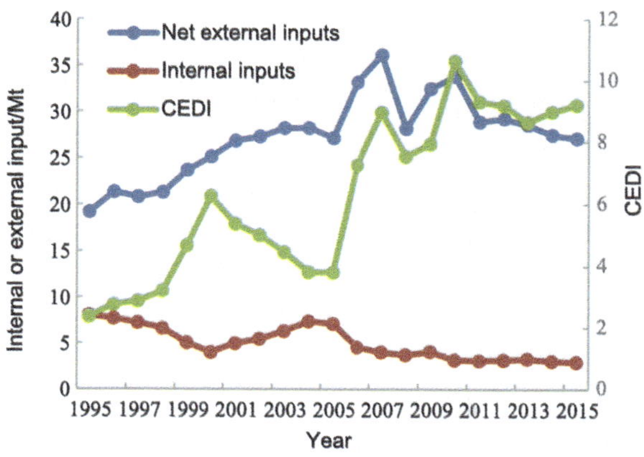

Fig. 9.5 Changes in the value of the carbon external dependence index (CEDI, which equals external inputs divided by internal inputs) for Beijing from 1995 to 2015

1995 (11% of the total), the contributions of both actors greatly changed during the study period. Retail and Catering increased significantly from 0.5% in 1995 to 8.7% in 2015, and Animal Husbandry decreased considerably from 10.0% in 1995 to 3.1% in 2015.

For Manufacturing, the initial net external input accounted for half of the total, and this proportion remained relatively stable until 2007. After 2007, Manufacturing's share of the total inputs began to decrease markedly, and in 2010 its contribution was exceeded by Energy Transformation and Production and Supply of Electric Power and Heat. In 2015, the inputs for Manufacturing decreased to less than 25% of its value from 2007 (and its proportion of the total decreased to 11%). This resulted mainly from a large decrease of inputs of non-renewable resources such as energy and minerals, which had the largest share of its inputs (accounting for 52% and 46%, respectively, of the total in 1995). By 2015, the inputs of energy and materials had decreased to 24% and 6% of their values from 1995, accounting for 48% and 8% of the total inputs, respectively. Renewable resources such as raw materials for production of paper and food accounted for less than 2% of the total inputs in 1995, but their flow increased gradually, reaching 8 times their value in 1995 and with their share of the total to 44%.

Net external inputs for Energy Transformation accounted for 28% of the total in 1995, which was less than those of Manufacturing and Energy Transformation. Its inputs increased continuously before 2010, reaching 1.6 times its value from 1995 and surpassing the contributions of Manufacturing and of Production and Supply of Electric Power and Heat. Thereafter, it decreased by 19% and its contribution was exceeded by that of Production and Supply of Electric Power and Heat. However, Manufacturing still accounted for 25% of total inputs by 2015. This was the consequence of the changes in the inputs of petroleum and coal products, which accounted for 63% and 37% of the total input, respectively. Oil products increased (by 96% compared with the level in 1995), whereas coal products decreased (by 97%). In 2015, these two inputs accounted for 94% and 0.6% of the total inputs, respectively. About 40% of the carbon input for Energy Transformation was exported to the external environment. The amount of exported carbon remained relatively stable during the study period, however, it increased sharply in 2015, reaching 5 times its level in 2014 (its proportion of the total also increased to 76%). Another part of the carbon, accounting for about 50%, flowed to Manufacturing, Transportation, Urban Households, and Production and Supply of Electric Power and Heat, and it fluctuated from 2007 to 2013. In 2010, it reached its maximum value and represented a 70% increase compared with the value in 2007. The last 3% of the carbon came from discharges during production processes.

Although the initial net external input of Production and Supply of Electric Power and Heat in 1995 accounted for only 9% of the total input, it increased to 24% (nearly 3 times its original value) by 2015. The net external inputs of Transportation and Urban Households accounted for 3% to 4% of the total in 1995, but rose to about half of the value for Production and Supply of Electric Power and Heat in 2015. The growth of net external inputs for Transportation was mainly affected by secondary energy from petroleum, such as gasoline, diesel, and kerosene, which accounted for

89% of the total energy consumption, and by 2015 increased to 5 times of the value from 1995. The input growth for Urban Households mostly resulted from increased consumption of food and plastics, reaching 12 and 10 times the values in 1995, respectively, and also from a 48% increase in energy consumption, which accounted for more than half of the total net external inputs. As can be seen in Fig. 9.4, Mining also contributed more than 20% of the carbon that was exported to the external environment. During the study period, about 63% of the output carbon was exported to the external environment and exported carbon decreased by 48%, with some fluctuation. The carbon (about 32% of the total) that was the output to Production and Supply of Electric Power and Heat, Energy Transformation, and Manufacturing also decreased by 72%, with fluctuation. The remaining 4% of the carbon was discharged during production by Mining. The amount of carbon that comprised external inputs was less than the amount of carbon output to the external environment. The carbon input mainly came from the internal environment and decreased by 64% from 1995 to 2015, with some fluctuation. This trend was mainly affected by the 55% decrease of the input energy, which accounted for more than 70% of total inputs. In addition, mineral inputs decreased by 83%, which accounted for less than 30% of inputs during the study period.

9.1.4 Comparison with Previous Research

Beijing's CII ranged from 4.5 to 8.4, and its minimum value was more than 2 times the global CII (1.8; Le Quéré et al., 2009) and 1.2 to 1.7 times the value for China as a whole (2.7 to 3.6; Piao et al., 2009), indicating that the city was a hotspot for carbon emission reduction. If we compare Beijing with other cities, the average value of CII in Beijing (6.5) was higher than that of Hangzhou (5.4; Zhao et al., 2010) and Miami (Dade County) (6.0; Escobedo et al., 2010). A possible reason for this difference could be dissimilarity in the accounting items. The Hangzhou and Miami (Dade County) studies only considered carbon sequestration by the Forest component. In contrast, the present analysis also included Other Forest, Cultivated Land, Grassland, and Wetlands, and added carbon emissions from Energy, Industrial Processes, and Waste Disposal, as well as carbon emissions from biological respiration, livestock rumination, and CH_4 release from Wetlands, and was therefore more comprehensive. In contrast, the Hangzhou study only considered carbon emission from industrial energy use, and the Miami (Dade County) study only accounted for carbon emissions from Energy and Waste Disposal.

If we use the same accounting categories as in previous studies, Beijing's CII ranged from 7.8 to 17.0 and 15.4 to 29.8 based on the Hangzhou and Miami (Dade County) accounting method, respectively. The minimum value of Beijing's CII was therefore 1.4 times that of Hangzhou, and 2.6 times that of Miami (Dade County). Compared with Hangzhou and Miami (Dade County), the average temperature in Beijing is lower, the weather is drier (with less rain), heat supply required to heat residences and workplaces is higher, and the carbon absorption capacity of the vegetation is weaker, leading to a larger carbon imbalance. In addition, Beijing has experienced faster economic development than Hangzhou, resulting in a larger carbon emission during this process. Moreover, Beijing has a lower energy efficiency than the more developed Miami (Dade County) and a higher carbon emission GDP per unit.

However, Beijing's post-industrialization process, which has consisted of transfer of industries that consume the most energy to other regions and strong growth of tertiary industries (particularly the services sector), has partially alleviated the carbon imbalance. Compared with three cities in the Republic of Korea (CII = 45.5 to 200.0; Jo, 2002), Chicago (CII = 333.3; McGraw et al., 2010), and Shenyang (CII = 384.6; Liu and Li, 2012), Beijing's CII was much lower. Since Chicago is an industrial center of the United States dominated by heavy industry, and Shenyang is a center of China's heavy industry, both cities have high energy consumption. Most of the energy requirement of various industries in the three cities of the Republic of Korean also came from coal-fired power plants, which have relatively large carbon emissions.

9.1.5 Explanations of the Research Results

The activities of Beijing's metabolic actors, and especially the key ones, led to heavy urban dependence on external resources, thereby creating a carbon imbalance. We summarized the key actors into three levels based on their CII (Table 9.1) and CEDI (Table 9.2) to help prioritize efforts to reduce carbon emission (Table 9.3). Production and Supply of Electric Power and Heat, Transportation, Urban Households, and Manufacturing, which had high values of both CII and CEDI, received the highest priority. In particular, Transportation's carbon emission was large and growing rapidly, mainly due to a rapid increase in private car ownership (which accounts for nearly half of Beijing's total vehicles). Although the government has implemented a series of measures to mitigate this problem, such as limiting the days on which a given vehicle can drive and limiting the number of vehicle permits, it has been unable to curb the growth of carbon emission by Transportation. Currently, to receive permission to buy a car in Beijing, drivers must participate in a registration lottery and only winners are granted the permission. Although the proportion of winners who were allowed to purchase a high-efficiency vehicle (e.g., electric, hybrid, or hydrogen) increased to 40% of the total registrations in 2017, the number of winners still cannot meet the demand. Therefore, development of electric and hybrid vehicles, as well as fuel cell vehicles, must be encouraged, and measures such as limits on the number of new licenses and the implementation of special roads reserved for use by these vehicles should be applied to encourage the replacement of old, energy-inefficient vehicles that consume fossil fuels.

Actors at grades II and III included Crop Cultivation, Rural Households, Energy Transformation, Mining, and Forest (absorption). These priorities included two actors with strong carbon absorption (Crop Cultivation and Forest). The reduction

Table 9.1 Key actors for Beijing from 1995 to 2015 based on their CII

	Key actors of carbon emission	Key actors of carbon absorption
High amount	Manufacturing, Production and Supply of Electric Power and Heat, Transportation, Other Services, Urban Households, Rural Households, Crop Cultivation	Forest, Crop Cultivation
Change trend	Increase	Decrease
First level	Transportation, Production and Supply of Electric Power and Heat, Urban Households, Other Services, Manufacturing, Crop Cultivation, Rural Households	Crop Cultivation
Second level	–	Forest
Third level	Retail and Catering, Waste Disposal, Mining	–

Note "Levels" include the major actors (first level) and minor but still significant actors (second and third levels). The table lists the key actors (the actors that together accounted for at least 70% of carbon emission) that most strongly affected trends in CII for the study period as a whole

Table 9.2 Key actors for Beijing from 1995 to 2015 based on their CEDI

	Key actors of net external input	Key actors of internal input
High amount	Manufacturing, Energy Transformation, Production and Supply of Electric Power and Heat, Transportation, Urban Households	Mining
Change trend	Increase	Decrease
First level	Manufacturing, Production and Supply of Electric Power and Heat, Energy Transformation, Transportation, Urban Households	Mining
Second level	–	–
Third level	Retail and Catering, Animal Husbandry	–

Note "Levels" include the major actors (first level) and minor but still significant actors (second level and third levels). Key actors (the actors that together accounted for 70% of carbon emission) were the actors that most strongly affected trends in CEDI for the study period as a whole

of the Crop Cultivation area and decreasing agricultural production per unit area (by 25% during the study period) resulted in considerably decreased agricultural carbon absorption. The main reason for this was rapid development of Beijing, leading to a conversion of agricultural land into built-up land. In order to counteract this decrease in the production, Crop Cultivation must become more modern and intensive; for example, real estate development in Farmland areas should be curtailed, and solutions such as three-dimensional agriculture (so-called "vertical farms") should be used to increase the production and agricultural carbon uptake per unit area. Importantly, Beijing's Forest actor has a considerable potential for additional development. Currently, 62% of the city's Forest is not healthy, however, if more than 90% of the

Table 9.3 Levels of the actors in Beijing's carbon metabolism

Key actor	CII	CEDI	Priority grade [a]	Regulation actor
First level	Crop Cultivation (absorption and emission), Manufacturing, Other Services, Production and Supply of Electric Power and Heat, Rural Households Transportation, Urban Households	Energy Transformation, Manufacturing, Mining, Production and Supply of Electric Power and Heat, Transportation, Urban Households	Grade I	Manufacturing, Production and Supply of Electric Power and Heat, Transportation, Urban Households
			Grade II	Crop Cultivation (absorption and emissions), Energy Transformation, Mining, Other Services, Rural Households
Second level	Forest (absorption)	–	Grade III	Forest (absorption)
Third level	Mining, Retail and Catering, Waste Disposal	Animal Husbandry, Retail and Catering	Grade IV	Retail and Catering
			Grade V	Animal Husbandry, Mining, Waste Disposal

We used the following principles for this division: actors at the first level based on both indexes receive a Grade I priority, actors at the first level based on only one of the indexes receive a Grade II priority, actors that are at the second level based on either index receive a Grade III priority, actors that are at the third level based on both indexes receive a Grade IV priority, and actors that are at the third level based on one or both indexes receive a Grade V priority. Grade I represents the highest priority for emission mitigation. CII, carbon imbalance index; CEDI, carbon external dependence index

Forest was restored to a healthy condition, it would become possible to increase the average carbon storage capacity of the trees by 100%, hence it is urgent to find means to improve forest health. This includes increased efforts to reduce air pollution, choosing species with higher pollution and stress resistance for new forests, and possibly fertilization and watering of existing forests.

Although actors at grades IV and V (Retail and Catering, Waste Disposal, Animal Husbandry, and Mining) had a smaller impact on the overall carbon metabolism, they should not be neglected. This is particularly true for Waste Disposal, which can decrease inputs of materials by their recycling and preventing outputs of materials as waste. Though such changes may be individually small, they can add up to a significant cumulative effect.

Figure 9.6 shows that while Beijing's population, GDP, GDP per capita, and built-up area increased continuously (a minimum growth of 73.5%), both the carbon emissions and net carbon inputs initially increased (by about 37%) and then decreased (by about 18%). This demonstrates that the efficiency of Beijing's carbon metabolism has improved, which is mainly a result of a significant decrease in the proportion of industries with a high carbon emission intensity. For instance, during the study period the proportion of Industry (including Manufacturing, Mining, Production and

Supply of Electric Power and Heat, and Energy Transformation) decreased from 35 to 16%, and simultaneously, the proportion of Other Services (i.e., service industries other than Transportation, and Retail and Catering) increased from 32 to 63%. This originated from upgrading and transformation of Beijing's industrial structure.

During the same period, the consumption of coal, which had a high carbon emission coefficient, decreased from 55 to 10% of energy consumption by Industry and Other Services, while the proportion of natural gas, which had a low carbon emission coefficient, increased from less than 0.3% to 20.0%, suggesting optimization and adjustment of Beijing's energy consumption structure. In fact, due to the 2008 economic crisis, emission of greenhouse gases decreased globally and in particular Beijing's carbon imbalance reduced due to decreasing carbon emissions, according to the current results. By comparing the trends for CII and CEDI with the trends for population, GDP, GDP per capita, and built-up area index, we can see that CII increased by 74% (with fluctuation) and CEDI increased to 4.0 times its value in 1995 during a period when GDP and GDP per capita increased to 15.0 and 8.8 times their values in 1995. However, the rates of population growth and expansion of Beijing's built-up area were lower than the increase in CEDI; the population grew exponentially, at about the same rate as CII (73.5%), and growth of the built-up area was about 2.6 times that of CII.

These results suggest that Beijing's carbon metabolism has not increased synchronously with carbon emission or carbon absorption while creating GDP and serving the city's population. We found that CEDI was strongly correlated with population, GDP, and GDP per capita ($r = 0.87$ to 0.90, $p < 0.05$), and moderately correlated with the city's built-up area ($r = 0.76$, $p < 0.05$), indicating that the population growth, economic development, and improved living standards have either increased Beijing's dependence on its external environment or failed to prevent this increase. Similarly, CII was moderately correlated with population, GDP, and GDP per capita ($r = 0.57$ to 0.64, $p < 0.05$), and strongly correlated with the built-up area ($r = 0.91$, $p < 0.05$), revealing that expansion of the built-up area has either increased the carbon imbalance or failed to prevent its increase.

9.2 Spatial Analysis for the Carbon Metabolism of an Urban Agglomeration

The IPCC (2006) noted that urban areas accounted for 78% of global CO_2 emission, of which about one-third resulted from land-use changes. Therefore, understanding how to reduce global carbon emission by regulating land-use changes and spatial adjustments has become an important area of research (Xia et al., 2018). For instance, researchers are exploring how to control urban spaces to promote low-carbon development (Tian et al., 2010). In 2017, China's CO_2 emission accounted for 28% of the world's total, making it the largest carbon emitter in the world, and the Beijing-Tianjin-Hebei (Jing-Jin-Ji) urban agglomeration accounts for > 20% of the

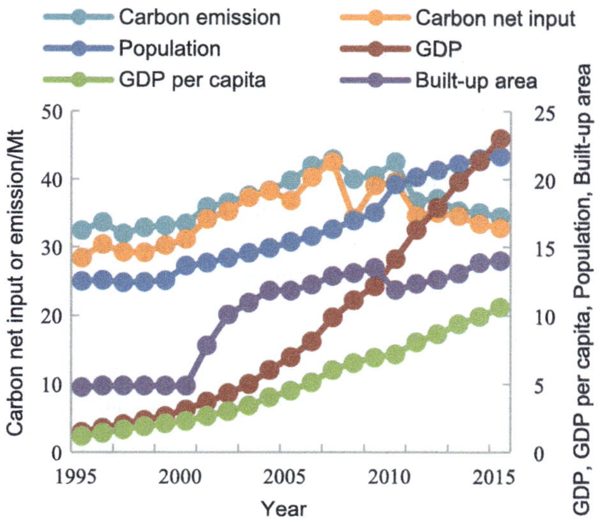

(a) The dynamic changes of the carbon input or emission versus the socioeconomic factors

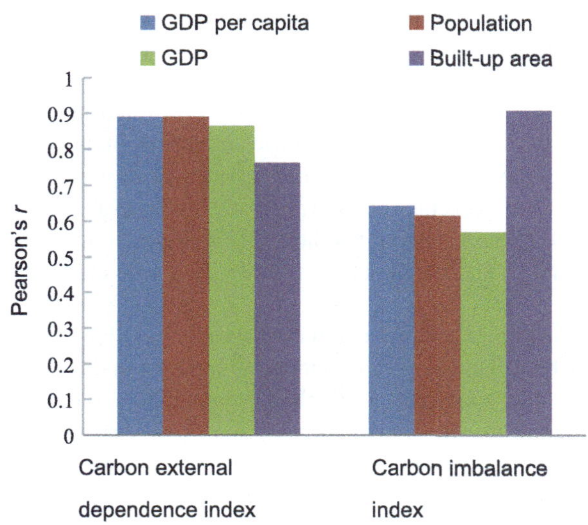

(b) The correlations (Pearson's r) between the socioeconomic factors and the carbon imbalance index (CII) and carbon external dependence index (CEDI)

Fig. 9.6 Relationships between the carbon metabolism indexes and socioeconomic factors (*Note* The unit of population is 10^6 people, the unit of GDP is 10^{10} RMB, the unit of GDP per capita is 10^4 RMB per capita, and the unit of built-up area is 10^2 km^2)

country's total emission. Land use within this agglomeration has changed frequently due to rapid socioeconomic development in recent decades (Chuai et al., 2015). Because these changes have serious consequences for CO_2 emission and absorption, improving our understanding of the underlying processes will help planners to adjust the spatial distribution of emission and absorption and thereby develop a regional approach to reduce carbon emission.

The emission and absorption capacities of the agglomeration's carbon metabolic actors vary, which leads to differences in the spatial distribution of the city's carbon metabolism and in the changes of this distribution over time. To provide deeper insights into these alterations, my research group calculated carbon emission and absorption in the agglomeration (Zhang et al., 2018a). We analyzed the spatial distribution of the agglomeration's carbon emission and absorption and examined the effects of land-use changes on the values of these indexes and their spatial patterns. The changes of land from an initial to a final use means that the land's emission and absorption capacities change to those of the final land use, which can lead to the identification of the changes in net absorption or net emission of carbon. Based on the spatial distributions of different land uses and their alterations over time, we can create a spatially explicit model of carbon emission and absorption. Our goal in creating this model was to provide empirical support for adjustment of the agglomeration's land use structure to optimize the distribution of its carbon emission and absorption and consequently to support efforts to reduce carbon emission (Fig. 9.7).

The spatial data used in this study were 1:100,000 land use vector maps from 2000, 2005, 2010, and 2015. The data were interpreted from Landsat TM images with a 30 m resolution. After image fusion, geometric correction, image enhancement, and splicing to create composite images, supervised classification was used to generate the land use maps. The mean classification accuracy for Cultivated Land, Urban Land, Rural Land, and Transportation and Industrial Land was > 85%, and the average classification accuracy for other land uses was > 75% (Xu et al., 2012; Liu et al., 2009).

To determine the impact of land-use change, we created a land transfer matrix that recorded the transfer of land between each pair of land use and cover type categories during each part of the study period. We then used the table to quantify the impacts of land-use change at 5-year intervals during our study period using the ArcGIS geographical information system (GIS) software. Further, we analyzed the spatial distribution of net carbon emission and absorption, and finally identified the key components of the metabolism that were responsible for this distribution.

In the rest of this section, I will discuss the implications of these results for regulating land use and its spatial distribution within the agglomeration and its 13 cities. To identify the major contributors to the carbon emission and absorption, I have divided the carbon emission and absorption rates into categories using the natural breakpoints method implemented by ArcGIS. Based on the maximum and minimum values during the study period, I used six grades for emission and absorption, from I to VI. Based on this approach, it is possible to identify the main components based on their contribution to the overall carbon emission and absorption, and to calculate the structure of their spatial distribution and their changes over time.

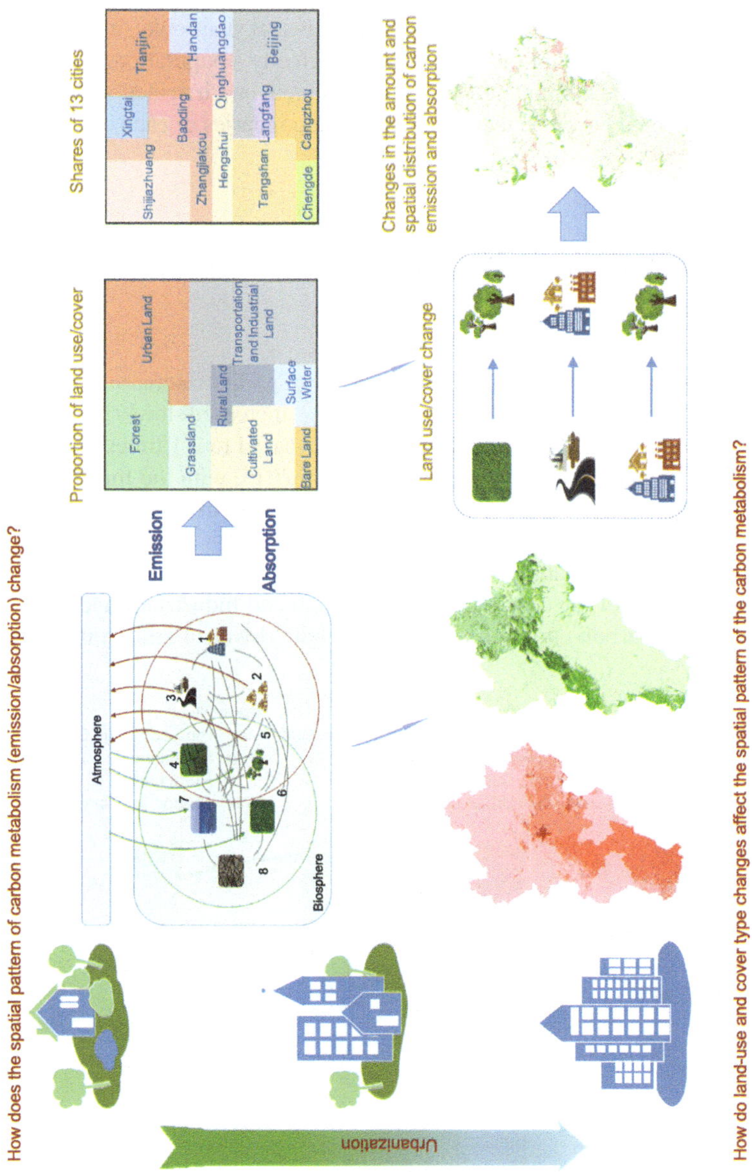

Fig. 9.7 Framework for analyzing the spatial variation of carbon absorption and emission from the carbon metabolism of the Beijing-Tianjin-Hebei urban agglomeration

The agricultural activities and energy consumption data came from the China Food Yearbook, China Energy Statistical Yearbook, Beijing Statistical Yearbook, Tianjin Statistical Yearbook, Hebei Economic Statistical Yearbook, and Hebei Cities Statistical Yearbook; however, the majority of other regions will have similar data available to support such an analysis. Using these statistical yearbooks, we collected information on energy consumption by industries, each city's population, number of livestock, fertilizer application quantities, cultivated area, and irrigated area for the cities, which we defined based on the administrative boundaries of each city (Zhang et al., 2018a).

9.2.1 Carbon Metabolism Accounting and Its Spatial Distribution

From 2000 to 2015, the values of carbon emission (C_E) first increased and then decreased, whereas the values of carbon absorption (C_A) increased slowly but steadily (Fig. 9.8). C_E reached its peak value in 2010, and increased to 2.1 times the values from 2000, and then decreased by 6.2% from 2010 to 2015, with the final rate still being 2.0 times the value from 2000. Therefore, the overall trend of C_E has been increasing.

The increase in C_E was largely caused by an increment in Transportation and Industrial Land (73.0% of the total C_E). Transportation and Industrial Land was the major contributor to carbon emission and thus the main driver of emission growth.

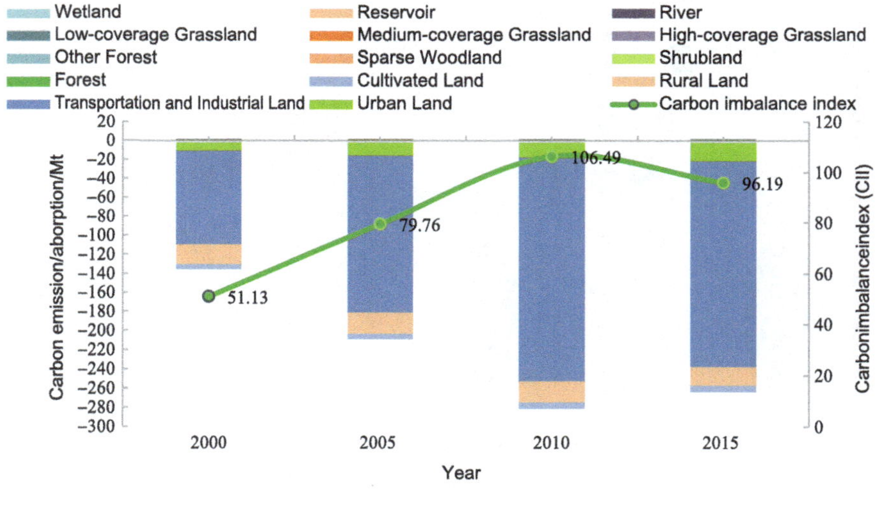

Fig. 9.8 Carbon emission (C_E) and absorption (C_A) and their components for the urban agglomeration, and the resulting carbon imbalance index (CII) (*Note* Positive values represent carbon absorption, while negative values represent carbon emission)

In 2010, the emission by this type of land increased by 138% compared with the value in 2000, which was 1.3 times the overall growth rate of C_E (108.1%). The emissions from Urban Land (about 7.0% of the total emission) increased by 70.6% from 2000 to 2010, which was 65% of the overall growth rate of C_E. Meanwhile, the contributions of Rural Land and Cultivated Land were mostly both < 10%, and their growth rates from 2000 to 2010 were 4 and 30%, respectively. Therefore, their effects on the overall growth of C_E were small. The decrease of C_E from 2010 to 2015 was also dominated by the contribution of Transportation and Industrial Land (a decrease of about 8.0%), whereas Urban Land showed the opposite trend, with an increase of 21.1%; yet, its impact on C_E was not obvious, because of the small growth rate.

The overall increase in C_A was relatively small, with an increase of only 3.8% from 2000 to 2015, mainly due to a strong increase in the Forest area (an increase of 9.8%), which accounted for 55.0% of C_A. The second contributor was High-coverage Grassland (accounting for about 11.0% of C_A), which decreased in absorption by 0.6% from 2000 to 2015, resulting in an insignificant overall increase of C_A. Although C_A of Rivers and Reservoirs increased by 24.0 and 49.3%, respectively, the C_A values of these two land uses were very low (<4% of the total), resulting in little effect on the overall increase of C_A.

During the study period, CII increased by 56.0% from 2000 to 2005, then further expanded by additional 33.5% from 2005 to 2010 (reaching 2.1 times its value from 2000), to finally decrease by 9.7% from 2010 to 2015 (Fig. 9.8). This final reduction still left the imbalance at about 1.9 times the value from 2000. C_A was only 1.0% of C_E in 2015, and exhibited an overall increasing trend. The CII increase from 2000 to 2010 resulted mainly from a 108.1% increase in C_E, while C_A changed little. The CII decrease from 2010 to 2015 stemmed from a 6.2% decrease of C_E and a 3.9% increase of C_A. However, the slow decrease of C_E from 2010 to 2015 (an annual reduction of about 1%), could not reverse the growth of C_E that occurred from 2000 to 2010 (an annual average increase of 10.8%).

9.2.1.1 Urban Contributions to Carbon Emission and Absorption

The Beijing-Tianjin-Tangshan metropolitan area and Shijiazhuang contributed the most to the overall carbon emission and absorption (Fig. 9.9). The contribution of Tangshan to C_E was relatively high, and increased during the study period, reaching its maximum (23.1% of C_E) in 2015, to 4 times its value from 2000. The contributions of Tianjin and Shijiazhuang to total C_E were not large, i.e., 19 and 9%, respectively, with a small fluctuation. Beijing's contribution to C_E gradually decreased from 18.7 to 7.6% during the study period. Conversely, the inputs of the 13 cities to C_A showed the opposite trend. Chengde made the biggest contribution, at about 33.3% of the overall C_A. Similarly, Zhangjiakou and Beijing's contributions were also relatively large, and increased steadily (by 8.1 and 6.4%, respectively); they were the highest in 2015, of 17.4 and 15.2%, respectively. Hengshui, Langfang, and Cangzhou contributed little to C_E and C_A, accounting for less than 6.0% of C_E and less than 0.8% of C_A, with

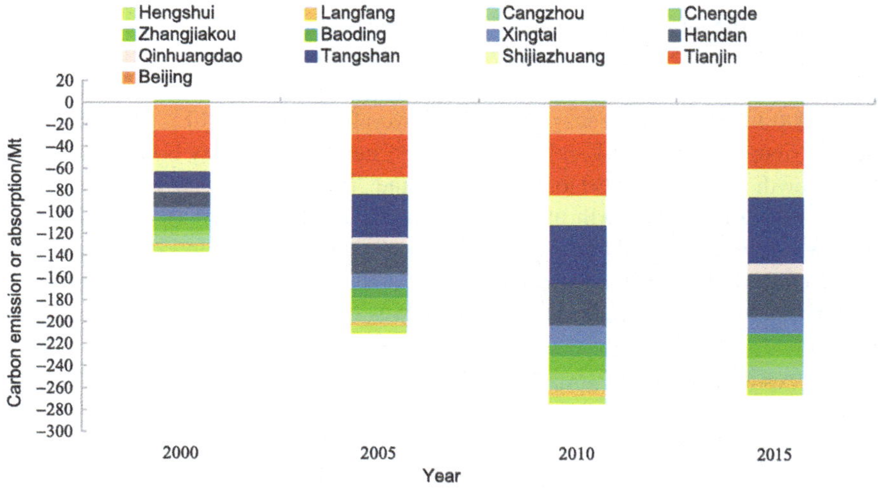

Fig. 9.9 Carbon emission (C_E) and absorption (C_A) of 13 cities in the Beijing-Tianjin-Hebei urban agglomeration (*Note* Positive values represent carbon absorption, while negative values represent carbon emission)

little fluctuation. Baoding contributed less than 8% of C_A and Qinhuangdao and Chengde contributed less than 4% of C_E.

9.2.1.2 Spatial Distribution of Carbon Emission

From 2000 to 2015, C_E and C_A showed strong spatial gradients (Fig. 9.10). Carbon emissions were low in the north and west, and high in the center, south, and southeast regions. Grade I carbon emissions (the largest values) in the urban agglomeration were mainly distributed along the coast of Bohai Bay (0.6% of the agglomeration's total area and 26% of the overall carbon emissions). Other areas with this grade were randomly spread in the northwest, where their area increased by 35% and their carbon emission increased by 20% during the study period. Gradually, the Grade I area in the Cangzhou region on the southern shore of Bohai Bay gradually decreased (by 27% in area and 10% in carbon emission). Although the high-emission area along the northern coast of Bohai Bay expanded inland from 2000 to 2010 (the area expanded by 42% and the carbon emission increased by 1.3 times), the area with high emission basically disappeared in 2015. Two centers appeared where the area of Grade I C_E increased; these areas accounted for 13 and 2% of total C_E, and were located in Beijing (at 0.1% of the area) and central Tangshan (at 0.2% of the area), respectively. The area with Grade I emission expanded into the periphery of these two areas. However, in 2010, the area of Grade I emission in Beijing's main urban area increased more obviously, to 4 times the level in 2005. C_E decreased by 66.7% in 2015, indicating that the carbon emission density (i.e., the emission per unit area) had concomitantly decreased, largely due to a reduction of the area of construction

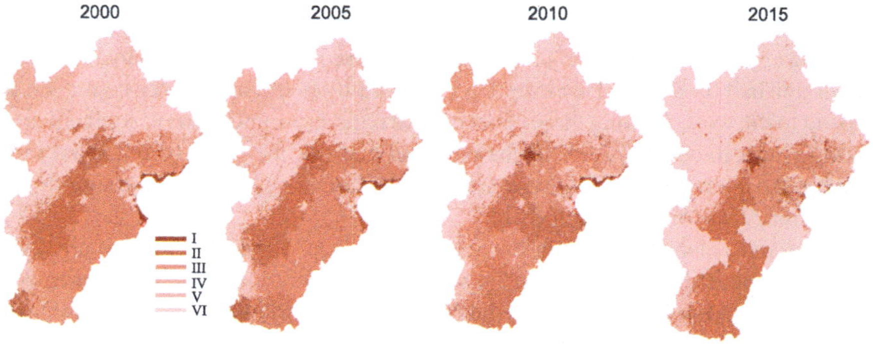

Fig. 9.10 Spatial distribution of carbon emission in the Beijing-Tianjin-Hebei urban agglomeration (*Note* Grade I represents the highest emission)

land. The area of Grade I emission in central Tangshan increased by only 12%, while C_E increased by 34% and 20%, since 2020 and by 2015, respectively, gradually forming a triangular zone (accounting for 1.2% of the agglomeration area, but 51% of the overall C_E). Inside this triangle, C_E of Grade I accounted for about 14% of the overall C_E.

Grade II carbon emission was mainly distributed east of the Taihang Mountains and in the Yanshan Mountains (which form the northwest and northern border of the study area) as well as in the northwestern part of the agglomeration (accounting for about 5% of the total area and about 30% of the overall carbon emissions). The area of this category in the central and central southern regions gradually decreased (by 40% since 2000), but carbon emission of areas with this grade increased by 130%, forming three major emission areas: Handan in the south, Baoding in the midwest, and the Tianjin-Tangshan area (accounting for 2.3% of the total area and 33% of the overall carbon emission). The area with Grade II emission in Beijing decreased (land changed to Grade I), and Grade II areas emerged on the southern shore of Bohai Bay (land change from Grade I). The area in this category on the southern shore of Bohai Bay has been decreasing and becoming more fragmented.

Grade III emission area was distributed from north to south along the eastern foothills of the Taihang Mountains (12% of the total area and 6% of the overall C_E), with sporadic distribution in the north and obvious reduction in the north (where the area decreased by 0.4%, and C_E increased by 140%). However, the southern and northern ends of this area shrank and the middle part expanded to the east, until 2015 (accounting for 17.8% of the area and 5.5% of the C_E). This area continued to merge with the southern Grade II areas (central and eastern Handan) in 2005, forming a north–south strip again that covered 18.1% of the area and accounted for 14% of C_E. The recurrence of this pattern also benefited from the northward extension of the southern Grade II zone, which covered several southern cities (an increase of 1.6% in area and 140% in C_E).

The area of Grade IV emission was mainly distributed in a strip southeast of Beijing, Tianjin, and Hebei (26% of the area and 4.5% of the C_E). Because emission in parts of Handan and Xingtai (8.3% of the area and 1.2% of C_E) and midwest Baoding (1.0% of the area and 0.3% of C_E) increased to Grade III, the area in Grade IV diminished by 83.5%. However, C_E increased by 670% as the area continued to expand to the west, forming three concentrated regions around eastern, southern, and northwestern Baoding (22.9% of the area and 6.3% of the C_E). By 2015, only the Tianjin-Tangshan region still had a concentrated area of Grade IV emission (9.6% of the area, 7.7% of the overall C_E).

There were also regions with low carbon emission. Grade V emission area was mainly distributed in the northern Jing-Jin-Ji agglomeration (13% of the area and 1% of the C_E). By 2015, the Zhangjiakou area in the northwest part of the agglomeration had a decreased Grade V area (because the emission decreased to Grade VI), however, C_E doubled. With continuous eastward expansion of the western border, only one concentrated area of Grade V emission remained in 2015, which was around Tangshan (a 91% reduction in area from the value in 2000 and a 20% reduction of C_E with this grade). Grade VI carbon emission was mainly distributed in the north, west, and on the coast of Bohai Bay, accounting for 43% of the total area but only 0.8% of C_E. In 2015, two areas with Grade VI emission (in Shijiazhuang and Cangzhou of the central region) constituted of 67% of the total area but only 2.2% of C_E, which represented an increase of 2800% compared to 2010.

9.2.1.3 Spatial Distribution of Carbon Absorption

Figure 9.11 shows the distribution of carbon absorption, with a division into six grades. C_A was highest in the north and west and lowest in the east and the south. In the north, areas with Grade I absorption were embedded in areas with Grade II and III absorption. Grade I was distributed in Zhangjiakou and Chengde, in the northern part of the agglomeration, and accounted for 19% of the total area and 61% of the total carbon absorption. By 2015, the western boundary had moved eastward and the area of Grade I carbon absorption had decreased by 70% (in Zhangjiakou it almost disappeared and became highly fragmented), and carbon absorption by land in this grade comprised for only 5.8% of the overall carbon absorption.

The area with Grade II carbon absorption was mainly distributed in a contiguous area in the north (4.8% of the area and 0.1% of carbon absorption) and in the west (5.7% of the area and 0.2% of carbon absorption), with fragmentation in the north and relative accumulation in the west. Over time, the northern area in this grade progressively reduced (48% of the decrease in the area and 29% of the decrease in the carbon absorption), while the southward expansion of the western area evidently increased. Nevertheless, carbon absorption decreased to Grade III near Shijiazhuang, and the western strip became fragmented. The western area of Grade II carbon absorption remained mostly unchanged (at about 5.5% of the area in 2015), but the overall area and carbon absorption still accounted for 8.7 and 25% of the total, respectively. The area with Grade III carbon absorption was mainly distributed in the Beijing-Tianjin-Tangshan area in the mid-east (only 0.1% of the area and 3.4% of carbon absorption), but gradually expanded to the north and west (accounting for

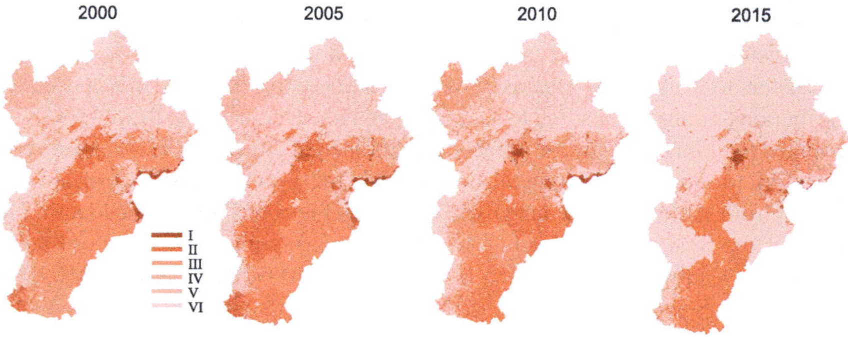

Fig. 9.11 Spatial distribution of carbon absorption in the Beijing-Tianjin-Hebei urban agglomeration (*Note* Grade I represents the highest absorption)

3% of the area and 5.3% of carbon absorption). By 2015, the area of this category in the north had considerably decreased (by 97%), accompanied by a 23% drop in carbon absorption, but it also reduced in the west (by 7.8% in area and 5.1% in carbon absorption).

The areas with Grade IV and V carbon absorption were randomly distributed in the central area (0.5% and 0.2%, respectively, for area, and 1.4 and 0.1% for C_A, respectively), and in the north (1.1 and 0.4% for area, respectively, and 2.1 and 0.4% for C_A, respectively). This mainly resulted from Grade VI areas in the north; by 2015, these regions had significantly increased (the area increased by 700% and 1600%, respectively, and C_A decreased by 23.6 and 8.1%, respectively) due to the decreased area of Grade I absorption. Overall, both the proportions of the total area and C_A increased (5.7% and 5.6% of the area, respectively, and 6.6% and 1.5% of the C_A, respectively). The area in Grade VI was widely distributed (61% of the area but only 0.6% of C_A), mainly in the east and south, with more fragmentation in the north and a sporadic distribution. During the study period, the spatial gradient did not change substantially, but by 2015, C_A had tripled compared with 2000 and accounted for 2.5% of C_A in the agglomeration.

9.2.2 *Impact of Land Use Changes on the Carbon Emission and Absorption*

Figure 9.12 shows the changes in carbon emission or absorption caused by land-use changes. Carbon emission increased steadily, mainly due to land transfers to Transportation and Industrial Land (accounting for about 90.0% of the total transfers leading to net emission); during the study period, the area of this type of land use increased to 1.3 times its value in 2000, and this increase was the main driver of the carbon net emission growth. The transfers to Urban Land and Rural Land accounted for < 6% of the total transfers leading to net emission. The net emission resulting from transfers to Urban Land increased by 28% during the study period, whereas the transfers to Rural Land decreased by 90%, leading to a decrease in Rural Land's contribution to transfers leading to net emission. However, because Rural Land's use accounted for a small proportion of the total, this decrease did not influence the overall increase in net emission. Among the transfers to Transportation and Industrial Land, transfers from socioeconomic land uses dominated the contributions (accounting for > 60% of the total transfers leading to net emission), with an increase of 88% in the transfers leading to net emission from the 2000–2005 period to the 2010–2015 period.

The transfers from Cultivated Land to Transportation and Industrial Land dominated the 2000–2005 and 2010–2015 periods, accounting for about 50% of the transfers that led to net emission, in comparison to only 15% in the 2005–2010 period. The net emission caused by changes in the carbon density (i.e., changes in the net emission per unit area) to the high value for Transportation and Industrial Land dominated the 2005–2010 period, accounting for about 50% of the overall transfers leading to net emission, versus 35% in the 2000–2005 period. However, the net emission resulting from self-transitions of Transportation and Industrial Land (i.e., changes between subcategories of these actors, such as from steel production to computer manufacturing) represented net absorption (i.e., reduced carbon emission or carbon density).

The contribution of net emission due to transfers from Rural Land to Transportation and Industrial Land increased greatly in the 2005–2010 period (to 24 times the corresponding transfers in the 2000–2005 period, reaching 11% of the total transfers leading to net emission in the 2005–2010 period). At the end of the study period

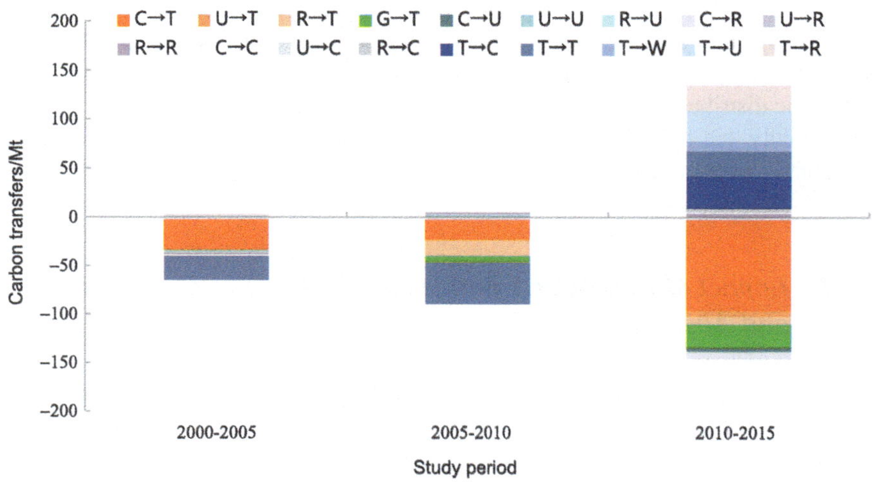

Fig. 9.12 Positive (i.e., leading to net carbon absorption) and negative (i.e., leading to net carbon emission) transfers among land uses from 2000 to 2015 (*Note* C-Cultivated Land; G-Grassland; R-Rural Land; T-Transportation and Industrial Land; U-Urban Land; W-Surface Water)

(2010–2015), net emission resulting from transfers from Rural Land to Transportation and Industrial Land were relatively insignificant, accounting for only 4.5% of the overall transfers leading to net emission. In addition, net emission resulting from transfers from Urban Land to Transportation and Industrial Land increased more evidently during the last part of the study (to nearly 500 times the value in 2000), accounting for 3% of the total transfers leading to net emission. The main natural component that became Transportation and Industrial Land was Grassland, comprising 60% of the transfers from natural components, with the transfer of Grassland increasing to 3 times the area in 2000 that was transferred, and accounting for 13% of the overall transfers leading to net emission.

The transfers leading to net absorption were relatively few and did not change significantly in the 2000–2005 and 2005–2010 periods. The main contribution resulted from transfers of Transportation and Industrial Land to Urban Land (T → U). However, the transfers leading to net absorption increased sharply in the 2010–2015 period, reaching 21 times the value of the 2005–2010 period. These transfers mainly resulted from self-transfers from Transportation and Industrial Land to the same land use (T → T, accounting for about 92% of the total transfers leading to net absorption and for a 12% decrease in the emission caused by Transportation and Industrial Land). Nonetheless, the small contribution of transfers from Rural Land to the same land use (R → R) accounted for only 2.4% of the total transfers leading to net absorption (a decrease in the carbon density), and transfers from Rural Land to Cultivated Land (R → C) accounted for only 2% of the transfers leading to net absorption. About 60% of the total transfers leading to net absorption were coming from Transportation and Industrial Land to socioeconomic land types, including transfers to Urban Land (16.5% of the overall transfers leading to net absorption), to Cultivated Land (17.3%), to Rural Land (13.2%), and to Transportation and Industrial Land (13.0%), with little difference among these transfer categories. Of all the transfers of Transportation and Industrial Land to the natural components of the system, Surface Water received most of them (4.8% of the total transfers leading to net absorption).

9.2.2.1 Contributions of Individual Cities to Carbon Emission or Absorption

Tianjin, Tangshan, Beijing, Shijiazhuang, and Handan contributed more than 76% of the total transfers leading to net emission (Fig. 9.13), however, the changes in these contributions differed among the cities during the study period. Overall, Tianjin and Tangshan contributed the most, but their shares of the total decreased from 46% in the 2000–2005 period to 36% in the 2010–2015 period. Their transfers leading to net emission decreased by 20% in the 2010–2015 period compared to the 2000–2005 period. These declining trends for Tianjin and Tangshan were mainly due to the increasing contributions from Beijing (to 5% of the total in the 2010–2015 period), Shijiazhuang (11%), and Handan (17%).

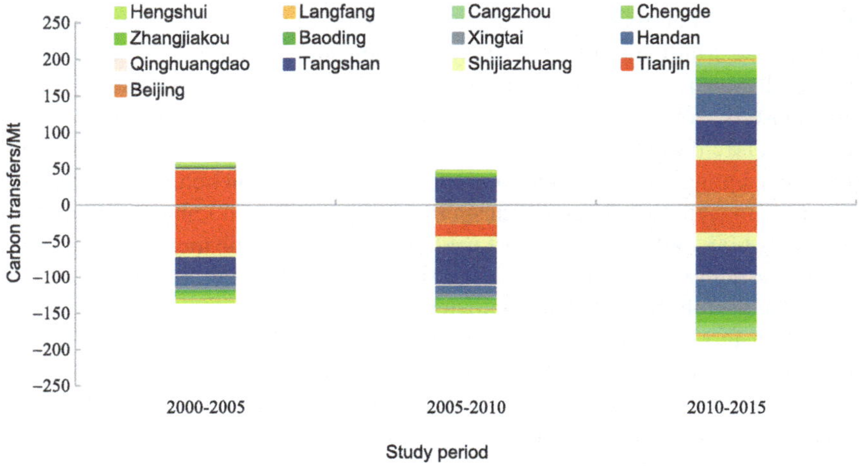

Fig. 9.13 Carbon transfers leading to net emission and absorption by 13 cities in the Beijing-Tianjin-Hebei urban agglomeration (*Note* Positive values represent carbon absorption, while negative values represent carbon emission)

Shijiazhuang increased its contribution throughout the study period, with emission reaching 2.5 times its value in the 2000–2005 period, whereas Beijing and Handan showed opposite trends. Handan's transfers leading to net emission first decreased (by 25% compared with the value in the 2000–2005 period) and then increased (to 1.9 times the value in the 2005–2010 period). Beijing's contribution initially increased in 2005–2010 (to 3 times its value in the 2000–2005 period) and then decreased in 2010–2015 (to 64% of its value in the 2005–2010 period).

Excluding Beijing, the cities that significantly contributed to transfers leading to net emission also made a considerable contribution to transfers leading to net absorption, accounting for more than 65% of these transfers. Generally, transfers leading to net absorption increased to 13 times the starting value during the study period. From 2000 to 2005, Beijing, Tianjin, and Shijiazhuang accounted for more than 15% of the total. From 2005 to 2010, Shijiazhuang accounted for the largest proportion (22%), while the remaining cities accounted for less than 10%. Finally, from 2010 to 2015, Tianjin, Tangshan, and Handan contributed the most, each accounting for more than 15% of the total.

9.2.2.2 Spatial Distribution of Carbon Transfers

Transfers within urban agglomeration leading to net emission increased gradually, and their spatial distribution shifted from dispersed to aggregated and then to dispersed again (Fig. 9.14). From 2005 to 2010, the area of these transfers decreased by 91%, whereas the net emission slightly increased (by 10%) compared to the 2000–2005 period. In the 2010–2015 period, both the transferred area and the net

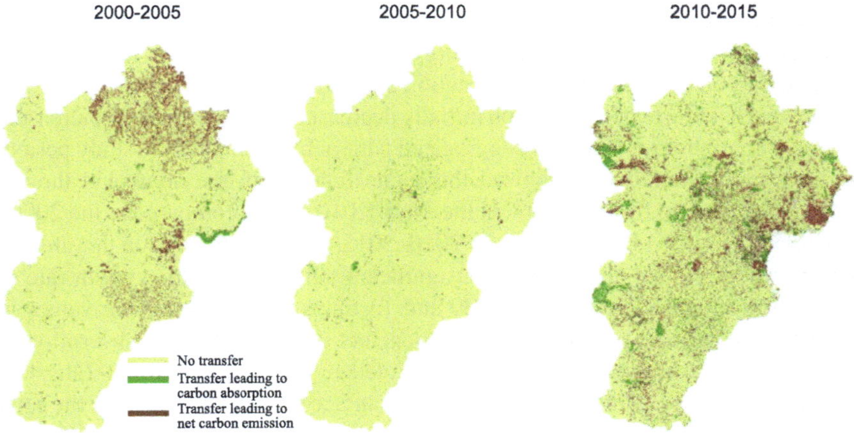

Fig. 9.14 Spatial distribution of the transfers leading to net emission (C_E) or absorption (C_A) caused by land-use change in the Beijing-Tianjin-Hebei urban agglomeration

emission all increased, but the transferred area grew faster than the net emission (which increased to 18.7 times the value in the 2005–2010 period and by 26%, respectively). In the 2000–2005 period, the transfers leading to net emission were mainly concentrated in the northeast (accounting for 5.5% of the total area). For instance, in Chengde and Cangzhou they accounted for 64.0% and 12.6% of the area with transfers leading to net emission, respectively, but the transfers leading to net emission accounted for < 1.6% of the total. The vast majority of the transfers leading to net emission were distributed in highly fragmented areas around Beijing's main urban zone (12% of the total transfers leading to net emission) and Tianjin (29% of the total transfers leading to net emission), constituting for 4.3% and 13.8%, respectively, of the areas with transfers leading to net emission.

In the 2005–2010 period, the pattern disappeared and the area of transfers leading to net emission decreased drastically. These transfers were only sporadically distributed in the Beijing-Tianjin-Tangshan and Shijiazhuang areas, because of China's hosting of the 2008 Beijing Olympic Games. To improve the environmental conditions before the games, the government requested to reduce the scale and frequency of many industrial activities in the Beijing-Tianjin-Tangshan and Shijiazhuang areas. As a result, the energy consumption by production activities, living activities, and livestock decreased, resulting in a reduction in the area of transfers leading to net emission.

In the 2010–2015 period, transfers leading to net emission became again more evident (accounting for 9.5% of the area of transfers leading to net emission, which was 19 times the value in the 2005–2010 period). The fragmented pattern of transfers leading to net emission spread throughout the Jing-Jin-Ji agglomeration and were mainly due to changes in the carbon density of the Transportation and Industrial Land. After the Beijing Olympics in 2008, the magnitude of the socioeconomic activities in various areas of the Jing-Jin-Ji urban agglomeration increased significantly. Therefore, the population and energy consumption has grown, leading to the increased carbon emission by Transportation and Industrial Land.

Transfers leading to net absorption initially decreased and subsequently increased, showing a continuous, moderately aggregated pattern by the end of the study period. The area of transferred land showed the same trend. Both the areas and the net absorption decreased by about 50% in the 2005–2010 period compared to the 2000–2005 period (Fig. 9.14). However, in 2010–2015, the growth rate of the area of transfers leading to net absorption was significantly higher than the growth rate of the net absorption (they increased to 20 and 13 times the corresponding values in the 2005–2010 period, respectively). The transfers leading to carbon absorption in the 2000–2005 period were mainly concentrated in the Beijing-Tianjin-Tangshan metropolitan areas. During this period, Tangshan accounted for 67.6% of the total area of these transfers, but the associated transfers accounted for only 5.4% of the total. The Beijing-Tianjin region accounted for < 20% of the total area of these transfers leading to carbon absorption, but the associated transfers accounted for > 80% of the total positive transfers. Gradually, the movement of these transfers towards the central and southern regions was mainly concentrated around the urban areas of Beijing and Shijiazhuang (accounting for 87% of the area and 67% of the net absorption).

In the 2010–2015 period, the transfers leading to net absorption became evident (6.9% of the area of transfers), and the fragmented distribution of these transfers covered most of the urban agglomeration. They were mainly concentrated in the Beijing-Tianjin-Tangshan area, and the Zhangjiakou and Shijiazhuang areas (68.8% of the area of transfers leading to absorption), accounting for 72% of the total net absorption.

9.2.3 Comparison of Carbon Spatial Variation with Other Studies

Previous research on urban carbon accounting generally focused on calculating carbon emission and absorption, but only few studies considered their spatial distribution. As a consequence, those studies provided limited guidance for the adjustment of urban land use patterns to minimize carbon emission (Xia et al., 2016). Analyzing the spatial patterns of urban carbon emission and absorption revealed areas where carbon emission and absorption were occurring, thereby providing a preliminary basis for more effective land-use management and for adjustment of the spatial distribution of a city's carbon emission and absorption (Zhang et al., 2014). For this reason, it is essential to understand how changes of land use affect the spatial pattern of carbon emission and absorption in order to provide a scientific basis for regulating land-use changes.

Some scholars have analyzed the spatial distribution of emission and absorption from a macroeconomic perspective. For example, Nowak and Crane (2002) confirmed that the carbon absorption by forests in cities of the northern, northeastern, south central, and southeastern United States was higher for cities in other areas. This difference resulted from the smaller scale of socioeconomic activities in these cities. Tao et al. (2015) found that Changzhou's Tianning district (the city's center, with high socioeconomic activity) had the highest carbon emission, whereas its Wujin and Xinbei districts (both close to the suburbs) had the lowest emission, which was caused by differences in economic development among the districts. We observed a similar pattern in the present study, i.e., carbon emissions were low in the north and west and high in the center, south, and southeast areas. This was mainly because the central, eastern, and southern parts of Beijing, Tianjin, and Hebei are mostly plains and construction land, with more complex socioeconomic activities and thus more carbon emissions. In contrast, the higher carbon absorption rates in the western and northern regions were related to the larger areas of forest and grassland.

Scientists have used software such as ArcGIS to analyze the sequential changes of carbon emission and absorption and to display the spatial distribution of the environmental effects (e.g., Harrison and Haklay, 2002). Hutyra et al. (2011) found gradients in carbon absorption and emission in buffer zones at different distances from the center of Seattle: the farther away from the city center, the lower the carbon emission. In contrast, Tao et al. (2015) discovered that the carbon emission density and carbon stocks gradually increased moving from urban land to the suburbs in Changzhou, China. This could be explained by the fact that industrial activity in Changzhou is concentrated in a ring around the city rather than at its center. Using GIS, researchers have drawn contour maps of carbon emission and absorption and used them to analyze the characteristics of the spatial distribution. For example, Zhang et al. (2016) observed that Beijing's carbon emission showed a single-center pattern, whereas carbon absorption exhibited a multicenter pattern; these patterns corresponded to the spatial concentration of industrial activities combined with the dispersed natural areas that are capable of absorbing carbon. Comparably, Chrysoulakis et al. (2013) found that Helsinki's carbon absorption showed a multicenter pattern. In line with this finding, we made a similar observation in the present study. Specifically, carbon emission showed multiple centers along the coast of Bohai Bay and in the Beijing-Tianjin-Tangshan area, and carbon absorption also displayed a multicenter pattern in the northern and western regions of the agglomeration.

9.2.4 Comparison of the Impact of Land Use Change on Carbon Throughput with Previous Research

Changes in land-use patterns will alter carbon emission and absorption within a region. Bolin (1977) studied the relationship between land-use change and the terrestrial carbon cycle, and concluded that changes from predominantly natural to predominantly socioeconomic components led to an increase of the atmospheric CO_2 content. Moreover, Houghton (2003) suggested that restoring forests and grassland could increase carbon absorption. Missing carbon sink by forests (Dixon et al., 1994) and the subsequent conversion of grassland and forest into farmland (Houghton and Goodale, 2004) indirectly increased carbon emission. Since the 1980s, many land types have changed to forest in Europe (Kauppi et al., 1992). In the United States, between 1850 and 2000, much farmland was returned to forest, thereby increasing carbon absorption (Houghton et al., 2000). In contrast, in China in the 1980s, a large number of planted forests and other natural land-use types were converted into farmland, which led to increased carbon emission (Houghton, 2003). On an urban scale, Svirejeva-Hopkins and Schellnhuber (2008) noted that many cities in the Asia–Pacific region were significant carbon sources, which was mainly caused by the conversion of forests to urban land.

Carbon emission resulting from land-use change is always closely related to the expansion of construction land, including the one being developed for Transportation, Mining, Urban Households, and Rural Households (Dhakal, 2009). Chuai et al. (2015) showed that the main land use conversion in China's Jiangsu coastal area from 1985 to 2010 was dominated by the conversion of cultivated land to construction land, leading to an obvious increase of carbon emission. Conversely, the transfer of Transportation, Industrial, and Mining land to other forms of construction land, such as Urban Land, Rural Land, and Cultivated Land, generally decreased carbon emission. We found similar results in the present study. The shift of cultivated land to Transportation and Industrial Land and changes (increases) in the carbon density of a given land type accounted for most of the carbon emission. In addition, Chuai et al. (2015) demonstrated that the transfer from grassland to construction land was important (accounting for about 7% of total transfers), which is consistent with our results, particularly in 2015 for the Jing-Jin-Ji agglomeration, where the transfer of Grassland to Transportation and Industrial Land, Urban Land, and Rural Land accounted for 15% of the transfers leading to net emission.

The analysis in Sect. 9.2 focused only on carbon emission and absorption based on changes in land use. However, additional work is needed to describe other aspects of these changes, such as the underlying processes responsible for the changes in emission and absorption. This will require more detailed data on the changes in the carbon stock of each land use type, and how that stock is changing through fluctuations in the carbon flows. In particular, we did not account for seasonal and longer-term alterations in the carbon cycles of natural components of the agglomeration, such as Forest and Grassland, which show both seasonal and long-term changes in canopy density and vegetation cover over periods as long as the 15 years of the present study; these changes would affect the estimates of their emission and absorption capacity. In future research, it will be necessary to use such improved estimates of the coefficients used to account for carbon in the current study and how these coefficients change over time.

9.3 Spatial Network Analysis of Beijing's Carbon Metabolism

One-third of carbon emission in the urban area results from land-use and land-cover changes (IPCC, 2006), and the absorption of carbon is also closely related to the land-cover status of the natural environment. In addition to the carbon absorption (inputs) and carbon emission (outputs) of the urban system and its atmosphere, the carbon transfer processes between a city's socioeconomic and natural subsystems and within the natural environments are directly related to land-use and land-cover changes. Beijing, the capital of China, has undergone rapid urbanization accompanied by dramatic changes in the land-use. Nearly 20% of the city's farmland had been converted to built-up land in the year 1992 to 2008 (Miao et al., 2011). This process leads to prominent development contradictions in Beijing due to the reduction in the available natural resources of the system that support socioeconomic development. According to literature (Xia et al., 2016), our research group analyzed Beijing's carbon conversion processes from the perspective of urban metabolism and explored the responsible interactions among the city's metabolic actors.

In order to support these analyses, a land-transfer matrix is created and used to analyze the changes in the land's carbon metabolism capacity by taking the changes in the carbon stock of the metabolic actors, and to quantify the exchanges between actors into account. To perform this analysis, a spatially explicit analytical framework (Fig. 9.15) is developed to model the carbon fluxes of the natural and socioeconomic components of an urban ecosystem based on their roles in the city's carbon metabolism. By using the model, the spatial distribution of the metabolic fluxes and their correlation is simulated. Such research can figure out the differences in the contributions ofrom the respective metabolic actors and can provide a more solid empirical basis for improving the use of the urban space and reducing its carbon emission.

The carbon emission and absorption rates (kg C/a) are calculated at 5-year intervals in the years 1990 to 2010. Here, we used government statistical data for Beijing, and the empirical coefficients for the carbon contents of each type of land-use are calculated for carbon emission (including human respiration system) and absorption. Further, the land-use transfer matrix is used to map the changes in carbon emission or absorption for each type of land-use and land-cover change. Based on the change in the emission or absorption intensity between the old and new land-use, the resulting carbon emission or absorption is calculated that resulted from land transfers between different metabolic actors. In addition, the network-flow analysis is used to calculate the integrated (i.e., direct plus indirect) network flows, to identify the different contributions to these flows due to the absorption and emission (inputs and outputs) by the different metabolic actors, and to simulate their spatial patterns and changes in their ecological relationships (Xia et al., 2017b, 2016). The results provide a more scientific basis for the regulation of the spatial distribution of urban carbon emission and absorption.

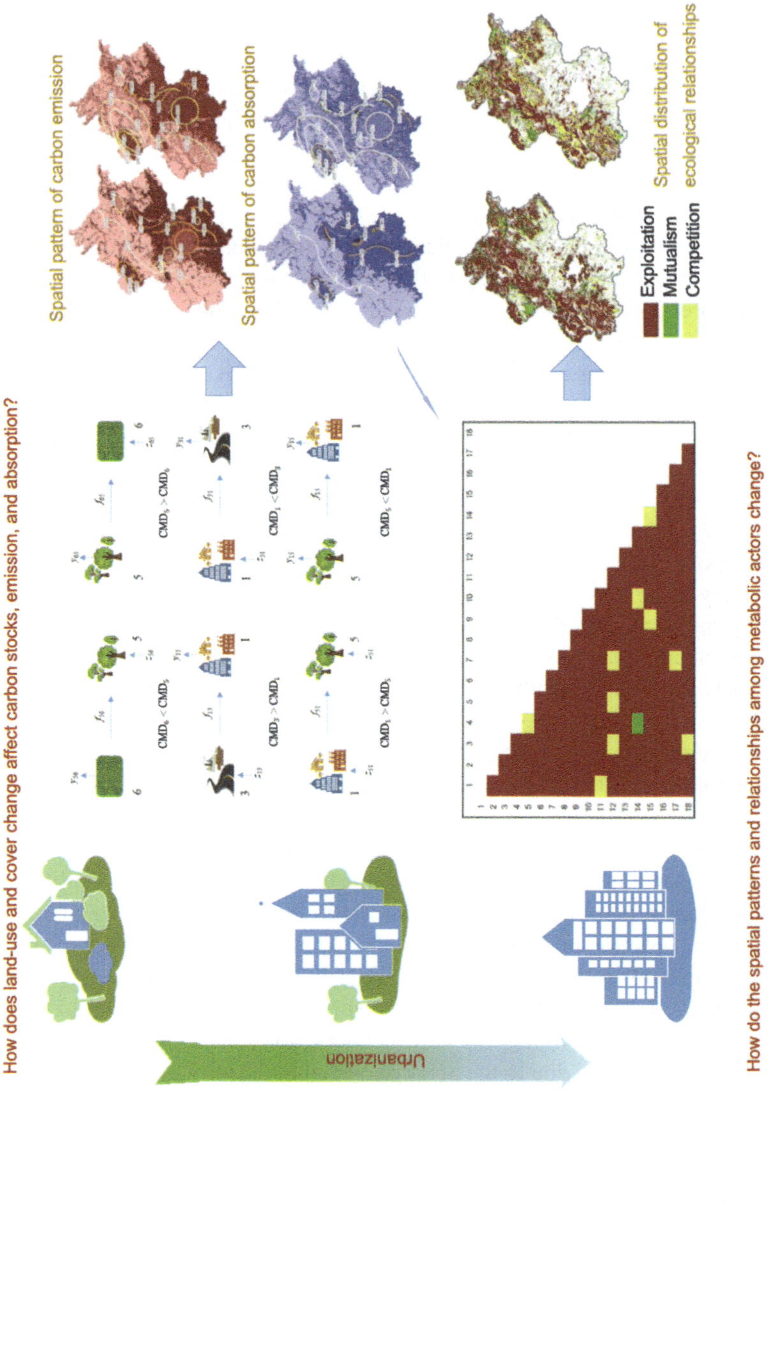

Fig. 9.15 Illustration of the analytical framework for spatially explicit analysis of the urban carbon metabolic network

9.3.1 General Spatial Characteristics

The total integrated carbon flows through Beijing's urban network had been decreased in the year 1990 to 2010 at an average annual rate of 7.5%. The total flows had been recorded as 41.3 × 10^8 kgC/a in the year 1990 to 1995 and 36.8 × 10^8 kgC/a in the year 1995 to 2000, respectively. The rate of land transfer between the system's components has been decreasing since 2000, resulted in the decreased integrated flows further. The total integrated flows had been recorded as 11.1 × 10^8 kgC/a in the year 2000 to 2005 and 8.8 × 10^8 kgC/a from the year 2005 to 2010, respectively. The spatial patterns were obtained by means of GIS analysis of the integrated flows. From the year 1990 to 2010, the GIS maps for the input and output flows revealed the spatial gradients from the city's center in the southeastern plains to the perimeter of the plains and from that western periphery to the northwestern mountains. Most input and output flows were centered on the southeastern plains with relatively small flows in the northwest. The average integrated output flow in the region (plains) with high-output was approximately 16 times the flow in the regions (mountains) with low-output. The average integrated input flow in the high-input region was approximately 12 times the flow in the low-input region (Xia et al., 2016).

The total area with the highest flow in the southeastern plains decreased in the year 2000 (Fig. 9.16). The distribution became more fragmented with many lower-flow patches dispersed within the area with the high-output flow. The main change in the integrated output flow occurred in the year 2005. From the year 1990 to 2005, the area with the highest output flows (1000 to 4500 kt C/a) was at the perimeter of the southeastern plains. There were multiple centers with high-output flows, and these flows were decreased in two directions, namely from the perimeter of the southeastern plains towards the center of the study area and towards the northwestern mountains. The output flow formed a different pattern in the year 2005 to 2010, decreasing to less than 1000 kt C/a at the perimeter of the southeastern plains. The area with the highest flow in the center of southeastern plains became an aggregated center with decreasing flows from the southeast towards the northwest.

The input flows were decreased significantly throughout the city after the year 1995 (Fig. 9.17). The input flows steadily decreased within the southeastern plains, but slightly increased in the northwest. In the year 1990 to 2010, the difference between the areas with high-input flows (plains) and low-input flows (mountains) was decreased from 20 times (1990 to 1995) to 4 times (2005 to 2010). Unlike the gradients of the output flows, the main change in the spatial distribution of the integrated input flows occurred in the year 1995. From 1990 to 1995, the flows in the highest category (1400 to 5000 kt C/a) occurred in the southeast plains, and the flows in the second-lowest category (76 to 235 kt C/a) occupied most of the mountains. The input flows therefore decreased moving from the southeastern plains to the northwestern mountains, except for one high-flow area in the mountains. After the year 1995, several patches in the high-flow category were distributed in the northeastern periphery of the plains and became the centers of spatial aggregation of these flows. The decrement resulted from the changes in the input flows in the

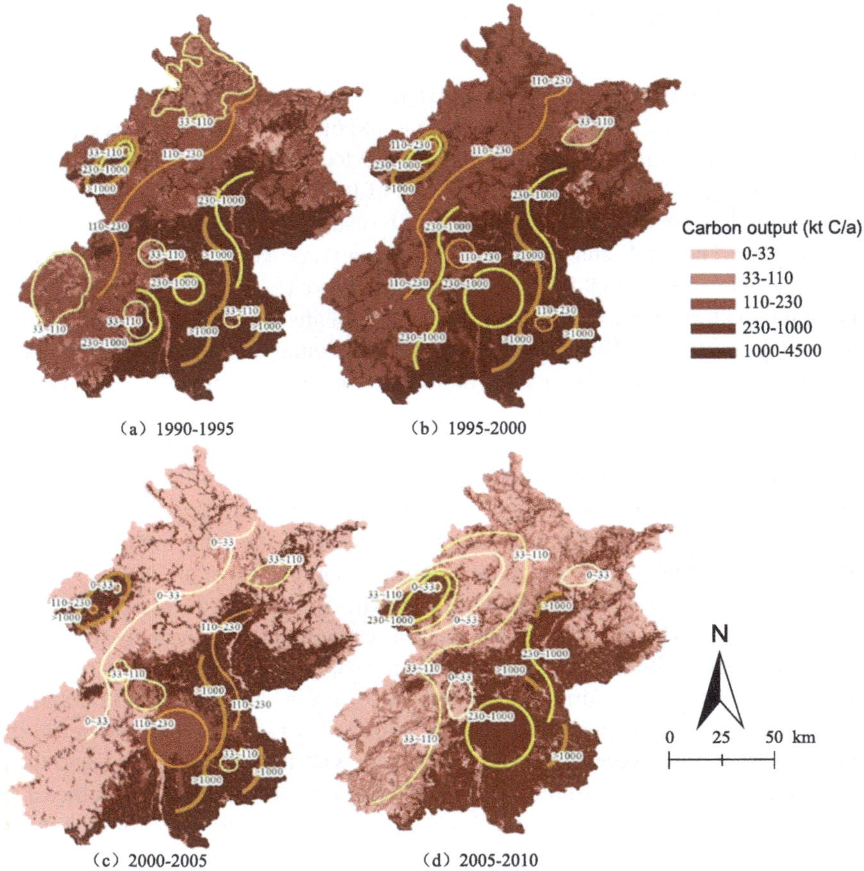

Fig. 9.16 Changes in the spatial distribution of integrated carbon outputs

two directions i.e., a particularly strong reduction from the eastern periphery to the built-up areas and mountains and a normal reduction from the built-up areas to the mountains.

The variations in the input flows occurred due to a series of policies, which were implemented in Beijing before in the year 1990, resulted in a unique change from 1990 to 1995. Beijing has been focusing on industrial and commercial development since the national government's "Reform and Opening-up Policy" was implemented in 1978. Large amounts of materials and energy resources were imported for industry and commerce, especially in urban areas, which were developed rapidly in the year 1990 to 1995. Simultaneously, the implementation of large-scale ecological policies such as the "Three-North Shelterbelt Project" and the "Beijing Forest Resources Protection Regulation" caused an increment in the input flows in the northwestern part of the study area. After 1995, regulation of urban growth and preservation of farmland have been increasing and emphasized under policies such as the "General

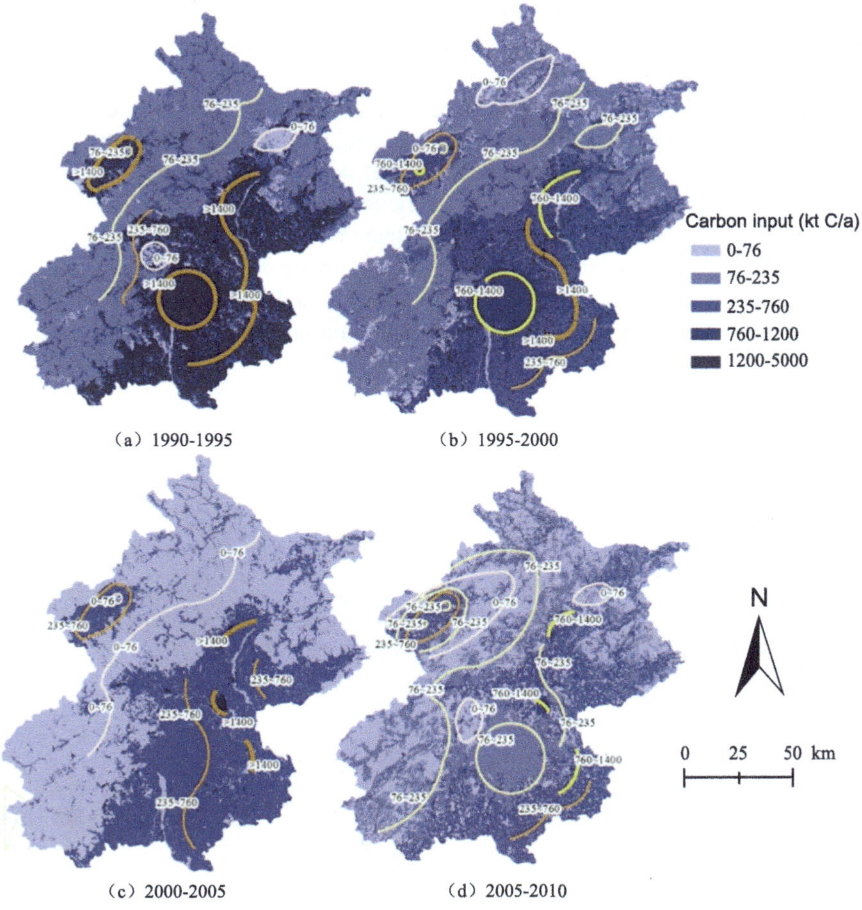

Fig. 9.17 Changes in the spatial distribution of the integrated carbon input flows

Plan for the Land Utilization of Beijing City from 2006 to 2020". The implementation of these policies has altered the center of the input flows from the core area to the southeastern periphery.

9.3.2 Ecological Relationships and Their Spatial Patterns

Figure 9.18 shows the distribution of all the correlated components in the model of Beijing's carbon flows. The spatial distribution has been changed over time as land-use changed, and the dominant components for each relationship also changed. The system's artificial components were dominated the exploitation relationships in the four-time periods, while the key components of the mutualism relationships changed from natural components and cultivated land to socioeconomic components with a low carbon density, and the competition relationships preferentially produced by the natural components and cultivated land (Xia et al., 2018).

The total number of the exploitation relationships had been decreased from 1990 to 2010 and the proportion of the exploitation relationships produced by artificial components increased i.e., accounting for 39.2, 47.0, 70.3, and 66.7% of the total number of the exploitation relationships in the four periods, respectively. In contrast, the number of the competition relationships increased continuously, and the proportion of the competition relationships produced by natural components and Cultivated Land increased i.e., accounting for 74.4, 90.7, 93.2, and 97.5%, respectively, in the four periods. Although the total number of the mutualism relationships increased

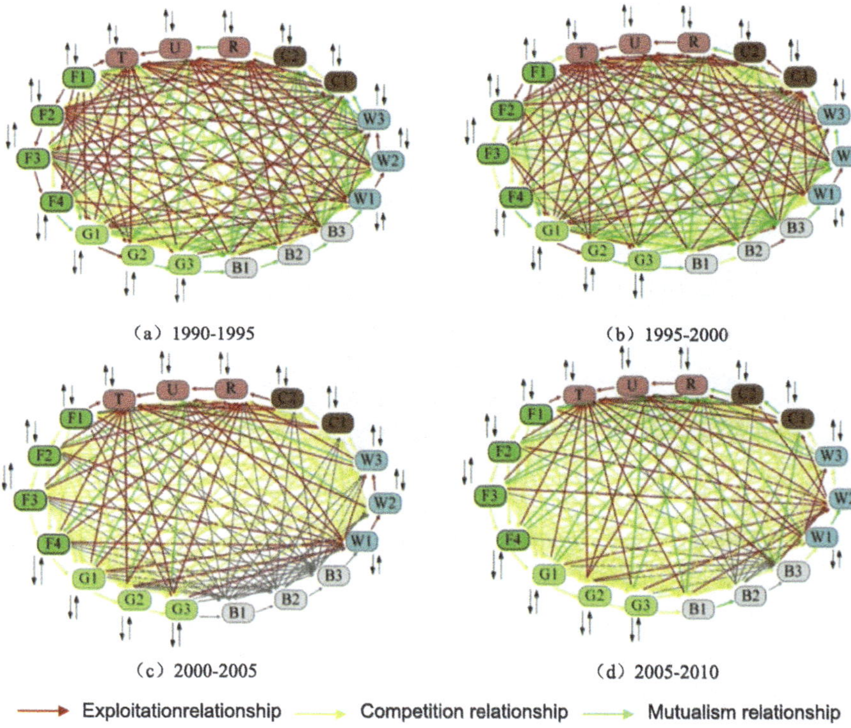

Fig. 9.18 Changes in the ecological relationships among the metabolic actors in the 18-node model of Beijing's carbon metabolism from 1990 to 2010 (*Note* Sand [B1], Barren Earth [B2], Bare Exposed Rock [B3], Paddy Land [C1], Dry Cultivated Land [C2], Forest [F1], Shrubland [F2], Sparse Woodland [F3], Other Forest [F4], High-coverage Grassland [G1, vegetation cover more than 50%], Medium-coverage Grassland [G2, vegetation cover between 20 and 50%], Low-coverage Grassland [G3, vegetation cover between 5 and 20%], Rural Land [R],Transportation and Industrial Land [T], Urban Land [U, urban built-up area], River [W1], Reservoir [W2], Wetland [W3])

slowly, and the proportion of mutualism relationships between natural and low-density artificial components (ex. Rural Land) increased by 22.5, 28.5, 100, and 97.7% of the relationships in the four periods, respectively.

During the studied period, spatial distribution of the ecological relationships based on the change in the land-use decreased, and the relationships became aggregated in and near the southeastern plains (Fig. 9.19). The exploitation, mutualism, and competition relationships formed an overlapping pattern in the mountainous area in the northwest, and the total area of all the relationships had been decreased sharply since 2000. However, the exploitation relationships occupied most of the area in the southeastern plains, and some mutualism relationships had been formed there since 2005. During the studied period, the exploitation relationships in the northwest mountainous area and the southeast plains showed large differences in the pattern between the northwest and southeast with the exploitation relationships occupying 50% of the total relationships. In contrast, the mutualism and competition relationships occupied a small area and spread slowly throughout the study area. Most of the relationships had been distributed throughout the southeastern plains since 2000, and approximately 90% of the exploitation relationships were distributed in the southeastern plains. This unique pattern reflected the exploitation of other components by the artificial components during urbanization.

The distribution of the mutualistic relationships had been gradually shrunk from the whole city to the southeastern plain in the year 1990 to 2010. The stable mutualism relationships mainly resulted from the natural and artificial components with a low metabolic density, which were distributed in the center of the southeastern plains. In the initial part of the studied period, Beijing underwent a relatively slow urbanization process when Cultivated Land was able to form mutualism relationships with artificial components that had low metabolic densities such as Rural Land and Urban Land. The mutualism relationships produced by Urban Land and Rural Land stabilized during the urbanization. Among these relationships, Rural Land formed relatively stable relationships with Wetland and Sand Land, except during the third period (2000–2005). The mutualism relationships between Rural Land and Wetland were located in the perimeter of the southeastern plains, and the mutualism relationships between Rural Land and Sand land were located at the northwest of the mountainous area and surrounded built-up area.

The competition relationships were primarily observed in the northwestern mountainous area and in the northern, eastern, and southern peripheries of the southeastern plains, but these relationships had been spread out to the northwestern mountainous area and gradually formed an evenly distributed pattern since 1995. The area with the competition relationships was relatively large, but it had a small carbon flow. The natural components and Cultivated Land had been observed as the predominant contributors to the competition relationships since 2000. The Cultivated Land was also one of the most important contributors to the carbon flow through competition relationships i.e., accounting for 35.8% of these relationships on an average during each period, and approximately 21.9% of these relationships represented the flows with negative utility. In addition, Grassland was a secondary contributor to the

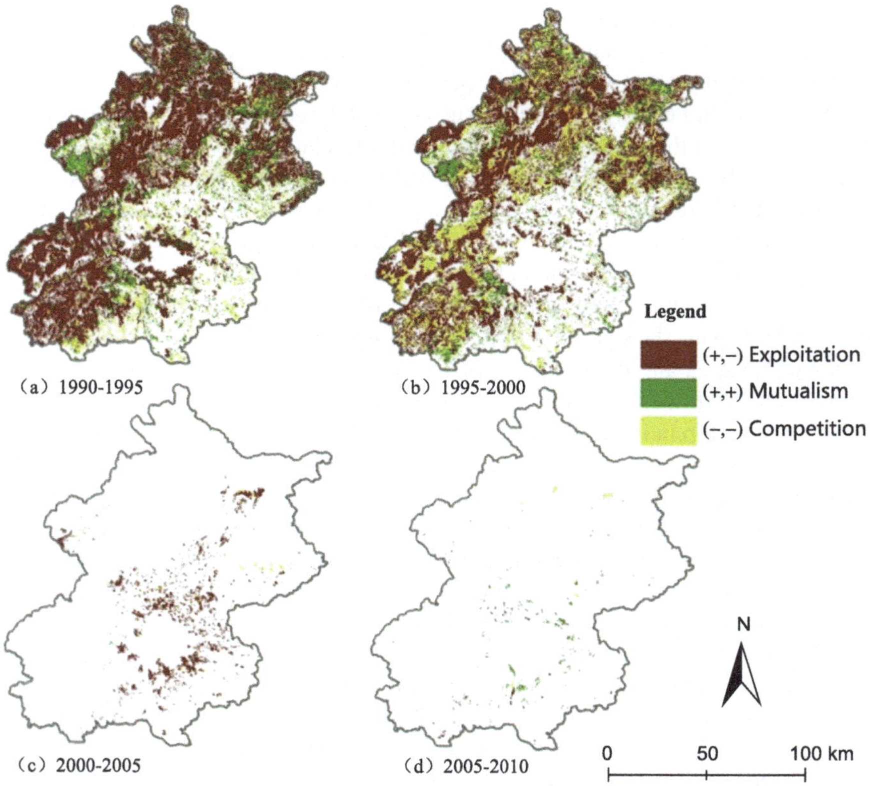

Fig. 9.19 Spatial distribution of the ecological relationships among the metabolic actors in the 18-node model of Beijing's carbon metabolism in the year 1990 to 2010

observed competition relationships, and the competition pairs associated with Grassland accounted for 38.9% of the total number of competition relationships, whereas carbon fluxes in the associated paths contributed to 19.2% of the total carbon fluxes with 18.6% being flowed with negative utility.

9.3.3 *Comparison with Previous Research on Spatial Distributions*

The adjustment of spatial planning and urban form has recently received a lot of attention. The spatial visualization of the interactions in an urban metabolism is important, and the development of GIS tools has made it possible to analyze the spatial patterns of urban metabolism. The GIS can integrate with the different types of data to reveal spatial relationships (Harrison and Haklay, 2002). As per our previous research (Xia et al., 2018), we identified the spatial patterns among the different relationships, but focused on the exploitation relationships because they were most common and had the most profound effects during urbanization. Our analysis revealed two hotspots, namely the southeast plains area and the northwest mountainous area (Zhang et al., 2014). Helsinki, Finland's capital, showed a similar spatial pattern in its carbon emissions, but unlike Beijing, Helsinki had multiple sub-centers and each sub-center showed a different gradient (Chrysoulakis et al., 2013). In addition, Leicester, a medium-sized city in the United Kingdom, research showed multiple carbon absorption centers (Davies et al., 2011).

The spatial patterns of an urban carbon metabolism are associated with the distribution of land uses. During Beijing's urbanization, the largest and most important carbon emission patches were centralized in the components of the system such as Transportation and Industrial Land (Zhang et al., 2014). Moreover, the area of high carbon absorption density depended mainly on the Forest component in the northwestern mountainous area. The research on the spatial distribution of the carbon metabolism of Toronto, Canada, showed that high emissions primarily occurred in areas associated with transportation and major housing areas while the high absorption concentrated in the large areas of green space (Christen et al., 2011). This reflected the influence of the changing urban form on the spatial patterns of carbon metabolism. Based on our previous research results (Zhang et al. 2014), the urban output and input fluxes usually occurred from multiple centers during a slow period of urban expansion, and the increase of carbon output fluxes was relatively slow. The concentration of Beijing's fluxes in and around built-up areas i.e., a small number of centers reveals the rapid development that has occurred. Therefore, Beijing's spatial structure should be adjusted to create a more diverse pattern with multiple centers.

9.3.4 Comparison with Previous Research on Ecological Relationships

The higher authority can use the spatial distribution of the ecological relationships between metabolic actors to locate areas that require more regulation to reduce carbon emissions. In this section of the chapter, the author showed the distribution of the exploitation relationships and reflected insight into the problems, which are created by the urban expansion. In Beijing's carbon metabolic system, the artificial components were the major contributors to the exploitation relationships, and the receptors of these relationships were the natural components and Cultivated Land. Both the areas that were transformed, resulting in carbon exchanges that resulted from the exploitation relationships associated with Cultivated Land and artificial land outweighed. The carbon exchanges related to the system's natural components was accounting for 48% on average of the total. This was consistent with the results of the previous studies. Liu et al. (2009) reported that the replacement of 40% of the Cultivated Land by artificial land-use occurred during the expansion of Beijing.

Moreover, Liu et al. (2010) showed that Cultivated Land was the main component, which was exploited by other components in their analysis of Beijing. In this section, we described a new result related to exploitation relationships, namely the existence of the periodic reversals, suggesting that the relationship between the pairs of components might change through this interaction. For example, Transportation and Industrial Land had been exploited by Cultivated Land from 1990 to 1995, but the reciprocal of an exploitation relationship is that the Transportation and Industrial Land was controlled by Cultivated Land in the following period. Zhang and Yang (2009) demonstrated the potential for the reversal of exploitation and control relationships in several Chinese cities. The results showed that the urban system's socioeconomic components exploited the natural components in Tianjin, Chongqing, and Guangzhou, whereas, in Beijing, Shanghai, and Shenzhen, the socioeconomic development was restricted (controlled) by the natural resources.

The conflict between land use and urbanization has become aggressively severe. The health of the urban ecosystem and its degree of mutualism are affected by the changes in the balanced activities of the metabolism. However, the analysis of the carbon metabolism network as described in this section can clarify the trends of the carbon flow, and the quantitative evaluation of this effect can support the targeted suggestions for spatial adjustments to improve the city's metabolism. So far, a key weakness in the analysis of the network approach has been the lack of spatial orientation, since the focus is still mostly on the specific land uses rather than the specific spatial plot. This creates weakness in the ability of the analysis to support the explicit recommendations spatially. Despite the crucial facts as described in this section, more work must be done to highlight the specific areas and build network models that can clearly indicate the to be needed adjustments of the city's metabolism (Xia et al., 2017a).

9.4 Path Analysis of the Carbon Involved in Trade Between the United States and China

China and the United States are not only the largest carbon dioxide (CO_2) emitters in the world, accounting for 45% of the global total annual CO_2 emission but also the two biggest economies, accounting for nearly 33% of global trade (IEA, 2015). The trade between China and United States accounted for roughly 25% of each country's total trade, demonstrating their close connections (WTO, 2017). On the other hand, CO_2 emissions associated with this trade account for about 26% of total global carbon emissions (Arce et al., 2016). These emissions have been neglected during the climate change negotiations and the formulation of the mitigation targets, and such emission greatly undermines the ability of these targets to mitigate the global climate change.

To provide insights into the emissions related to this trade, my research group performed the research that is explained in this section (Zhang et al., 2019, 2018b).

By using the Eora Global Supply Chain Database (https://worldmrio.com/), which covers 189 countries and 26 industrial sectors, we developed a framework for the upstream and downstream linkages between the sectors involved in this trade (Fig. 9.20). Further, we used this framework to calculate the total CO_2 transfers, resulting from the final consumption in United States and China, and analyzed the long-term dynamic changes and structural characteristics of these transfers in the year 1993 to 2013. Based on the results of our analysis, we identified the key domestic and foreign sectors in the United States and China as well as the major domestic-domestic, foreign-foreign, and domestic-foreign transfer paths (Zhang et al., 2019, 2018b). The results of this analysis can help to quantify the impact of the final consumption in China and the United States on the global CO_2 emission, to clarify the effects of the two countries on the other countries and regions of the world, and to provide support for the development of better regulations and plans to control CO_2 emission by allocating reasonable emission targets.

9.4.1 CO_2 Transfers in Imports and Exports

Herein, the three-letter abbreviations are created for each form of production and consumption. The first letter represents F = foreign or D = domestic intermediate production. The second letter represents F = foreign or D = domestic final production. The third letter represents the location of the consumption (foreign or domestic). The CO_2 transfers in the imports by the United States had been increased from 1993 to 2013, and then decreased [Fig. 9.21(a)]. The value in the year 2013 represented an increase of 124% as compared with the value in the year 1993. The proportions of the total accounted for the domestic final productions created from the foreign intermediate production and imported for the domestic consumption (FDD) and the foreign final production created from the foreign intermediate products and imported for the domestic consumption (FFD) were stable, at around 50% of the imports. The total transfers in imports by the United States had been increased by 1.68 Gt CO_2 (2.2 times its value in the year 1993) from 1993 to 2003 and reached a maximum value of 2.45 Gt in 2003 mainly due to FDD, which accounted for 56% increment. The total value had been decreased by 30% (1.72 Gt CO_2) in the year 2003 and 2013, and this decrease was driven by both FDD and FFD. The increment and reduction of the domestic final productions created from the domestic intermediate products and exported for the foreign consumption (DDF) and the foreign final productions created from the domestic intermediate products and exported for the foreign consumption (DFF) were largely canceled each other out, leaving emissions embodied in the exports by the United States stable at around 0.75 Gt. Also, the ratio of DFF to DDF was also stable at nearly 2:1. The export transfers were decreased by 19% (0.59 Gt CO_2) only as compared in the year 1998 due to the roughly equal reductions in the DDF and DFF.

The export transfers by China [Fig. 9.21(b)] showed the same order of magnitude as the transfers in imports by the United States (reaching a value greater than 2

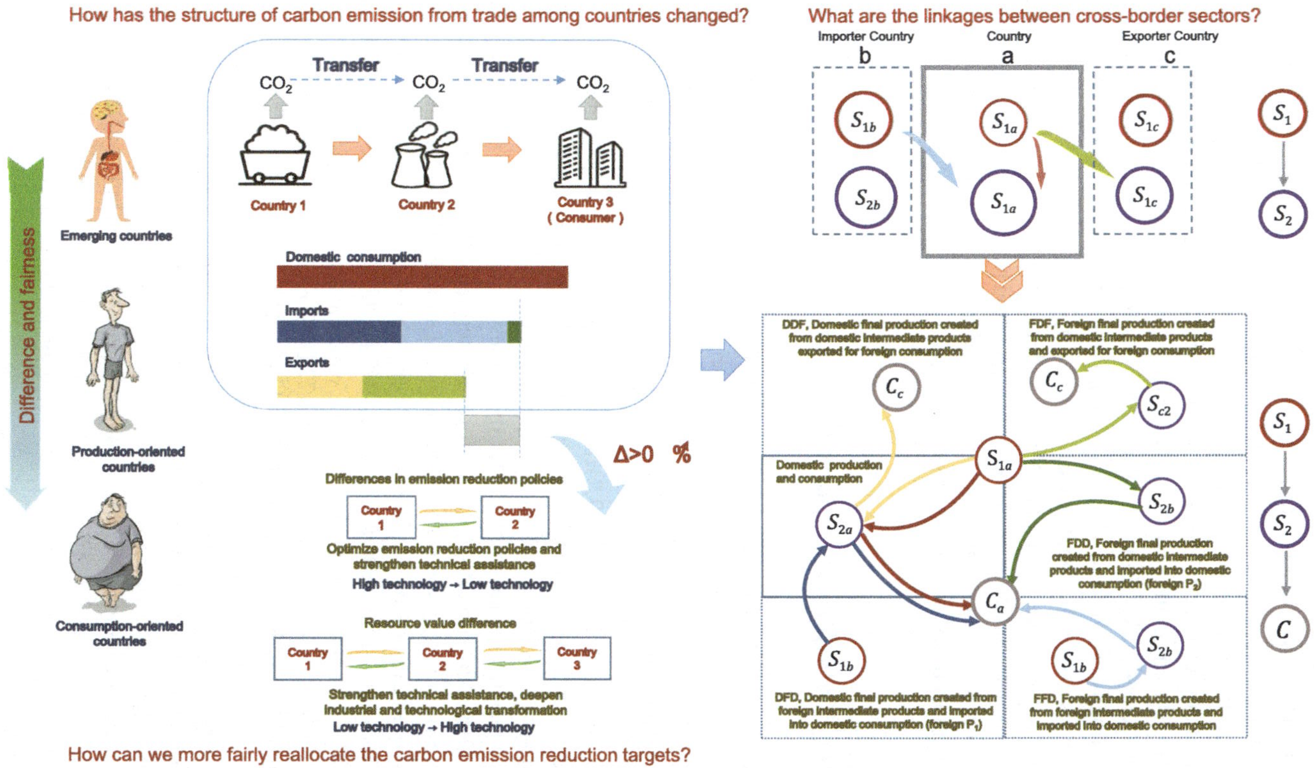

Fig. 9.20 Illustration of the analytical framework for carbon transfers during international trade and for the path analysis framework (Definitions: a, b, and c represent three countries, consumer [C], sector [S], Sector that produces intermediate products [S_1], Sector that produces final products [S_2])

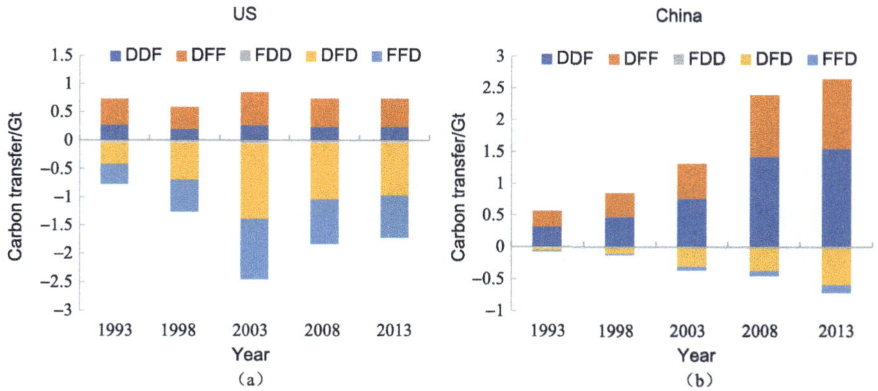

Fig. 9.21 CO_2 transfers during trade between the United States and China (*Note* Positive values represent imports, negative values represent exports. Domestic final production created from domestic intermediate products and exported for foreign consumption [DDF], foreign final production created from domestic intermediate products and exported for foreign consumption [DFF], foreign final production created from domestic intermediate products and imported for domestic consumption [DFD, foreign S_2], Domestic final production created from foreign intermediate production and imported into domestic consumption [FDD, foreign S_1], foreign final production created from foreign intermediate products and imported for domestic consumption [FFD])

Gt CO_2 in 2013). However, China's export transfers increased by roughly 400% as compared to that of value in the year 1993, reaching 2.4 Gt CO_2 in the year 2008 and 2.7 Gt CO_2 in the year 2013 [Fig. 9.21(b)]. DDF contributed 60% of the increment and the gradual increment observed in the year 2008 to 2013 (about 10% increment), and the shares of DDF and DFF remained stable at around 58 and 42%, respectively, during the studied period. China's import transfers value was about 12% of the corresponding export value, but this increased to 25% as the import value increased to 8.8 times that of value (0.71 Gt CO_2) in the year 1993 to 2013. FDD always accounted for more than 75% of the total increment and contributed about 78%.

In a comparative study, we analyzed the transfers from an additional 23 regions, which accounted for at least 70% of the total of both exports and imports (Fig. 9.22). The spatial distribution of the countries that exchanged imports and exports with the United States and China were the same. The number of countries linked with the United States was only 70% of the number linked with China, and the United States' links were more centralized. For China, most of the linked countries were in Asia with a few links in Europe and North America. In addition, Canada (2.6% of the total number of links), France (2.6%), and Singapore (2.1%) were the key linked countries for China because these countries are together accounted for more than 30% of the total Chinese CO_2 exports.

The location of the United States have strongly affected its CO_2 transfers in the imports and exports. Canada and Mexico, whose border area meet with the

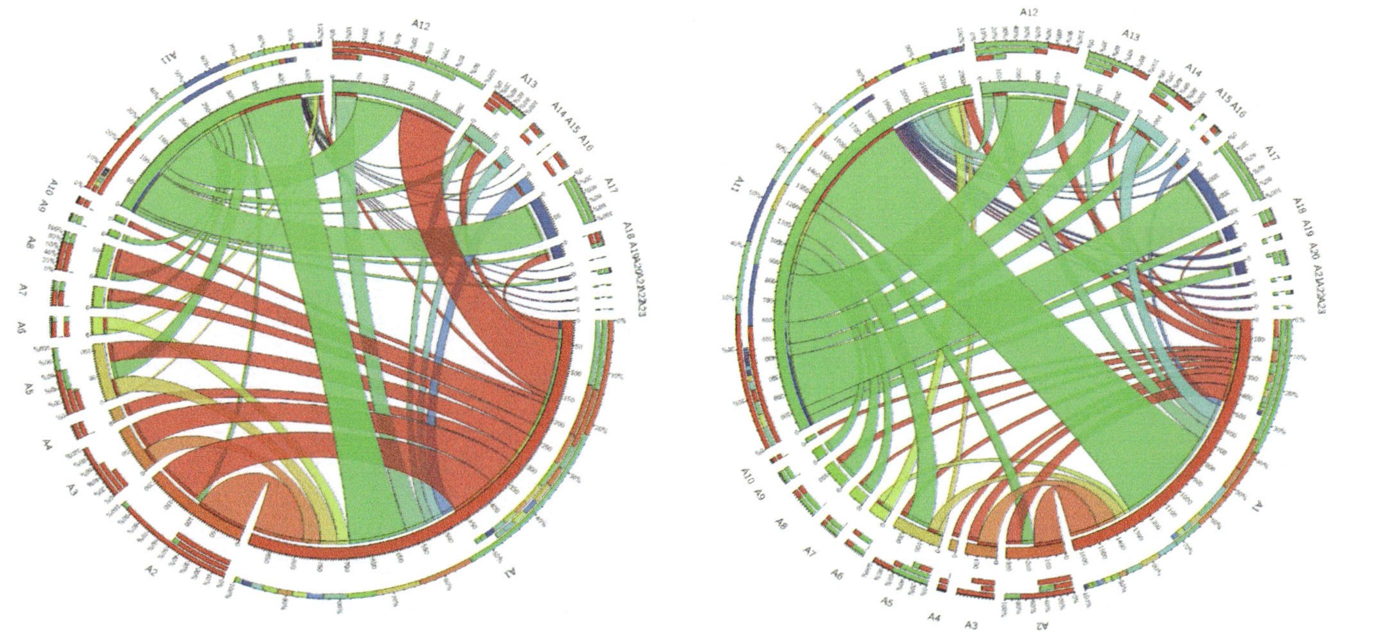

(a) United States (b) China

Fig. 9.22 Key international CO_2 transfer pathways for the United States and China in 2013 (*Note* United States [A1]; Canada [A2]; Mexico [A3]; Guyana [A4]; Germany [A5]; Russia [A6]; France [A7]; United Kingdom [A8]; Italy [A9]; Brazil [A10]; China [A11]; Japan [A12]; the Republic of Korea [A13]; India [A14]; Singapore [A15]; Taiwan, China [A16]; Hong Kong, China [A17]; Indonesia [A18]; Malaysia [A19]; Thailand [A20]; Kazakhstan [A21]; Saudi Arabia [A22]; South Africa [A23]. The unit of path flux is Mt. The pathways sent or received by different metabolic bodies are shown in different colors)

United States, accounted for around 10 and 7% of both imports and exports, respectively. However, transfers from more distant countries were also large. The exports to Guyana also reached 5% in the year 2013 after increasing by 50%. The transfers from China to the United States remained the largest transfers throughout the studied period and accounted for about 33% of United States imports by the year of 2013 after increasing 3.4 times that of value in the year 1993. The imports from India accounted for only 3% of the total imports by the United States, but these imports also increased rapidly (1.5 times that of the value in the year 1993).

The number of key North American and European countries linked to China was relatively small, but their transfers were large. Among them, 7 countries are accounted for 60% of the total trade. The exports from China to the United States stabilized at 22% of total exports of the United States, which was the largest share, and imports from the United States accounted for 8% of China's total import. The exports to and imports from Europe accounted for around 15% of China's total transfer. Among these countries, the proportion of imports from Russia (9%) was similar to that from the United States, and Germany accounted for 4% to 5% of the total import. Although the CO_2 transfers embodied in the bilateral trade between China and Japan was increased during the studied period, the proportion of China's exports to Japan was decreased up to 10% of China's total export, but its share of imports from Japan remained stable (at about 9%). In contrast, imports from India rose dramatically (to 23 times its value in the year 1993) and accounted for 6% of China's total imports in 2013.

9.4.2 Import Links Among Sectors in the United States and China

Foreign sectors with an S_1 role (i.e., foreign intermediate production imported for the domestic consumption) and domestic sectors with an S_2 role (i.e., foreign final production created from the domestic intermediate products and imported for the domestic consumption) that participated in the imports by the United States and China were similar, but their shares of the total clearly separate (Fig. 9.23). The service sectors with an S_2 role in the United States accounted for a remarkable share of the linkages. The CO_2 transfers related to imports by the Public Administration sector accounted for the largest share of the total market (25% in the year 2013) after increasing to 3.4 times its value in the year 1993, whereas the shares of the financial intermediation sector and the Education and Health sector stabilized at around 11 and 7%, respectively. The shares of the Transportation Equipment sector, Electrical and Machinery sector, and Petroleum, Chemical and Non-metallic Mineral Products sector accounted for nearly 20%, and the Petroleum, Chemical and Non-metallic Mineral Products sector also accounted for 6–9% of the total share. The Construction sector accounted for 9% of the FDD value in the United States.

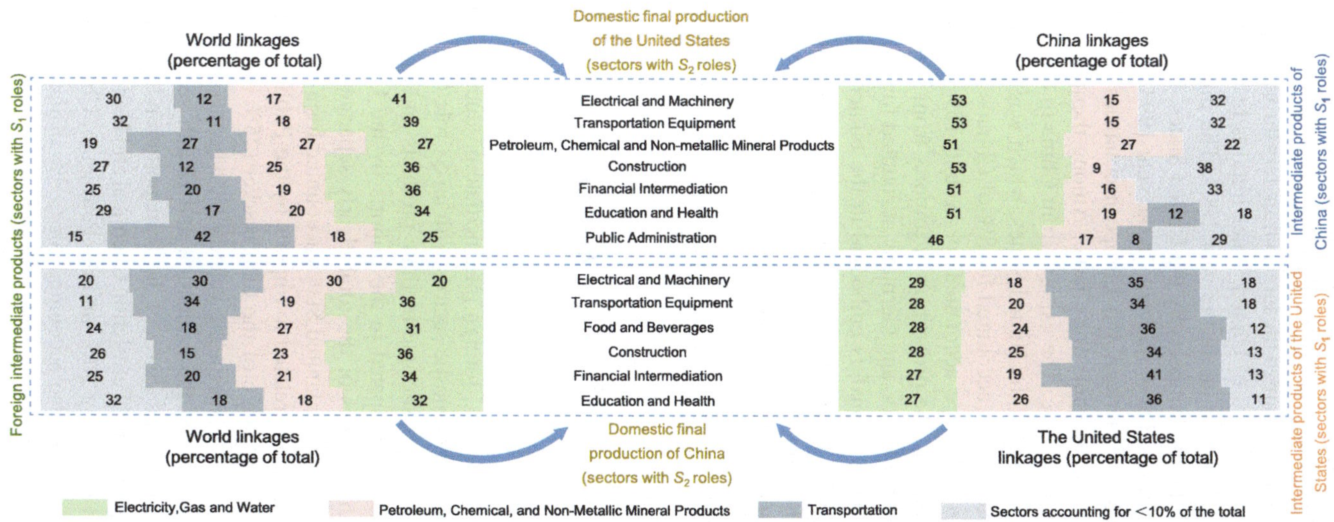

Fig. 9.23 Average proportions of the total number of linkage paths for final production sectors of the United States and China from 1993 to 2013 (*Note* Sectors at the center of the figure represent the key final production sectors of the United States and China, the sectors at the left side of the figure represent the key intermediate production sectors [global total], and sectors at the right side of the figure represent the key intermediate production sectors in China and United States. The numbers for each sector represent their percentage of the total transfers and only nodes with a percentage larger than 10% are included)

In contrast, the Construction sector accounted for the largest share of China's total final production sectors (>33%). Out of the 30% share by the final production sectors, the Electrical and Machinery sector and the Transportation Equipment sector accounted for 15% and 10%, respectively, with the remaining 5% accounted for by the Food and Beverage sector. China's Education and Health sector (about 6% of the total) accounted for 75% only of that of the United States (8%), whereas the share of the Financial Intermediation sector in both countries was similar (about 5%). However, the import values of these two service sectors in China were significantly increased (17 and 14 times their value in the 1993, respectively). Therefore, it represents an important concern because of this rapid fast increment.

The key foreign intermediate production sectors for both countries were all related to the traditional energy supply and consumption sectors. Even though the United States and China had different economic development levels, they both rely heavily on the energy flows and inputs from other countries. The key final production sectors in the United States had linkages with foreign Energy, Gas and Water, Petroleum, Chemical, and Non-Metallic Mineral Products, and Transportation sectors, which matched/agreed with the pattern for China's final production sectors, but with different percentages. The foreign intermediate production sectors and their shares of the Financial Intermediation sector and the Education and Health sector in the United States were the same. Among them, the share of Energy, Gas and Water was more than 33%, whereas China's Energy, Gas and Water sector had a larger share (nearly 50% in 2013). The foreign Transportation sectors (intermediate production sectors) accounted for more than 40% of the United States Public Administration sector's imports, whereas China's Energy, Gas and Water sector passed the Transportation Equipment sector to occupy the imports by the Public Administration sector (46% of the total in 2013) in the United States, which reflected the support and contribution to China's energy sectors to the service sectors in the United States. The foreign intermediate production sectors linked to the Transportation Equipment sector and Electrical and Machinery sector of the United States were mainly the Energy, Gas and Water sector, the Petroleum, Chemical, and Non-Metallic Mineral Products sector, and the Transportation sector (which together accounted for nearly 65% of the total). In addition, the foreign Mining and Quarrying sector accounted for 35% of the total for the Petroleum, Chemical, and Non-Metallic Mineral Products sector. These values reflected the importance of the foreign construction raw materials, logistics, and energy flows to support the various infrastructures in the United States.

The types and shares of the foreign intermediate production sectors linked with each final production sector in China were similar to those in the United States. Among them, the share of the foreign Energy, Gas and Water sector was stable (about 30%), whereas the Petroleum, Chemical, and Non-Metallic Mineral Products sector, and the transportation sector accounted for 20–30% and 15–20%, respectively, of the imports by each final production sector. However, the shares of the foreign Transportation sector in China's Electrical and Machinery sector and Transportation Equipment sector were different (30 and 10%, respectively). The Transportation Equipment sector's share was low because the foreign Electrical and Machinery sector contributed about 18% of China's Transportation Equipment sector, which was greater than the value of 10% for the same sector in the United States, revealing China's need for the foreign information technology and equipment.

The key intermediate production sectors in the United States agreed with those in other countries, but their shares differed in the total. The transportation sector in the United States had the largest share (more than 30%) of each of China's key final production sectors, indicating that the contribution to the logistics from the United States was much greater than that from other countries. In addition, the Energy, Gas and Water sector and the Petroleum, Chemical, and Non-Metallic Mineral Products sector were stable at around 25% of the total, which decreased the Electrical and Machinery sector's share of China's Transportation Equipment sector up to less than 10%. Interestingly, the key intermediate production and final production sectors in the United States were all different, but some of the key intermediate production and final production sectors in China were the same, reflecting the need for vertical integration and upgraded technology in China.

9.4.3 Export Links Among Sectors in the United States and China

In this section, we chose the sectors that together accounted for at least 70% of the transfers to clarify the key intermediate production sectors of the United States and China and most strongly linked foreign final production sectors with them. The domestic key intermediate production sectors were the same in the United States and China, including the Transportation Equipment sector, the Energy, Gas and Water sector, the Electrical and Machinery sector, and the Petroleum, Chemical, and the Non-Metallic Mineral Products sector. The share of Energy, Gas and Water increased to 5.9 times its value in the year 1993 and accounted for the largest share of China's DFF (51% in 2013). The Electrical and Machinery sector, the Petroleum, Chemical, and Non-Metallic Mineral Products sector, and the Transportation sector had lower increment (to 4.1, 2.8, and 2.6 times of those values in 1993, respectively) with their shares of the total reaching 23, 10, and 7%, respectively, in the year 2013. The values of the key intermediate production sectors in the United States did not vary greatly. The transfers by the Transportation sector, the Energy, Gas and Water sector, and the Petroleum, Chemical, and Non-Metallic Mineral Products sector increased by 10 to 15% with their shares of the total reaching 50, 25, and 17% of the total, respectively, whereas the share of the Electrical and Machinery sector decreased by 20 to 3% only of the total.

The final production sectors that China's key sectors linked to were also similar, but they were widely distributed as compared to the domestic intermediate production sectors. Each domestic key sector was mainly linked to the foreign Electrical and Machinery sector, Education and Health sector, and Construction sector, and they altogether accounted for 30–50% of the total. Among the foreign sectors linked to the Chinese sectors, the Electrical and Machinery sector and the Construction sector both accounted for more than 10% of the total (Fig. 9.24). The Electrical and Machinery sector had the highest share of each of China's key intermediate production sectors (>13%), and it accounted for 30% of China's Electrical and Machinery sector. The foreign Construction sector also accounted for more than 10% of each Chinese intermediate production sector. For the final production sectors in the United States, the distribution was different from the foreign final production sectors, but the shares of the Electrical and Machinery sector and the Construction sector remained high i.e., at 7–9 and 11%, respectively. The foreign Transportation sector and Public Administration sector linked to the Transportation sector in the United States, and the foreign Transportation Equipment sector basically linked to the Energy, Gas and Water sector and the Electrical and Machinery sector in the United States, and these altogether accounted for more than 10% of the export value for the intermediate production sectors.

Among the key final production sectors linked with intermediate production sectors in the United States, the Construction's share remained above 10% and was even higher for the Petroleum, Chemical, and Non-Metallic Mineral Products sector (15%). However, the foreign Transportation Equipment sector that were linked with the Electrical and Machinery sector and the Energy, Gas and Water sector in the United States remained stable with values near 10 and 16%, respectively. For China's final production sectors, the Electrical and Machinery sector's share was mostly high (17% for Energy, Gas and Water) with respect to the 5 to 8% only for Electrical and Machinery in the United States. In addition, China's Electrical and Machinery sector was linked with foreign Electrical and Machinery sector, and China's Transportation sector was linked with the Food and Beverages sector of the United States, which was in whole had a share of more than 30% of the total. The foreign Construction sector accounted for 15% of the transfers from China's Petroleum, Chemical, and Non-Metallic Mineral Products sector. Other Chinese final production sectors had different sectoral distributions from the foreign final production sectors, and mostly had a share of 5 to 8%.

By comparing the intermediate production sectors of the United States and China, the results revealed that the foreign final production sectors linking their intermediate production sectors including Transportation sector, Electrical and Machinery sector, and the Petroleum, Chemical, and Non-Metallic Mineral Products sector were all the same, and their shares of the transfers to the foreign Petroleum, Chemical, and Non-Metallic Mineral Products sector were around 33% for both countries. The foreign final production sectors linking the Transportation sector and the Petroleum, Chemical, and Non-Metallic Mineral Products sector in the United States were the same. However, the shares of the Transportation sector and the Petroleum, Chemical, and Non-Metallic Mineral Products sector in the United States linking with the final production sectors (17 and 11%, respectively) were larger than those for China (9 and 7%, respectively). It indicated that the Transportation sector, the Electrical and Machinery sector, and the Petroleum, Chemical, and Non-Metallic Mineral Products sector in the United States and China strongly supported the same foreign sectors.

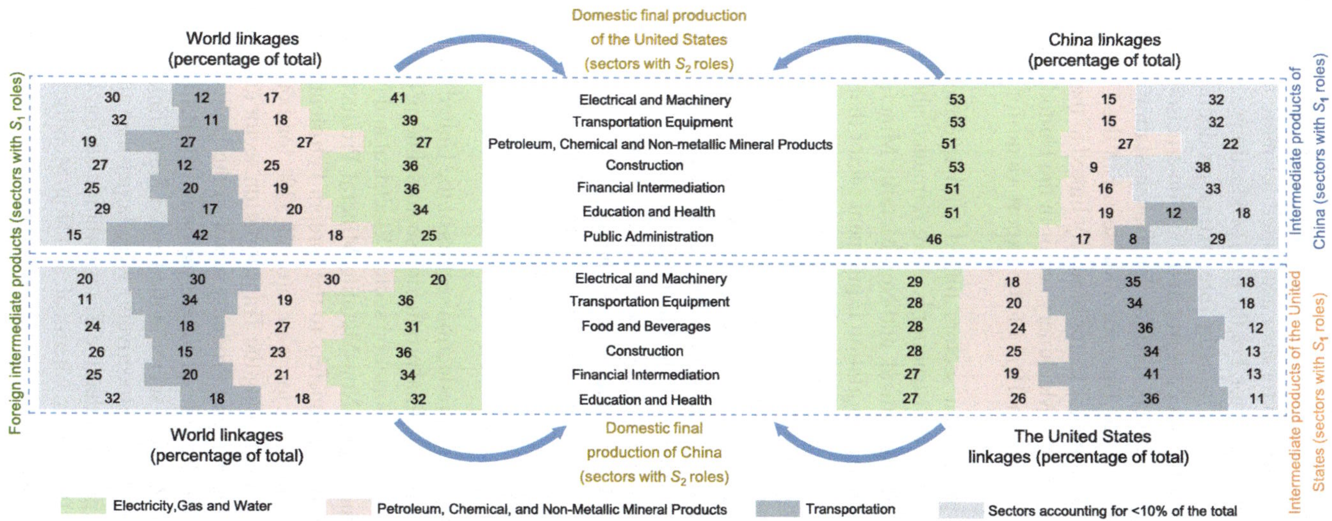

Fig. 9.24 Average proportions of the linkage paths for intermediate production sectors of the United States and China from 1993 to 2013 (*Note* The sectors at the center of the figure represent the key intermediate production sectors of the United States and China, sectors at the left side of the figure represent the key final production sectors of the world, and sectors at the right side of the figure represent the key final production sectors from China and the United States. The numbers for each sector represent their percentage of the total transfers, only nodes with a percentage larger than 5% are included)

9.4.4 Adjustment of the Carbon Mitigation Targets to Account for CO_2 Transfers in Trade

Based on the above-mentioned results in the previous section, we can propose the revisions to the United States and China's emission reduction targets specified under the 2015 Paris Agreement. The intended nationally determined contribution for the United States aims for 26% fewer emissions until the year 2025 in comparison to 2005. According to the Eora Global Supply Chain Database (2013) CO_2 emission data, only a 9% reduction has been achieved. However, based on the net imports described earlier, an additional 7% reduction is needed for a total of 33%. This is because the net CO_2 imports embodied in trade by the United States are equivalent to 14% of the emission in 2005, so it is assume that if the United States would account for its share of the responsibility (50%), we can propose a revised carbon reduction target of 33% (i.e., 26% + 14%/2). Table 9.4 summarizes this data and the sources of the transfers from other countries. Following by, we can allocate the other half of the 14% (i.e., 7%) to the other countries based on their technological and economic levels (i.e., their ability to achieve the specified reduction). The United States imported embodied carbon emissions from the 10 key countries, which would increase the share of the United States for the emissions in those countries. Among them, China is needed to transfer the largest share of its emissions to the United States, accounting for 68% of the 7% allocated increment of the United States, followed by India (11% of the 7% increment), and the remaining countries each accounted for less than 6%. As compared with these countries, the carbon intensity (t CO_2 per unit GDP) of the United States was low, and the per capita GDP was high [Fig. 9.25(a)]. Therefore, to further reduce the responsibility for the domestic emissions, the United States should prioritize providing technical support to develop low-carbon production to the countries it trades with, especially China and Russia, which combined high emission per unit GDP with low per capita GDP. This would improve their production technology and reduce their carbon emissions per unit GDP, thereby reducing the carbon emission that results from exports to the United States. Among the other countries, only Singapore combined a high per capita GDP with a low carbon emission intensity. As a source of imports by the United States, Singapore can effectively reduce carbon emissions. The United States had high net exports to four countries, namely Guyana (5.9% of total net exports), Singapore (2.9%), the United Kingdom (2.5%), and Brazil (1.8%). These countries are also responsible for carbon emissions related to trade with the United States. Because the United States was a net importer, it would bear additional responsibility for the reduction in carbon emission and should reduce the amount of carbon transferred to the other countries through its own final production technology.

Based on the principle of common and differentiated responsibilities that was proposed by the UNFCCC (http://unfccc.int/resource/docs/convkp/conveng.pdf), China's submission of the intended nationally determined contribution did not involve actual emission reduction tasks, but instead emphasized that emissions should decrease from now to 2030 while the carbon intensity should decrease to 60% of that

Table 9.4 The adjusted carbon emission mitigation index (i.e., the change in the intended nationally determined contribution) and their associated countries and regions

Country	Adjusted mitigation index (%)	Associated countries and regions (percentage of responsibility)/%				
		China	U.S	India	Canada	Russia
U.S	+ 7	68.22	–	11.04	5.52	3.50
China	– 9	–	21.89	2.18	3.14	– 1.09
Country	Adjusted mitigation index (%)	Associated countries and regions (percentage of responsibility)/%				
		Japan	Germany	The Republic of Korea	France	U.K
U.S	+ 7	1.00	0.98	0.03	– 1.58	– 2.48
China	– 9	9.05	6.02	3.03	3.34	4.27
Country	Adjusted mitigation index (%)	Associated countries and regions (percentage of responsibility)/%				
		Singapore	Brazil	Mexico	Guyana	Italy
U.S	+ 7	– 2.88	– 1.85	5.74	– 5.89	–
China	– 9	3.04	–	–	–	3.20
Country	Adjusted mitigation index (%)	Associated countries and regions (percentage of responsibility)/%				
		Thailand	Malaysia	Indonesia	Saudi Arabia	South Africa
U.S	+ 7	–	–	–	–	–
China	– 9	2.36	1.30	0.98	0.29	0.11
Country	Adjusted mitigation index (%)	Associated countries and regions (percentage of responsibility)/%				
		Kazakhstan	–	–	–	–
U.S	+ 7	–	–	–	–	–
China	– 9	– 0.52	–	–	–	–

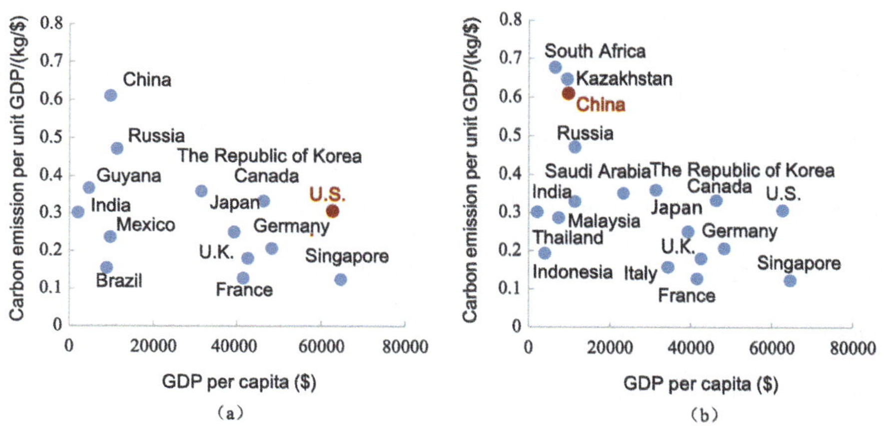

Fig. 9.25 Relationships between the CO_2 emission intensity (emission per unit GDP) and per capita GDP for key countries linked to the United States and China

value in year 2005. According to China's Ministry of the Ecological Environment (MEE, 2018), China's carbon emission intensity in 2017 had decreased by about 46% as compared with 2005, so its responsibility is nearing completion. However, based on the difference between China's net export and import values, 9% of China's emission reduction responsibility should be borne by other countries. According to the assumption of equally shared responsibility that we used for the United States, we allocated the emission reduction targets to the 19 key countries that trade with China. China was a net exporter to 17 of these countries. Therefore, these countries should bear the responsibility for some of China's emissions. Based on their shares of China's net exports, the United States would be responsible for 22% of China's carbon emissions. Among the 17 countries, the per capita GDP of India, Indonesia, and Thailand was lower than that of China [Fig. 9.25(b)], but their carbon emission intensity was also lower than that of China. Therefore, China should improve its industrial technology level to further reduce its carbon transfers in exports. The remaining 14 countries had a low carbon intensity and a high per capita GDP. In particular, the United States, which accounted for 20% of the total, should strengthen its technical assistance and support to China to improve China's production technology. Russia (1.0% of total net exports to China) and Kazakhstan (0.5%) were net import sources for China, and both had carbon emission intensity similar to or greater than that of China. By combined with their low per capita GDP, indicating that China's strengthening of imports from these countries can mitigate global carbon emission.

9.4.5 The Importance of the Research Perspective

Some scholars have pointed out that CO_2 transfers to satisfy domestic or foreign final consumption have become an important part of the global CO_2 emissions (Mi et al., 2017; Kanemoto et al., 2014; Davies and Caldeira, 2010). Such transfers have attracted the attention of researchers in the context of the potential adjustments of the national mitigation targets. When analyzing CO_2 transfers in trade, researchers have emphasized the importance of sectoral analysis to explain export transfers, whereas others have emphasized the need to examine both export and import behavior (Steininger et al., 2018). Through this approach, the net export or import values can be captured and used to reveal a nation's role in CO_2 emission and its responsibility for mitigation (Peters et al., 2011; Peters and Hertwich, 2008). In addition to focusing on a nation's total contribution, Mi et al. (2017) also stressed that path analysis for the cross-border CO_2 transfers can support the calculation of the transfers of mitigation responsibility among the nations. Based on this approach, Meng et al. (2018) proposed an upstream–downstream framework for the cross-border CO_2 transfers embodied in a trade that made it possible to analyze CO_2 transfer paths among sectors. In Sect. 9.4, a framework is utilized to study trade-induced CO_2 transfers by the United States and China and analyzed the export and import values and their structural features at a global scale. Being an expert, I also described the characteristic differences of the linkage paths for domestic intermediate production and final production sectors between the two countries. By combining national distribution characteristics that support sectoral trade in the United States and China, it was possible to propose specific mitigation measures.

In the future, it can be possible to integrate the sectoral linkage path results to derive mitigation targets for the specific sectors. This can follow the results of the sectoral transfer paths identified earlier in this section to clarify capital, technical, and cooperative action plans. The trade-induced CO_2 transfers described in this section were based on the total consumption, and this approach can be extended to reveal which specific consumption type is most strongly motivated consumption (e.g., governments vs. households). In addition, the identification of key sectors was based on the 26 Eora sector types. In future research, these sectors can be disaggregated to allow an examination of the specific industrial activities based on the more than 2000 kinds of CO_2 emission inventories contained in the Eora data, including inventories for the agricultural transportation equipment, agricultural machinery energy use, and straw burning in agriculture.

9.4.6 Comparison with Previous Research

Previous analyses of the sectoral contributions to CO_2 transfers in exports and imports by the United States and China differed somewhat from our results (Table 9.5). For example, Xu et al. (2011) and Lin and Sun (2010) calculated the shares of key export sectors based on 45- and 16-sector input–output tables for China in the year 2008 and 2005, respectively. Their results were similar in the Machinery Processing (the total of the Electrical and Machinery sector, the Transportation Equipment sector, and the Food and Beverages sector) that had the largest share (47.3% and 42.0% for the 16- and 45-sector tables, respectively), whereas the shares of the Petroleum, Chemical and Non-metallic Mineral Products sector, and the Textiles and Wearing Apparel sector were around 11%. In the present study, the sectors with the largest share of China's DFF in 2008 were the Electricity, Gas, and Water sector (44.2%) and the Petroleum, Chemical and Non-metallic Mineral Products sector (19.8%). Both the previous results and our results highlighted the importance of the Electricity, Gas, and Water sector, but calculated different percentages, it may be attributed to the differences in the creators of the input–output tables for single countries or regions accounted for the export data for each sector.

In an analysis of transfers between China and the United States, Guo et al. (2010) and Du et al. (2011) examined input–output tables for China in the year 2005 and 2007, respectively, and found that the Electrical and Machinery sector accounted for 25.5–37.2% of China's exports to the United States. In contrast, our results showed that the key Chinese export sectors were Electricity, Gas, and Water (44.2%) and Petroleum, Chemical, and Non-metallic Mineral Products (19.8%). These differences may be caused by the different sectoral divisions. For example, China's 42-sector

Table 9.5 Comparison of the sectoral contributions to exports in previous research and the present study

	Year	Methodology (no. of sectors)[a]	Share of sector (percentage of total) [a]/%						Source
			EM	TE	FB	PCNM	EGW	TWA	
China's exports	2005	SRIO (16)	47.3			10.4	0.2	9.7	Lin and Sun (2010)
	2008	SRIO (45)	23.0	190		14.0		12.0	Xu et al. (2011)
	2008	MRIO (26)	19.0	2.0	2.0	15.0	27.0	11.3	Present study
China's exports to the United States	2007	SRIO (42)	25.5	204		9.6		20.4	Du et al. (2011)
	2005	SRIO (42)	37.2	1.7		28.5		16.7	Guo et al. (2010)
	2008	MRIO (26)	10.2	1.3	0.6	19.8	44.2	8.0	Present study

Abbreviations: Electricity, Gas, and Water (EGW); Electrical and Machinery (EM); Food and Beverages (FB); multi-region input–output analysis for 26 sectors (MRIO); Petroleum, Chemical, and Non-metallic Mineral Products (PCNM); single-region input–output analysis for 16, 42, or 45 sectors (SRIO); Transportation Equipment (TE); and Textiles and Wearing Apparel (TWA) single-region input–output table separated Transportation into sea, land, and air transportation, and combined Electrical and Machinery with Transportation Equipment, which differs from the 26 sectors in the Eora data that I used in this section.

In addition, Arce et al. (2016) estimated that the main destinations for China's CO_2 transfers in exports were the United States (24.3%), Japan (9.2%), Germany (5.3%), and the United Kingdom (4.2%). This was similar to our results with the United States (23.9%), Japan (10.8%), Germany (5.5%), and the United Kingdom (4.0%) being important destinations. The differences among these values may result from the different data sources (the Global Trade Analysis Project in the Arce et al. study and Eora in the present study). However, our study went beyond the previous research by identifying the key intermediate production and final production sectors for China and the United States. We found that in the FDD, the final production sectors in both the United States and China were linked with the foreign Transportation sector and Electricity, Gas, and Water sector, whereas, in the DFF, the intermediate production sectors of both countries were mainly associated with the foreign Construction sector and Electrical and Machinery sector.

References

Arce G, López LA, Guan D (2016) Carbon emissions embodied in international trade: the post-China era. Appl Energy 184(12):1063–1072

Bolin B (1977) Changes of land biota and their importance for the carbon cycle. Science 196(4290):613–615

Christen A, Coops NC, Crawford BR et al (2011) Validation of modeled carbon-dioxide emissions from an urban neighborhood with direct eddy-covariance measurements. Atmos Environ 45(33):6057–6069

Chrysoulakis N, Lopes M, San José R et al (2013) Sustainable urban metabolism as a link between bio-physical sciences and urban planning: the BRIDGE project. Landsc Urban Plan 112(2):100–117

Chuai X, Huang X, Wang W et al (2015) Land use, total carbon emissions change and low carbon land management in coastal Jiangsu, China. J Clean Prod 103(17):77–86

Davies ZG, Edmondson JL, Heinemeyer A et al (2011) Mapping an urban ecosystem service: quantifying above-ground carbon storage at a citywide scale. J Appl Ecol 48(5):1125–1134

Davis SJ, Caldeira K (2010) Consumption-based accounting of CO_2 emissions. Proc Natl Acad Sci USA 107(12):5687–5692

Dhakal S (2009) Urban energy use and carbon emissions from cities in China and policy implications. Energy Policy 37(11):4208–4219

Dixon RK, Solomon AM, Brown S et al (1994) Carbon pools and flux of global forest ecosystems. Science 263(5144):185–190

Du H, Guo J, Mao G et al (2011) CO_2 emissions embodied in China-US trade: input-output analysis based on the emergy/dollar ratio. Energy Policy 39(10):5980–5987

Escobedo F, Varela S, Min Z et al (2010) Analyzing the efficacy of subtropical urban forests in offsetting carbon emissions from cities. Environ Sci Policy 13(5):362–372

Guo J, Zou LL, Wei YM (2010) Impact of inter-sectoral trade on national and global CO_2 emissions: an empirical analysis of China and US. Energy Policy 38(3):1389–1397

Harrison C, Haklay M (2002) The potential of public participation geographic information systems in UK environmental planning: appraisals by active publics. J Environ Planning Manage 45(6):841–863

Houghton R A (2003) Revised estimates of the annual net flux of carbon to the atmosphere from changes in land use and land management 1850–2000. Tellus Ser B—Chemical & Phys Meteorol 55(2):378–390

Houghton RA, Goodale CL (2004) Effects of land-use change on the carbon balance of terrestrial ecosystems//DeFries R S, Asner G P, Houghton R A. In: Washington USA (ed) Ecosystems and land use change. American Geophysical Union, pp 85–98

Houghton RA, Hackler JL, Lawrence KT (2000) Changes in terrestrial carbon storage in the United States. 2: The role of fire and fire management. Glob Ecol Biogeogr 9(2):145–170

Hutyra LR, Yoon B, Alberti M (2011) Terrestrial carbon stocks across a gradient of urbanization: a study of the Seattle, WA region. Glob Chang Biol 17(2):783–797

IEA (International Energy Agency) (2015) CO_2 emission from fuel combustion (2015-11-20) [2022-1-12]. http://www.iea.org/statistics/

IPCC (Intergovernmental Panel on Climate Change) (2006) Guidelines for national greenhouse gas inventories (2006-4-12) [2022-1-1]. https://www.ipcc-nggip.iges.or.jp/public/2006gl/index.html

Jo HK (2002) Impacts of urban green space on of carbon emissions for middle Korea. J Environ Manage 64(2):115–126

Kanemoto K, Moran D, Lenzen M et al (2014) International trade undermines national emission reduction targets: new evidence from air pollution. Glob Environ Chang 24(1):52–59

Kauppi PE, Mielikainen K, Kuusela K (1992) Biomass and carbon budget of European Forests, 1971 to 1990. Science 256(5053):70–74

Le Quéré C, Raupach MR, Canadell JG (2009) Trends in the sources and sinks of carbon dioxide. Nat Geosci 2(12):831–836

Li J, Zhang Y, Liu NY et al (2018) Flow analysis of the carbon metabolic processes in Beijing using carbon imbalance and external dependence indices. J Clean Prod 201(21):295–307

Lin B, Sun C (2010) Evaluating carbon dioxide emissions in international trade of China. Energy Policy 38(1):613–621

Liu CF, Li XM (2012) Carbon storage and sequestration by urban forests in Shenyang, China. Urban for & Urban Green 11(2):121–128

Liu G, Yang Z, Chen B (2010) Extended exergy-based urban ecosystem network analysis: a case study of Beijing, China. Procedia Environ Sci 2(1):243–251

Liu JY, Zhang ZX, Xu XL et al (2009) Spatial patterns and driving forces of land use change in China in the early 21st century. J Geog Sci 64(12):1411–1420 (in Chinese)

McGraw J, Haas P, Young L et al (2010) Greenhouse gas emissions in Chicago: emissions inventories and reduction strategies for Chicago and its metropolitan region. J Great Lakes Res 36(S2):106–114

MEE (Ministry of Ecological Environment) (2018) China's policies and actions for addressing climate change (2018-11-27) [2021-12-1]. http://english.mee.gov.cn/Resources/Reports/reports/201812/P020181203546772849924.pdf

Meng B, Peters GP, Wang Z et al (2018) Tracing CO_2 emissions in global value chains. Energy Economics 73(5):24–42

Mi ZF, Wei YM, Wang B et al (2017) Socioeconomic impact assessment of China's CO_2 emissions peak prior to 2030. J Clean Prod 142(2):2227–2236

Miao LJ, Cui XF, Luan YB et al (2011) Similarities and differences of Beijing and Shanghai's land use changes induced by urbanization. J Meteorol Sci 31(4):398–404

Nowak DJ, Crane DE (2002) Carbon storage and sequestration by urban trees in the USA. Environ Pollut 116(3):381–389

Peters GP, Hertwich EG (2008) CO_2 embodied in international trade with implications for global climate policy. Environ Sci Technol 42(5):1401–1407

Peters GP, Minx JC, Weber CL et al (2011) Growth in emission transfers via international trade from 1990 to 2008. Proc Natl Acad Sci USA 108(21):8903–8908

Piao SL, Fang JY, Ciais P et al (2009) The carbon balance of terrestrial ecosystems in China. Nature 458(7241):1009–1013

Satterthwaite D (2008) Cities' contribution to global warming: notes on the allocation of greenhouse gas emissions. Environ Urban 20(2):539–549

Steininger KW, Munoz P, Karstensen J et al (2018) Austria's consumption-based greenhouse gas emissions: identifying sectoral sources and destinations. Glob Environ Chang 48(1):226–242

Svirejeva-Hopkins A, Schellnhuber J (2008) Urban expansion and its contribution to the regional carbon emissions: using the model based on the population density distribution. Ecol Model 216(2):208–216

Tao Y, Li F, Wang R et al (2015) Effects of land use and cover change on terrestrial carbon stocks in urbanized areas: A study from Changzhou, China. J Clean Prod 103(17):651–657

Tian GJ, Wu JG, Yang ZF (2010) Spatial pattern of urban function in the Beijing metropolitan region. Habitat Int 34(2):249–255

WTO (World Trade Organization) (2017) World Trade Report 2017 (2017-8-10) [2021-11-12]. https://www.wto.org/english/res_e/publications_e/wtr17_e.htm

Xia LL, Fath BD, Scharler UM et al (2016) Spatial variation in the ecological relationships among the components of Beijing's carbon metabolic system. Sci Total Environ 544(3):103–113

Xia LL, Liu Y, Wang XJ et al (2018) Spatial analysis of the ecological relationships of urban carbon metabolism based on an 18 nodes network model. J Clean Prod 170(1):61–69

Xia LL, Zhang Y, Sun XX et al (2017a) Analyzing the spatial pattern of carbon metabolism and its response to change of urban form. Ecol Model 355(13):105–115

Xia LL, Zhang Y, Wu Q et al (2017b) Analysis of the ecological relationships of urban carbon metabolism based on the eight nodes spatial network model. J Clean Prod 140(1):1644–1651

Xu M, Li R, Crittenden JC et al (2011) CO_2 emissions embodied in China's exports from 2002 to 2008: a structural decomposition analysis. Energy Policy 39(11):7381–7388

Xu XL, Liu JY, Zhuang DF (2012) Remote sensing monitoring methods of land use/cover change in national scale. J Anhui Agric Sci 40(4):2365–2369 (In Chinese)

Zhang Y, Li J, Fath BD et al (2015) Analysis of urban carbon metabolic processes and a description of sectoral characteristics: a case study of Beijing. Ecol Model 316(7651):144–154

Zhang Y, Li YG, Hubacek K et al (2019) Analysis of CO_2 transfer processes involved in global trade based on ecological network analysis. Appl Energy 233–234(1):576–583

Zhang Y, Li YG, Liu GY et al (2018a) CO_2 metabolic flow analysis in global trade based on ecological network analysis. J Clean Prod 170(1):34–41

Zhang Y, Wu Q, Zhao XY et al (2018b) Study of carbon metabolic processes and their spatial distribution in the Beijing-Tianjin-Hebei urban agglomeration. Sci Total Environ 645(23):1630–1642

Zhang Y, Xia LL, Fath BD et al (2016) Development of a spatially explicit network model of urban metabolism and analysis of the distribution of ecological relationships: case study of Beijing, China. J Clean Prod 112(2):4304–4317

Zhang Y, Xia LL, Xiang WN (2014) Analyzing spatial patterns of urban carbon metabolism: A case study in Beijing, China. Landsc Urban Plan 130(5):184–200

Zhang Y, Yang ZF (2009) A method of analyzing the interactions in an urban metabolism system. J Environ Sci 29(1):217–224

Zhao M, Kong ZH, Escobedo FJ et al (2010) Impacts of urban forests on offsetting carbon emissions from industrial energy use in Hangzhou, China. J Environ Manag 91(4):807–813

Chapter 10
Analysis of the Urban Nitrogen Metabolism

Nitrogen is an essential component of an urban metabolism, since it is a key ingredient of the fertilizers used to sustain crop growth, of the food produced using that fertilizer, and of the waste generated as a result of preparation and consumption of that food. It is also a key element in many industrial processes, and is released into the atmosphere during combustion of fossil fuels. As a result, it's important to understand the urban metabolism of nitrogen. Similar to carbon in Chapter 9, I will examine the role nitrogen plays in an urban metabolism using the tools developed in the first part of this book. As in the previous chapter, Beijing is considered to demonstrate how these tools can be applied, and the insights they can provide.

10.1 Accounting for Nitrogen Metabolism and Its Key Influencing Factors in Beijing

The production of global anthropogenic reactive nitrogen increased by nine times its original value in the past 100 years from the 1890s to the 1990s, while natural reactive nitrogen decreased by 11% (from 100 TgN/year to 89 TgN/year). Therefore, we can say that the global nitrogen cycle has been greatly disturbed by human activities (Galloway and Cowling 2002), such as energy consumption, production and consumption of food and inorganic fertilizers, industrial production, and animal husbandry. Furthermore, cities have accounted for a significant proportion of global nitrogen consumption and emission because of their attributes like intensive production and consumption and large-scale populations. For example, global NO_x emission from both energy and food production has increased by more than 5 times its 1890 level between 1890 and 1990, with significant effects on human health and regional sustainable development. To control urban nitrogen pollution, it is necessary to reduce nitrogen consumption at the source, so adopting the metabolic perspective to link resource consumption and pollution emissions has become an important subject in nitrogen research.

Y. Zhang, *Urban Metabolism*,
https://doi.org/10.1007/978-981-19-9123-3_10

In the research on which this section is based (Zhang et al. 2020), my research group developed indexes that describe reactive nitrogen inputs and anthropogenic nitrogen consumption, and used them to characterize the most important nitrogen sinks and sources and the socioeconomic factors that affected them.

We analyzed the changing trend and structural characteristics of Beijing's total anthropogenic nitrogen consumption from 1995 to 2015 and used the logarithmic mean Divisia index method (I have described this method in Chapter 6) to construct a factor decomposition model including six factors: ① the nitrogen content of the material flows, ② the material intensity (i.e., the material consumption per unit of the output value), ③ the industrial structure, ④ the per capita GDP, ⑤ the material consumption structure, and ⑥ the population (Zhang et al. 2020). These data were examined, and we identified the main factors that promoted or inhibited the effect on Beijing's consumption of reactive nitrogen using these data (Fig. 10.1). Overall, we aimed to provide scientific support for formulating policies related to nitrogen consumption control and to help the development of Beijing's nitrogen metabolism to become healthier and more stable (Zhang et al. 2017).

10.1.1 Analysis of the Total Input of Reactive Nitrogen

Figure 10.2 shows that during the two decades from 1995 to 2014, the input of reactive nitrogen (red line) in Beijing generally increased, with peaks in 2001 and 2010. From 1995 to 2000, the rate of increase of the reactive nitrogen input was slow, with an annual increase of about 0.6%, but the growth rate increased greatly (to 2.2% per year) from 2000 to 2010. The growth rate reached 40% from 2000 to 2001, which resulted in the second peak in 2001. The input reached another peak in 2010, at 641.2 Gg, about 1.2 times the value in 1995, then slowly decreased to 94.4% of the 2010 value by 2015.

The curves in Fig. 10.2 show that the consumption of food and energy, which are large sources of reactive nitrogen input, increased to 2.6 and 1.6 times the 1995 values, respectively, with some fluctuation, while the reactive nitrogen input from fertilization decreased to 0.5 times the 1995 value. Before 2000, the input of reactive nitrogen from fertilization was the largest one, and initially increased slowly, but decreased steadily from 1995 to 2000, and became smaller than energy consumption in 2001, and fell continuously at an average annual rate of 4.0% from 2000 to 2015, reaching about 50% of the 1995 value by 2015.

After 2000, the input of reactive nitrogen from energy consumption was the largest input, and increased to about 1.8 times its 2000 value by 2015. The upward trend for energy consumption closely followed the total input curve. Although the nitrogen input in food consumption increased steadily (to about 2.6 times the 1995 level), it was much smaller than the input from energy and fertilizer use. Nitrogen in livestock feed was also significant, with an increase to about 2 times its 1995 value by 2003, followed by a decrease to about 50% of the 2003 level by 2015. The other inputs of reactive nitrogen, including chemical product consumption, biological nitrogen

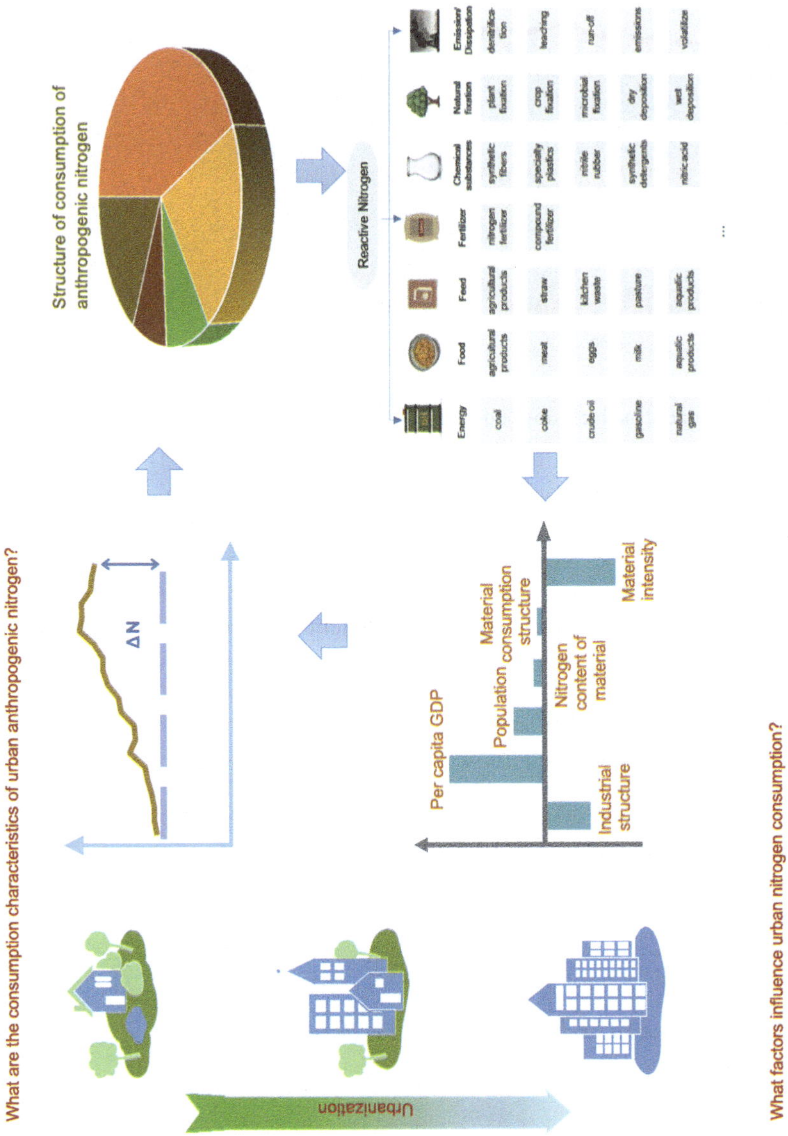

Fig. 10.1 Analytical accounting framework for an urban nitrogen metabolism and identification of its key influencing factors

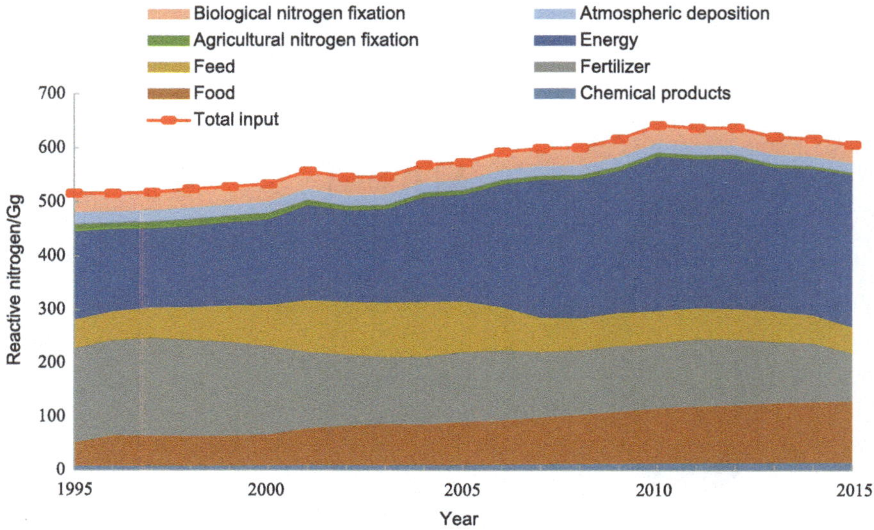

Fig. 10.2 Total input of reactive nitrogen by Beijing and its sources

fixation, and atmospheric deposition, were small and showed little change during the study period.

Figure 10.3 shows the balance between natural and anthropogenic nitrogen production and the balance between internal and external sources of reactive nitrogen. The proportion of anthropogenic nitrogen was greater than 88% throughout the study period. The natural reactive nitrogen input decreased continuously, so the proportion of the total accounted for by anthropogenic nitrogen increased steadily, from 88.9% in 1995 to 91.5% in 2015, with only a slight decrease after 2010. Similarly, the total input of reactive nitrogen from the external environment increased steadily as internal production decreased, so the proportion of total nitrogen input increased from 62.3% to 75.9%, and the quantity input from the external environment to the urban system increased to about 1.4 times the 1995 level, indicating that a rapidly growing metropolis such as Beijing depends increasingly on material inputs from its external environment.

10.1.2 Analysis of the Structural Characteristics of the Reactive Nitrogen Inputs

Food and energy consumption were two principal sources for the increased input of reactive nitrogen in Beijing. Thus, I will discuss the changes of the consumption structure of food and energy as well as their influence on the total input of reactive nitrogen. First, the fluctuation of consumption and of the consumption structure of several key food categories in Beijing from 1995 to 2015 is analyzed. Figure 10.4

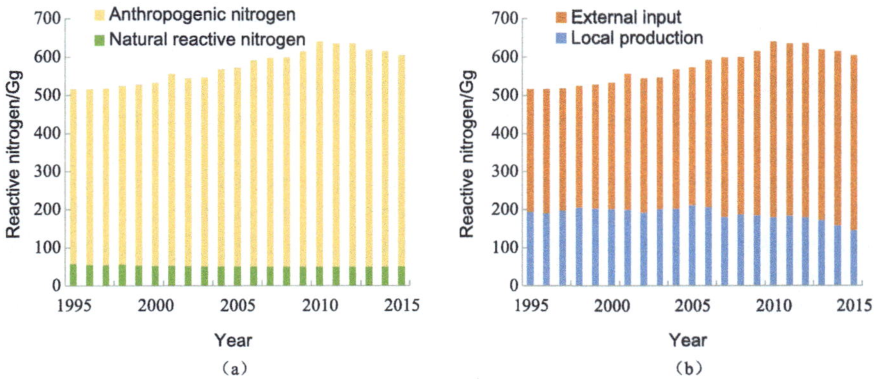

Fig. 10.3 Total inputs of reactive nitrogen by Beijing and their sources

shows that the total consumption of food in Beijing rose significantly, increasing to nearly 1.9 times the 1995 value by 2015. However, this increase was smaller than the increase in reactive nitrogen caused by increased food consumption. The reactive nitrogen input caused by food consumption increased to about 2.1 times the 1995 value because the food consumption structure changed.

The consumption of agricultural products (including grains, vegetables, soybeans, and fruit) was far greater than consumption of meat (pork, beef, mutton, poultry, eggs, milk, aquatic products) during the study period, with vegetable consumption greatest towards the end of the period. However, the proportion of total consumption accounted for by agricultural products fell from 84.7% in 1995 to 76.0% in 2015 because the proportion of meat products with high nitrogen content was increasing.

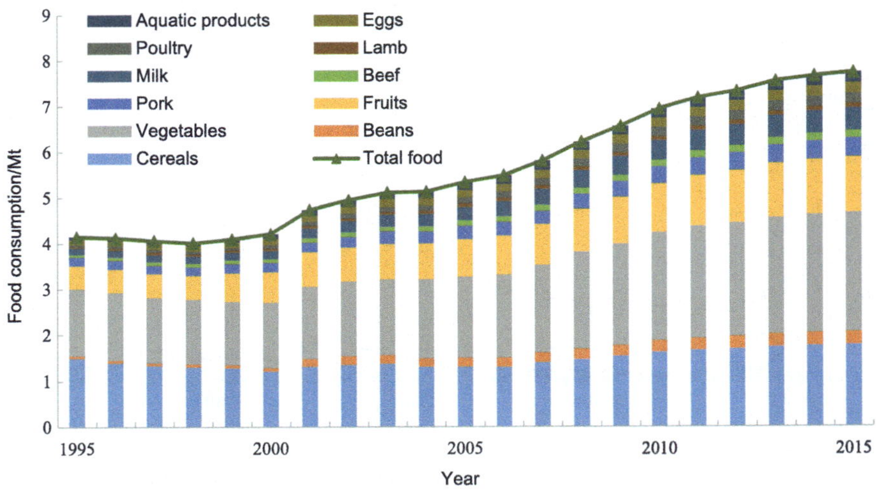

Fig. 10.4 Consumption structure for Beijing's food categories

The consumption of pork and milk were the largest, with these foods increasing to 2.1 and 3.4 times the 1995 level, respectively, by the end of the study period. Consumption of poultry and aquatic products increased even more, to 5.3 and 4.2 times the 1995 level, respectively, but they accounted for only a small proportion of the total ($\leq$3.5%).

Because Beijing's energy consumption is still dominated by high-nitrogen fossil fuels, it's important to account for this nitrogen source. Figure 10.5 shows the total energy consumed by each energy consumption sector from 1995 to 2015 (converted to units of tonnes coal-equivalent, tce). The total energy consumption was dominated by the Industry, Household Consumption, and Transportation sectors, particularly towards the end of the study period. The energy consumed by the Industry sector steadily decreased, with fluctuation, to about 26.9% of the 1995 value by 2015. In contrast, the energy consumed by the Household Consumption and Transportation sectors increased, with fluctuation, reaching 2.1 and 9.9 times the 1995 value, respectively, by the end of the study period. Because of the size of this component of the metabolism, the increase for the Transportation sector was the most significant one. An important contributor to this trend was the high emission of nitrogen oxides (NO_x) from burning the main fuels consumed by the Transportation sector, which had higher emission coefficients than the energy sources used by the other sectors. Hence, the Transportation sector's energy consumption contributed the most to the total emission of NO_x in Beijing.

From 1995 to 2001, the three sectors with the largest energy consumption were Industry > Household Consumption > Transportation. The energy consumed by the Industry sector was an order of magnitude larger than that of the Household Consumption and Transportation sectors. Energy consumption by the Household

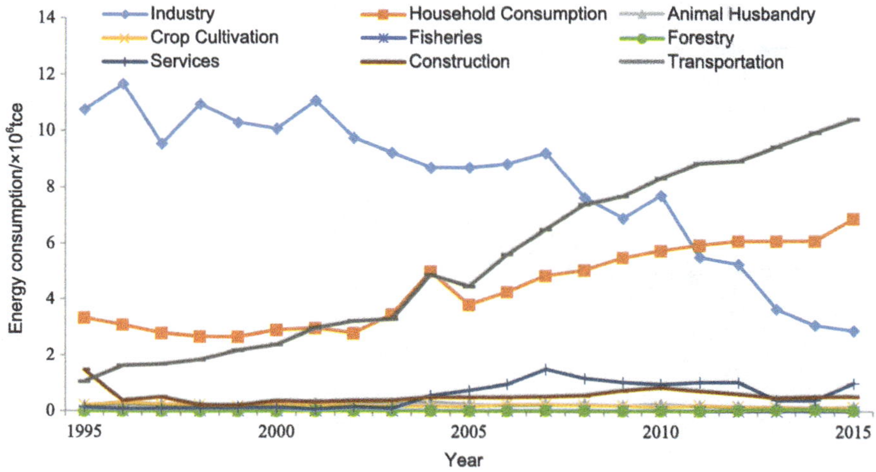

Fig. 10.5 Energy consumption by Beijing's sectors from 1995 to 2015 (*Note* Values are in standard tonnes coal-equivalent)

Consumption sector did not change significantly until about 2003, remaining relatively constant at about 2.8 × 10^6 tce, then increased steadily during the rest of the study period, whereas consumption by the Transportation sector increased steadily and rapidly, with an average annual increase of 12.2%. From 2001 to 2005, the energy consumed by the Transportation sector was comparable to consumption by the Household Consumption sector, whereas the Industry sector decreased continuously, at an average annual rate of 5.8%. From 2005 to 2008, the ranking changed to Industry > Transportation > Household Consumption. Although the two latter sectors had comparable energy consumption during this period, Transportation consumed about 9.8% more energy overall, and increased much faster, at an annual average rate of 18.3%. From 2008 to 2011, the ranking of energy consumption was Transportation > Industry > Household Consumption, with energy consumption by the Transportation sector increasing continuously at an annual average rate of 6.2% and exceeding consumption by the Industry sector in 2009. During the rest of the study period, the ranking changed as Transportation > Household Consumption > Industry, because the total energy consumption by the Industry sector decreased at an average annual rate of 14.9%, while the average annual increase of the Transportation sector decreased to 4.1%.

The other sectors consumed less energy. The consumption by Animal Husbandry, Crop Cultivation, Fisheries, and Forestry remained less than about 0.32 × 10^6 tce respectively, and had similar values throughout most of the study period. Energy consumption by the Services sector remained low and relatively constant until 2003, after which it increased continuously to about 2007, reaching 10.6 times the 1995 value, with an average annual increase of 22.0%. Subsequently, the sector's energy consumption decreased at an average annual rate of 11.5%.

Based on aggregated data, my research group compared the consumption of different types of energy by different sectors. Figure 10.6 reveals great differences between the energy consumption structures of Beijing's nine sectors. Among the sectors, coal was generally still the largest energy category, accounting for more than 30% of the total, except for the Services sector, the Transportation sector, and the Construction sector. Figure 10.6 declares that in the Industry Sector, the two main energy sources were raw coal and coke, with the proportion of raw coal consumption increasing from 36% of the total to 58% during the study period, while the proportion of coke consumption decreased sharply, from 44.9% of the total to 0.1%. The Household Consumption sector was initially dominated by raw coal consumption, but the proportion of gasoline consumption increased steadily from 5.5% to 65.3% during the study period, and in 2006, replaced raw coal as the principal energy consumed. In addition, the gasoline produced high emission of NO_x, with a value of 16.7 kg N/t, and therefore contributed greatly to NO_x production by burning fossil fuel in the Household Consumption sector. For the Transportation sector, which experienced a large increase in the total energy consumed, kerosene was the principal energy consumed, and its proportion of total consumption increased from 58.2% to 76.0%. The NO_x emission factor from burning kerosene was the largest for all sectors, with a value of up to 27.4 kg N/t, leading to the largest increase of NO_x emission from energy consumption by any sector. The consumption of diesel oil by the Transportation sector also increased dramatically, to 7.0 times the 1995 value, but this energy source accounted for only 16.5% of the total in 2015. In other sectors with less total energy consumption, such as Crop Cultivation, Forestry, Animal Husbandry, and Fisheries, raw coal and diesel oil dominated the energy consumption.

10.1.3 Analysis of Anthropogenic Nitrogen Consumption

Anthropogenic nitrogen consumption is an indicator that represents the nitrogen consumption and nitrogen sink status which result from human activities. Figure 10.7 shows that Beijing's consumption of anthropogenic nitrogen can be divided into three distinct periods: a period with a slow increase (1995–2000), a period with an accelerated increase (2000–2010), and a period with decreasing nitrogen consumption (2010–2015). Energy nitrogen was the largest component of the total after 2002, accounting for more than 33% of the total and continuing to grow throughout the study period. The proportions of energy nitrogen and fertilizer nitrogen were roughly equal in 1995 (at about 38% of the total), but energy nitrogen increased to a maximum of 51% of the total by the end of the study period and fertilizer nitrogen decreased to about 20% of the total. The growth of total anthropogenic nitrogen consumption also resulted from food nitrogen, which increased throughout the study period, from 9.7% of the total in 1995 to 20.7% by 2015. The reduction of total anthropogenic nitrogen consumption after 2010 was mainly caused by the large reduction of fertilizer nitrogen. The share of fertilizer nitrogen, which decreased throughout the study period, was similar to that of food nitrogen, at about 20% of the total in 2015. However, feed nitrogen reached a peak around 2004, then decreased to around its 1995 level by the end of the study period. The chemical products nitrogen changed little, accounting for 2% to 3% of the total.

There was a decrease in total nitrogen consumption by 33.6 Gg during the 12th Five-Year Plan period from 2011 to 2015 due to the reduction of feed, fertilizer, and energy nitrogen consumption by 30.7 Gg, 12.5 Gg, and 3.7 Gg, respectively (Fig. 10.7). At the same time, the government further planned to adjust the industrial structure, hoping that the proportions of primary and secondary industries would be reduced. Tertiary industries would increase to more than 78%, with a significant decrease in the fertilizer nitrogen, feed nitrogen, and industrial energy nitrogen as demands of the primary and secondary industries and Beijing's total nitrogen consumption.

Overall, the trend for anthropogenic nitrogen consumption, in which there were two crucial inflection points in the years 2001 and 2010, respectively, followed the trend for reactive nitrogen input (Figs. 10.3, 10.7). There was a significant increase in total nitrogen consumption in 2011, which was mainly driven by increased consumption of food nitrogen, feed nitrogen, and energy nitrogen, accounting for 13.9%, 19.6%, and 35.8%, respectively, due to the increasing demand for food (including animal products) with Beijing's population constantly increasing. In

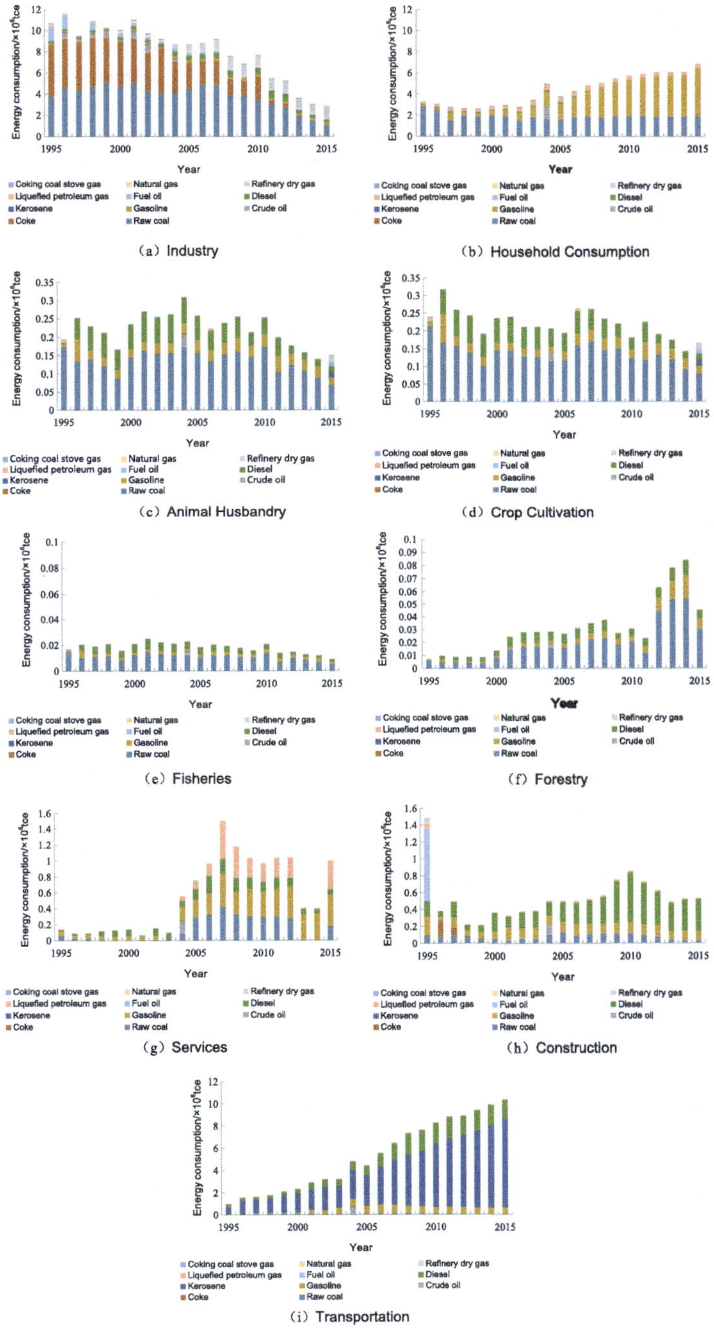

Fig. 10.6 The energy consumption structures of Beijing's nine sectors from 1995 to 2015

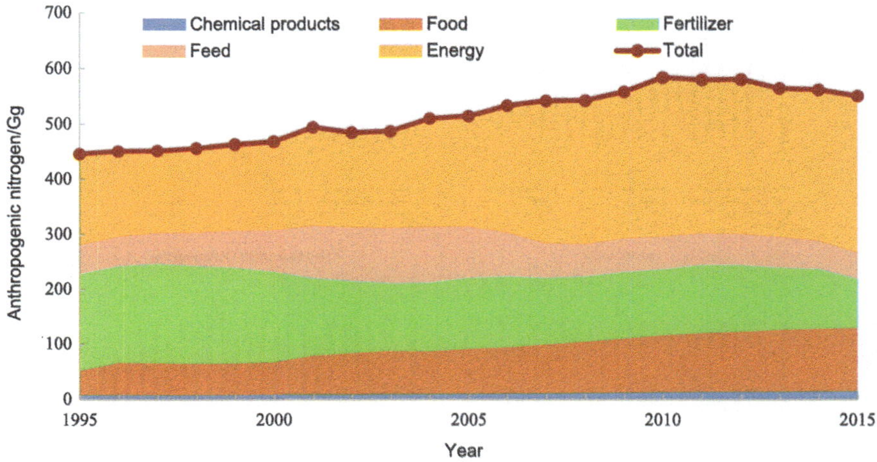

Fig. 10.7 Total amount and structure of Beijing's anthropogenic nitrogen consumption from 1995 to 2015

addition, Beijing's population increased dramatically in 2000 (Fig. 10.8) by 8.5% compared with an average of 0.9% for the previous four years (Zhang et al. 2020). Household Consumption had a similar increasing rate to population. Meanwhile, per capita income was also improving, so there was a tremendous increase in high-nitrogen food consumption, such as meat.

The growth rate of Beijing's total consumption of livestock and poultry from 2000 to 2001 was 33.2%, which was much higher than the average value for the previous five years (14.6%), and then there was a peak in 2011 (Fig. 10.9). As can be seen, this increase contributed to the rapid growth of total food nitrogen consumption in 2001 (Fig. 10.8) at 20.0%, which was also much higher than the average value

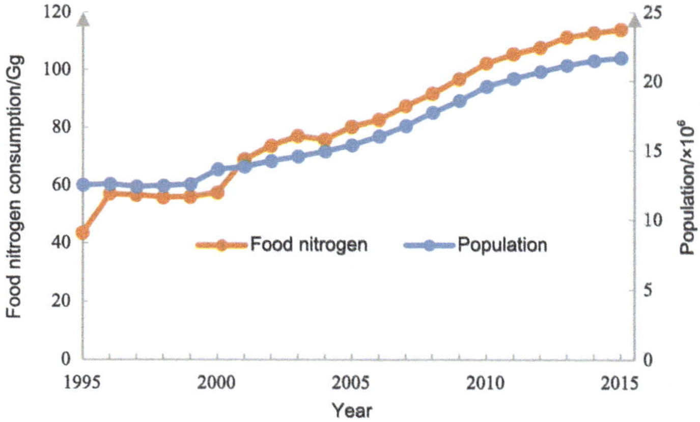

Fig. 10.8 Trends in Beijing's population and food nitrogen consumption from 1995 to 2015

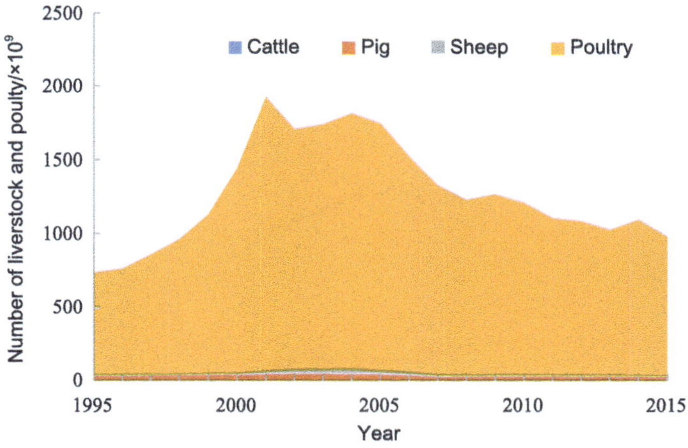

Fig. 10.9 Number of livestock and poultry slaughtered in Beijing from 1995 to 2015

for the previous five years (7.2%). Meanwhile, the increased livestock and poultry production (Fig. 10.9) hugely increased feed nitrogen consumption, which was 27% higher than that in 2000 (Zhang et al. 2020). Poultry overwhelmingly dominated this production during the whole study period, with a proportion of more than 94% of the total livestock.

Besides the rise in total consumption of livestock and poultry, there was an increase in total energy nitrogen consumption during the early period of the 10th Five-Year Plan (2001–2005) with a planned target of 9% GDP growth, when the increase of energy nitrogen consumed by the Industry and Transportation sectors accounted for a relatively large proportion (Fig. 10.10). In summary, there was a peak in Beijing's anthropogenic nitrogen due to multiple factors. For industrial energy, nitrogen consumption, it was decreased significantly. Still, the rate was slower than Beijing's proportion of the secondary industry, which came up to 28.9% during the 11th Five-Year Plan period because of the preparation for the 2008 Olympic Games launching in 2002 and the industrial restructuring proposed by the Chinese government.

From 2002 to 2010, there was accelerated growth at an average rate of 2.6% of Beijing's anthropogenic nitrogen consumption, which was much higher than that of 1.0% in the previous period. Therefore, there was a peak in 2010 (Fig. 10.7). It was interesting that it was closely related to the growth of energy nitrogen, which afterward had a dramatic increasing demand during the period of the 11th and 12th Five-Year Plans (2005–2010 and 2010–2015, respectively) due to the accelerating construction of urban transportation infrastructure. In addition, compared to the year 2001, the energy nitrogen consumption of the Transportation sector had increased by 2.2 times by 2010. At the same time, Beijing's population grew continuously by 41.6% in 2010 compared to the value in 2001. Therefore, the consumption of Beijing's energy nitrogen increased rapidly and constantly to 156% of the value in 2001.

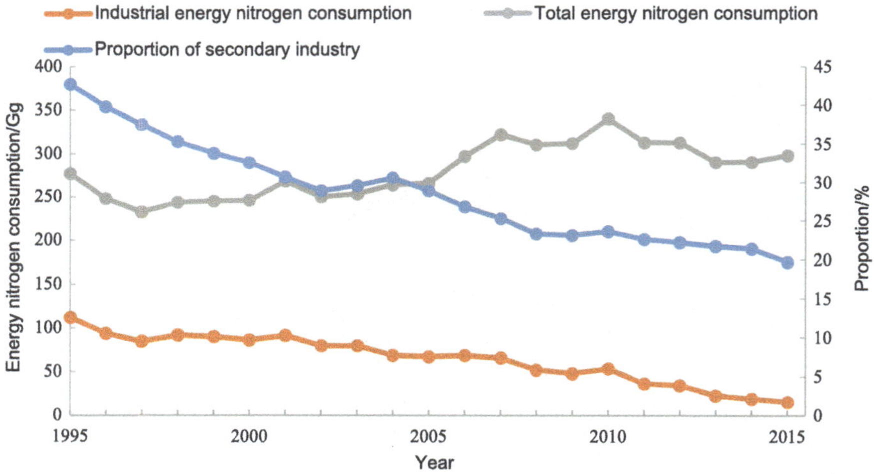

Fig. 10.10 Trends in Beijing's proportion of secondary industry and energy nitrogen consumption from 1995 to 2015

From 2011 to 2015, there was a downturn in total nitrogen consumption of a total amount of 33.6 Gg, which was mainly caused by the reductions in feed (30.7 Gg), fertilizer (12.5 Gg), and energy nitrogen consumption (3.7 Gg), respectively (Fig. 10.7). And the further adjustment of industrial structure helped form this trend of decreasing consumption of Beijing's fertilizer nitrogen, feed nitrogen, industrial energy nitrogen, and total nitrogen. During the 12th Five-Year Plan period (2011–2015), the government wanted to reduce the proportions of primary and secondary industries to make a higher proportion of more than 78% of tertiary sectors.

10.1.4 Contributions of Influencing Factors

As a global megacity, Beijing has a high concentration of socioeconomic activities and a large population. The per capita GDP and the population must therefore be considered to account for the intensity of these activities. In addition, Beijing has gone through a critical period of industrial transformation, in which industries with high energy consumption were shut down or moved to other parts of the country, combined with upgrading the remaining industries through technological innovation aiming to reduce energy and material consumption; therefore, I should take the industrial structure into account. Further in this section, I focused on material consumption structure, and the reasons are as follows. First, it is too high intensity of urban socioeconomic activities and profound industrial transformation that had increased severe pressure by resource consumption and its environmental impacts, and the material consumption reduction has been very urgent. The parameter which can show the material consumption reduction is material intensity, meaning the consumption of a material per unit output. In addition, the variety of urban industries, the complex dietary structure of residents, and differences in consumption levels all affect the material consumption structure (Zhang et al. 2020).

In the research on which this section is based (Zhang et al. 2020), my research group calculated the relative contributions of the scale (population), material intensity (material consumption per unit output), and industrial structure effect on nitrogen consumption. Overall, the scale effect (i.e., the growing population) promoted the rise of anthropogenic nitrogen consumption between 1995 and 2010 (Fig. 10.12), but thereafter, consumption decreased despite the increasing population. The ratio of population to total N consumption was up to 5.5% from 1995 to 2000, versus 12.1% from 2001 to 2010, followed by a decrease to 8.4% from 2011 to 2015, demonstrating that the impact on Beijing's consumption of anthropogenic nitrogen brought by population growth was so enormous that we couldn't ignore it even when the effect began to decrease. However, here presented the opposite pattern of the intensity effect on anthropogenic nitrogen consumption, with decreases during all three periods but the most significant reduction from 2001 to 2010. The results of per capita GDP and material intensity had a proportion of 20.0% of the total from 1995 to 2000, then decreased to 4.3% from 2011 to 2015 since some inhibitory factors, such as material intensity, continuously increased. Some promoting factors, such as the per capita GDP, had gradually weakened. For example, the contribution of material intensity had risen from 21.6% (from 2000 to 2010) to 36.6% (from 2010 to 2015), while the assistance of the per capita GDP decreased from 41.5% of the total (from 1995 to 2000) to 32.3% (from 2010 to 2015). The overall structural effect (the sum of the nitrogen content, material consumption, and industrial structure) consistently inhibited the growth of anthropogenic nitrogen consumption. Still, the magnitude of the inhibition decreased over time (Zhang et al. 2020). This was because the industrial structure's inhibitory effect (more than 15% of the total) was much greater than the promoting effect of both the material consumption structure and the nitrogen content of materials (less than 7.5%). Overall, there was a significantly decreasing trend of consumption by industrial design from the second period to the third period. At the same time, there was an increasing trend in the sum of material consumption structures and nitrogen content.

Of the six factors that influenced consumption, only the material intensity and industrial structure showed inhibitory effects (Fig. 10.12); the other four showed blatant nitrogen consumption promotion. The material intensity and the industrial system had similar inhibitory effects from 1995 to 2000, accounting for more than 20% of the overall contribution. However, the inhibitory effect of material intensity increased dramatically during the second period. In contrast, the effect of the industrial structure weakened significantly during the third period, decreasing to less than half of the contribution of material intensity from 2000 to 2010.

Per capita GDP was the main driving force behind the growth of Beijing's anthropogenic nitrogen consumption (Table 10.1). The contribution of this factor (41.5%) during the first period was close to the combined contributions of material intensity and industrial structure. Although the contribution of per capita GDP decreased after that, it remained greater than 32.0% of the total donation throughout the study period.

It was followed by the industrial structure, which fell from 26.9% during the first period to 15.4% in the last period. The material consumption intensity increased from 21.6% to 36.6% during the study period. The population was also a significant driving factor of anthropogenic nitrogen consumption in Beijing. However, the strength of the population (an overall average of 9.6%) was lower than those of several other factors and varied fairly widely during the study period. There was a continuous decrease in the material consumption structure by two orders of magnitude.

The population factor had a positive but volatile contribution to the growth of Beijing's anthropogenic nitrogen consumption during the study period, as it first nearly doubled and then decreased to about 70.0% of its peak. Meanwhile, the increase of the promotion effect also resulted from the amount of nitrogen content in the materials (ΔN_F, whose contribution increased from 2.1% to 7.3%) (Zhang et al. 2020). Its contribution in the third period increased by 2.4 times its starting value, but it still did not reach 10.0%. For the material consumption structure, its promotion effect decreased continuously by two orders of magnitude. Overall, the above two factors had a relatively small impact on Beijing's anthropogenic nitrogen consumption.

The increasing effect of Beijing's anthropogenic nitrogen consumption was mainly because of the impact of GDP and population, and the growth rate of the former was much larger than that of the latter. However, the ratio of GDP growth to population growth has presented a continuous decreasing trend over time, which was 11.5% during the first period, 10.4% during the second period, and 5.7% during the third period. In fact, this reduction was due to the 12th Five-Year Plan (2010–2015), when China's economy entered a new period with a gradual slowing pace of increasing the quality of the economic development was paid more attention than its quantity and the financial structure should be adjusted. At the same time, there was an increase in the inhibitory effect of the material intensity (that is, the material consumption per unit of GDP) accompanied by the growth of material utilization efficiency with two actions. First, Beijing's Municipal Government promoted moving industries with high energy consumption to other regions and technological improvements, which led to a significant decrease in Beijing's energy consumption per unit GDP from 2.3 tce/10^4 RMB in 1995 to 0.3 tce/10^4 RMB in 2015. Second, to make progress in scientific and technological innovation and improve energy and material utilization efficiency, Beijing's government, issued a series of planning documents in 2017 to promote the development of high-tech industries such as information technology, integrated circuits design and manufacturing, and the production of new energy-efficient automobiles (Zhang et al. 2020). Overall, it was convinced that there would be a sustained growth of the inhibitory effect of material intensity.

Accompanied by changes in Beijing's industrial structure (representing a decreasing trend of primary and secondary industries) the total N consumption of Industry was decreasing, especially the fertilizer and feed nitrogen, which is required by both Agriculture and Animal Husbandry, reducing by 48.9% and 9.9%, respectively. Meanwhile, there was also a decreasing trend of energy nitrogen, which was applied in industrial production, and the chemical product's nitrogen (Fig. 10.7).

Table 10.1 Contributions (C) of the factors (ΔN)[a] that affected Beijing's consumption of anthropogenic nitrogen from 1995 to 2015

Period	ΔN_P	C/%	ΔN_R	C/%	ΔN_{IS}	C/%	ΔN_{ME}	C/%	ΔN_{MS}	C/%	ΔN_F	C/%
1995–2000	39.29	5.46	298.63	41.52	−193.54	−26.91	−155.10	−21.56	17.35	2.41	15.41	2.14
2000–2010	187.81	12.12	584.53	37.74	−279.22	−18.03	−437.58	−28.25	3.14	0.20	56.67	3.66
2010–2015	57.52	8.42	220.86	32.34	−104.92	−15.36	−250.28	−36.64	−0.16	−0.02	49.57	7.26
1995–2015	284.63	9.64	1104.03	37.41	−577.68	−19.57	−842.96	−28.56	20.33	0.69	121.66	4.12

[a] Definitions of changes in anthropogenic nitrogen consumption: ΔN_P, due to changes in the population; ΔN_R, due to changes in per capita GDP; ΔN_{IS}, due to changes in industrial structure; ΔN_{ME}, due to changes in material intensity; ΔN_{MS}, due to changes in the material consumption structure; ΔN_F, due to changes in material nitrogen content

Unlike the primary and secondary industries, the tertiary industries were significantly increasing, especially the service sector (including tourism, catering, and public transit), its proportion of Beijing's GDP rising from 52.5% of in 1995 to 79.7% in 2015. This led to significant consumption of anthropogenic nitrogen and furthermore, the consumption of food and transportation energy nitrogen, rising by 0.6 and 7.1 times the 1995 level, respectively (Figs. 10.7, 10.10). Overall, there was a decrease of the strength of Beijing's anthropogenic nitrogen consumption by the industrial structure factor that it had a proportion of 26.9% in the first period, and then 14.6% during the second period (Fig. 10.11). Contrastly, for the factor of the nitrogen content of materials, its contribution was consistently less than 10%, and we could owe this phenomenon to changes of composition of some materials, including energy, feed, inorganic fertilizer, and chemical products. For example, there were major changes to the dietary structure of Beijing's Household Consumption because of the standard of living improving with the growth of per capita GDP. The proportion of food with a high nitrogen content (meat, eggs, and milk) increased from 15.5% in 1995 to 41.5% in 2015 (Zhang et al. 2020). The proportion of the nitrogen content of materials was increasing steadily, which had the same developing trend of consumption, from 2.1% in the first period to 3.7% in the second period and then to 7.3% in the third period (Fig. 10.11). Meanwhile, there was also a continuously small contribution of the material consumption structure, which was less than 3.0% of the total and then less than 1.0% during the second period. This demonstrated that there was little contribution of the changing material consumption structure to the consumption of Beijing's total anthropogenic nitrogen.

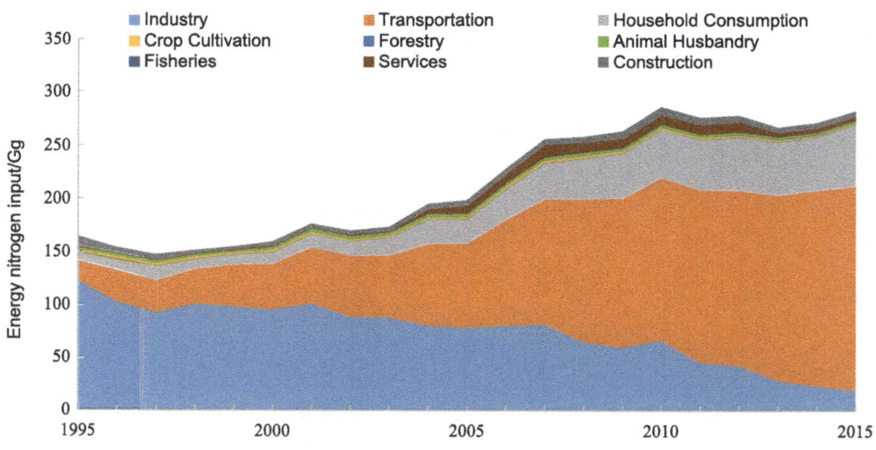

Fig. 10.11 Changes of Beijing's energy nitrogen consumption structure from 1995 to 2015

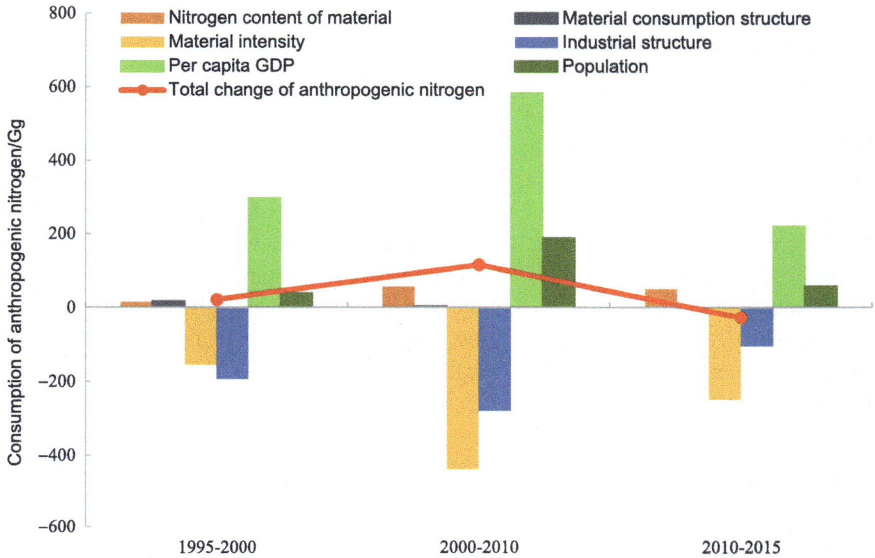

Fig. 10.12 Contributions of the factors that affected Beijing's nitrogen consumption from 1995 to 2015

10.1.5 Comparison with Previous Research on the Total Characteristics of Urban Nitrogen Metabolism

We should have the ability to estimate the inputs, accumulation, and outputs of nitrogen related to human activities (Keeney 1979), and subsequently, the scholars agreed with this proportion. Actually, there were many apparent differences between the human systems and natural ecosystems, and human systems have significantly affected the global nitrogen cycle (Vitousek et al. 1997). Among the world's human systems, cities have become the most full sink of nitrogen because of the large demand for matter and energy for themselves (Kaye et al. 2006). It is universally acknowledged that the initial point to knowing the nitrogen cycle of urban ecosystems is a detailed nitrogen budget (Baker et al. 2001). Indicators of anthropogenic nitrogen flows can characterize these flows from the source through consumption and emission, so nitrogen accounting based on these indicators can quantify these flows (Zhang et al. 2020).

We listed the per capita anthropogenic nitrogen consumption in this study and the related data in urban areas in previous studies (Table 10.2), among which there were three studies conducted in China, including this study, with the result of per capita consumption of more than 32.0 kg. However, there was only one study in Phoenix comparable, with only around 20% of Beijing's population and per capita consumption annually at 29.60 kg, which was at least 8% lower than that of studies in China. Furthermore, the nitrogen consumption of food, fertilizer, and energy consumption was also small in Phoenix, with a proportion of less than 20% of Beijing's corresponding consumption. Shanghai's per capita anthropogenic nitrogen consumption was the lowest of the Chinese values, at 32.04 kg annually, which is slightly lower than the Chinese average level of 32.12 kg annually in 2005 (Zhang et al. 2020), probably because inorganic fertilizer nitrogen was not considered in the accounting processes of Shanghai's nitrogen. Compared with the average per capita anthropogenic nitrogen consumption in Chinese cities, the value of the whole world (about 29.32 kg annually in 1990) was a little lower than that, while the value of Asia (about19.92 kg annually in 1995) was much lower than that, due to lower food and energy nitrogen consumption in Asian countries accompanied by a relatively low level of urbanization in Asian countries. For the average per capita anthropogenic nitrogen consumption in Brazil, the value at the national level was nearly equivalent to that at the urban level in 1995 at 30.86 kg annually. However, this value had increased to 53.71 kg annually by 2002, much higher than the value of any other studies. This was possible because, under Brazil's unique geographical and climatic conditions, the plants proliferate, and organic matter decomposes rapidly. The poor soils need extra nitrogen input by supplemental nitrogen fertilizer. The nitrogen consumption of Brazil's agricultural products had a tremendous increase and a significant proportion of 94% of the total consumption, which was as twice the value in China, despite 2005 Brazil's population being less than 15% of China's population.

Due to data limitations, we used interpolation to estimate some missing data in the study on which this section was based (Zhang et al. 2020). The accuracy of the data also needs to be improved. For example, the consumption of certain chemical substances in some years was not available. We, therefore, estimated the anthropogenic nitrogen consumption for these missing chemical substances by calculating

Table 10.2 Comparison of per capita anthropogenic nitrogen consumption in different regions of the world

Area	Time	Annual per capita anthropogenic N consumption/kg	Source
Global	1990	29.32	Galloway et al. (2004)
	2005	28.77	Cui et al. (2013)
Asia	1995	19.92	Galloway and Cowling (2002)
Brazil	1995	30.86	Filoso et al. (2006)
	2002	53.71	
China	2005	32.12	Galloway et al. (2004)
Phoenix	1996	29.60	Baker et al. (2001)
Hangzhou	2004	34.02	Gu et al. (2009)
Shanghai	2004	32.04	Gu et al. (2012)
Beijing	1996	35.82	Present study
	2004	34.16	

the overall per capita nitrogen consumption for chemical products (i.e., by dividing Beijing's total chemical product nitrogen consumption by the population in a given year), and then multiplying the per capita value by the population in each focal year. There was also inadequate data for the city's recycling system. If the actual chemical consumption data can be obtained through the future empirical research, then the accuracy of the results will improve. Given the importance of such studies and the frequency with which they are being conducted, we also propose greater transparency and consistency of methods and data collection to facilitate comparisons among case studies. We could also provide more suggestions through the improved data on identifying industries that consumed the maximum nitrogen to take targeted actions to reduce nitrogen consumption.

10.1.6 Comparison with Previous Research on the Structural Characteristics of Urban Nitrogen Metabolism

From the global perspective, the reactive nitrogen in the terrestrial ecosystem has supplied doubled mainly due to the extra demand for food and energy since the year 1860 and the problem that extensive nitrogen accumulated appeared (Galloway 1998). Among the different components of anthropogenic nitrogen inputs, the two critical parts are food and energy (Yu et al. 2012; Galloway et al. 2004). We listed the existing variation in different studies as much as possible (Table 10.3), and we found the main reason—other calculation methods, which are described below. Therefore, as we perform more and more case studies of nitrogen accounting, researchers should adopt more consistent standardized practices to facilitate comparisons between studies (Zhang et al. 2020).

There was an exciting result that food nitrogen consumption in Beijing was as 1.2 times as that in Phoenix in 1996 since the calculation methods of the two studies were different. The research in Phoenix utilized the protein content needed by other age groups to calculate the food nitrogen. In contrast, our research in Beijing utilized actual patterns of food consumption and estimated accounting coefficient to calculate the food nitrogen. For feed nitrogen consumption, its value in Beijing was more than three times that value in Phoenix because of the difference in calculation methods. The research in Beijing only estimated the amount of feed nitrogen for the production of dairy products, while the study in Beijing took more sources of feed nitrogen into accounts, such as the feed nitrogen for the production of meat, eggs, and dairy.

For nitrogen consumption in agricultural products, only biological nitrogen fixation was considered in Hangzhou, while all food nitrogen except biological nitrogen fixation was considered in Beijing, such as meat, eggs, and milk, and therefore food nitrogen consumption in Hangzhou in 2004 was only about 70% of that value in Beijing. In contrast, the number of livestock in Hangzhou was 2.6 times that value in Beijing, while Hangzhou's feed nitrogen consumption was only more than half that in Beijing because we took poultry and fish into account in the study in Beijing

Table 10.3 Comparison of the per capita inputs of food nitrogen, feed nitrogen, and energy nitrogen in the present study of Beijing and in cities around the world

City	Year	Urbanization rate/%	Per capita food N/(kg/a)	Per capita feed N/(kg/a)	Per capita energy N/(kg/a)	Source
Paris	2006	–	8.07	–	–	Billen et al. (2012)
Phoenix	1996	77.0	3.67	1.30	13.46	Baker et al. (2001)
Toronto	2001	–	6.40	–	–	Forkes (2007)
	2004	–	6.35	–	–	
Shanghai	2004	81.16	8.33[a]		13.78	Gu et al. (2012)
Hangzhou	2004	43.40	3.55	10.0	5.03	Gu et al. (2009)
Xiamen	2008	68.28	7.21	–	16.41	Huang et al. (2016)
Beijing	1996	76.06	4.53	4.30	12.20	Present study
	2004	79.53	5.08	6.86	13.06	
	2006	84.33	5.16	5.05	14.25	
	2008	84.90	5.18	3.37	14.58	

[a] This value equaled the sum of food and feed nitrogen

and the population in Beijing was 30% more than that of Hangzhou. The per capita food nitrogen consumption values for Xiamen, Shanghai, Toronto, and Paris were all more than 1.0 kg annually, more significant than Beijing's value in the same year, mainly due to the dietary structure of their Household Consumption sectors (Zhang et al. 2020). We could take Paris as an example. The amount of nitrogen in fish, meat, and eggs (high-nitrogen foods) was 59% more than in Beijing in 2006. The study conducted in Toronto used protein consumption to achieve the result of per capita food nitrogen consumption rather than a total diet.

The per capita annual consumption of energy nitrogen was in the range of 12–17 kg in most cities (Table 10.3), but there was only an exception—Hangzhou, accounting for 38.5% of Beijing's per capita energy nitrogen consumption. However, the same parameter in Shanghai was 10% higher than that in Beijing. These differences relate to differences in the energy consumption structure of each city (Zhang et al. 2020). High-nitrogen fuel oil consumption had a proportion of 7.0% of Beijing's total energy consumption in 2004, which was as nearly double the balance in Hangzhou (3.9%). Furthermore, the low-nitrogen raw coal consumption had a relatively large proportion of 89.1%, which led to per capita energy nitrogen consumption is much lower. Compared with Shanghai, Beijing consumed less energy provided by oil but more high-nitrogen coke and the proportion were 17% and 130% (Beijing accounting for 32.2% of the total; while Shanghai accounted for 24.4% of the total), respectively, so Beijing's consumption of high-nitrogen coke considerably narrowed the gap in per capita energy nitrogen consumption. For Xiamen, its per capita energy nitrogen consumption was at annually 16.41 kg, which was the highest among listed cities in 2008, while the total value was 80% less than the total value of Beijing at the same time, largely because Xiamen's population in 2008 was less than 20% of Beijing's population. These two effects compensated for each other to some extent, thereby reducing the gap between the two cities (Zhang et al. 2020).

With Beijing's urbanization rate at the range from 76 to 80%, there was an obvious increase in the per capita food nitrogen consumption. Contrastly, there was also an obvious increase in per capita energy nitrogen consumption when the urbanization rate was between 80 to 86%, and there was a peak at the urbanization rate of 84.5% (Fig. 10.13). For the per capita food and nitrogen consumption, the values in Beijing were higher than those of Phoenix but lower than those of Xiamen and Shanghai. Therefore, consumption-dominated cities, represented by Shanghai and Xiamen mentioned above, could absorb some experiences from some regulations made by other cities' governments, such as slowing down urbanization rate and per capita nitrogen consumption and reversing energy consumption structure. According to the research of Hangzhou (Gu et al. 2009), the per capita nitrogen consumption increased with the urbanization rate increasing, but the growth rate of per capita nitrogen consumption was different among the different urbanization rates. For example, when Hangzhou's urbanization rate increased from 25.0% to 43.4%, the per capita food and energy nitrogen consumption increased by about 0.06 kg annually for every 1% increase in the urbanization rate (Gu et al. 2009; Zhang et al. 2020). However, the relationship between the urbanization rate and per capita annual nitrogen consumption was different in Beijing that with the range of urbanization rate of Beijing from 83.6% to 86.5%. When the urbanization rate increased by 1%, there was an increase of 0.12 kg/a of its per capita food nitrogen consumption while there was only an increase of 0.03 kg/a of its per capita annual energy nitrogen consumption.

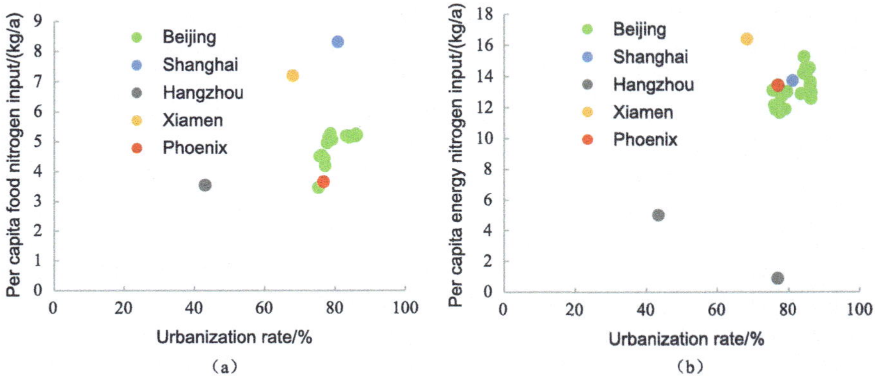

Fig. 10.13 Comparison of per capita food nitrogen and energy nitrogen input in each city (*Note* Data and their sources are shown in Table 10.3)

10.1.7 Comparison with Previous Nitrogen Metabolism Research

Analyzing the factors significantly related to the consumption of anthropogenic nitrogen has become a very important part of guiding us to reduce and control nitrogen consumption (Liu et al. 2014). Based on the previous sections, we studied anthropogenic nitrogen consumption and its structural characteristics in order to identify and quantify the effects made by socioeconomic drivers behind anthropogenic nitrogen consumption and advance some scientific suggestions to support the healthier development of urban ecosystems. And in the existing research related to nitrogen consumption was the Logarithmic Mean Divisia Index method, which could divide driving factors into effects of scale, structure, and intensity. The scale effect usually leads to increasing in nitrogen consumption and emission (Pang et al. 2013; Li et al. 2012a; Zhang et al. 2020), while the economic scale factor usually contributes more than 40% due to the main responsibility of consumption and emission of anthropogenic nitrogen (Ding et al. 2017; Jia et al. 2017; Wang 2017).

It was the economic development needs of artificial systems that leaded to high proportion of the scale effect (Sect. 10.1.4), which could be supported by the above analysis that it always accounted for 5% to 15% of the total contribution as a driver of the growth of anthropogenic nitrogen consumption in Beijing. This is because the population factor was chosen as the scale effect, and fortunately per capita GDP factor contained both the population and the economic scale. In our study in Beijing, the contribution of per capita GDP drove the growth of more than 32% of the total anthropogenic nitrogen consumption in the whole study period. Therefore, it was the main factor (an intensity factor). In addition, it was proved that there was an inhibitory effect of both technological factors (Ding et al. 2017; Li et al. 2012a; Pang et al. 2013) and intensity factors (Jia et al. 2017; Wang 2017) on the growth of nitrogen consumption and emission in most previous studies, and the result in this study was in line with previous studies. However, there was also a high inhibitory effect of the material intensity factor, with a contribution of more than 20% of the total. There was not only promoting effect but also an inhibitory effect of structural factors on reactive nitrogen consumption for a different phase in previous research (Ding et al. 2017; Wang 2017) but there was only the inhibitory effect of the industrial structural factors in our present study in Beijing, which was because of Beijing's adjustment of industrial structure to downsize the proportion of industries with high nitrogen consumption. This approach may be unique to Beijing's political and cultural status and may not be possible in all cities (Zhang et al. 2020).

Recently, the rapid growth of Beijing's population brought harm to sustainable urban development, which was realized by the Beijing Municipal Government, representing a series of control measures implemented. Hence, the growth of Beijing's population has slowed since 2010. Furthermore, on 11 August 2016, Beijing implemented a points-based domestic registration system to strengthen further population control (Zhang et al. 2020). Actually, decreasing the population will be very difficult because of Beijing's status as the capital of China and too large population. It is predicted that population will be positive but gradually weaken the driving force of Beijing's anthropogenic nitrogen consumption for a long time. It will be another difficulty that we should promote the recommendations for Beijing on a global scale if we want to strengthen the world's nitrogen management.

Inevitably, there were only six factors with available and comprehensive data in our factor-decomposition model (Sect. 10.1). We can further rigorously analyze more industrial sectors and more types of nitrogen-containing materials by recounting the impact of factors, including the industrial structure, based on the condition that more detailed data on additional factors can be collected. Using these additional data, we can find a more detailed model to compare the development patterns of different cities. It can improve our understanding of differences in the developing processes promoting urban ecology.

10.2 Analysis of Beijing's Nitrogen Metabolism Network

The socioeconomic sectors were regarded as a whole black box at an early stage, in which only nitrogen exchanges with the external natural environment or major metabolic processes of nitrogen were considered, such as the nitrogen metabolic processes in the food system. However, the metabolic perspective can cover the above shortages as it can describe the overall processes, not only those related to the external environment. The concept of urban metabolism was proposed in 1965, and it subsequently provided a research framework for urban nitrogen metabolism (Wolman 1965). Then there have been increasing researchers focusing on urban nutrient metabolism since the early twenty-first century, especially urban metabolic processes related to nitrogen and phosphorus (Faerge et al. 2001).

From this perspective, a city is seen as a giant organism in which the metabolic processes of nitrogen are analyzed that it absorbs nitrogen as both natural and anthropogenic inputs, and then circulates, transformations nitrogen among different sectors, and finally releases nitrogen outside as outputs of nitrogen. However, most studies of these nitrogen metabolisms have regarded the natural environment solely as the support for urban socioeconomic subsystems, and did not include this environment as an equal metabolic actor in their models when they analyzed the interactions among the sectors involved in nitrogen metabolism from a network perspective.

In the research on which this section is based (Zhang et al. 2016a, 2016b), my research group analyzed the processes in Beijing's nitrogen metabolism from a network perspective to identify the key nodes in the urban system (i.e., the ones that had the most frequent relationships with other nodes), whether these actors were socioeconomic sectors or part of the city's natural ecosystem, and the key pathways among the components of the overall system (those that had the largest nitrogen flows). Figure 10.14 shows the results of analytical framework. We also used network-flow and utility analysis methods to simulate the integrated flow (i.e., the total of the direct plus indirect flows) and the distribution of utility among the network's nodes. The results provide the capability to deeply characterize the relationships among the nodes and the indirect effects caused by the frequent exchanges within the network. This, in turn, let us identify the nodes that controlled the network or that represented bottlenecks. Using the concepts of a food web and a trophic hierarchy, we then quantitatively simulated the hierarchical structure of the urban nitrogen metabolism processes, determined their health and stability, and identified key metabolic sectors that influenced the state of the whole system (Zhang et al. 2018). The results provide a combination of empirical and theoretical support for identifying the most important sectors to reduce nitrogen consumption, improve nitrogen-use efficiency, and maintain the stability of the city's nitrogen metabolism to promote sustainability.

10.2.1 Direct-Flow Analysis of Nitrogen Metabolism

Figure 10.15 shows the direct flows (paths with a length of 1) between network nodes from 1995 to 2015 in Beijing. Nitrogen flows from the external environment and self-feedback flows within a sector have been omitted to simplify the diagrams. From 1995 to 2015, Beijing's direct flow network illustrated heterogeneous characteristics clearly (i.e., the flow diagrams showed wide variation of the flows along the different paths among Beijing's metabolic actors). The difference between the maximum and minimum direct flows was 7 orders of magnitude. In the direct-flow network, only 20% of the paths had a relatively large flow (greater than 30 Gg). The number of paths was greatest for direct flows less than 10 Gg (black line), which accounted for about 60% of the total network paths (33 paths). There were 10 paths with a direct flow between 10 and 30 Gg (pink lines), which represented 20% of the total network paths.

Before 2000, the paths with the largest direct flows were concentrated in the upper right side of the network, including the triangle formed by the vertices of node 2 (Industry), node 12 (Atmosphere), and node 15 (Cultivated Land), and the path from node 15 to node 4 (Crop Cultivation). As time passed, a triangle from node 2 to node 9 (Transportation) and then to node 12 also gradually emerged. Nodes and paths with a flow greater than 30 Gg accounted for about 10% of the total network input throughout the study period, and can be regarded as important nodes and paths. Node 2 had the largest number of output paths and the largest output flow. The largest receiver of its output flows was node 12, but the flow along this path decreased over time, reaching only 17% of the 1995 flow by 2015. At the same time, the proportion of the total output flows accounted for by node 2 also decreased, from 40.6% in 1995 to 6.6% in 2015, reflecting the gradual reduction of Beijing's industrial NO_x emissions.

Node 12 (Atmosphere) not only received NO_x from the combustion of energy by node 2 (Industry), but also received emissions from volatilization of organic fertilizer and denitrification of organic fertilizer in node 15 (Cultivated Land) that should not be ignored. The proportion of total input flows accounted for by node 12 has decreased,

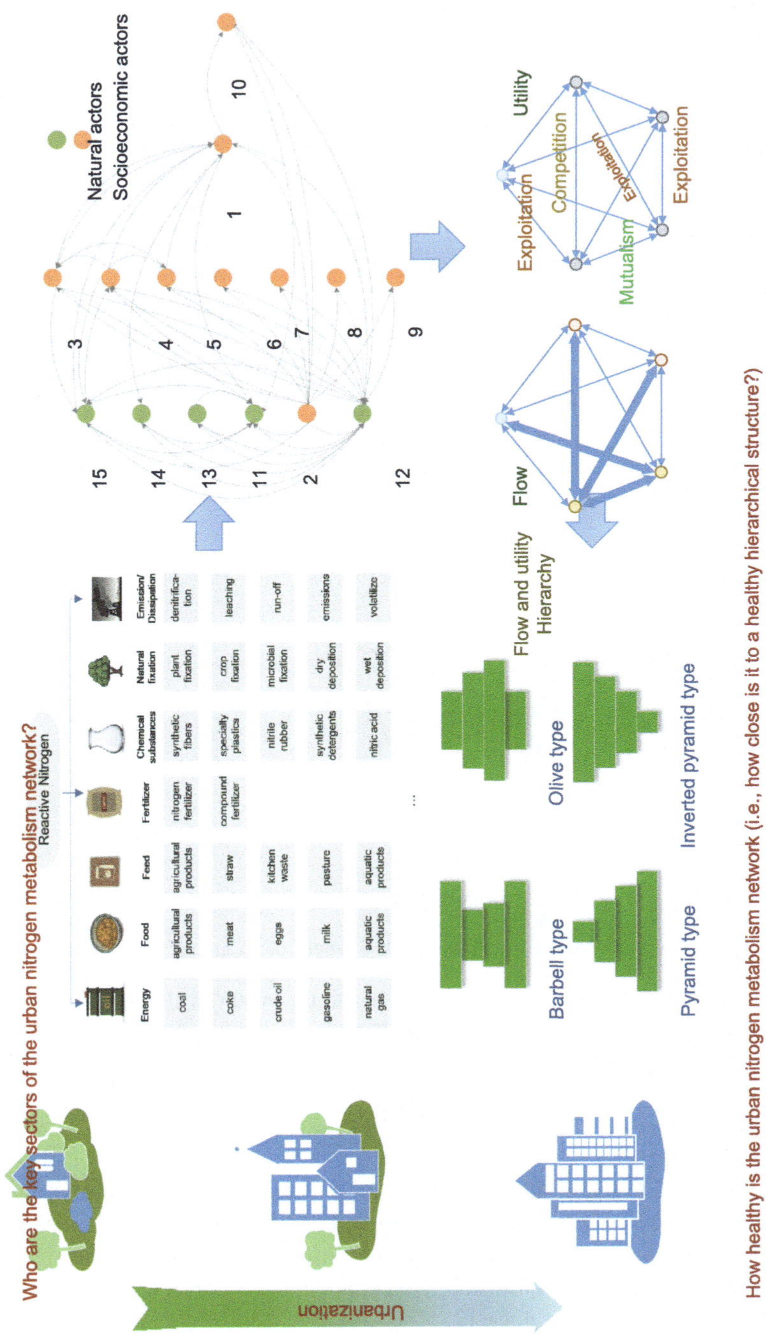

Fig. 10.14 Analytical framework for an urban nitrogen metabolism network and its hierarchical and ecological characteristics

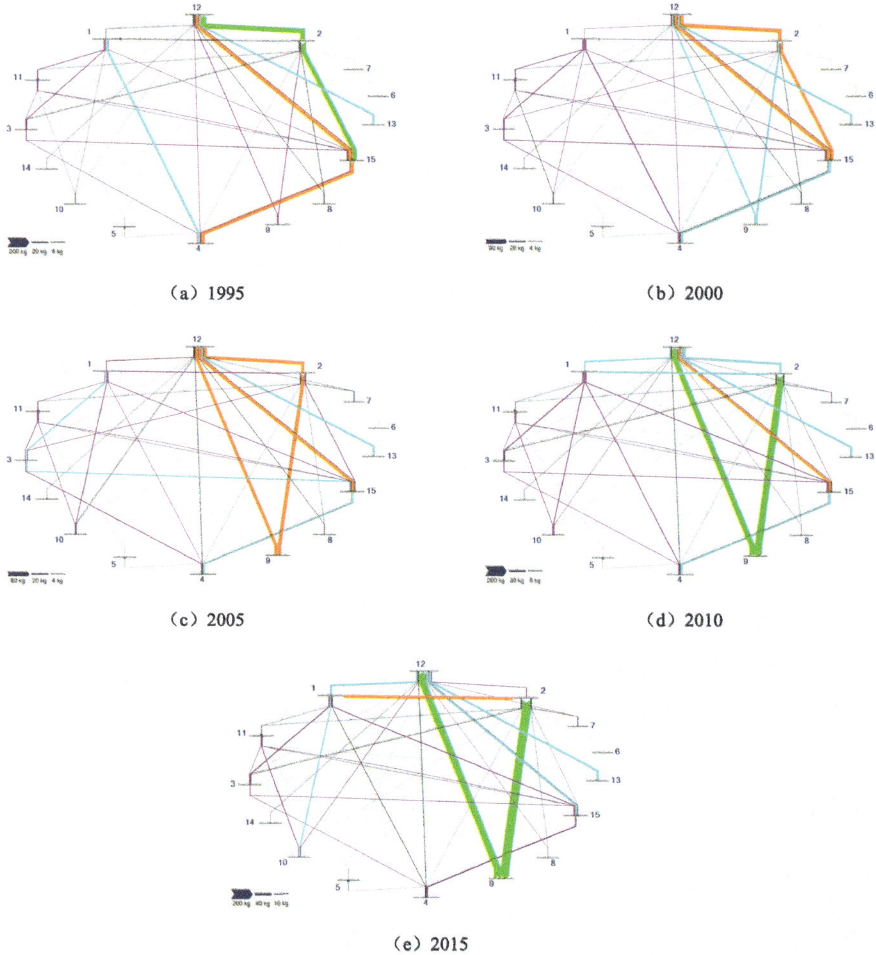

Fig. 10.15 Direct flows of nitrogen (i.e., paths with a length of 1) among the components of Beijing's nitrogen metabolic network from 1996 to 2015 (*Note* the *green*, *orange*, *blue*, *cyan*, and *gray lines* represent flows of >90 Gg, 60 Gg to 90 Gg, 30 Gg to 60 Gg, 10 Gg to 30 Gg, and 0 to 10 Gg, respectively. Nodes: 1-Household Consumption, 2-Industry, 3-Animal Husbandry, 4-Crop Cultivation, 5-Fisheries, 6-Forestry, 7-Services, 8-Construction, 9-Transportation, 10-Sewage Treatment, 11-Surface Water, 12-Atmosphere, 13-Forest, 14-Grassland, 15-Cultivated Land. The pathways sent or received by different metabolic bodies are shown in different colors)

from 67.3% in 1995 to 23.0% in 2015. In addition to having the largest input flow, node 12 had the largest output flow, mainly to node 13 (Forest). Nitrogen obtained from node 12 was captured by node 13 through nitrogen fixation and atmospheric deposition, but this output flow has stabilized at about 45 Gg, accounting for about 75% of the total output flow of node 12.

Some nodes had smaller flows that gradually increased to more than 30 Gg. For instance, the flows of chemical products from node 2 (Industry) to node 1 (Household Consumption) were less than 30 Gg from 1995 to 2006. Thereafter, the flow was greater than 30 Gg, which represents an increase to 5.9 times the 1995 value by 2015. At the same time, the waste discharge by node 1 also increased, and the average annual growth rate of nitrogen output to node 10 (Sewage Treatment) and node 12 (Atmosphere) reached 10.2 and 9.6%, respectively, by 2015. The discharges by node 1 increased above 30 Gg in 2008 and 2011. The energy nitrogen consumption by node 9 (Transportation) increased significantly during the study period. Since 1997, it has exceeded 30 Gg, and the NO_x output flow to node 12 also began to exceed 30 Gg. The energy that node 9 obtained from node 2 and the NO_x output to node 12 both increased to 5.4 times the 1995 level by 2015.

10.2.2 Integrated Flows of Nitrogen

Indirect flows pass through at least one intermediate node before they reach their final destination, and adding these flows to the direct flows creates the total integral flow. Figure 10.16 shows that the integral exchanges among the nodes in the flow network were more frequent than the direct flows, increasing by about 3 times, from 51 direct paths to 159 integrated paths, and the flows along these paths also increased. The integral flows along most paths were still dominated by flows less than 10 Gg, which accounted for about 74% of the total number of flow paths. The number of paths with an integral flow of 10 Gg to 30 Gg ranged from 19 to 24 during the study period, which is about twice the number of direct flow paths, and accounted for 6.3% to 15.1% of the total integral flow. The difference between the maximum and minimum flows in the integral flow network was 6 orders of magnitude, and high variability in the flows was still obvious.

The paths whose integral flow was always above 30 Gg were all related to node 12 (Atmosphere), and the sum of their input and output flows accounted for more than 8% of the total flow along all paths in the network. Output flows from node 12 were mainly supplied to node 2 (Industry), which accepted about 30% of the output flow. However, the change in these output flows was not obvious, in contrast with the downward trend for direct flows from node 12. In addition, the contribution of node 15 (Cultivated Land) to node 12 (Atmosphere) was relatively large, but fluctuated over time. In 2015, it decreased to 66% of the 1995 value, and the proportion of atmospheric input flows also dropped from 16.8% to 7.1%, which is the same trend observed for the direct flows. The input and output flows of node 12 were both large, and the main output flow was transferred to node 13 (Forest), accounting for about 15% of the total atmospheric output flow. The difference between the integral flow and the direct flow from node 12 to node 13 in each year was small, and the changes over time were small. The paths with a large integral flow to or from node 12 also had large direct flows, with values always greater than 30 Gg.

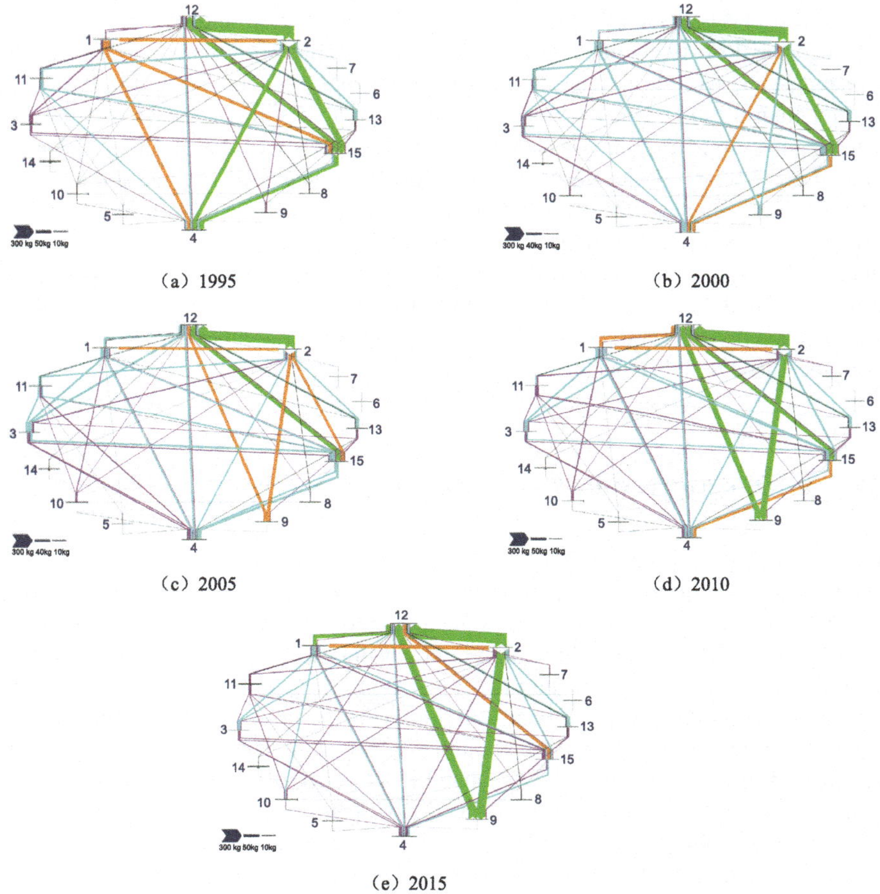

Fig. 10.16 Integral (direct plus indirect) nitrogen flows among the sectors of Beijing's nitrogen metabolic network (*Note* the *green*, *orange*, *blue*, *cyan*, and *gray lines* represent flows of >90 Gg, 60 Gg to 90 Gg, 30 Gg to 60 Gg, 10 Gg to 30 Gg, and 0 to 10 Gg, respectively. Nodes: 1-Household Consumption, 2-Industry, 3-Animal Husbandry, 4-Crop Cultivation, 5-Fisheries, 6-Forestry, 7-Services, 8-Construction, 9-Transportation, 10-Sewage Treatment, 11-Surface Water, 12-Atmosphere, 13-Forest, 14-Grassland, 15-Cultivated Land. The pathways sent or received by different metabolic bodies are shown in different colors)

In addition, the integral flow of node 1 (Household Consumption) became increasingly significant over time, mainly due to the contribution of node 2 (Industry) and node 4 (Crop Cultivation), with the contribution from node 2 to node 1 accounting for about 16% of the input flow of node 1 in all years. The direct flow along this path increased considerably during the study period, but the change of the integral flow was relatively small. The integral flow from node 4 to node 1 fluctuated frequently, but it remained at around 71 Gg from 1995 to 1998. From 1998 to 2003, there was a substantial decrease (by about 50%), but the flow then doubled from 2003 to 2008.

Although the trend for flows from node 4 to node 1 was the same as the trend for direct flows, the magnitude of the change was smaller than that for direct flows. Node 2 (Industry) was the main supplier of nitrogen to all other nodes, and it had the largest number of output paths. At the same time, the sum of the input and output flows for this node accounted for about 32% of the total network flows.

In addition to the important flows for node 12 (Atmosphere), node 1 (Household Consumption), node 2 (Industry), and node 3 (Animal Husbandry) also had large input flows, mainly from node 4 (Crop Cultivation) and node 15 (Cultivated Land). The input flow of node 4 accounted for about 22% of the input flow for node 3 (Animal Husbandry) during the whole period of the study. The integral input flow from node 15 to node 3 was the only indirect flow between these two nodes.

Changes in the paths in the integral flow network are also worthy of attention. Some have changed from small paths to important paths with a flow >30 Gg. The nodes with the most obvious growth rates included node 10 (Sewage Treatment) and node 9 (Transportation). The average annual growth rates of the total input and output flow for these two nodes were 8.6% and 11.3%, respectively. The integral flow from node 1 (Household Consumption) to node 10 increased to about 5.9 times the 1995 value. The integral flows from node 2 (Industry) to node 9 and from node 9 to node 12 (Atmosphere) increased to 5.6 and 5.1 times the 1995 value, respectively. The integral flows of these three paths were all close to the direct flows. Some flows were greater than 30 Gg in the early phase, and then decreased again to <30 Gg. The most obvious change was observed for the integral output flow from node 2 to node 3 (Animal Husbandry), node 4 (Crop Cultivation), and node 11 (Surface Water). The integral flows along these paths decreased to about 37% of their 1995 levels by 2015. At the same time, the direct flows remained less than 30 Gg throughout the study period, and the indirect flows for these nodes made the integral flows unusually large in some years.

10.2.3 The Relationships Between Metabolic Actors

The mutualism index (M) that was described in Chapter 5 can be calculated from the integrated utility sign matrix, sgn(U), to support additional analysis of the distribution of ecological relationships among the nodes of a metabolic network. Figure 10.17 shows the changes in Beijing's M from 1995 to 2015, and reveals periods with decreases and increases. M ranged between 0.7 to 0.9 during the study period, indicating that negative relationships dominated the system (i.e., $M < 1$). There was a slight early increase (by 0.9%/a) between 1995 and 1997, followed by a steady decrease (by 3.5%/a) between 1997 and 2001. M then increased rapidly (by 3.5%/a) to reach the 1997 level by 2004. From 2004 to 2009, M showed a reduction once more. Thereafter, it fluctuated around the value of 0.8. The proportion of mutualism relationships decreased slowly during the study period, whereas the proportion of competition relationships remained relatively stable and the proportion of exploitation relationships slowly increased.

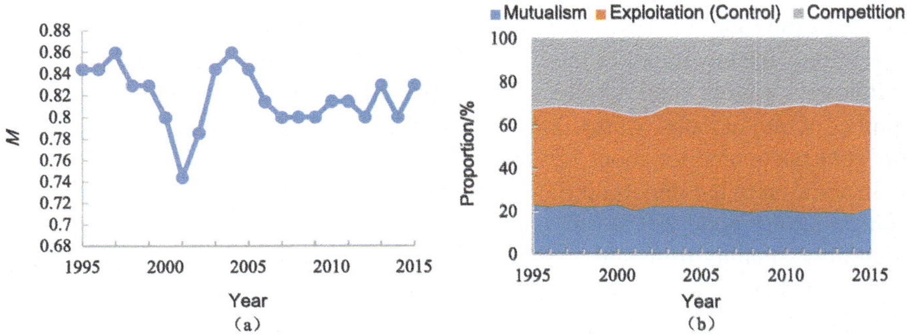

Fig. 10.17 Changes in the value of the mutualism index (M) and the proportions of the ecological relationship in each category for Beijing's nitrogen metabolic system

The fluctuation of M was influenced by changes in the 225 pairs of relationships that were defined in Beijing's urban nitrogen metabolic system (Zhang et al. 2018). There was an average proportion of 44.4% for exploitation relationships between 1995 and 2002, while the value was transformed to 47.9% between 2003 and 2015. For exploitation relationships, there was an average proportion of 50.7% in 2013 and 2014, while the value was lower than 50.0% in all other years. There was a proportion in the range between 29.0% and 36.0% for competition relationships during the whole study period. Contrastly, the mutualism relationships accounted for a maximum of 23.6% during the entire study period as the lowest proportion of the three types of relationships, which was approximately equal to half of the exploitation relationships, and this proportion indicated the lack of mutualistic relationships because of its significance for the health of the system. Overall, there were a slight increase in exploitation relationships and a slight decrease in competition and mutualism relationships, and three types of relationships remained stable in the whole study period from 1995 and 2015, which expressed the meaning that changes in their proportions of them were small. Furthermore, we counted the unchanged relationships between nodes in at least 15 years, including competition, exploitation (hereafter, including control), and mutualism relationships, as can be seen in Fig. 10.18. For the bidirectional relationships between any two sectors, there remained a stable trend during the whole study period, and exploitation relationships had the largest proportion of more than 42.0% of the total unchanged relationships, while mutualism had the smallest proportion. Therefore, the contradictions in Beijing's nitrogen metabolic processes were becoming more obvious (Zhang et al. 2020), and we should pay more attention to ecological relationships among the sectors in urban systems.

The values of M for the relationships were between 0.7 to 0.9 during the whole study period, but it was different from the values for 15 individual nodes. During the whole 21 years, there were several nodes whose values of M were greater than 1, such as Surface Water, Atmosphere, and Forests, and it meant that there were mostly positive relationships between them. In addition, for a part of the study period, there were also several nodes whose values of M were greater than 1, such as Animal Husbandry, Crop Cultivation, and Fisheries, most of which are natural sectors or

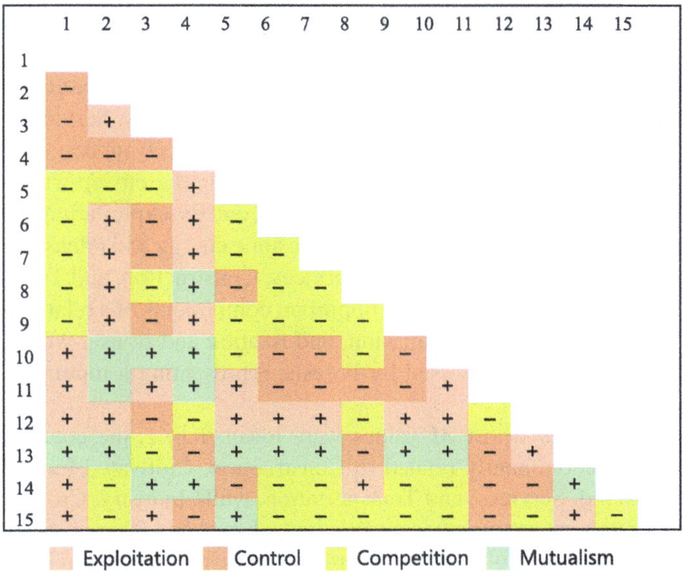

Fig. 10.18 Distribution of stable ecological relationships (those with values that did not change for at least 15 of the 21 years) between sectors within Beijing's nitrogen metabolic system from 1995 to 2015 (Nodes: 1-Household Consumption, 2-Industry, 3-Animal Husbandry, 4-Crop Cultivation, 5-Fisheries, 6-Forestry, 7-Services, 8-Construction, 9-Transportation, 10-Sewage Treatment, 11-Surface Water, 12-Atmosphere, 13-Forest, 14-Grassland, 15-Cultivated Land)

industries since they use raw materials produced by the natural sectors. Furthermore, we can explain differences in the M values by summarizing the type of ecological relationships between each pair of nodes, and the relationship sign matrix sgn(U) is a useful tool to achieve our goal of its symmetrical characteristics. We only analyzed half of the relationships below the diagonal, resulting in a total of 105 relationships (Zhang et al. 2020). We defined that when there was an unchanged·trend lasting for at least 15 years in a relationship between pairs of nodes, it was regarded as a stable relationship (Fig. 10.18).

For the nodes whose M values were greater than 1, they still had a control relationship with other nodes and received more benefits, while for the nodes whose M values were lower than 1, they were exploited by other nodes. Animal Husbandry, Surface Water, and Atmosphere were three nodes that received more benefits from other nodes, and the main type of ecological relationship related to them was exploitation (control), with a proportion of more than 40.0% of all relationships with these nodes. The mutualism relationships were also important, and the proportion of mutualism relationships related to Animal Husbandry and Surface Water (excluding self-mutualism for the flows within a node) above came up to 14.0% of all the relationships related to these two nodes, and it was the reason why they had higher M value. In addition, Animal Husbandry received industrial feed and farm products from Industry and Crop Cultivation. Surface Water mainly received runoff from Forests, grasslands, and Cultivated Land (Zhang et al. 2020), resulting in their higher value of *M*. The node of the Atmosphere received materials in the form of gas, mainly including nitrogen produced by denitrification from Sewage Treatment gases and nitrogenous flue gases from fossil fuel combustion by Forestry, Services, Construction, and Transportation. The node of Forests also has relatively more positive benefits than other nodes. Still, the primary type of ecological relationship was mutualism, with a proportion of 57.0% of the total for this node, and the exploitation relationships had a proportion of only 7.0%. Some nodes consume energy and release nitrogenous gas to the atmosphere, such as Forestry, Services, Construction, and Transportation, and these bidirectional processes were important components of a relatively virtuous nitrogen cycle of consumption, emission, and fixation and re-use, which was realized by nitrogen fixation processed by Forests, representing a mutually beneficial relationship.

For the other nodes whose *M* values were lower than 1, the competition relationships were predominantly related to Construction, Cultivated Land, Fisheries, Forestry, Grassland, Services, and Transportation, while Industry, Crop Cultivation, and Sewage Treatment are mainly controlled by other nodes. There was a proportion of more than 35.0% for exploitation (control) relationships related to Industry, Crop Cultivation, and Sewage Treatment of all the relationships, while more than 20.0% for competition relationships related to Industry and Sewage Treatment, and only 7.0% for relationships related to Crop Cultivation. Industry and Crop Cultivation supply food and energy sources for humans and animals in the city. Sewage Treatment has the capacity to treat nitrogen waste and participates thereby in the recycling of urban nitrogen, but it was mostly exploited by other nodes. The main relationship type for Fisheries, Forestry, Services, Construction, Transportation, Grassland, and Cultivated Land was competition, which accounted for more than 40.0% of all the relationships related to these nodes; exploitation relationships for these nodes accounted for more than 14.0% of the total, but less than 21.0%. This is mainly because Forestry, Services, Construction, and Transportation are all energy-consuming sectors that simultaneously receive energy nitrogen from Industry, thereby competing for this nitrogen. Grassland and Cultivated Land had no direct links with Forestry, Services, Construction, and Transportation, but competed between grassland and farmland through Industry and Atmosphere as intermediate nodes.

10.2.4 The Utility of the Metabolic Sectors

The results described in this section are based on previous work by my research group (Zhang et al. 2018). The synergism index (*S*) that was discussed in Chapter 5 reflects the actual benefits received by the system and its nodes. Figure 10.19 shows the nature of the flows among the nodes of the network (i.e., self-beneficial, positive, and negative). *S* fluctuated between 7 and 11, and there was no obvious long-term trend. The magnitude of these benefits fell by 29.5% between 1995 and 1999. Between 2003 and 2010, the *S* value increased to almost the same level as in 1995 (>10).

Nevertheless, there was a gradual decrease in the level of benefits from 10.27 in 2010 to 7.76 in 2015, and during the whole study period, none of the nodes had uniformly positive benefits. Our study showed that *S* values for individual nodes, which were the same as the sum of the values of all cells in the ***U*** matrix, dropped in the range from −2.56 to +2.04. Among these nodes, there were some nodes with a decreasing *S* value, such as Household Consumption, Industry, Animal Husbandry, Crop Cultivation, Fisheries, Construction, Grassland, and Cultivated Land, and the *S* value of Construction had the biggest decrease in the *S* value of all nodes with both positive and negative benefits that it dropped to 20% of the *S* value in 1995 from 1995 to 2015. The positive and negative benefits of Fisheries were almost equal, though the negative benefit slightly outweighed the positive, leading to a long-term decrease in its *S* value (Zhang et al. 2020). There was also a decreasing trend of *S* values of Animal Husbandry, Crop Cultivation, Cultivated Land, Grassland, Household Consumption, and Industry due to the fact that their negative benefits increased. However, there was a general increasing trend of *S* values of the remaining nodes, such as Forestry, Services, Transportation, Sewage Treatment, and Surface Water, because their negative benefits decreased and the positive benefits increased. And the *S* values of some nodes had an overall increase, such as Atmosphere and Forests, due to the fact that increase of the positive benefits was clearly larger than that of the negative benefits.

The factor which could greatly affect the size of the *S* value of each node was related to the changes in the flow of benefits between nodes. Throughout the study period, there was an increase in the number of negative benefit paths from 105 to 117, of which the distribution was changed from multiple paths with a large amount to fewer paths with a lower amount, while there was a decreasing trend of the number of positive benefit paths, of which the distribution was changed from homogenization to aggregation. In addition, there was a relatively stable trend of the number of paths that had relatively large absolute benefits with $|u| > 0.5$ throughout the study period, with their proportions in fluctuating between 3.0% and 5.0% of the total number of paths. The number of paths with a large negative benefit was 7 in 1995, and it was lower than 1, that is, the number of paths with a positive negative benefit in the same year, but in other years, the result was opposite that the paths with a large positive benefit with $|u| > 0.5$ were dominant and had a large proportion within the range between 55% and 67.0% of the total paths. However, the sum of the positive and negative benefits for paths with large $|u|$ increased over time (Zhang et al. 2020).

There was a large increase of the sum of positive benefits by nearly five times the 1995 value in the whole study period, while there was only a slight increase of the sum of the absolute values of the negative benefits, which came up to 1.1 times the 1995 value in 2015. For the paths which were with large benefits, the sum of the values with a large negative benefit was 8.2 times that with a large positive benefit in 2015, and the two sums in all other years were similar.

Nodes of Fisheries, Forestry, Services, Construction, Transportation, and Sewage Treatment achieved large benefits from their corresponding flows, but the *S* values had different changing trends that there was a decrease in Construction and Fisheries, while there was an increase in other nodes. For the Construction node, there were

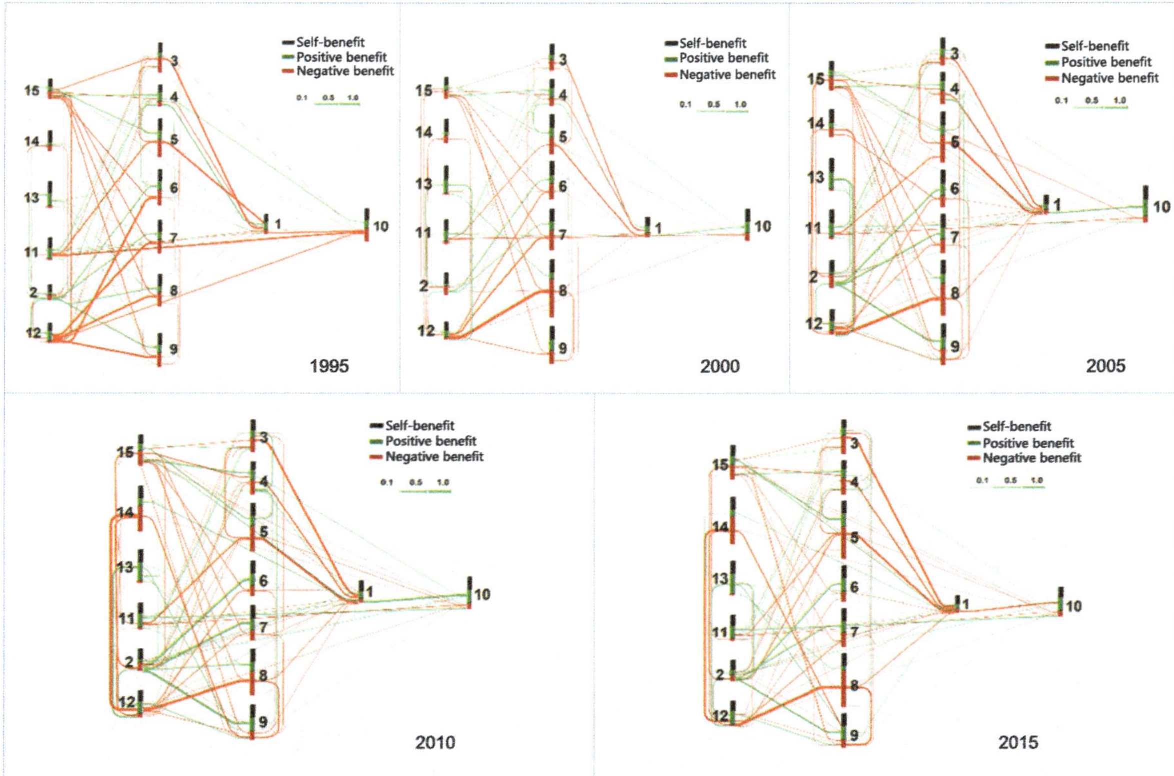

Fig. 10.19 Diagram of the utility flows among the 15 nodes of Beijing's nitrogen metabolic network (Nodes: 1-Household Consumption, 2-Industry, 3-Animal Husbandry, 4-Crop Cultivation, 5-Fisheries, 6-Forestry, 7-Services, 8-Construction, 9-Transportation, 10-Sewage Treatment, 11-Surface Water, 12-Atmosphere, 13-Forest, 14-Grassland, 15-Cultivated Land)

no large flows of positive benefit, which had a total decreased by 0.426, and the negative benefits mainly came from Transportation and Atmosphere, accounting for a proportion of more than 65.0% of its total negative benefits during the study period, with two of the related flows increasing by 0.676 and 0.729, respectively. Moreover, the self-benefit value for Construction increased during the study period, but only by 1.9%, so this change had little effect on the overall S value (Zhang et al. 2020). For the Fisheries node, there were a number of negative flows from the Household Consumption node continuously increasing, with a large proportion of 45.0% of the total negative flows received by Fisheries, and there was an increase by 0.580 of the absolute value of the total negative benefits. In contrast, there was a steady positive flow for the Fisheries node from the Cultivated Land node, with a proportion of more than 38.0% of the total positive benefit flows, leading to the total positive benefits increasing by 0.543 and the node's self-benefits decreasing by 3.7%, and therefore the S value of Fisheries decreased. For the Sewage Treatment node, there were positive benefits from the Household Consumption node, with a proportion of more than 60.0% of its total received positive flows, while there were negative benefits from the Surface Water node, with a proportion of around 45.0% of its total received negative benefits. For the nodes of Forestry, Services, and Transportation, there were positive benefits mainly from Industry, with a proportion of more than 70.0% of received positive benefit flows of each node, while there were negative benefits from Atmosphere, with a proportion of more than 59.0% of received negative benefit flows of each node. The inputs of larger positive benefits flow increased the total positive benefits and decreased the negative benefits for each node (Zhang et al. 2020). In the whole study period, there the S values of the above nodes all increased due to a relatively low proportion of less than 5.0% for the self-benefit of each node. Furthermore, for the S values of the above nodes, the increase was greatest for the Sewage Treatment node by 0.890 because of its largest decrease of negative benefits by 0.664.

10.2.5 The Structure of the Flow Hierarchy

From 1995 to 2015, the flow hierarchy structure of Beijing's nitrogen metabolism network maintained an inverted pyramid shape (Fig. 10.20); that is, the upper levels of the hierarchy had larger weights (the contributions of the components to the whole network) than the lower levels (Zhang et al. 2018). In 2003, the difference in weights between the bottom level and the top level was the largest (0.386). The weights of levels 1 and 2 decreased continuously and the gap between the two levels gradually decreased. The gap was 0.038 in 1995, and by 2004 the gap was only 0.002. Subsequently, between 2004 and 2011, the weight of the lower levels increased once more. The weights of levels 1 and 2 both fluctuated around 0.1, with little difference. After 2011, the weight of level 1 began to increase, and the weight of level 2 began to decrease; however, since no data was available after 2015, it's not possible to conclude whether this trend continued.

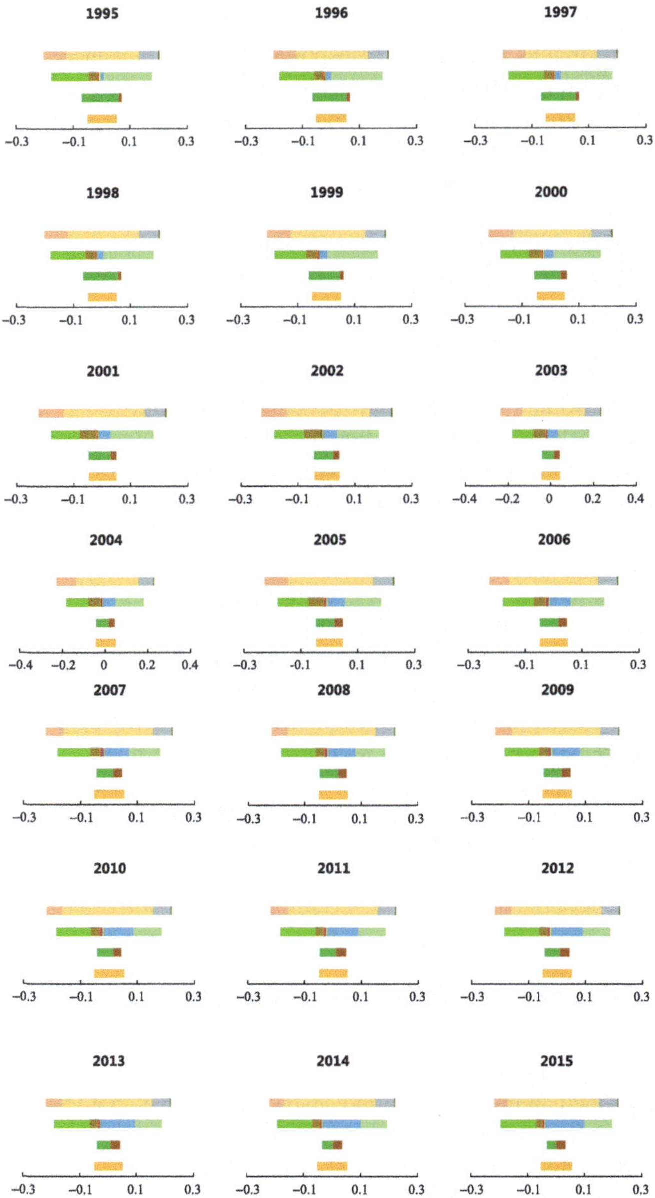

Fig. 10.20 The hierarchical structure of Beijing's nitrogen metabolism network from 1995 to 2014 (*Note* From bottom to top—level 1 comprises node 2; level 2 comprises nodes 4 and 10; level 3 comprises nodes 1, 3, 5, 6, 7, 8, 9, and 15; level 4 comprises nodes 11, 12, 13, and 14. The length of the color block represents the weight *W* of each node. Nodes: 1-Household Consumption, 2-Industry, 3-Animal Husbandry, 4-Crop Cultivation, 5-Fisheries, 6-Forestry, 7-Services, 8-Construction, 9-Transportation, 10-Sewage Treatment, 11-Surface Water, 12-Atmosphere, 13-Forest, 14-Grassland, 15-Cultivated Land)

The weights of levels 3 and 4 were greater than the weights of levels 1 and 2 throughout the study period. The weight of level 3 increased slowly, from 0.352 in 1995 to 0.392 in 2015, due to the strong contributions from the nitrogen-consuming sectors. Level 4 showed a similarly strong increase, from 0.407 in 1995 to 0.471 in 2003, followed by a decrease to about 0.430 from 2008 to 2015. This level was mainly composed of natural metabolism sectors. Level 2 showed the opposite trend to level 4, and its proportion decreased from 0.139 in 1995 to 0.083 in 2003, resulting in the largest difference in weight between the top and bottom levels in 2003. Thereafter, level 2 maintained a weight of around 0.090 until 2011, after which its weight decreased to only 0.064 in 2015. The weight of level 1 did not change significantly during the study period, and remained at about 0.110. Levels 1 and 2 both supply nitrogen, whereas level 3 consumes nitrogen and level 4 represents the post-consumer nitrogen emission role. Figure 10.20 shows that the weight of the suppliers remained relatively stable, whereas the weight of the nitrogen consumers increased. Thus, a large and growing proportion of the nitrogen consumption depended on an external supply, and more and more nitrogen-containing pollutants were discharged into the natural sectors.

The weight change of each level of the hierarchy is determined by changes in the weights of the sectors that make up each level. The weight reduction of level 2 mainly resulted from decreasing contributions from node 4 (Crop Cultivation). Although the weight of node 10 (Sewage Treatment) has gradually increased, the 9.7% decrease of node 4 was greater than the 2.1% decrease of node 10. The increase in the weight of level 3 was mainly due to the increased contribution of node 9 (Transportation). Its weight increased from 0.013 in 1995 to 0.138 in 2015. The pressure of the nitrogen flows in Beijing has increased sharply, but at the same time the weight of node 15 (Cultivated Land) showed a significant downward trend, resulting in slower growth of this level. The decrease in the weight of node 15 resulted from a gradual decrease in the production of local agricultural products in Beijing, which changed simultaneously with the weight of node 4 in level 2. The growth of level 4's weight from 1995 to 2003 was mainly due to the increased weights of node 11 (Surface Water), node 12 (Atmosphere), and node 13 (Forest), these three nodes had relatively large proportions of the total weight, and the subsequent decrease was caused mainly by the influence of node 11.

10.2.6 The Structure of the Utility Hierarchy

In addition to the flow level, the nitrogen net benefits of nodes (dimensionless values) can be calculated according to the network utility matrix, and then the utility hierarchy can be constructed. During the study period, the overall hierarchical structure of Beijing's nitrogen metabolism changed from a pyramid to an irregular barbell (Fig. 10.21). This structural change resulted from a gradual increase in the weight of level 4 of the hierarchy, a decrease of the weight of level 3 of the hierarchy, and a decrease of the weight of level 2 in the hierarchy. Figure 10.21 shows that the weight of level 1 in the hierarchy (producers) remained greatest throughout the study period, despite some fluctuation. Its overall proportion did not change significantly (mean $\pm$ SE, 0.560 $\pm$ 0.080). In contrast, level 2 of the hierarchy (primary consumers) showed the opposite trend, with an overall decreasing trend during the study period; its weight decreased from 0.258 to 0.086, with its weight in 2015 equaling 33% of its weight in 1995.

In addition, there was a relatively stable weight of 0.090 $\pm$ 0.013 of secondary consumers, that is, level 3 of the hierarchy, and only the Household Consumption node was included. It could result from the overall utilization of reactive nitrogen by urban residents having little change. For Level 4 in the hierarchy (decomposers), only the Sewage Treatment node was included, and there was a continuous increase of its weight by 1.4 times in 2015 (the end of the study period, 0.177), compared to the value in 1995, and this trend showed the ability to decompose nitrogen waste of Sewage Treatment node steadily increasing.

The hierarchical structure was also impacted by changes in the positive benefits of each node. For level 1 of the hierarchy structure (producers), Forests, Surface Water, and Atmosphere contributed most to its weight, their flows with a proportion of more than 30.0%, 26.0%, and 17.0% of the positive benefits of level 1, respectively. In fact, the benefits of these three nodes were always positive and increased throughout the study period (Zhang et al. 2020). For the Grassland node, it had negative benefits in 2010 and 2015, but the negative effect was relatively small, with a value of less than 1, and the positive benefits were more than 6. Therefore, the greatest weight belonged to level 1 during the whole study period. For the weight of level 2, the changing trend depended on the number of nodes with negative benefits. The components that remained positive benefits were Forestry, Transportation, and Services sectors, with a proportion of 14.0%, 14.0%, and 7.0% of the positive benefits for this level, respectively, but the negative benefits of Level 2 were larger than the positive benefits due to the continuously increasing number of nodes with negative benefits, resulting the decrease of the overall weight of level 2. Only Animal Husbandry in level 2 (a primary consumer) showed a small negative benefit in 1995 (–0.050); by 2000, it had recovered positive benefits (Zhang et al. 2020). The weight of the Construction node in Level 2 had a significant decrease of 6.4% due to its relatively large negative benefits of -1.280, and its negative benefits decreased to -0.84 in 2005. At the same time, there were also some small negative benefits for the node of Fisheries (-0.260). Therefore, there was a slight increase of 4.5% in the weight of level 2 in 2005. Afterward, in 2010, due to the negative benefits of nodes of Fisheries and Construction further decreasing to -0.330 and -0.250, respectively, and the other nodes' positive benefits in Level 2 increasing, such as Forestry, Services, and Transportation, the weight of level 2 increased and came up to its maximum at 0.300. In 2015, there were large number of nodes with negative benefits, such as Animal Husbandry, Construction, Crop Cultivation, and Fisheries, and the weight of level 2 had a sharp decrease to its minimum value of 0.086.

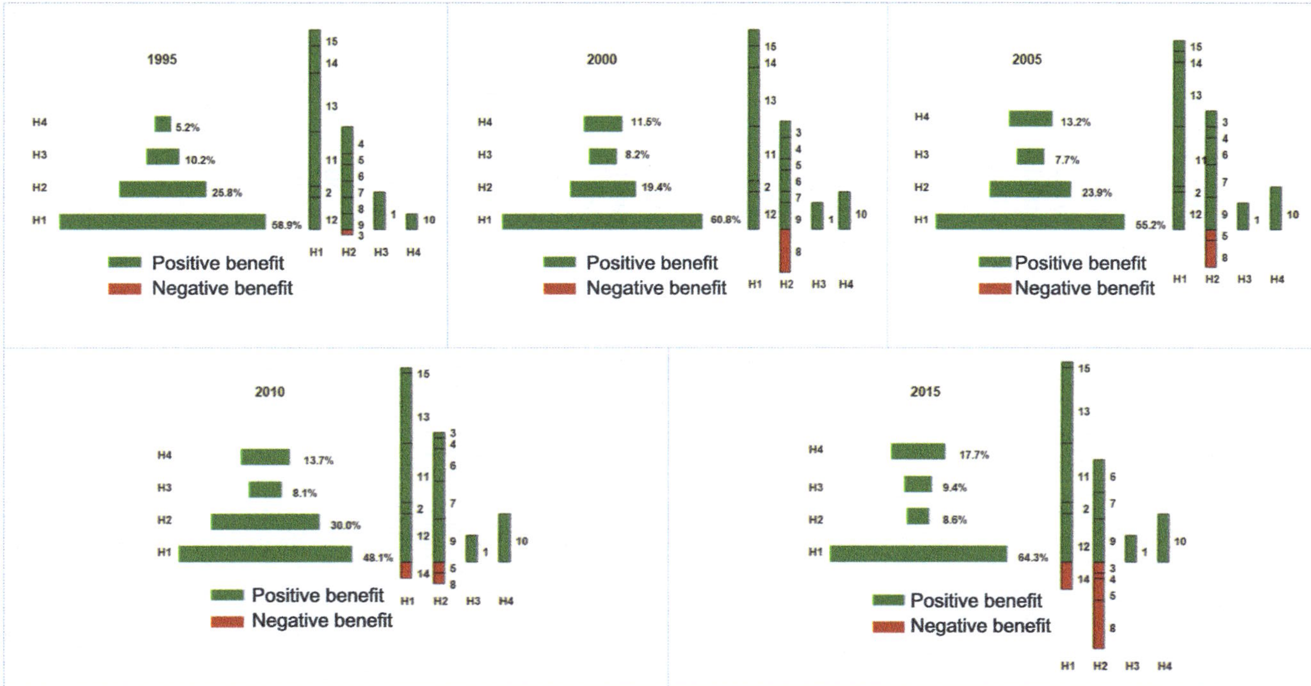

Fig. 10.21 Illustration of the net benefit hierarchy for the structure of Beijing's nitrogen metabolic system (*Note* Graphs on the left show the overall weights of each level (H1 to H4) in the hierarchy. Graphs on the right show the weights for each node in the network. Nodes: 1-Household Consumption, 2-Industry, 3-Animal Husbandry, 4-Crop Cultivation, 5-Fisheries, 6-Forestry, 7-Services, 8-Construction, 9-Transportation, 10-Sewage Treatment, 11-Surface Water, 12-Atmosphere, 13-Forest, 14-Grassland, 15-Cultivated Land)

10.2.7 Significance of the Network Analysis

In investigating the influence of human activities on nitrogen cycling, most previous researchers calculated the nitrogen inputs and outputs to clarify the sources and fates of nitrogen, as well as the environmental effects (Cui et al. 2013; Gu et al. 2012, 2009; Galloway et al. 2004). However, such research cannot clarify the indirect effects of multi-level transmission and transformations because the research did not quantify indirect flows between urban nodes through multiple paths with a length greater than 1. To effectively identify the problems within an urban nitrogen metabolism, the research on which this section was based (Zhang et al. 2016a) relied on an urban metabolic network model. My research group started by quantifying the direct and indirect flows of nitrogen so that we could quantify the ecological relationships among the network's nodes and identify the reasons for changes over time in the mutualism and synergism levels.

Under the urban metabolism framework, we can track the flows of nitrogen through a city's metabolism and describe the processes that drive those flows. In the network model, the city obtains nitrogen from the environment, and discharges nitrogen pollutants into the environment after human production and consumption activities. Models based on flow analysis and utility analysis provide an effective research perspective and research methods that make it possible to simulate the integrated flows and distribution of utility among network nodes. The approach provides deep insights into the status and roles of network nodes, and create opportunity to quantify the relevance, advantages, and disadvantages of each network node.

From a network perspective, clarifying urban nitrogen transfer processes, the quantities transferred, and the resulting net benefits can provide scientific support for efforts to regulate nitrogen flows and improve the structure of the urban nitrogen metabolism, thereby mitigating the environmental impacts of reactive nitrogen consumption. However, there is still room for improvement of these models. In particular, we need more high-quality and high-resolution data to allow us finely divide model nodes (e.g., to move from sectors to the individual industries within a sector) and obtain insights at the level where changes can be implemented (e.g., a specific problematic industry rather than the whole sector it belongs to). Eventually, we will need to apply our models to more cities to test their effectiveness under a range of environmental, social, and economic constraints. This will also facilitate comparisons between cities that provide additional, new insights to guide urban development.

10.2.8 Comparison with Previous Research on the Ecological Components and Their Relationships

Ecological network analysis is an effective method to study the functional relationships among the components of an urban ecological network. The results of research on natural ecosystems revealed that mutualism relationships are ubiquitous in nature (Fath and Patten 1999). Table 10.4 compares the characteristics of the ecological network models that were developed in previous research. Mutualism relationships accounted for less than 20.0% of the total relationships in most socioeconomic systems, whereas exploitation (control) relationships accounted for 46.0 to 76.0% of the total; competition was more variable, accounting for 9.0% to 44.0% of the total. In contrast, the mutualism proportion for the Beijing eco-industrial park's carbon metabolic system was 33.3% and that in the Lubei eco-industrial park's sulfur metabolism was 22.2%, both of which are higher than the 20.0% maximum for the other studies. This high level of mutualism resulted from the dominant characteristics of an eco-industrial park, namely the emphasis on recycling energy and materials to maximize the efficiency of the flows. This leads naturally to a high proportion of mutualism relationships. In addition, Chinese scholars have studied the virtual water metabolism system of the Lake Baiyangdian basin and the virtual water metabolism system of the Heihe River Basin and found that the mutualism relationships accounted for about 20.0% of the total, which is relatively high for a hybrid natural- socioeconomic system. This result could be closely related to the fact that both studies involved watersheds that have a strong natural component. Other scholars have analyzed the social metabolism of China (14.3% mutualism), the energy metabolism of Beijing (16.7% mutualism), and the materials metabolism of Beijing (14.3% mutualism). In these studies, mutualism accounted for a smaller proportion of the total, possibly because of the smaller roles of natural components in the study systems, which would have been dominated by the socioeconomic components. Nonetheless, in each case, the study systems can be viewed as highly organized systems, with up to 15 metabolic nodes in the network model.

In the example with the smallest number of socioeconomic nodes (a study of the Jing-Jin-Ji urban agglomeration and 27 Chinese provinces, mutualism relationships in the virtual energy metabolism accounted for only 4.0% of the total. This low proportion can be explained by the fact that the network nodes were based on administrative regions and therefore lacked a dominant Industry sector; as a result, there was less resemblance to an organism or ecosystem, and the proportion of mutualisms was correspondingly small. In the present study, the mutualism relationships for Beijing's nitrogen metabolism accounted for the highest proportion in any of the studies that did not involve an eco-industrial park (23.0%), mainly due to the important roles of Surface Water, Forest, Grassland, Crop Cultivation, and Fisheries and their interactions with the socioeconomic nodes, particularly in terms of the mutualisms between three socioeconomic sectors (Forestry, Transportation, and Construction) and two natural sectors (Forest and Grassland). Therefore, increasing the share of a system's natural components and improving the metabolic system's recycling component can increase the proportion of symbiotic relationships in the system, as in the case of the eco-industrial parks.

As the political and economic center of China, Beijing has a large proportion of socioeconomic components, and a small proportion of natural components (e.g., forest, grassland) with a small amount of green space existing and most green space now fully occupied, and therefore Beijing needs to improve the ecological potential

Table 10.4 Comparison of the proportions of the different ecological relationships for different metabolic systems

Metabolic system	n	n_1	n_2	Proportion of total relationships/%		
				Mutualism	Exploitation (Control)	Competition
Chinese social metabolism (Zhang et al. 2012)	6	1	5	14.3	76.2	9.5
Heihe Basin virtual water metabolism (Fang and Chen 2015)	6	0	6	19.0	67.7	13.3
Jing-Jin-Ji agglomeration and 27 provinces (virtual energy metabolism) in China (Zhang et al. 2015a)	5	0	5	4.0	53.0	44.0
Beijing's material metabolism (Li et al. 2012b)	7	0	7	14.3	61.9	23.8
Beijing's energy metabolism (Zhang et al. 2010a)	4	0	4	16.7	66.6	16.7
Lake Baiyangdian basin's virtual water metabolism (Mao and Yang 2012)	5	1	4	20.0	70.0	10.0
Beijing industrial park's carbon metabolism (Lu et al. 2015)	9	0	9	33.3	57.2	9.5
Lubei eco-industrial park's sulfur metabolism (Zhang et al. 2015b)	12	0	12	22.2	55.6	22.2
Beijing's nitrogen metabolism (Present study)	15	5	10	23.0	46.0	31.00

Note n is the total number of nodes, n_1 is the number of natural nodes, and n_2 is the number of socioeconomic nodes

of existing urban green space. For example, we can improve the canopy density of the vegetation and the nitrogen-fixation capacity of the green space to increase the potential of Beijing's limited green lands. In addition, we can expand the urban green space by creating smaller patches of green space in the unused land. Since there is not too much free space to expand the green land in the horizontal direction, we can expand the green land in the vertical direction. For example, we can create green borders in unused space on sidewalks, in the medians between roads, and along the sides of railways and other transportation corridors, and these green borders above can be built with the promise of urban planners in the places such as roofs, slopes, walls, and the interior of buildings.

Actually, it is only a part of the solution that we add some natural components to the urban system, and we should do strategic planning related to reducing high nitrogen consumption and increasing the efficiency of nitrogen utilization to improve nitrogen recycling. Especially city managers should pay more attention to the nodes which had a large nitrogen consumption or emission and find approaches to add nitrogen recycling links from the existing examples of eco-industrial parks. Therefore, nitrogen utilization efficiency of these nodes could be improved. In fact, it may be possible to construct eco-industrial parks in or near the city to take advantage of the ability of these parks to simulate the recycling characteristics of natural ecosystems (Zhang et al. 2020).

In the socioeconomic components of the urban system, Exploitation and Control were the main types of ecological relationships (Table 10.4). Especially in the research of Chinese society and Lake Baiyangdian Basin, the Exploitation and Control relationships had a large proportion of 76.2% and 70.0%, respectively, which could be related to the role of natural nodes in the models. In the research of Chinese society, exploitation (control) relationships with natural nodes had a proportion of 25.0% of the total, while this value was only 20.0% in the research of Lake Baiyangdian Basin. Similarly, there was a high proportion of exploitation (control) relationships in the research of Beijing's energy metabolism, Beijing's material metabolism, and Heihe River Basin virtual water metabolism, as 66.6%, 61.9%, and 67.7%, respectively. However, the value of the proportion of exploitation (control) relationships in our research was just 46.0%. This was because of our subdivision of the natural and socioeconomic nodes, and then mutualism relationships and competition relationships had a relatively high proportion of 23.0% and 31.0%, respectively. For the competition relationships, they existed both between two natural nodes (flows among the Atmosphere, Grassland, and Cultivated Land nodes) and between the natural and socioeconomic nodes (flows among the Grassland, Cultivated Land, Forestry, and Services nodes), accounting for 2.6% and 12.1%, respectively. In contrast, the most competition relationships were found in the virtual energy metabolism of 27 provinces and regions in China (44.0%), in which China's regions compete aggressively for resources (Zhang et al. 2020). Actually, there was a proportion of less than 25.0% for competition relationships in other studies. For our study, the proportion of competition relationships was an intermediate value of 31.0%.

10.2.9 Comparison with Previous Research on Metabolic Utility and the Hierarchical Structure

The mutualism (*M*) and synergism (*S*) indexes in different studies were summarized in Table 10.5. For mutualism (*M*) indexes, most of them had an increasing trend with the number of nodes increasing. For systems with less than ten nodes, their *M* values were greater than one due to their fewer competition relationships, with a proportion of less than 25.0% of the total relationships, and the number of mutualism relationships was more than that of competition relationships. However, the main reason for a high *M* in Beijing's material metabolism was self-mutualism (i.e., all elements of the diagonal of the utility matrix were +) (Zhang et al. 2020). The proportion of self-mutualism relationships was 10% lower than that of competition relationships, and therefore only self-mutualism relationships could not offset the effects of competitive relationships if there were no strong mutualism in this urban system. But the situation was special in Lubei industrial park with nodes of more than ten and *M* value greater than 1, due to its special design as tight recycling of products, byproducts, and waste, to increase the industrial mutualisms to the greatest extent, and the proportion of mutualism relationships was as the same as that of competition relationships (22.2%). In the cases of the energy metabolism models in four Chinese municipalities and Shandong Province, with the number of nodes of 17 and 19, respectively, their proportion of competitive relationships was high, accounting for more than 30.0% of the relationships, and therefore their *M* values were more than 1. For our study of the 15-node nitrogen metabolism, the proportion of mutualism relationships (<15.0%) was much lower than that of competition relationships (31.0%), and the *M* value (0.82) was also less than 1, possibly due to the strict classification standards of nodes.

There are different types of hierarchical structures of metabolic systems, including pyramid (balanced), barbell, inverted pyramid (highly unbalanced), and irregular types, which have a close relationship to the characteristics of urban development, the weights (contributions) of the nodes, the hierarchical relationships among the nodes, and the changes of the three factors above. Our research group studied the energy metabolism with the system divided into five nodes, and the system included four municipalities which were directly managed by China's Central Government (Zhang et al. 2010b), with the result of the distribution of different hierarchical structures that Beijing, Shanghai, and Tianjin had the hierarchical structures of inverted pyramids, while only Chongqing, which held rich coal resources, had the hierarchical structure of a pyramid shape. Therefore, Chongqing has the highest energy production weight (i.e., the strongest producer level) of these cities. Furthermore, a 17-node model for the energy metabolism of the same four municipalities was developed (Zhang et al. 2010a). In the 17-node model, we found that Beijing's hierarchical structure was an inverted pyramid type and proposed the reason that Beijing had colossal energy consumption and relatively low energy production due to its political and socioeconomic status. All the other cities had a pyramid structure due to the presence of large energy storage nodes at the bottom of the hierarchy (Zhang et al.

Table 10.5 Comparison of the mutualism index (*M*) and synergism index (*S*) values among studies of different metabolic systems

Metabolic system	*M*	*S*	Number of nodes
Chinese society's metabolism (Zhang et al. 2012)	1.45	–	7
Beijing's material metabolism (Li et al. 2012b)	1.30	5.64	7
Lubei eco-industrial park's sulfur metabolism (Zhang et al. 2015b)	1.25	6.67	12
Beijing industrial park's carbon metabolism (Lu et al. 2015)	2.06	7	9
Shandong Province's energy metabolism (Zheng and Shi 2016)	0.63	–	19
Energy metabolism system of four Chinese cities (each with five nodes) (Zhang et al. 2010b)			
Beijing	2.2	–	5
Shanghai	1.78	–	5
Tianjin	1.78	–	5
Chongqing	0.92	–	5
Energy metabolism systems of four cities (each with 17 nodes) (Zhang et al. 2010a)			
Beijing	0.74	–	17
Shanghai	0.70	–	17
Tianjin	0.61	–	17
Chongqing	0.29	–	17
Beijing's nitrogen metabolism (Present study)	0.82	7–11	15

2020), and numerous studies of different cases, such as China's social metabolism (Zhang et al. 2010a), Beijing's water metabolism (Zhang et al. 2010c), Beijing's material metabolism (Li et al. 2012b), and the carbon metabolism of Beijing's eco-industrial park (Lu et al. 2015), had the same result of hierarchical structure. In this research of different metabolic systems, industries of the biological resource, non-living resources, and recycling all can be regarded as producers, but the specific functions were different. The first two sectors could support both the internal and external environments, and the last sector could provide raw materials to the other nodes.

In our research, the decomposers (Sewage Treatment) were regarded as a component of level 4 in the structural hierarchy due to their position downstream from producers and consumers, but at the same time, they played the role of producers that the wastes were recycled in the decomposers and were returned to the sectors in levels 1 to 3 of the hierarchy as inputs. So our result did not show the trend of an increasing imbalance of the hierarchical structure, although the weight of level 4 increased in the whole study period, and it possibly demonstrated an increasingly healthy metabolic system. Meanwhile, the direct or indirect consumption of flows of many socioeconomic sectors was from producers, and therefore they acted as primary and secondary consumers, which led to a smaller weight at higher consumer levels and a higher weight at lower consumption levels. We obtained a similar hierarchical structure in Sect. 10.2.6 early in the study period, although the structure has changed over time (Zhang et al. 2020) because the natural and industrial nodes functioning in the processes were actually combined, and there was a finding that during the end of the study period, a barbell-shaped structure was developing, and there was a large decrease of many primary consumer nodes including Animal Husbandry, Crop Cultivation, Fisheries, and Construction, and this was mainly because of their negative benefits.

There was a fact should be noticed that the calculating methods of hierarchical weights strongly decided the specific hierarchical structure. For example, the weights in the hierarchy could be described using the index—the ratio of the direct flows of each node in each layer to the total flow of the system, but it did not highlight the relevance of the indirect flows (Lu et al. 2015), while my research group described the hierarchical structure using integrated flows, which was added indirect flows based on the direct flows, but the index of integrated flows also has a disadvantage that the related result is difficult to be applied to propose a specific solution due to the obvious differences in the hierarchical structure under the condition of integrated flows. In our research, we used the index of net benefits for a node to accurately describe the actual contribution of each node within the system and furthermore show the urban system's hierarchical level. This will improve the scientific basis for decision-making to improve the health of a city's nitrogen metabolism (Zhang et al. 2020).

References

Baker LA, Hope D, Xu Y et al (2001) Nitrogen balance for the Central Arizona-Phoenix (CAP) ecosystem. Ecosystems 4(6):582–602

Billen G, Barles B, Chatzimpiros P et al (2012) Grain, meat and vegetables to feed Paris: Where did and do they come from? Localising Paris food supply areas from the eighteenth to the twenty-first century. Reg Environ Change 12(2):325–335

Cui S, Shi Y, Groffman PM et al (2013) Centennial-scale analysis of the creation and fate of reactive nitrogen in China (1910–2010). Proc Natl Acad Sci USA 110(6):2052–2057

Ding L, Liu C, Chen KL et al (2017) Atmospheric pollution reduction effect and regional predicament: an empirical analysis based on the Chinese provincial NO_x emissions. J Environ Manage 196(13):178–187

Faerge J, Magid J, de Vries FWTP (2001) Urban nutrient balance for Bangkok. Ecol Model 139(1):63–74

Fang D, Chen B (2015) Ecological network analysis for a virtual water network: a case study of the Heihe River Basin. Environ Sci Technol 49(11):6722–6730

Fath BD, Patten BC (1999) Review of the foundations of network environ analysis. Ecosystems 2(2):167–179

Filoso S, Martinelli LA, Howarth RW et al (2006) Human activities changing the nitrogen cycle in Brazil. In: Martinelli LA, Howarth RW (eds) Nitrogen cycling in the Americas: natural and anthropogenic influences and controls. Springer, Dordrecht, the Netherlands, pp 61–89

Forkes J (2007) Nitrogen balance for the urban food metabolism of Toronto, Canada. Resour Conserv Recycl 52(1):74–94
Galloway JN (1998) The global nitrogen cycle: changes and consequences. Environ Pollut 102(1):15–24
Galloway JN, Cowling EB (2002) Reactive nitrogen and the world: 200 years of change. Ambio 31(2):64–71
Galloway JN, Dentener FJ, Capone DG et al (2004) Nitrogen cycles: past, present, and future. Biogeochemistry 70(2):153–226
Gu B, Chang J, Ge Y et al (2009) Anthropogenic modification of the nitrogen cycling within the Greater Hangzhou area system, China. Ecol Appl 19(4):974–988
Gu B, Dong X, Peng C et al (2012) The long-term impact of urbanization on nitrogen patterns and dynamics in Shanghai, China. Environ Pollut 171(1):30–37
Huang W, Cui S, Bing Gao YY et al (2016) Urban nitrogen metabolism in Xiamen City, China (2016–12–15) [2022–8–16]. http://agronomyaustraliaproceedings.org/images/sampledata/ini2016/pdf-papers/INI2016_Huang_Wei.pdf
Jia JS, Jian HY, Xie DM et al (2017) Multi-perspectives' comparisons and mitigating implications for the COD and NH_3-N discharges into the wastewater from the industrial sector of China. Water 9(3):1–18
Kaye JP, Groffman PM, Grimm NB et al (2006) A distinct urban biogeochemistry? Trends Ecol Evol 21(4):192–199
Keeney DR (1979) A mass balance of nitrogen in Wisconsin. Trans Wis Acad Sci, Arts, Lett 67(1):95–102
Li CJ, Lei HJ, Pan CZ et al (2012a) Decomposing the influencing factors of COD and NH_4-N discharged in industrial water environment. J Beijing Norm Univ (nat Sci) 48(5):476–482 (in Chinese)
Li SS, Zhang Y, Yang ZF et al (2012b) Ecological relationship analysis of the urban metabolic system of Beijing, China. Environ Pollut 170(1):169–176
Liu C, Fei JL, Hayashi Y et al (2014) Socioeconomic driving factors of nitrogen load from food consumption and preventive measures. Ambio 43(5):625–633
Lu Y, Chen B, Feng KS et al (2015) Ecological network analysis for carbon metabolism of eco-industrial parks: a case study of a typical eco-industrial park in Beijing. Environ Sci Technol 49(12):7254–7264
Mao XF, Yang ZF (2012) Ecological network analysis for virtual water trade system: a case study for the Baiyangdian Basin in Northern China. Ecol Inform 10(3):17–24
Pang J, Shi YC, Hu T et al (2013) Structural decomposition analysis of pollutants emission change embodied in exports of China. China Environ Sci 33(12):2274–2285 (in Chinese)
Vitousek PM, Aber JD, Howarth RW (1997) Human alteration of the global nitrogen cycle: sources and consequences. Ecol Appl 7(3):737–750
Wang LQ (2017) Investigation of NO_x emissions reduction and its strategies over China's 30 provinces based on LMDI. Acta Sci Circumstantiae 37(6):2394–2402 (in Chinese)
Wolman A (1965) The metabolism of cities. Sci Am 213(3):178–190
Yu Y, Cui SH, Zhao SN et al (2012) Changes of residents' nitrogen consumption and its environmental loading from food in Xiamen. Acta Ecol Sin 32(19):5953–5961 (in Chinese)
Zhang Y, Li SS, Fath BD et al (2010a) Analysis of an urban energy metabolic system: comparison of simple and complex model results. Ecol Model 223(1):14–19
Zhang Y, Liu H, Li YT et al (2012) Ecological network analysis of China's societal metabolism. J Environ Manag 93(1):254–263
Zhang Y, Lu HJ, Fath BD et al (2016a) A network flow analysis of the nitrogen metabolism in Beijing, China. Environ Sci Technol 50(16):8558–8567
Zhang Y, Lu HJ, Fath BD et al (2016b) Modelling urban nitrogen metabolic processes based on ecological network analysis: a case of study in Beijing, China. Ecol Model 337:29–38
Zhang Y, Lu HJ, Zhang XL (2017) Analysis of nitrogen metabolism processes and a description of structure characteristics. Ecol Model 357(15):47–54

Zhang Y, Yang ZF, Fath BD (2010b) Ecological network analysis of an urban water metabolic system: model development, and a case study for Beijing. Sci Total Environ 408(20):4702–4711

Zhang Y, Yang ZF, Fath BD et al (2010c) Ecological network analysis of an urban energy metabolic system: model development, and a case study of four Chinese cities. Ecol Model 221(16):1865–1879

Zhang XL, Zhang Y, Fath BD (2020) Analysis of anthropogenic nitrogen and its influencing factors in Beijing. J Clean Prod 244(2):118780

Zhang Y, Zhang XL, Zhao XY (2018) Analysis of the ecological relationships and hierarchical structure of Beijing's nitrogen metabolic system. Ecol Indic 94(SI): 39–51.

Zhang Y, Zheng HM, Fath BD (2015a) Ecological network analysis of an industrial symbiosis system: a case study of the Shandong Lubei eco-industrial park. Ecol Model 306(12):174–184

Zhang Y, Zheng HM, Yang ZF et al (2015b) Multi-regional input-output model and ecological network analysis for regional embodied energy accounting in China. Energy Policy 86(11):651–663

Zheng SS, Shi L (2016) Analysis of Shandong Province energy metabolism network based on ecological network analysis. Front Environ Sci 6(6):159–170

Chapter 11
Analysis of Metabolic Processes in Eco-Industrial Parks

In the previous chapters, I have shown how the urban metabolism tools described in the first part of this book can be used to examine the flows of materials and energy within a complex urban system. However, the examples, including studies of energy metabolism (Chapter 8), carbon metabolism (Chapter 9), and nitrogen metabolism (Chapter 10), involved relatively simple networks, typically with 15 or fewer metabolic actors. The limited size of these networks was partially due to the difficulty obtaining sufficient reliable and comparable data to conduct a higher-resolution analysis. However, urban metabolism analysis can be conducted for substantially more complex networks if sufficient data is available. In this final chapter of the book, I will provide examples of a far more complex network, with up to 48 nodes. Because the previous examples have focused on urban-scale metabolisms, which are inevitably unbalanced compared with a natural ecosystem, I will also explore eco-industrial parks, where the park's design is intended to increase cycling of materials within the park to make the metabolism more ecologically sustainable and thus, make the system more closely resemble a natural ecosystem or a living organism.

11.1 Symbiotic Metabolic Processes in an Eco-Industrial Park

China launched a pilot project for developing eco-industrial parks in 1999. Through these eco-industrial parks, recycling the industrial waste as inputs for different enterprises could tremendously reduce the environmental burden which was created by industrial waste. China's State Environmental Protection Administration confirmed the Guangxi Guigang Eco-Industrial Park as the first national ecological industrial demonstration park. Shortly afterward, Zhejiang, Jiangsu, Shandong, Liaoning,

Y. Zhang, *Urban Metabolism*,
https://doi.org/10.1007/978-981-19-9123-3_11

Tianjin, Xinjiang, and Guangdong, among other provinces, cities, and autonomous regions, carried out their own pilot projects to build eco-industrial parks. In 2011, China's Ministry of Environmental Protection, Ministry of Commerce, and Ministry of Science and Technology jointly issued the document "Guiding Opinions on Strengthening the Construction of the National Eco-Industrial Demonstration Park" (MEEC 2011), which proposed that 50 additional demonstration parks, with specific distinctive features and desired results, should be built during the 12th Five-Year Plan period (2011-2015). These eco-industrial parks will represent the third generation of these parks in China, after previous generations that focused on economic and technological development zones and then on high-tech development zones. Nonetheless, each generation of eco-industrial park focused on the construction of an industrial network that resembled an ecosystem in terms of the circular flows designed to maximize the utilization and reuse of resources, thereby reducing the emission of pollutants at the source.

In the research on which this section is based (Zhang et al. 2013), my research group examined 10 typical Chinese and foreign eco-industrial parks as examples, and analyzed the quantities of the flows and the paths they flowed along. We then described the network structure using an analytical framework (Fig 11.1) that let us analyze the topological relationships and connectivity among the network nodes. Our goal was to gain insights into the details of the networks. We used social network analysis to analyze the degrees of connectedness of the network and of the subnetworks formed by exchanges of products, byproducts, and waste within these symbiotic networks and to clarify the relationships among their nodes. The social network analysis approach that we used was developed based on the sociometry techniques of Moreno and Jennings (1938). Their approach analyzes the relationships among actors within a system by focusing on the relationships among actors rather than on the actors themselves (e.g., individuals, groups, and organizations). The approach therefore focuses on social phenomena and structures. We analyzed the network's density and centrality to identify the dominant nodes and the core-periphery structure of the networks and sub-networks. The results can reveal the operational problems caused by inefficiencies of the industrial symbiotic network structure, thereby providing an empirical basis for improving the completeness and efficiency of the symbiotic network, while also enriching the theory and methods that support research on industrial symbiosis.

We chose 10 representative eco-industrial parks for our research: Kalundborg (Denmark; Mihelcic et al. 2014), Choctaw (Oklahoma, USA; Potts 1998), Kitakyushu (Japan), Styria (Austria; Schwarz and Steininger 1997), Guigang (Guangxi Zhuang Autonomous Region, China; Zhu et al. 2007), Lubei (Shandong Province, China; Yang et al. 2004), Changsha Huangxing (Hunan Province, China; Wang 2005), Shihezi (Xinjiang Uygur Autonomous Region, China; Wu et al. 2004), Wujing (Shanghai City, China; ODEHCIAP 2011), and the Tianjin Economic-Technological Development Area (TEDA; Tianjin City, China; Shi et al. 2010). We

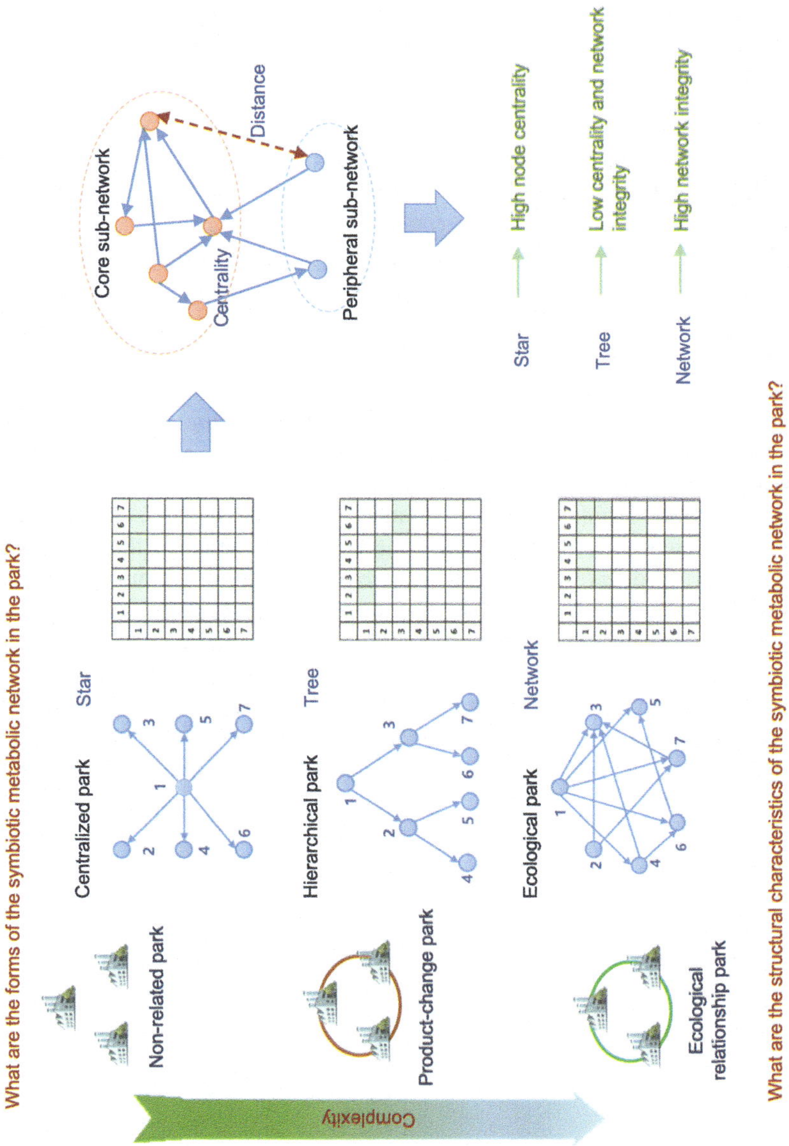

Fig. 11.1 Analytical framework for studying the connectivity within the symbiotic metabolic networks in an eco-industrial park

obtained data from these published references in order to describe these networks. Although these 10 networks describe symbiotic development of all kinds of types of industries, and cannot be compared directly because of their diversity, social network analysis still can provide insights into their operations. In addition, It is suitable to adopt this method to identify and describe their various network topologies, meanwhile, the method also can be used to look forward to different ways of optimizing the exchanges among nodes to improve the network structure.

11.1.1 A Symbiotic Metabolic Network Model

According to the characteristics of the members of the network and the flows of materials, energy, and information among them, we established networks of flows for each of the eco-industrial parks. (Figure 11.2). First, we identified the members of the parks, and it is assumed that all nodes were active enterprises, including members engaging in production, circulation, consumption, and service activities. The ecological industrial parks are not isolated from the surrounding environment, though we used the administrative boundary of each park as the network boundary. On the contrary, ecological industrial parks are inevitably connected with rivers and other bodies of water, farmland, residential areas, local governments, and business organizations located outside of the park. Therefore, we expanded the networks, including critical members outside of the immediate geographic boundary of the park but nonetheless exchanged essential resources with members within the park to build a relatively complete model truly reflecting the symbiotic network. Then, we defined the relationships among networks' members, which include the exchanges of products, byproducts, waste, and energy. In order to choose the industrial symbiosis linkages to construct the adjacency matrices correctly, considering all of the exchange of byproducts and waste, not just exchanges of products, was also important.

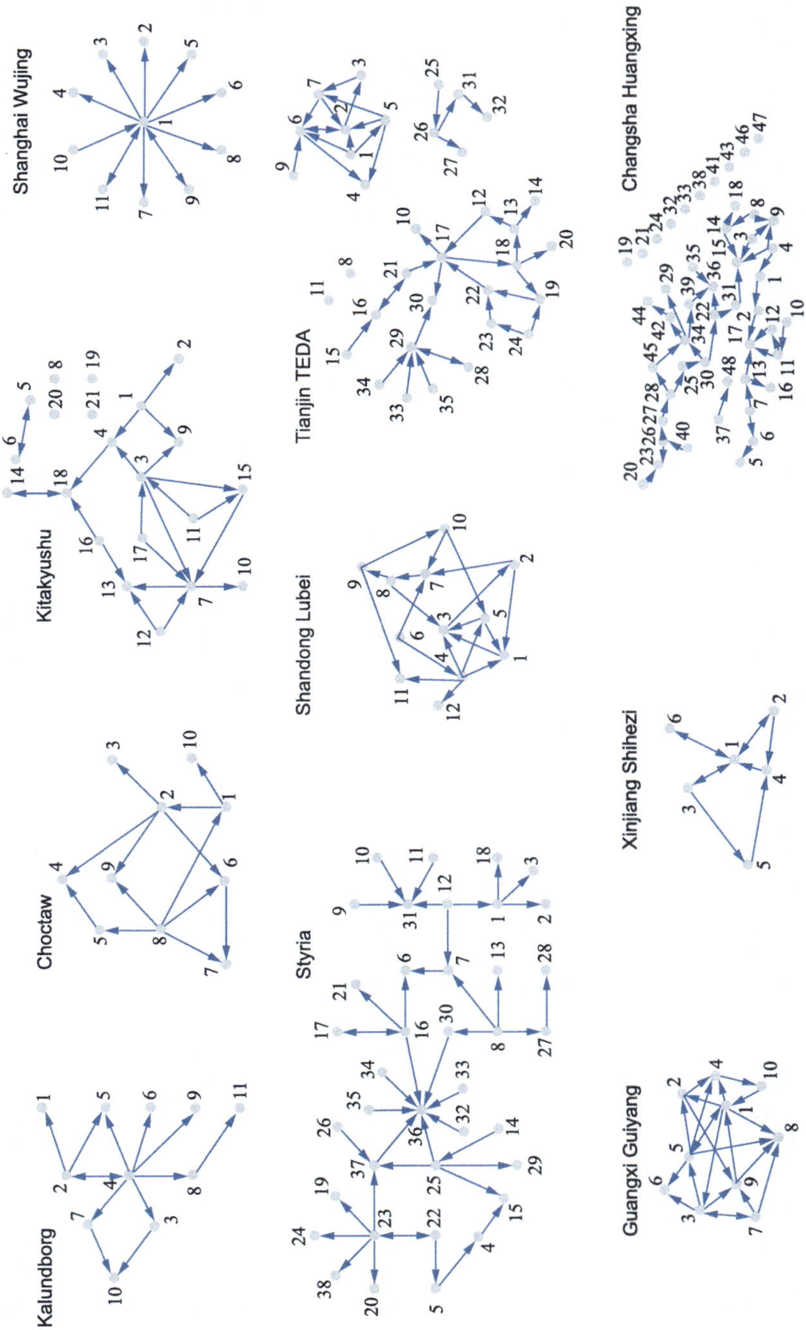

Fig. 11.2 Models of the symbiotic metabolic networks in the 10 eco-industrial parks

Nodes For Kalundborg: 1-liquid fertilizer; 2-Statoil refinery; 3-Asnaes fish farm; 4-Asnaes power plant; 5-Gyproc gypsum board plant; 6-Alborg cement plant; 7-Novo Nordisk biopharmaceutical plant; 8-Kalundborg City; 9-recovered nickel and vanadium; 10-farm; 11-A-S Bioteknisk Jordens.For Choctaw: 1-tire shredding plant; 2-tire pyrolysis plant; 3-hard rubber tire manufacturers; 4-carbon black processing plant; 5-ink cartridge production and recovery plant; 6-plastics plant; 7-plastic products plant; 8-wastewater treatment plant; 9-greenhouse; 10-crushed steel recovery plant.For Kitakyushu: 1-plastic bottle recycling plant; 2-car disassembly factory; 3-waste office equipment plant; 4-home appliance recycling factory; 5-PCB treatment facilities; 6-composite core facility; 7-construction waste treatment plant; 8-medical equipment plant; 9-fluorescent lamps plant; 10-empty cans plant; 11-reused computer plant; 12-recreational machine plant; 13-waste wood and plastics plant; 14-cooking oil plant; 15-styrofoam plant; 16-ink cartridges plant; 17-scrap car plant; 18-organic solvent and waste plastics plant; 19-wastepaper plant; 20-kitchen waste; 21-tofu and food waste.For Styria: 1-paper product industry 3; 2-pressboard plant; 3-paper product industry 4; 4-scrap material dealer; 5-wastewater treatment plant; 6-mining company; 7-paper product industry 1; 8-wastepaper dealer; 9-textile plant 1; 10-textile plant 2; 11-chemical plant; 12-sawmill; 13-paper product industry 6; 14-iron scrap dealer; 15-construction materials plant 1; 16-power plant 1; 17-region of Voitsberg; 18-stone and ceramic industry 2; 19-cement plant 6; 20-construction materials plant 2; 21-cement plant 3; 22-region of Graz (Austria); 23-power plant 2; 24-cement plant 4; 25-iron manufacturing industry; 26-used tire dealers; 27-paper product industry 5; 28-plastics plant; 29-pigment plant; 30-paper product industry 2; 31-stone and ceramic plant 1; 32-used oil dealer 3; 33-used oil dealer 2; 34-used oil dealer 1; 35-fuel producer; 36-cement plant 2; 37-cement plant 1; 38-cement plant 5.For Guangxi Guigang: 1-sugar refinery; 2-alcohol plant; 3-pulp and paper plant; 4-compound fertilizer plant; 5-power plant; 6-wastewater treatment plant; 7-alkali recovery plant; 8-cement plant; 9-light calcium plant; 10-sugarcane planting system.For Shandong Lubei: 1-ammonium phosphate plant; 2-sulfuric acid plant; 3-cement plant; 4-thermal power plant; 5-chlorine plant; 6-aquaculture plant; 7-bromine plant; 8-salty gypsum production plant; 9-raw salt production; 10-chlor-alkali plant; 11-potassium and magnesium salt production; 12-seawater.For Changsha Huangxing: 1-food plant; 2-cellulose enzyme plant; 3-tea-leaf plant; 4-aloe deep-processing plant; 5-cosmetics plant; 6-camellia oil refinery; 7-camellia oil plant; 8-citrus plant; 9-beverage plant; 10-beer plant; 11-nucleic acid extraction; 12-antiviral medicine plant; 13-tea cake processing plant (which processes the residues after oil extraction); 14-orange peel deep-processing plant; 15-medicine plant; 16-daily-use chemical plant; 17-fertilizer plant; 18-food additives plant; 19-integrated circuit (IC) design; 20-IC manufacturing; 21-motor plant; 22-agricultural production plant; 23-IC packing plant; 24-liquid crystal display (LCD) plant; 25-purifying agent plant; 26-plastics manufacturing plant; 27-household appliances plant; 28-Yuanda air conditioner plant; 29-metal recovery plant; 30-wastewater treatment plant; 31-rice husk plant; 32-optical fiber plant; 33-Ni-MH battery plant; 34-environmental protection equipment plant; 35-green (i.e., ecologically friendly) paint plant; 36-green building materials plant; 37-intelligent metals plant; 38-Liyuan new materials plant; 39-green adhesives plant; 40-plastics plant; 41-antimicrobial products plant; 42-ceramics plant; 43-fire retardant plant; 44-building bricks plant; 45-equipment parts plant; 46-fiber plant; 47-food packaging plant; 48-building materials plant.For Xinjiang Shihezi: 1-achnatherum cultivation system; 2-paper-making system; 3-livestock breeding system; 4-wastewater treatment system; 5-animal-products processing system; 6-eco-tourism industry.For Shanghai Wujing: 1-Shanghai coking plant; 2-Jinghua chemical plant; 3-titanium white plant; 4-Morgan carbon products plant; 5-Liancheng chemical plant; 6-hydrogen peroxide plant; 7-Wujing chemical plant; 8-Zhongxing chemical plant; 9-Linde CO2 plant; 10-Chlor-alkali plant; 11-Card Brantec chemical plant.For Tianjin TEDA: 1-TEDA water treatment plant; 2-industrial, commercial, and residential users; 3-TEDA wastewater treatment plant; 4-construction companies; 5-Guahua cogeneration plant; 6-Binhai energy cogeneration plant; 7-TEDA New Water Source Co.; 8-TEDA eco-landscaping Co.; 9-desalination plant; 10-FAW Resource Recovery Co.; 11-other Toyota-family firms; 12-Toyota FAW Dies Co.; 13-Tianjin Rainbow Hills Cast Iron Co.; 14-other auto die makers; 15-Tianjin AW Automatic Transmission Co.; 16-Tianjin Toyotsu aluminum smelting; 17-Tianjin FAW Toyota motor Co.; 18-Tianjin Toyota Resource Management Co.; 19-Takaoka Lioho industries; 20-Tianjin pipe corporation; 21-Tianjin FAW Toyota Engine Co.; 22-Toyota-family auto part makers; 23-CMW Industrial Co.; 24-steel scrap contractors; 25-refineries; 26-Cabot Chemical Co.; 27-TEDA Chemical Park Corporation; 28-Tianjin Tong Tee Industrial Co.; 29-Tianjin Tobo Lead Recycling Co.; 30-Tianjin cement mill; 31-Kumho Tire Co.; 32-Tianjin Aoxing Rubber Co.; 33-Tianjin Motorola China; 34-Tianjin Yuasa Batteries Co.; 35-various lead-acid battery users

11.1.2 Topology of the Symbiotic Metabolic Networks

Figure 11.2 shows that the 10 networks have a variety of topologies, including star, network, tree, and mixed types. Among them, Shanghai Wujing and Xinjiang Shihezi symbiotic networks belong to the star type. In a star network, one node is the central node, and the other nodes (branch nodes) are connected to the central node but generally not to other branches. The star topology is relatively simple, with the central node controlling the material and energy flows. The advantage of this topology is that the flow rate from the center node to the periphery is fast, and the network is easy to monitor and manage, but there is a problem of over-reliance on central nodes, which leads to a heavy burden on the metabolic network. From the perspective of the park's ecosystem, this type can be interpreted as a form of parasitic symbiosis. The parasitic (branch) nodes in the network obtain all kinds of raw materials for their own production from the host (central) nodes so as to reduce the environmental pressure created by pollution, and they obtain the resources they need by outsourcing the supply of those resources to the host nodes to obtain the benefits. The host nodes can process a certain amount of waste materials from both their own production activities and transfers from the branch nodes, and have obvious advantages in resource use, production processes, and production design. Generally, a host node can supply multiple parasitic nodes and provide a stable resource supply. The parasitic relationship is relatively stable; the host nodes gain the benefits of reducing waste emissions, but the value-added by this reduction is not obvious. Materials and value are redistributed from the host nodes to the parasitic nodes, with the flows primarily in one direction (towards the parasitic nodes).

The Choctaw, Kitakyushu, Shandong Lubei, and Guangxi Guigang symbiotic networks belong to the network type of topology. The status of their nodes is more balanced due to the close relationships among the nodes. Because of this, the overall stability of the network is strong. If one path breaks, materials and energy can still be transferred through other paths, and the utilization rate of materials and energy is high. The distributed structure of this network type is prominent, and sharing resources is easier than in a star topology. The disadvantage is that the network's structure is more complex, which makes it more difficult to control and supervise. From the perspective of the park's ecosystem, the network topology can be interpreted as mutualism. Two or more nodes in the network form a community of interests through the mutual benefits and complementary advantages they obtain from their relationship, and all of the nodes can benefit from the exchanges of materials and energy. The basic characteristic of this topology is that the links between nodes are relatively stable, and form an approximately closed circulatory system; the nodes may be more or less active in the network, but don't automatically play active or passive roles defined by the topology because they are more likely to play equal and indispensable roles.

The Kalundborg and Styria symbiotic networks belong to the tree type of topology. The network has branches, and each branch has sub-branches. Compared with the star type, this type represents a hierarchical centralized control network. Its advantage is that the flow between nodes is relatively convenient, and it is easy to establish new path connections. At the same time, if the leaf nodes (i.e., nodes at the end of branches) and their connected paths are damaged or destroyed, the network will not be fatally damaged because other paths remain intact. However, there is also the problem of excessive dependence on the root (central) node. If the source node for any branches of the tree fails, the network will not operate normally.

The Tianjin TEDA and Changsha Huangxing symbiotic networks belong to the mixed type of topology, in which some of the network nodes form a network structure, while most others form a hierarchical tree structure. In contrast, most nodes in the Changsha Huangxing park form a tree-type hierarchical structure, but a few nodes are relatively independent and do not exchange material and energy with many other nodes.

11.1.3 Core-Periphery Structure Analysis

Networks are composed of core nodes, which are at the center of the network and are connected to many other nodes, and peripheral nodes, which lie at the edges of the network and are connected to fewer nodes. My research group used the UCINET software (http://www.analytictech.com/archive/ucinet.htm) to analyze the core-periphery structure of the 10 symbiotic networks described in the previous section. First, we used the software to determine the class memberships (i.e., whether each node was a core or peripheral node), then analyzed the number of relationships and the proportions of core and peripheral nodes in each structure (Table 11.1). Next, we divided them among four types of relationships: core-core, core-periphery, periphery-core, and periphery- periphery.

The core members outnumbered peripheral members in the Changsha Huangxing, Styria, Kitakyushu, and Kalundborg parks. Core members accounted for more than 60% of all members in each case, with more than 90% core members in the Styria park. In the Shandong Lubei park, core members were roughly balanced with peripheral members. In the remaining parks, peripheral members outnumbered the core members, with at least 65% peripheral members.

According to our analysis of the core-periphery structure (Table 11.2), the exchanges of resources in Xinjiang Shihezi, Shanghai Wujing, Guangxi Guigang, and Choctaw ecological industrial Parks were mainly between core and periphery members. On the contrary, exchanges in the Shandong Lubei, Changsha Huangxing, Styria, Kitakyushu, and Kalundborg ecological industrial parks were mainly between core members, rather, the exchanges within the Tianjin TEDA park were mainly between peripheral members.

11.1.3.1 Core-Core and Periphery-Periphery Relationships

Most of the resources were exchanged between core members instead of peripheral members in the Kalundborg and Changsha Huangxing networks (Table 11.1). Besides, all of the relationships originally came from core members, which means that core members provide all resources. In the Kitakyushu network, all the resources were exchanged between core members, which means that core members were not only providers but also consumers of products, byproducts, and wastes within the network.

In the Shandong Lubei network, 55% of all the exchanges of resources originally came from core members, while 45% came from peripheral members (Table 11.1). Besides, 55% of the exchanges of resources were received by core members, while 45% by peripheral members, which indicates that core members received more from and provided more to peripheral members. Exchanges of resources between core members were 40% of the total, of which between core members and peripheral members only were 30% in the proportion of all of the relationships. And the remaining 30%, were between peripheral members. Therefore, exchanges of

Table 11.1 The number of relationships between core (C) and peripheral (P) nodes and their proportions of the total relationships

Kalundborg						Choctaw					
Number of relationships			Proportion (percentage of total)/%			Number of relationships			Proportion (percentage of total)/%		
	C	P		C	P		C	P		C	P
C	8	5	C	61.54	38.46	C	4	8	C	30.77	61.54
P	0	0	P	0	0	P	0	1	P	0	7.69
Kitakyushu						Changsha Huangxing					
Number of relationships			Proportion (percentage of total)/%			Number of relationships			Proportion (percentage of total)/%		
	C	P		C	P		C	P		C	P
C	23	0	C	100	0	C	40	1	C	97.56	2.44
P	0	0	P	0	0	P	0	0	P	0	0
Styria						Guangxi Guigang					
Number of relationships			Proportion (percentage of total)/%			Number of relationships			Proportion (percentage of total)/%		
	C	P		C	P		C	P		C	P
C	42	1	C	95.45	2.27	C	5	11	C	20.83	45.83
P	1	0	P	2.27	0	P	2	6	P	8.33	25.00
Xinjiang Shihezi						Shanghai Wujing					
Number of relationships			Proportion (percentage of total)/%			Number of relationships			Proportion (percentage of total)/%		
	C	P		C	P		C	P		C	P
C	0	3	C	0	33.33	C	1	8	C	8.33	66.67
P	3	3	P	33.33	33.34	P	3	0	P	25.00	0
Shandong Lubei						Tianjin TEDA					
Number of relationships			Proportion (percentage of total)/%			Number of relationships			Proportion (percentage of total)/%		
	C	P		C	P		C	P		C	P
C	8	3	C	40	15	C	15	0	C	34.88	0
P	3	6	P	15	30	P	0	28	P	0	65.12

Note Relationship directions are defined as originating in the first column of the table and flowing to the second and third columns.

Table 11.2 Types of sub-network relationships for the 10 symbiotic networks

Network type	Eco-industrial park
Core-core	Changsha Huangxing, Kalundborg, Kitakyushu, Shandong Lubei, Styria
Core-periphery	Choctaw, Guangxi Guigang, Shanghai Wujing, Xinjiang Shihezi
Periphery-periphery	Tianjin TEDA

resources were mostly between core members. However, when compared with other eco-industrial parks, the resource exchanges between core and peripheral members were relatively more frequent.

According to the core-periphery analysis of the Styria eco-industrial park, 97.7% of the exchanges of resources originally came from core members and only 2.3% from peripheral members (Table 11.1). The proportion of exchanges of resources between core members was 95.4% of the total, which shows that exchanges of resources were mainly provided by core members for other members and also indicates that 97.7% of the resources have flowed to the core members and almost all resources have flowed from core members; only 2.3% of flows of resources were from core members to peripheral members. It indicates that core members are on duty for the inputs and outputs of resources in this park, while peripheral members contribute little to all of the flows of resources.

In Tianjin TEDA eco-industrial park, 34.9% of all the exchanges originally came from core members, while 65.1% were from peripheral members (Table 11.1). Besides, core members received 34.9% of the exchanges of resources, while peripheral members received 65.1%. Therefore, peripheral members are involved in most of the exchanges, which shows that peripheral members were both producers and consumers of resources, and also energetically exchanged resources with others. There were far more peripheral members than core members, so the peripheral members participated in most exchanges, which also reflects that the peripheral members actively participated in establishing relationships, contributing significantly to the formation of the industrial symbiotic network of the park.

11.1.3.2 Core-Periphery Relationships

In the Guangxi Guigang eco-industrial park, 66.7% of the relationships originally came from core members, and only 33.3% were originally from peripheral members. Besides, 70.8% of the exchanges flowed to peripheral members, more than one-third of which were exchanged among peripheral members. The remaining 29.2% of the exchanges flowed to core members, nearly one-third of which flowed from peripheral members. It shows that peripheral members mainly used by-products and wastes by raw materials received from core members. Exchanges of resources between peripheral members accounted for 25.0% and are far lower than that between core members and peripheral members (54.2%). The exchange rate between core members was the lowest (20.8%). Thus, the main form of symbiotic relationship was resource exchanges between core members and peripheral members.

In the Shanghai Wujing eco-industrial park, 75.0% of the exchanges originally came from core members, while only 25.0% were from peripheral members. Besides, core members received 33.3% of the exchanges of resources, while peripheral members received 66.7%. The proportion of exchange resources between core members was only 8.3% of the total, while the remaining exchanges were all conducted between core members and peripheral members, accounting for 91.7%. Therefore, the exchanges were primarily between core and peripheral members. The peripheral members used the by-products and wastes from the core members as raw materials to support their production, but there was no exchange between the peripheral members, so all of the exchanges that were received by the peripheral members totally came from the core members, which shows that the peripheral members are at the end of all ecological industrial chains of the park and are the final destination for wastes and byproducts.

In the Choctaw network, 92.3% of exchanges of resources originally came from core members, while only 7.7% of exchanges were from peripheral members, which indicates that core members devoted significant resources needed for the network. Besides, 30.8% of these relationships were between core members, and only 7.7% of these relationships were between peripheral members, which indicates that 61.5% of the exchanges of resources flowed to peripheral members from core members and leading exchanges of resources were between core members and peripheral members.

In the Xinjiang Shihezi network, peripheral members devoted 66.7% of exchanges of resources, while core members only devoted 33.3%, which shows that peripheral members provided primary resources. In addition, 33.3% of the relationships were between peripheral members, with 0% between core members, which indicates that 66.7% of exchanges of resources were between core and peripheral members.

11.1.4 Degree of Connectedness

11.1.4.1 Network Degree of Connectedness

Table 11.3 shows that the network density of the 10 eco-industrial parks was the highest for Xinjiang Shihezi, followed by Guangxi Guigang, and was the lowest for Styria and Changsha Huangxing. In an actual social network graph, the maximum density is typically larger than 0.5 (Liu 2004; Mayhew and Levinger 1976). However, when this approach is applied to an eco-industrial park, the results in Table 11.3 show a much lower density for all ecological industrial parks, in which all values are equal to or even less than 0.3, which means that the exchanges of resources among members of parks are relatively sparse and it has considerable potential to strengthen the symbiotic relationships.

Table 11.3 The network density (*D*) and number of paths for the 10 eco-industrial parks

Name of parks	*D*	Number of paths
Xinjiang Shihezi	0.300	9
Guangxi Guigang	0.267	23
Shandong Lubei	0.152	20
Choctaw	0.145	13
Kalundborg	0.118	13
Shanghai Wujing	0.109	12
Kitakyushu	0.055	23
Tianjin TEDA	0.036	43
Styria	0.031	43
Changsha Huangxing	0.018	41

Note Parks are presented in order of decreasing *D*

The members of Xinjiang Shihezi ecological industrial park were closely connected, so the park had the highest density of 0.300, and its network was relatively complete. The network of Changsha Huangxing had the lowest density only, accounting for 0.018, followed by Stiria (0.031). It indicates that the members of them are relatively unconnected, so their networks of them were a little sparse. In these parks, it is difficult for them to form symbiotic relationships, because resources cannot be used by all of the members of a complex system. After a comparative study of these 10 networks, it is found that the network density of Changsha Huangxing ecological industrial park and Styria ecological industrial Park is low because of the large scale of these two networks (48 and 38 members, respectively), indicating that in these two giant networks, the degree of attention between members is lower than that of parks with fewer members. As a matter of fact, the degree of competence of an industrial symbiotic network mainly depends on the frequency of interactions among members of the network.

The centrality of the network is generally used to analyze network density except for different scales. Both the Xinjiang Shihezi ecological industrial park and Guangxi Guigang ecological industrial park had the highest densities partly as a result of the high degrees of centrality (also referred to as nodal degrees, which represent the number of nodes directly connected to a given node). The nodes in the symbiotic network were all balanced in the Xinjiang Shihezi ecological industrial park, with high input and output nodal degrees. However, the nodes in the Guangxi Guigang ecological industrial park mainly were either of high input degree or high output degree, so they were imbalanced. On the contrary, Changsha Huangxing and Stiria parks had instead a low degree of density. In summary, both the Changsha Huangxing eco-industrial park (with 48 nodes) and Stiria eco-industrial park (with 38 nodes) had relatively few direct relationships among all of the nodes and were in large network scale, low nodal degree and the imbalanced degree of the input and output, so they were both relatively of the lowest density.

It is revealed that the completeness of eco-industrial parks in China was relatively high by comparing the density of eco-industrial parks in China and abroad, which is in accordance with their characteristics of development. Lots of eco-industrial parks in China are joint ventures, with one or several large enterprises at the center. Therefore, the industrial symbiosis network is generally dominated by large-scale enterprises and their affiliated organizations. They belong to a single economic entity and interact with each other firmly, just like Xinjiang Shihezi ecological industrial parks and Guangxi Guigang ecological industrial park, whose completeness of networks is relatively higher. On the contrary, central enterprises were still independent, whose symbiotic relationships were established spontaneously by means of contracts between owned and operated enterprises referring to the development of foreign ecological industrial parks. This is different from the joint-enterprise types of parks in China, in which symbiotic relationships are quickly established. On the contrary, Stiria has many members in its network without a dominant central one. As a result, interactions between members were less, and the networks were relatively spare, forming a lower density and degree of completeness of networks.

11.1.4.2 Sub-Network Degree of Connectedness

Based on the number of sub-network and their distribution in each eco-industrial park, the intensity of the relationships among the sub-networks can be described, but it was impossible to directly determine the degree of connectedness of the sub-network due to different network scales (i.e., the sub-networks had different numbers of nodes). Therefore, in order to analyze the degree of connectedness, we calculated the density of the sub-network using a block adjacency matrix (i.e., an adjacency matrix separated into blocks that comprise the sub-networks) (Table 11.4).

Based on the distribution of sub-networks in Table 11.4, the high degree of connectedness of the Xinjiang Shihezi symbiotic network is caused by the frequent interactions between core members and peripheral members, while the high degree of connectedness of the Guangxi Guigang symbiotic network results from the interactions between core members and between core and peripheral members. On the contrary, as a result of the low frequency of interactions between core members and peripheral members, the symbiotic networks of Changsha Huangxing and Stiria have the lowest degree of connectedness (i.e., low core-core and low periphery-periphery).

We can identify the dominant sub-network based on the degree of connectedness of sub-networks, so that all of the ecological industrial parks can be divided into three types, including parks dominated by core-core relationships, by core-periphery relationships, and by periphery-periphery relationships (Table 11.5).

The structure of the core-core networks is relatively compact, and core members exchange frequently with each other, forming an advantage of an organic symbiotic network. This advantage is conducive to reduce the risk that breaking a linkage will seriously damage the entire network. From the perspective of the connectedness of the sub-networks composed of core members, Guangxi Guigang has the

Table 11.4 The degree of connectedness of the sub-networks of the symbiotic networks for the 10 eco-industrial parks

Kalundborg			Choctaw		
	Core	Periphery		Core	Periphery
Core	0.190	0.179	Core	0.333	0.333
Periphery	0	0	Periphery	0	0.033
Styria			Guangxi Guigang		
	Core	Periphery		Core	Periphery
Core	0.033	0.014	Core	0.833	0.524
Periphery	0.014	0	Periphery	0.095	0.143
Xinjiang Shihezi			Shanghai Wujing		
	Core	Periphery		Core	Periphery
Core	0	0.600	Core	0.500	0.444
Periphery	0.600	0.150	Periphery	0.167	0
Kitakyushu			Changsha Huangxing		
	Core	Periphery		Core	Periphery
Core	0.075	0	Core	0.030	0.002
Periphery	0	0	Periphery	0	0
Shandong Lubei			Tianjin TEDA		
	Core	Periphery		Core	Periphery
Core	0.400	0.086	Core	0.268	0
Periphery	0.086	0.143	Periphery	0	0.400

Note Matrix rows represent the senders in a relationship and matrix columns represent the recipients

Table 11.5 The dominant sub-networks within the overall network

Dominant sub-network type	Eco-industrial park
Core-core dominant	Changsha Huangxing, Choctaw, Guangxi Guigang, Kalundborg, Kitakyushu, Shandong Lubei, Shanghai Wujing, Styria
Core-periphery dominant	Xinjiang Shihezi
Periphery-periphery dominant	Tianjin TEDA

highest degree of core connectedness (0.833), followed by Shanghai Wujing, Shandong Lubei, Choctaw, Kalemburg, Kitakyushu, Styria and Changsha Huangxing. In the Guangxi Guigang and Shanghai Wujing Parks, the degrees of connectedness among core members are high because there are few core members, but they interact frequently. The network connectedness of Stiria Park and Changsha Huangxing Park is relatively low, which is due to a large number of core members of these two parks and the relatively small interaction. Although the relationship between the core members of Stiria and Changsha Huangxing eco-industrial park is dominant, the connectedness of the three subsidiary networks was lower than other parks.

Therefore, the future development of these parks should establish core-core and core-periphery relationships to rapidly improve the completeness of the network in the short term, and then add periphery-periphery relationships to make the network more stable in the long term.

In the network dominated by core-periphery relationships (Xinjiang Shihezi), the periphery members were highly reliant on the core members. Therefore, if the core members make changes due to process adjustments, material or scale changes, or management environment changes, it will be incredibly harmful to the stability of the relationships between the core industries and downstream industries.

In the one sub-network (Tianjin TEDA) dominated by periphery-periphery relationships, the interactions between peripheral members are relatively frequent. Core members did not drive the development of the network, so network members are all more equal, but they cannot give total energy to their vitality. In order to make the network more compact and stable, raising the status of core members and adding relationships among them or between core-peripheral members will be necessary in future development.

11.1.5 Importance of Research Modalities

The degree of connectedness of three types of sub-networks (core-core, core-periphery, and periphery-periphery) can be analyzed by combining density and core-periphery structure analyses. The structural attributes can be studied at levels from the respective of the whole network to the sub-network levels. Meanwhile, in order to determine how specific network structures can improve or damage the performance of industrial symbiotic networks, it was possible to compare the similarities and differences between different eco-industrial parks in China and abroad by combing density and core-periphery structure analysis. Identifying some problems and proposing measures based on these differences was beneficial to improve the degree of connectedness of the networks and eliminate or mitigate the inefficiencies (These issues will be discussed in three of the parks in the next section.) Through the analysis and research of this combined method, the eco-industrial parks with different origins, compositions and structures can be described effectively, which is a unique advantage.

We can learn about how social network analysis permits exploratory research on symbiotic networks by these results, but it is necessary to adopt further study and discussion for some issues. We did not fully describe these flows, but simply dealt with the information flow and interpersonal relationship between actors in the network (i.e., as a hidden part of material or energy flow, rather than as an independent flow part). Therefore, it is necessary to find a more practical and reliable method to quantify these factors and incorporate them into the analysis of future research. Social network analysis is a larger and more complex method than I introduced in this section. For instance, because the network analysis was mainly used to demonstrate the value of concepts primarily, I only used some fundamental indicators in network analysis. Therefore, it is necessary to improve the characteristics of paths between nodes and the overall characteristics of the networks. Meanwhile, it is also necessary to further analyze the relationships between network structure and characteristics of notes because some networks may have star or symmetric structures, as well as the relationship between network structure and overall network characteristics. These improvements will enable us to analyze the structural patterns of symbiotic networks in more detail.

11.1.6 Findings and Recommendations

To illustrate the insights provided by social network analysis, we can select three typical parks (Changsha Huangxing, Xinjiang Shihezi, and Tianjin TEDA) from the 10 symbiotic networks, and propose ways to improve their network connectivity and eliminate network inefficiencies to resolve or mitigate existing problems. Similar analyses can be conducted for the other parks.

In the Changsha Huangxing Ecol-industrial park, the following exchanges could be implemented: the intelligent metals plant could be connected with the Liyuan New Material Industry of the park (a relationship between core and peripheral members) and other industries using the byproducts of the metal plant. For instance, if connections can be established with the components (i.e., a relationship between the core components) and the building materials plant can recycle their metal waste (a relationship between the core components and the peripheral components), the metal recycling plant could use scrap metal byproducts. Increasing the establishment of such communication mechanisms would improve the density of the network and efficiency of resource utilization, thus improving the sustainability of parks' operations.

In the Xinjiang Shihezi eco-industrial park, it is necessary to increase the exchanges among the surrounding members in the future in order to weaken their dependence on core members and increase the stability of the network. For instance, the water from the wastewater treatment system could be delivered to the livestock breeding system to clean livestock facilities, to the animal-products treatment system to beautify the landscape around the facilities, and to the papermaking system as low-quality industrial water by expanding the use of treated water.

In the Tianjin TEDA eco-industrial park, the connection between these two cogeneration plants and other members of the park using their products is weak. The steam from the Guahua cogeneration plant could also be used by the desalination plant (a relationship between core members), whose bottom ash could be delivered to the construction companies (a relationship between core members). The slag waste from two plants could be used in the construction industry. Increasing the visits to the TEDA eco-landscaping company, including visits to the TEDA water treatment plant and the TEDA new water source company, is beneficial to make the public and other enterprises further understand further production processes (a relationship between core members and peripheral members).

11.2 Analysis of the Sulfur Metabolism in an Eco-Industrial Park

The Shandong Lubei ecological industrial park has been developing rapidly and steadily since it was approved for construction in 2003, which is located in the Yellow River Delta, adjacent to the Bohai Sea and Huanghua Port in Hebei Province. The park is the leader in the development of China's ecological industrial parks, whose ecological and economic benefits have been significantly improved. It is operated by Shandong Lubei Chemical Limited Liability Company and is a typical representative of China's ecological industrial parks in specific industries. Lubei Park is an essential basis of the inorganic and salt chemical industry in China, which has the biggest ammonium phosphate, sulfuric acid, and cement joint-production plants in China and be regarded as a typical ecological industrial park of China by the United Nations Environment Programme.

Most resources (including industrial waste and byproducts) could be reused or recycled within Shandong Lubei eco-industrial park; In this way, the efficiency of resources utilization can reach 95.6%, and the efficiency of clean energy utilization can reach 85.9%. Shandong Lubei ecological industrial park mainly has three industrial chains: the stepped utilization of seawater chain, the ammonium phosphate-sulfate acid-cement chain, and the cogeneration of salt, alkali, and electricity chain. The industrial processes utilizing sulfur can be tracked through each of these three chains (Yang et al. 2004). Therefore, tracking these processes in the metabolism will provide an essential basis for understanding the symbiotic status of the networks of the park and identifying the critical enterprises and processes in the park.

In this section, in order to provide an example that can gain insights through comprehensive analysis, I will focus on the flows of sulfur in the metabolic processes of the Shandong Lubei eco-industrial park. Then, from the perspective of considering the relationship between pairs of linked nodes and flows of resources the network analysis method, I will use network analysis to analyze the internal characteristics of the park; to be specifical, it is precisely in the research based on this section, my research group examined the structural distribution and functional attributes of the park's network to reveal the characteristics of enterprises (nodes) that are engaged in the regular operation of the park (Zhang et al. 2015a). Our goals were to identify the core and peripheral nodes, to study the paths along which materials and energy are exchanged between nodes, and to identify the key paths (Zhang et al. 2015b). Our ultimate goal was to identify the different roles of nodes and the different processes related to sulfur flows and reveal the similarities and differences among nodes and paths by combining the characteristics of the structure and function of the network. Such analysis can be used to reveal the shortcomings of some nodes and paths in order to explore the methods to improve the development of the Shandong Lubei ecological industrial park by smoothing and stabilizing the sulfur metabolism processes (Fig. 11.3).

The quantitative data about the exchanges within the park or between the park and its external environment is from the research literature (i.e., Chen 2003; Feng 2003a, 2003b), and from the Shandong Lubei Enterprise Group General Company(http://www.lubei.com.cn/page-6833.html).

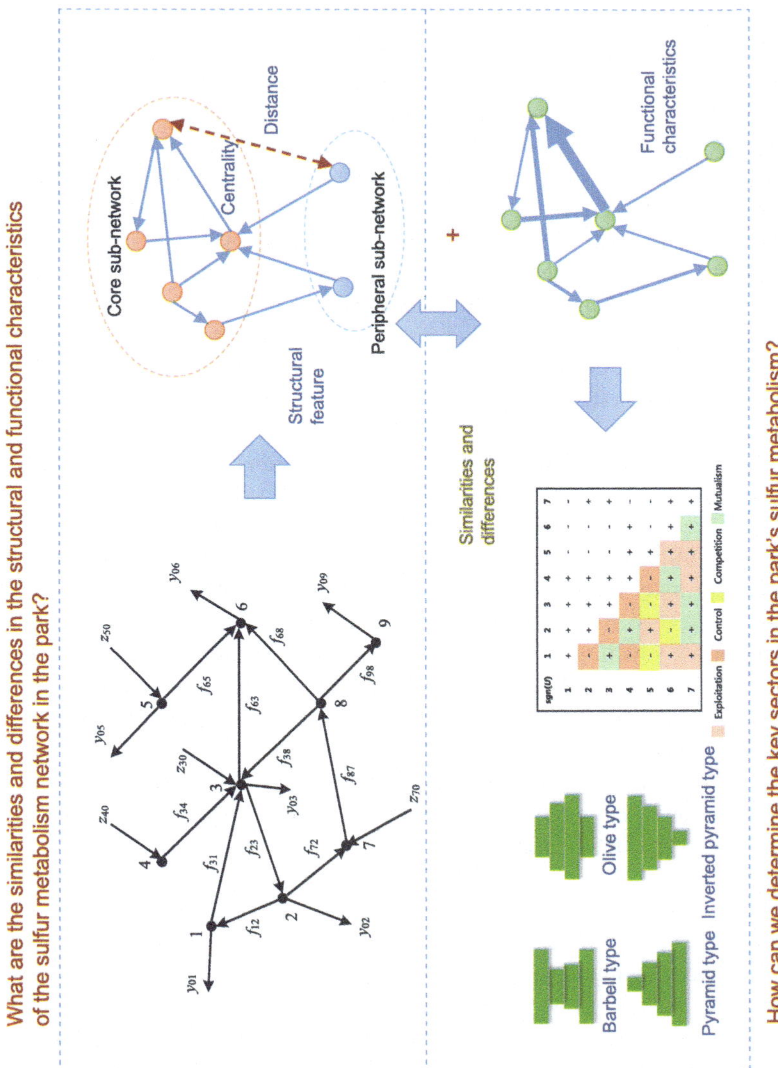

Fig. 11.3 Analytical framework for the sulfur metabolism processes in the Shandong Lubei eco-industrial park (*Note* In the flow diagram, f_{ij} represents the flow of materials from node *j* to node *i*; *y* represents outputs to the external environment, and *z* represents inputs from the external environment. (The diagram has been simplified compared with the actual network structure developed for this analysis.) sgn(U) represents the sign utility matrix)

11.2.1 Construction of the Network Model for the Park's Sulfur Metabolism

Shandong Lubei eco-industrial park has 12 enterprises engaging in the materials and energy exchanges, including aquaculture plants, bromine plants, salt gypsum production plants, raw salt production plants, chlor alkali plants, potassium magnesium salt production plants, ammonium phosphate plants, sulfuric acid plants, cement clinker plants, cement products production plants, ammonia plants and thermal power plants (Fig. 11.4). These enterprises establish symbiotic exchange networks by exchanging materials or energy. Aquaculture plants, raw salt production plants, and chlor alkali plants in the network model provide materials for other nodes but do not exchange sulfur with other nodes, so they are not included in the sulfur analysis. The remaining of this section only analyzes enterprises engaged in sulfur exchanges, so aquaculture plants, raw salt production plants, and chlor alkali plants are excluded because they do not participate in sulfur exchanges. In order to further improve the description of parathion metabolism processes, the cement plant is divided into two sections with different characteristics, including cement clinker production and cement product production (Chen 2003). The processes of sulfur metabolism consist of three subsidiary chains, each of which is represented by a different color. The nine metabolic actors that exchange sulfur form three chains: the stepped utilization of seawater chains, the ammonium phosphate-sulfuric acid-cement, and their derivatives chains, and the cogeneration of salt, alkali, and electricity chain.

In the stepped utilization of the seawater chain, raw seawater can be evaporated to produce primary brine containing sulfate ions, which is delivered from the bromine plant to the system, and then the primary brine can be further evaporated into secondary brine, which is then delivered to the salty gypsum production plant. Part of the secondary brine can be delivered to the potassium and magnesium salt production plant. Finally, the sulfate ions are incorporated into potassium and magnesium salt products. The secondary brine that passes through the production process of the salty gypsum production plant is incorporated into the salt gypsum products, and these products are finally delivered to the cement clinker production plant and the cement products production plant.

In the ammonium phosphate-sulfuric acid-cement chain, phosphogypsum and sulfur-containing coal are delivered to the cement clinker production plant, where they are divided into three branches: one involves loss during the production process, another involves the cement clinker products, and the third involves a substance that is transformed into the gas produced by the cement kiln that is used to supply the sulfuric acid plant (Zhang et al. 2015a). The sulfuric acid plant then divides all its sulfur into four branches: one is related to the loss during the production of sulfuric acid, the other is related to the raw materials for the production of liquid SO_2 and bromine gas (Br_2) in the chlor alkali plant, and the other two branches are related to the production of sulfuric acid products (Zhang et al. 2015a). Most of these products are sent to the ammonium phosphate plant to react with phosphate rock, and a small number of sulfuric acid products are exported to the external

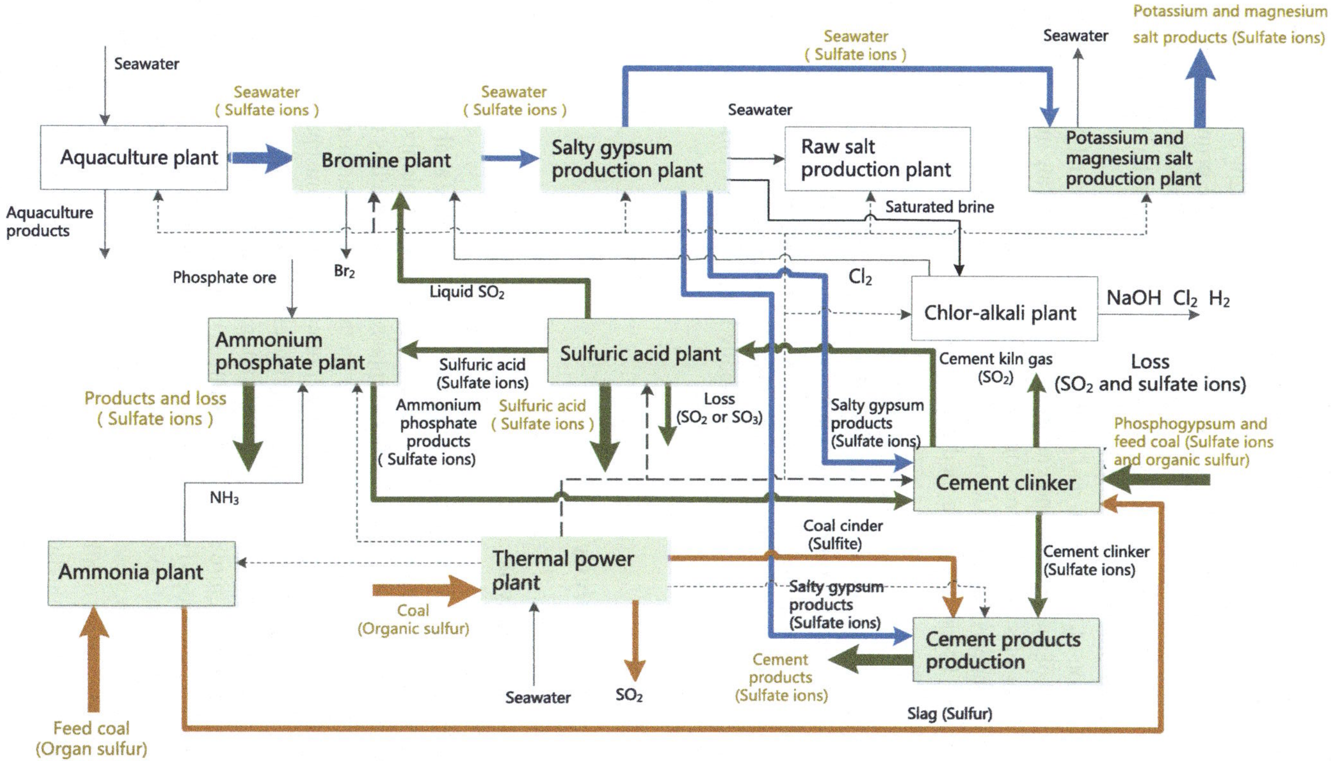

Fig. 11.4 Sulfur metabolic processes in the Lubei eco-industrial park (*Note* The solid lines represent exchanges of materials or energy; the dashed lines represent the sharing of infrastructure; the white rectangles represent the enterprises that are not involved in the park's sulfur metabolic processes but that nonetheless exchange materials with other enterprises in the park; the colored rectangles represent the enterprises that are involved in the sulfur metabolic processes; the three line colors represent different sulfur metabolic paths; and the thick lines are inputs from or outputs to the external environment)

environment. Some sulfur in the ammonium phosphate unit is lost in the production process, and some sulfur is included in phosphate rock products and phosphogypsum by-products. Phosphogypsum can also be used as raw material for cement clinker production. Finally, the sulfur in the cement clinker production section is included in the products delivered to the cement product production section (Zhang et al. 2015a).

In the ammonium phosphate-sulfuric acid-cement chain, as the primary energy resource of the ammonia plant, the processes of feed coal's metabolism have generated the sulfur, which can be transformed into slag waste, and then be delivered to the cement clinker production plant to mix with others.

In the cogeneration of salt, alkali, and electricity chain, the thermal power plant is the producer both of the thermal and energy of the park, whose raw materials are coal containing a lot of sulfur. It is reasonable to use lime to desulfurize raw coal to produce a coal cinder, and the cinder is exactly the auxiliary material for cement product production. The sulfur in the feed coal is transformed into SO_2, which is a waste sulfur resource.

In order to analyze the inherent characteristics of the processes of sulfur metabolism in the Shandong Lubei ecological industrial park, the metabolism processes are supposed to be abstracted as a network model, including enterprises as nodes and exchanges among enterprises as paths. The nine enterprises are considered to be involved in the processes of sulfur metabolism as nodes in this model. What's more, the paths between pairs of nodes can be defined on the basis of the exchange relationships between enterprises (Fig. 11.5).

11.2.2 Structure Analysis

11.2.2.1 Centrality Analysis

Table 11.6 concludes the results of the centrality analysis. In this analysis, I have used three indexes of centrality: the nodal degree, which is the number of nodes directly connected to the given node; the betweenness degree, which is the number of nodes whose flows pass through a given node on their way to another node; and the closeness degree, which represents the number of nodes a flow must pass through before reaching a given node. Node 3 (cement clinker production) had the highest node degree, medium degree, and closeness, indicating that this node occupied a core position in the network structure and lots of resources related to sulfur are controlled by this node. The nodal degree of node 8 (salt gypsum production plant) was also high (25.0%). The nodal degree of node 2 (sulfuric acid plant) and node 6 (cement product production) is next, but both were less than 20%, which was only 60% of the value of node 3. The degree of nodes of all other nodes was less than 50% of the value of node 3. Node 8 (salt gypsum production plant) and node 6 (cement products production) had relatively high dielectric values, and were equal to 71.5% and 50.0% of node 3, respectively; The median value of all other nodes was much smaller, including four nodes with a value of 0. Compared with the other two indicators, the proximity of all

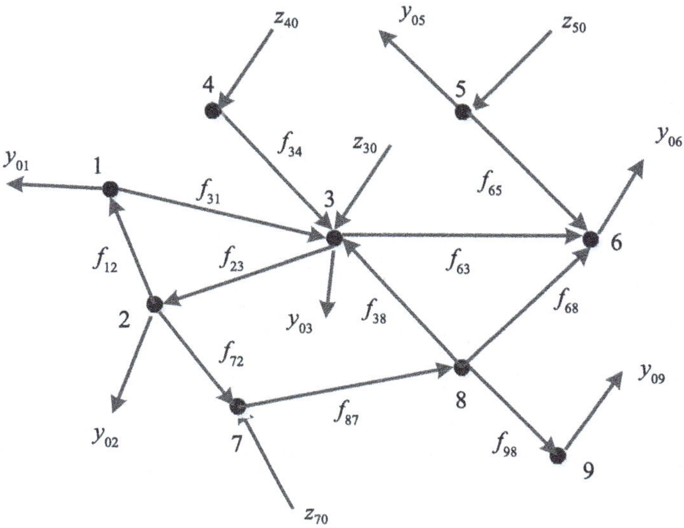

Fig. 11.5 Network model of the sulfur metabolic processes in the Shandong Lubei eco-industrial park (*Note* Flow f_{ji} represents the flow from node i to node j; z_{j0} represents the flow from the external environment (node 0) to node j; and y_{0i} represents the flow from node i to the external environment. Three of the original 12 nodes shown in Figure 11.4 have been omitted because they don't participate in sulfur flows) Nodes: 1-ammonium phosphate plant; 2-sulfuric acid plant; 3-cement clinker production; 4-ammonia plant; 5-thermal power plant; 6-cement products production; 7-bromine plant; 8-salty gypsum production plant; 9-potassium and magnesium salt production plant.

nodes was more significant than or equal to 40%, and the value of six nodes in nine nodes was more significant than or equal to 50%. This result shows that the distance between any two nodes was relatively short.

11.2.2.2 Core-Periphery and Density Analysis

Our core-periphery analysis (Table 11.7) showed that Node 1 (ammonium phosphate plant), Node 2 (sulfuric acid plant), Node 3 (cement clinker production), Node 6 (cement product production) and Node 8 (salt gypsum production plant) occupy the core position of the network. Nodes 4 (ammonia plant), 5 (thermal power plant), 7 (bromine plant), and 9 (potassium magnesium salt production plant) are located at the periphery (Zhang et al. 2015a). Table 11.7 indicates that more than half of the exchanges (54.6%) are among core nodes, of which half (27.3%) are from core nodes to peripheral nodes and the rest of which are from peripheral nodes to core nodes. There were no exchanges between peripheral nodes. In all exchanges, 81.8% came originally from core nodes, and 72.7% flowed to core nodes. Therefore, the flows delivered to and received by peripheral nodes only accounted for a small section totally.

Table 11.6 Node characteristics in the Shandong Lubei eco-industrial park

Node	Nodal degree/%	Betweenness degree/%	Closeness degree/%
1	12.500	0	50.000
2	18.750	7.143	53.333
3	31.250	50.000	72.727
4	6.250	0	44.444
5	6.250	0	40.000
6	18.750	25.000	61.538
7	12.500	3.571	50.000
8	25.000	35.714	66.667
9	6.250	0	42.105

Nodes 1-ammonium phosphate plant; 2-sulfuric acid plant; 3-cement clinker production; 4-ammonia plant; 5-thermal power plant; 6-cement products production; 7-bromine plant; 8-salty gypsum production plant; 9-potassium and magnesium salt production plant.

Table 11.7 The proportions of flows within the four sub-network types, and the degree of connectedness for exchanges within the four sub-networks

	Proportion for the whole network		Degree of connectedness (sub-networks)	
	Core	Periphery	Core	Periphery
Core	54.55	27.27	0.30	0.15
Periphery	18.18	0	0.10	0

Note Flows are from the node type in the first column of each part of the table to the node types in the next two columns.

The density of the network totally was 0.139 (i.e., only a tiny proportion of the theoretically possible paths). On the contrary, the connectedness within the four sub-networks (core-core, core-periphery, periphery-core, periphery-periphery) was considered a little higher than others, where the connectivity of core-core nodes was up to 0.30 (Table 11.7). When there are many paths between each pair of nodes in the network, the connectedness rate is in the high value. The five nodes at the core positions have six exchanges ($f_{12}, f_{23}, f_{31}, f_{38}, f_{68}$, and f_{63}, where f_{ij} represents the flow from node j to node i), and the density of this sub-network is more than twice that of the overall network (Zhang et al. 2015a). All core nodes are in the ammonium phosphate-sulfate acid-cement chain except node 8 (salty gypsum production plant), which shows that enterprises in this chain have made outstanding contributions to the sulfur exchange of the park. The dependence on the core nodes of this park is prominent, whose structure may increase the importance of core members, but the contribution of peripheral members may decline. Since the linkages of core-periphery and periphery-core are different in their directions, they are likely to be combined into a single core-periphery subsidiary network if we concentrate on the existence of exchanges instead of their directions. The average connectedness of the subsidiary network was 0.125, which was similar to the total density. In the subsidiary network relationships f_{34}, f_{65}, f_{72}, f_{87}, and f_{98}, most peripheral nodes are linked with node 8 (salty gypsum production plant).

11.2.2.3 Average Distance

The average distance of all of the nodes is 2.553, which means that, generally, any pair of nodes have to rely on more than two intermediate nodes to connect with each other. Table 11.8 demonstrates the distance between any pair of nodes in the network. The maximum distance between any two nodes was 5, and this was the length from node 1 to node 9 and from node 4 to node 9. In these two pairs of nodes, sulfur was gradually used by exchanging along the paths of five nodes. The paths from node 1 (ammonium phosphate plant) to node 9 (potassium magnesium salt production plant) were 1→3→2→7→8→9, and the delivered resources were phosphogypsum, cement kiln gas, liquid SO_2, and seawater. These pathways connect the ammonium phosphate sulfate cement chain and the seawater chain for stepped utilization (Zhang et al. 2015a). Liquid SO_2 is an essential sulfur resource for bromine factories and a critical raw material for bromine production. The paths that were from node 4 (ammonia plant) to node 9 (potassium and magnesium salt production plant) were 4→3→2→7→8→9, which connected all three chains.

There were several routes with a path length of 4 between the two nodes: from node 7 to node 1, from node 1 to node 8, from node 4 to node 8, and from node 3 to node 9. It is founded that all four pairs of routes generally share two primary paths by checking their respective specific paths, which are the paths from node 3 to

Table 11.8 Distances between pairs of nodes in the Shandong Lubei sulfur metabolism network

	1	**2**	**3**	**4**	**5**	**6**	**7**	**8**	**9**
1	0	1	2	3	0	0	4	3	0
2	2	0	1	2	0	0	3	2	0
3	1	2	0	1	0	0	2	1	0
4	0	0	0	0	0	0	0	0	0
5	0	0	0	0	0	0	0	0	0
6	2	3	1	2	1	0	2	1	0
7	3	1	2	3	0	0	0	3	0
8	4	2	3	4	0	0	1	0	0
9	5	3	4	5	0	0	2	1	0

Note Distances are defined from the node in the top row of the table to the node in the first column of the table Nodes: 1-ammonium phosphate plant; 2-sulfuric acid plant; 3-cement clinker production; 4-ammonia plant; 5-thermal power plant; 6-cement products production; 7-bromine plant; 8-salty gypsum production plant; 9-potassium and magnesium salt production plant.

node 2 and from node 7 to node 8. The results of this and the path with a length of 5 show that these two paths play an essential role in connecting network members and exchanging resources, which are the representatives of relationships about the recycling and reusing between upstream and downstream members of the network and also can connect the three major industrial chains.

Some paths are circular. For example, node 1 can directly link with node 3, node 3 delivers resources to node 2, and then node 2 delivers resources back to node 1. Node 6 (cement products production) and node 9 (potassium and magnesium salt production plant) do not deliver materials to other nodes; this means that when sulfur is delivered to these enterprises, most of the sulfur is transformed into products or lost during the production processes.

11.2.3 Functional Analysis

11.2.3.1 Flow Analysis

Based on the direct and indirect flows for each node, we can analyze the different ecological roles that result from this flow distribution: basal (producer) species (nodes that accept sulfur as their inputs), intermediate (consumer) species (nodes that receive and deliver surfur), and top (consumer) species (nodes that receive and output sulfur). According to the direct and indirect flows of each node, the various ecological roles caused by this flow distribution can be specifically analyzed (Fig. 11.6).

The direct inputs of the external environment of node 7 (bromine plant) were the largest, while the direct inputs of node 4 (ammonia plant) were the smallest. The maximum value (114,558 tons) was approximately 180 times the minimum value (650 tons). So the bromine plant was an essential primary species. Node 9 (potassium magnesium salt production plant) had the highest environmental outputs, which was the core top species of node 1 (ammonium phosphate plant), and its value was 10 times the minimum. The most enormous direct flows focused on the stepped utilization of the ammonium phosphate-sulfate acid-cement chain and the seawater chain, and it means that the direct flows in these two chains play an important role in exchanging and must be studied to improve the efficiency of the circulation.

The indirect flows between nodes were usually more extensive and more frequent than the direct flows (Fig. 11.6). There were solely 11 direct paths but 54 indirect paths of all nodes. The maximum of indirect flows was among node 3 (cement clinker production), node 2 (sulfuric acid plant), and node 1 (ammonium phosphate plant), with a value ranging from 5.8×10^5 t to 7.7×10^5 tons. Therefore, the magnitudes of indirect flows are 4 times larger than those of direct flows. The direct flows between the same three nodes were incredibly high, which indicates that they can make exchanges with each other directly, however, they adopted lots of circulating flow paths to transform sulfur indirectly. These observations show that the combined flows (direct and indirect) between the three nodes are also at the highest level. The indirect flows are mainly in the ammonium phosphate-sulfate acid-cement chain,

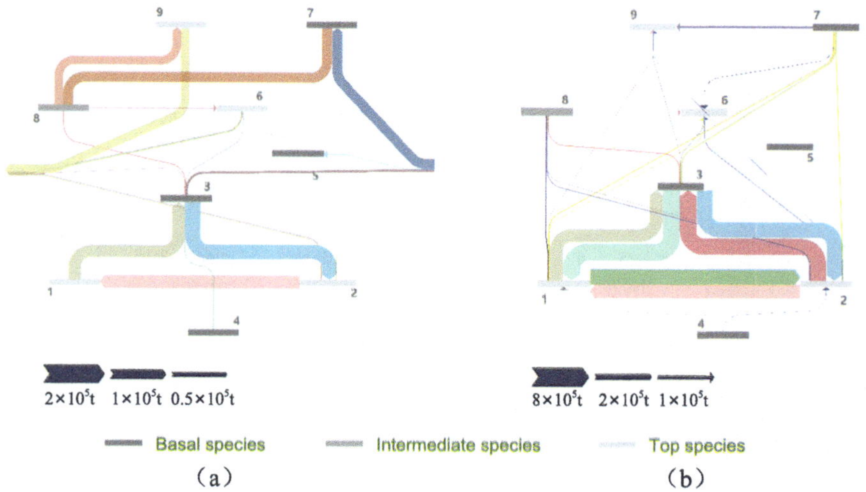

Fig. 11.6 Direct and indirect flows among different nodes (*Nodes* 1-ammonium phosphate plant; 2-sulfuric acid plant; 3-cement clinker production; 4-ammonia plant; 5-thermal power plant; 6-cement products production; 7-bromine plant; 8-salty gypsum production plant; 9-potassium and magnesium salt production plant. The pathways sent or received by different metabolic bodies are shown in different colors)

strengthening the importance of enterprises in the chain. The minimum of indirect flows are from node 4 (ammonia plant) to node 6 (cement product production), then to node 7 (bromine plant), node 8 (salt gypsum production plant), and node 9 (potassium magnesium salt production plant) (Zhang et al. 2015a). These flows are all less than 100 t. The indirect flow from Node 5 (thermal power plant) and Node 6 (cement product production) to other nodes is less than 1 t, so these were too small to be shown in Fig. 11.6.

Some pairs of nodes were dominated by direct flow, including the flows from Node 7 (bromine plant) to Node 8 (salty gypsum production plant), the flows from Node 8 to Node 6 (cement products production), and the flows from Node 8 to Node 9 (potassium and magnesium salt production plant). Others were dominated mainly by indirect flows that were much larger than the direct ones. These are the flows from nodes 1 (ammonium phosphate plant), 2 (sulfuric acid plant), and 3 (cement clinker production) to other nodes (Fig. 11.6). The path from node 5 (thermal power plant) to node 6 (cement products production) is totally controlled by direct flows, which indicates that node 5 is at the end of each chain and is controlled by node 6 (Zhang et al. 2015a). Through cyclic exchange between nodes, nodes that cannot be directly connected are in the high indirect flows, for example, from node 1 (ammonium phosphate plant) to node 2 (sulfuric acid plant), from node 2 to node 3 (cement clinker production), and from node 3 to node 1.

As mentioned earlier, the indirect flows were significantly larger than direct flows, so the relative abilities to absorb sulfur through indirect and direct flow paths were further analyzed (Fig. 11.7). The ratios of indirect flows to direct flows for node 1

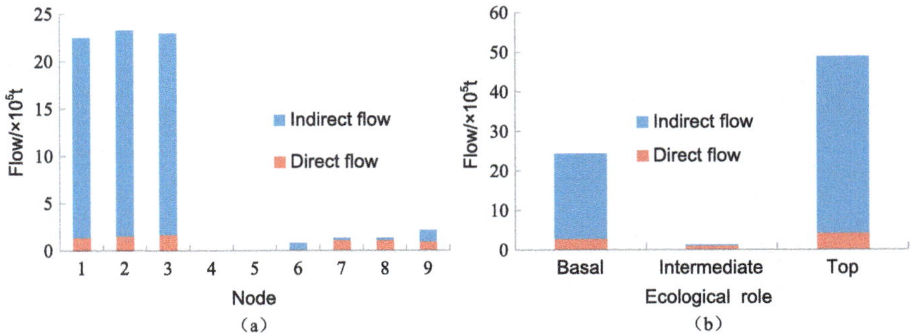

Fig. 11.7 Direct and indirect flows for the nodes and the three ecological roles (*Nodes* 1-ammonium phosphate plant; 2-sulfuric acid plant; 3-cement clinker production; 4-ammonia plant; 5-thermal power plant; 6-cement products production; 7-bromine plant; 8-salty gypsum production plant; 9-potassium and magnesium salt production plant)

(ammonium phosphate unit), node 2 (sulfuric acid unit), and node 3 (cement clinker production) were more extensive than 10, whose integrated flows (total direct flow plus total indirect flow) were also the highest. The indirect flow accepted by node 6 (production of cement products) was 3 times the direct flow, but the direct and indirect flows for this node were both smaller than those of nodes 1 to 3. For these four enterprises, absorbing sulfur flows from other members of the park can provide support for their stable production processes. The indirect flows at node 9 (potassium magnesium salt production plant) were slightly higher than the direct flows (the ratio is 1.2). The direct flows absorbed by node 7 (bromine plant) and node 8 (salt gypsum production plant) were 5 times the indirect flows, which shows that the direct flows were the primary source of sulfur. Both of node 4 (ammonia plant) and node 5 (thermal power plant) were dependent on direct flows, but these flows were only 650 tons and 5183 tons, respectively; Therefore, they are too small to be seen in this chart. The flow of the top species is the largest, about twice the total flow of the base species, while the flow of the intermediate species is very low. The indirect flow of top and bottom species is about 10 times the direct flow, while the direct flow of intermediate species is about 5 times of the indirect flow.

11.2.3.2 Ecological Relationships

There were 36 pairs of ecological relationships, excluding the mutualisms that resulted from flows within each node. Of this total, 8 pairs represented mutualism and 8 pairs represented competition, each accounting for 22.2% of the total number (Fig. 11.8). Control and exploitation can be regarded as a single relationship. Together, they account for 55.6% of the remaining 20 pairs of relationships, because they are mutually related (i.e., only the direction of the relationship is different, which is represented by the positive or negative sign in the sign matrix), which are the dominant relationship in the system.

sgn(***U***)	1	2	3	4	5	6	7	8	9
1	+	+	+	+	-	-	-	+	-
2	-	+	+	+	-	-	+	+	-
3	+	-	+	+	-	-	+	+	-
4	-	+	-	+	+	+	-	-	+
5	-	+	-	-	+	-	-	-	+
6	+	-	+	+	+	+	+	+	-
7	+	+	+	+	+	+	+	-	+
8	-	+	-	-	-	-	+	+	-
9	-	+	-	-	-	-	+	+	+

Exploitation Control Competition Mutualism

Fig. 11.8 The utility sign matrix sgn(***U***) and resulting ecological relationships for the flows of sulfur in the Shandong Lubei eco-industrial park (*Note* Signs are for the flows between the nodes in the top row of the table and the nodes in the first column of the table) Nodes: 1-ammonium phosphate plant; 2-sulfuric acid plant; 3-cement clinker production; 4-ammonia plant; 5-thermal power plant; 6-cement products production; 7-bromine plant; 8-salty gypsum production plant; 9-potassium and magnesium salt production plant.

Of all the mutualism relationships, four of them were related to node 7 (bromine plant): $(su_{27}, su_{72}) = (+, +)$, $(su_{37}, su_{73}) = (+, +)$, $(su_{67}, su_{76}) = (+, +)$, and $(su_{97}, su_{79}) = (+, +)$. It shows that mutual relationships have been established in the bromine factory with the sulfuric acid plant, cement clinker production, cement products production, and potassium and magnesium salt production. Only one of them flowed directly (the sulfuric acid plant transfers resources to the bromine plant); the rest were controlled by indirect flows. The cement clinker production makes interactions with the bromine unit by the sulfuric acid plant. What's more, the bromine plant has a mutually beneficial relationship with the production of cement products. Among the other two pairs of mutualism relationships, two were connected with node 2 (sulfuric acid plant): $(su_{24}, su_{42}) = (+, +)$ and $(su_{28}, su_{82}) = (+, +)$. The ammonia plant (node 4) was linked with the sulfuric acid plant (node 2) through the delivery of resources from cement clinker production (node 3). The sulfuric acid plant makes interactions with the salty gypsum production plant by flowing into the bromine plant. The results show that lots of reciprocal relationships are caused by indirect flows among network members.

The number of competition relationships was the same as the number of mutualism relationships. Node 9 (potassium and magnesium salt production plant) competes with three nodes: $(su_{19}, su_{91}) = (-, -)$, $(su_{39}, su_{93}) = (-, -)$, and $(su_{69}, su_{96}) = (-,$

−). The indirect flows were important because there were no direct flows between the potassium and magnesium salt production plant and the other three nodes. Flows from the sulfuric acid plant, bromine plant, and salty gypsum production plant permit interactions with the potassium and magnesium salt production plant while using cement clinker production as an example. Besides, three members of the network compete with node 5 (thermal power plant): $(su_{15}, su_{51}) = (-, -)$, $(su_{35}, su_{53}) = (-, -)$, and $(su_{58}, su_{85}) = (-, -)$. The thermal power plant does not receive sulfur resources from other members of the park, which indicates that the thermal power plant competes with these other three members when it provides resources to the same downstream receiver. For example, the thermal power plant and cement clinker production compete to provide resources for cement product production. Node 2 (sulfuric acid plant) and node 6 (cement products production) also compete, as $(su_{26}, su_{62}) = (-, -)$; both receive the same resources from cement clinker production and compete for the sulfur it produces.

Control and exploitation were dominant in all of the relationships in the system, which means that unequal relationships dominated the sulfur flows in Lubei park. Most members of the network are controlled by or exploit other members, which indicates that these members rely on each other for resource utilization. Node 6 (cement products production) exploits the resources of three members: $(su_{61}, su_{16}) = (+, -)$, $(su_{63}, su_{36}) = (+, -)$, and $(su_{65}, su_{56}) = (+, -)$. Through cement clinker production, the delivery relationship between cement product production and the ammonium phosphate plant is established. Cement clinker production gains ammonium phosphate from the ammonium phosphate plant and uses it to condense the cement clinker, then the cement clinker becomes the main raw material for the production of cement products.

The exchanges of resources among these three nodes were all related to the finished products of each node. The cement products production receives slag from the thermal power plant and treats it as a blending material. Cement products production also exploits sulfur from the thermal power plant. These exchanges of byproducts and waste reflect the ability of cement products production to fully reuse the waste and resources produced by other members of the park. Node 7 (bromine plant) also exploits three members of the network: $(su_{71}, su_{17}) = (+, -)$, $(su_{74}, su_{47}) = (+, -)$, and $(su_{75}, su_{57}) = (+, -)$. The bromine plant interacts with all three other members through indirect flows, and thus exploits their resources indirectly. Node 1 (ammonium phosphate plant) also exploits three members: $(su_{21}, su_{12}) = (-, +)$, $(su_{41}, su_{14}) = (-, +)$, and $(su_{81}, su_{18}) = (-, +)$. The ammonium phosphate plant receives sulfur products from the sulfuric acid plant, and also interacts with the other two members through indirect flows. Node 8 (salty gypsum production plant) controls three members: $(su_{81}, su_{18}) = (-, +)$, $(su_{83}, su_{38}) = (-, +)$, and $(su_{86}, su_{68}) = (-, +)$. The salty gypsum production plant delivers cement kiln gas containing sulfate ions directly to cement clinker production and salty gypsum products to cement products production.

We can calculate that the mutualism index $(M) = 1.25$ using the data in Fig. 11.8, which is higher than 1. So the overall system is mutualistic, and positive signs dominate. This mainly results from the positive self-feedbacks within each node, because the eight pairs of mutualism relationships and the eight pairs of competition relationships can offset each other. The numbers of positive are equal to the numbers of negative signs in the exploitation (control) relationships. The synergism (S) value was 6.67, indicating that the network's final benefit was positive, and that the positive flows between nodes are greater than the negative flows, which again demonstrates the primarily mutualistic state of the network. Among the nine nodes, node 6 (cement products production) had the largest benefit (1.39), followed by node 9 (potassium and magnesium salt production plant, 1.34). Only node 4 (ammonia plant) and node 5 (thermal power plant) had a benefit less than 0.10, and node 5 had a negative benefit (–0.12).

11.2.3.3 Hierarchy Analysis

By examining the integral utility values and the ecological relationships, the hierarchy of the network members can be determined. In an eco-industrial network, a given node may play different ecological roles when it interacts with the other members of the network. For example, it may act as a producer for the flows to another node, but become a consumer for the flows from one node. I will choose the position in the hierarchy which appeared most often as a node's level to resolve such conflicts.

The pairs $(su_{83}, su_{38}) = (-, +)$, $(su_{81}, su_{18}) = (-, +)$, $(su_{89}, su_{98}) = (-, +)$, and $(su_{86}, su_{68}) = (-, +)$ indicate that node 8 (salty gypsum production plant) is exploited by node 1 (ammonium phosphate plant), node 3 (cement clinker production), node 6 (cement products production), and node 9 (potassium and magnesium salt production plant). These nodes are therefore all higher in the hierarchy than node 8. Based on the pairs of $(su_{23}, su_{32}) = (+, -)$, $(su_{52}, su_{25}) = (+, -)$, $(su_{95}, su_{59}) = (+, -)$, and $(su_{19}, su_{91}) = (-, -)$, node 2 (sulfuric acid plant) exploits node 3 (cement clinker production), node 5 (thermal power plant) is located higher in the hierarchy than node 2 (sulfuric acid plant), node 9 (potassium and magnesium salt production plant) is higher than node 5 (thermal power plant), and node 9 (potassium and magnesium salt production plant) and node 1 (ammonium phosphate plant) compete, and are therefore at the same level in the hierarchy. (Zhang et al. 2015a).

The relationships $(su_{61}, su_{16}) = (+, -)$, $(su_{63}, su_{36}) = (+, -)$, $(su_{65}, su_{56}) = (+, -)$, $(su_{64}, su_{46}) = (+, +)$, and $(su_{67}, su_{76}) = (+, +)$ mean that node 6 (cement products production) is higher in the hierarchy than node 1 (ammonium phosphate plant), node 3 (cement clinker production), and node 5 (thermal power plant), and is at a different position in the hierarchy than node 4 (ammonia plant) and node 7 (bromine plant). Therefore, node 6 is located between node 7 and node 4. Because $(su_{49}, su_{94}) = (+, -)$ and $(su_{74}, su_{47}) = (+, -)$, node 4 (ammonia plant) is higher in the hierarchy than node 9 (potassium and magnesium salt production plant), and node 7 (bromine plant) is higher than node 4 (ammonia plant). (Zhang et al. 2015a).

Based on these relationships, we can divide the hierarchy into eight layers (Fig. 11.9). It is necessary to analyze the interactions among the members further to see whether any layers can be combined, because this complex hierarchy is somewhat cumbersome to work with, and does not provide a clear parallel with the levels in a natural system. We can divide the nodes into producers, primary consumers, and secondary consumers based on the analogy of a natural ecosystem. First, both nodes have more exploitation relationships than the other nodes, so we can combine nodes 6 and 7 as secondary consumers, which places them higher in the hierarchy. Second, these three nodes are exploited mainly by other nodes and belong to more control relationships than exploitation relationships, so we can combine nodes 2, 3, and 8, which can be treated as producers, because they are the leading providers of resources to other nodes. Finally, nodes 1, 4, 5, and 9 not only receive resources from the lower level of the hierarchy but also provide resources to the higher level of the hierarchy, so they have complex relationships with the other nodes, which therefore function as primary consumers. We present the resulting simplified hierarchy structure in Fig. 11.9.

We can calculate the weight of the integral flows provided by each node to the network and incorporate this information in the hierarchy structure. Figure 11.9 shows that the integral flow weights of sulfur form a pyramidal hierarchy. The influence of the nodes at a given level in the hierarchy decreases moving from the producer level to the two consumer levels. This structure, with the highest weight for the producer level, means that the members of the producer level contribute the most sulfur to the other members (the consumers); thus, upstream enterprises can support and promote the sulfur utilization of downstream enterprises (Zhang et al. 2015a). The balanced pyramidal pyramid structure suggests that the distribution of resources is favorable for the long-term development of Lubei park. The integral flow weights were most significant for nodes 1, 2, 3, and 6 among all the members of the system, which means that they contribute more sulfur to the other members in the system.

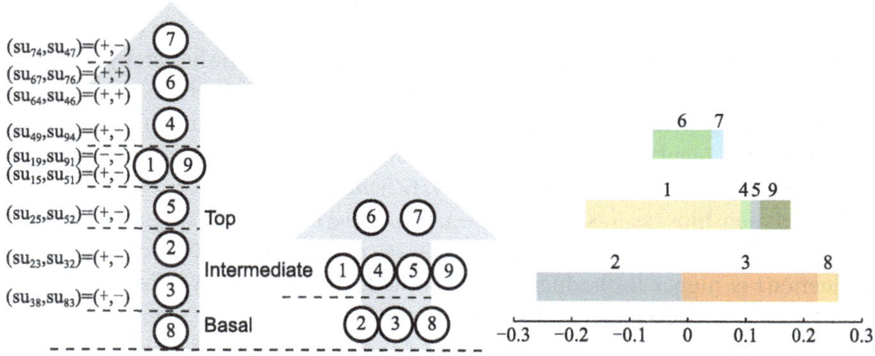

Fig. 11.9 Detailed and simplified hierarchies of the Shandong Lubei eco-industrial park system based on the values in the integral utility matrix for flows of sulfur (*Nodes*: 1-ammonium phosphate plant; 2-sulfuric acid plant; 3-cement clinker production; 4-ammonia plant; 5-thermal power plant; 6-cement products production; 7-bromine plant; 8-salty gypsum production plant; 9-potassium and magnesium salt production plant. Nodes 2, 3, and 8 are producers; nodes 1, 4, 5 and 9 are primary consumers; and nodes 6 and 7 are secondary consumers. The bar lengths represent the weights of each node based on the integral flow it receives from other nodes)

Therefore, they play a more critical role in the network compared to other nodes and represent core members of the system.

Node 2 (sulfuric acid plant) provides the most sulfur to the other nodes (25.1% of the total) in the producer level of the hierarchy, including delivery to the other members of the system directly and indirectly. It is followed by node 3 (cement clinker production), which contributes 20.3% of the total. The weight of the integral utility of node 8 (salty gypsum production plant) is the smallest of the producers, as it only accounts for 3.7% of the total. In the primary consumer layer of the hierarchy, due to the essential circular flow structure among nodes 1, 2, and 3, node 1 (ammonium phosphate plant) accounts for the most significant proportion of the flows (26.8%) and supplies more sulfur to upstream members of the network. Node 1 not only provides sizeable direct sulfur flows to node 3 but can also supply node 2 with resources through indirect flows that pass through node 3. Node 1 acts as a consumer, but it can help to recycle sulfur for the producers. Node 4 (ammonia plant) and node 5 (thermal power plant) provide the most minor sulfur of the other three primary consumers, which both account for only 1.7% of the flows. Compared with the four core members, they contribute little sulfur to other members of the network and only serve as peripheral members. Among the secondary consumers, node 7 (bromine plant) only provides 2.0% of the flows of sulfur in the system.

11.2.4 Comparison with Previous Research

11.2.4.1 Comparison with Previous Research on Node Characteristics

By combining the structural characteristics of the node analysis (Table 11.6, Table 11.7, and Table 11.8), the characteristics of the nodes in the Shandong Lubei network can be described in considerable detail. In previous studies, Ashton (2008) used network analysis to analyze the relationship between industrial network members and found that the centrality index they used fully reflected whether network members were controlled by one or more nodes (Zhang et al. 2015a). It shows that the center is the enterprise that sells the most goods to other parks of such a specific eco-industrial park. Based on this approach, Doménech and Davies (2011) used the centrality theory to determine the importance of the core members of the Karenburg Eco-industrial Park (Zhang et al. 2015a). It is found that the core members not only were connected with more members, but also played key roles in improving the connectedness degree of the overall network.

In the research on which this section was based, in order to divide all nodes into core and surrounding subsidiary networks, my research group used core-periphery analysis. Nodes at the core frequently interacted with other nodes, and these companies were linked continuously by more than one exchange flow; On the contrary, peripheral nodes had not communicated directly with other peripheral nodes and had less frequent communication with core nodes. In many cases, they were just recipients of garbage flows. By combining centrality and core-periphery analysis, we found that node 3 (cement clinker production) and node 8 (salt gypsum production) were the core members of the whole network, which greatly affected resource delivery and control resources of the network. On the contrary, node 4 (ammonia plant) and node 5 (thermal power plant) were obviously peripheral members.

It can be found that node 1 (ammonium phosphate plant), node 2 (sulfuric acid plant), and node 3 (cement clinker production) were key members of the sulfur delivery network, and the throughput of the overall metabolic system was involved in these three members by considering the flows along each path and analyzing the network functions. The results of functional analysis and structural analysis are both similar and different. Node 3 (cement clinker production) was a crucial node in these two analyses because it made interactions frequently with others and exchanged a large amount of sulfur with others, which played an essential role in the primary species of the whole system. Node 1 (ammonium phosphate unit) and node 2 (sulfuric acid unit) did not play key roles in structural analysis, but they were critical nodes in flow analysis, which increases their importance of functions. Another similarity between functional analysis and structural analysis is that node 4 (ammonia plant) and node 5 (thermal power plant) were both peripheral nodes, which played a weak role in structural and functional analysis. As essential species, they were the source of sulfur in the system and the basis for the expansion of sulfur metabolism processes. So their contributions to critical nodes should not be ignored.

In the process of identifying key nodes, the similarities and differences between the structural attributes and functional attributes of the network are caused by the different weights of each path. In structural analysis, all paths have the same weight, and their positions and roles are evaluated only based on whether they exchange materials with other nodes. On the contrary, all paths have corresponding quantities in the functional analysis, so the capacity and weight of each node can be quantitatively described.

11.2.4.2 Comparison with Previous Research on Path Characteristics

Figure 11.10 indicates the relationship of the distances between pairs of nodes and the direct flows along these paths. It is found that there was a statistically significant linear relationship between the two parameters, including the longer the distance between the two nodes and the higher the direct flows, which is caused by the differences between the socioeconomic system and the natural system of the ecological industrial park; Enterprises with high trophic levels can obtain resources from some of the bottom levels and directly from the external environment, which will usually increase rather than decrease the flows after each step along a linked series of paths. But some relationships deviated from the linear relationship. For example, the two circles in Fig. 11.10 show that two paths should be used to move from node 4 (ammonia plant) to node 6 (cement product production), while only one path can be used to move from node 2 (sulfuric acid plant) to node 7 (bromine plant) (Fig. 11.5). Both of these flows occur over a short distance and involve small direct flows, so they represent the “bottleneck” of the whole process of the metabolism of sulfur. In future analysis

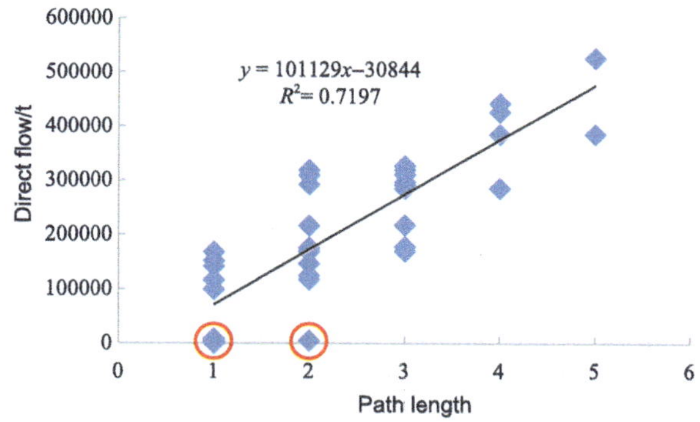

Fig. 11.10 Relationships between the distance between nodes and the direct flows that occur over these distances (*Nodes* 1-ammonium phosphate plant; 2-sulfuric acid plant; 3-cement clinker production; 4-ammonia plant; 5-thermal power plant; 6-cement products production; 7-bromine plant; 8-salty gypsum production plant; 9-potassium and magnesium salt production plant. The two circled points deviate from the linear relationship. However, the regression equation includes the values of both circled points)

and research, sulfur can be fully utilized and transferred in the park, which is likely to be the potential to increase the importance of these two paths by expanding the production scale.

We found that all of these flows followed the path from node 3 (cement clinker production) to node 2 (sulfuric acid plant), and from node 7 (bromine plant) to node 8 (gypsum production plant) by analyzing the flows that take 4 or 5 routes to reach the destination. Therefore, these two paths carry large direct flows (1.677×10^5 t and 1.162×10^5 t, respectively) and are essential intermediate nodes between other network members. They are not only critical ways to connect multiple enterprises, but also play an essential role in transporting sulfur from primary species to intermediate or top species.

It was found that the members in the core-core sub-network interacted frequently and that the flow paths were concentrated in this sub-network (Zhang et al. 2013). We found similarities in the structural and functional analysis after adding the magnitude of the flows. For example, the paths from node 2 (sulfuric acid plant) to node 1 (ammonium phosphate plant), from node 3 (cement clinker production) to node 2, and from node 1 to node 3 are key paths for the overall system, because they carried large direct and indirect flows, and the cycling among them greatly influenced sulfur transfers in the network. This agrees with the results of Chen (2003), who found that the direct flows among the ammonium phosphate plant, sulfuric acid plant, and cement clinker production plant contributed significantly to sulfur recycling in Lubei park. It is also found no direct interactions among peripheral nodes in the periphery-periphery sub-network, but the indirect flows between node 7 (bromine plant) and node 9 (potassium and magnesium salt production plant) increased the interactions among peripheral members.

11.2.4.3 Comparison with Previous Research on Ecological Relationships

In a word, the Lubei system is mutualistic, as one might expect from the ecological industrial park, whose design goal is to achieve such a symbiosis. This mutualistic status of the system is controlled by the number of positive relationships, including the positive elements in the diagonal of the sgn($\boldsymbol{U}$) matrix; that is, the positive self-feedback effects were important. In this system, the relationship of exploitation accounted for the majority, which indicates that members of this socioeconomic system tend to exploit resources from others because it helps to promote the exchange of finished products, by-products, or wastes between members. The network can produce positive benefits by passing these materials on to other members of the park. In Lubei park, these relationships ultimately contribute net positive benefits for the overall system, because the integral utility of the exploitation (control) relationships is positive, which also means that in a symbiotic system, the relationships of exploitation (control) are not necessarily negative. It shouldn't be judged whether a system is mutually beneficial or estimate the degree of reciprocity only based on the number of reciprocal relationships. In a sense, exploitation relationships resemble the obligatory symbiosis described by Liwarska-Bizukojc et al. (2009). These authors stated that when a byproduct of one enterprise becomes an input material for another enterprise, both benefit: the receiver obtains cheaper input materials, while the provider minimizes its waste disposal problems, which results in the coupling of the two enterprises and improves the overall resource utilization in industrial symbiosis. This relationship can also expand the manufacturing scale of enterprises. If the export of products increases, them this may generate more waste, but if one or more downstream enterprises can receive waste and treat them as useful resources, these enterprises mitigate the environmental effects of their production (Liwarska-Bizukojc et al. 2009).

Multiple types of relationships will co-exist in a real-world system. Liwarska-Bizukojc et al. (2009) believed that although managers do not expect negative effects (exploitation, control, or competition) among members, this is inevitable in a free market economy. At the same time, a market-oriented economy and society also requires fair competition among enterprises (Liwarska-Bizukojc et al. 2009), and these relationships are necessary for network stability and symbiosis. In previous research by my group (Zhang et al. 2012) we also found that the competitive relationships promoted socioeconomic development to a certain extent. Drawing on the factors that control socioeconomic development, a certain degree of competition can be seen to promote technological improvements by enterprises, improve the internal organizational structure of the enterprises, and increase the network's ecological efficiency. Effective competition is neither excessive nor insufficient; it is just enough to enhance communication and contacts between companies, thereby establishing new symbiotic relationships, improving equipment or processes, and improving the output quality of byproducts and waste. Of course, for long-term stable development of the park, its managers must take measures to ensure that predominantly positive relationships are formed between enterprises to promote a coupling effect that leads to greater economic and environmental benefits (Liwarska-Bizukojc et al. 2009).

11.2.5 Planning for Future Development

Node 3 (cement clinker production) is clearly the key enterprise in the park from both structural and functional perspectives, and is at the center of many flow paths. However, the relationships among the peripheral members of the network should be enhanced. Establishing connections among these members could mitigate the network's dependence on the core members, thereby lowering the risk that would result from adverse changes in any of the core members, and could also reveal unexpected information or opportunities for the peripheral members to cooperate by encouraging exchanges of innovative technical and scientific ideas. Therefore, the peripheral members would play a more important role in the cycling of currently unused or underutilized resources and should be the subject of careful analysis during future planning of the park's development (Granovetter 1973).

Since 2005, the managers of the Lubei eco-industrial park have done much work to improve utilization of the waste heat produced by the sulfuric acid plant and the two components of the cement plant. Previously, engineers reduced excessive pressures and temperatures in the steam lines of these plants by releasing the energy as steam. To capture some of this energy, managers of the plants have introduced an 1800 kW turbine that is driven by a fan, and the work performed by this turbine will lower the consumption of electricity produced outside the park by these plants. The turbine therefore replaced the previous method and also produces low-temperature and low-pressure steam that can be used by other enterprises in the park. This enables sulfuric acid plants to produce more sulfur, reduce carbon dioxide emissions, and reduce energy use. The low-temperature and low-pressure steam can also be delivered to other enterprises, for example, the cement plants and the salty gypsum production plant, or to the residential living area near the park. Now, it is also more economical than it used to be to deliver the more than 100 t of sulfur that is produced to ammonium phosphate plant or to export this material.

Emissions of SO_2 from the thermal power plant can also be reused by transforming this resource into sulfate and sulfite ions. This would not only reduce waste SO_2 but also improve the recycling efficiency of sulfur. These two measures would decrease the reliability on node 3 (cement clinker production) and increase the overall sulfur utilization efficiency.

Managers should also motivate the step-by-step utilization of materials in chains of transformation processes. For instance, the managers of the ammonium phosphate plant have improved the utilization rate of phosphoric acid production equipment. They improved the utilization of different grades of phosphoric acid, in which the wet purification of phosphoric acid based on defluorination, desulfurization, and iron removal can produce compound fertilizer, sodium tripolyphosphate, and potassium phosphate. So the project improved the socioeconomic benefits of the ammonium phosphate plant (i.e., produced better fertilizer that will improve agricultural quality), which can not only improve the utilization of phosphoric acid resources but also will not impose additional environmental impacts.

References

Ashton WS (2008) Understanding the organization of industrial ecosystems: A social network approach. Journal of Industrial Ecology 12(1):34–51

Chen DJ (2003) Research on integration and complexity of industrial ecosystem analysis. Tsinghua University, Beijing, China (in Chinese)

Doménech T, Davies M (2011) Structure and morphology of industrial symbiosis networks: The case of Kalundborg. Procedia Social and Behavioral Sciences 10(1):79–89

Feng JT (2003) Lubei Enterprise Group's practice and exploration of developing an ecological industrial chain. China Population Resources and Environment 13(1):110–112 (in Chinese)

Feng JT (2003) A case study of Lubei eco-industrial park. China's Population Resources and Environment 13(4):98–102 (in Chinese)

Granovetter MS (1973) The strength of weak ties. American Journal of Economics and Sociology 78(6):1360–1380

Liu J (2004) Introduction to Social Network Analysis. Social Sciences Literature Press, Beijing, China

Liwarska-Bizukojc E, Bizukojc M, Marcinkowski A et al (2009) The conceptual model of an eco-industrial park based upon ecological relationships. Journal of Cleaner Production 17(8):732–741

Mayhew BH, Levinger R (1976) Size and the density of interaction in human aggregates. American Journal of Sociology 82(1):86–110

MEEC (Ministry of Ecology and the Environment of the People's Republic of China), 2011. Guiding Opinions on Strengthening the Construction of the National Eco-Industrial Demonstration Park.(2011-12-5) [2022-4-5]. http://www.mee.gov.cn/gkml/hbb/bwj/201112/t20111208_221112.htm. (in Chinese)

Mihelcic JR, Zimmerman JB, Auer MT (2014) Environmental engineering: Fundamentals, sustainability, design. Wiley, Hoboken, NJ

Moreno JL, Jennings HH (1938) Statistics of social configurations. Sociometry 1(3/4):342–374

ODEHCIAP (Office Development Experts and Heavy Chemical Industry in Anhui Province), 2011. Investigation Report of Shanghai Chemical Industry Park.(2011-7-20) [2021-12-4]. https://wenku.baidu.com/view/b0d9887201f69e31433294e4.html. (in Chinese)

Potts CAJ (1998) Choctaw eco-industrial park: An ecological approach to industrial land-use planning and design. Landscape and Urban Planning 42(2–4):239–257

Schwarz EJ, Steininger KW (1997) Implementing nature's lesson: The industrial recycling network enhancing regional development. Journal of Cleaner Production 5(1–2):47–56

Shi H, Chertow M, Song YY (2010) Developing country experience with eco-industrial parks: A case study of the Tianjin Economic-Technological Development Area in China. Journal of Cleaner Production 18(3):191–199

Wang RX (2005) Research on the planning and design of the Huangxing National Eco-industrial Park in Changsha, China. Northeast Normal University, Changchun, China (in Chinese)

Wu YP, Duan N, Qiao Q et al (2004) Research on the industrial symbiosis chain network structure of a brand-new eco-industrial park. China Population Resources and Environment 14(2):125–130 (in Chinese)

Yang L, Hu SY, Liang RZ et al (2004) Eco-industrial model of Lubei, China. Chinese Journal of. Process Engineering 4(5):467–474 (in Chinese)

Zhang Y, Liu H, Li YT et al (2012) Ecological network analysis of China's societal metabolism. Journal of Environmental Management 93(1):254–263

Zhang Y, Zheng HM, Chen B et al (2013) Social network analysis and network connectedness analysis for industrial symbiotic systems: Model development and case study. Frontiers of Earth Science 7(2):169–181

Zhang Y, Zheng HM, Fath BD (2015) Ecological network analysis of an industrial symbiosis system: A case study of the Shandong Lubei eco-industrial park. Ecological Modelling 306(12):174–184

Zhang Y, Zheng HM, Yang ZF et al (2015) Analysis of the industrial metabolic processes for sulfur in the Lubei (Shandong Province, China) eco-industrial park. Journal of Cleaner Production 96(11):126–138

Zhu QE, Lowe EA, Wei Y et al (2007) Industrial symbiosis in China: A case study of the Guitang Group. Journal of Industrial Ecology 11(1):31–42

The manufacturer's authorised representative in the EU is Springer Nature Customer Service Centre GmbH, Europaplatz 3, 69115 Heidelberg, Germany. If you have any concerns regarding our products, please contact ProductSafety@springernature.com

Printed and bound by CPI Group (UK) Ltd, Croydon, CR0 4YY
30/09/2026
02225244-0017